Alpine, S[illegible] Minx & Rapier Owners Workshop Manual

by J H Haynes
Member of the Guild of Motoring Writers

and M S Daniels

Models covered

Talbot Alpine GL, LS, LE, S, GLS and SX; 1294 cc, 1442 cc, 1592 cc
Talbot Solara GL, LS, LE, GLS and SX; 1294 cc, 1592 cc
Talbot Minx; 1592 cc
Talbot Rapier; 1592 cc

Covers all Special and Limited Edition models

ISBN 1 85010 304 6

Printed in England *(337-7N4)*

ABC

Haynes Publishing Group
Sparkford Nr Yeovil
Somerset BA22 7JJ England

Haynes Publications, Inc
861 Lawrence Drive
Newbury Park
California 91320 USA

British Library Cataloguing in Publication Data

Daniels, Marcus S.
Talbot Alpine & Solara owners workshop manual
4th ed. – (Owners Workshop Manual)
1. Sunbeam Alpine automobile 2. Solara automobile
I. Title II. Daniels, Marcus S. Talbot/
Chrysler Alpine and Solara, Minx and Rapier
1975-85 owner's workshop manual III. Series
629.28'722 TL215.S85
ISBN 1-85010-304-6

Acknowledgements

Our thanks are due to Chrysler/Talbot UK Ltd for the supply of technical information and certain illustrations. The Champion Sparking Plug Company supplied the illustrations showing the various spark plug conditions.

Special thanks are due to F W B Saunders of Yeovil for their co-operation in loaning us a Talbot Solara.

Lastly, thanks are due to all those people at Sparkford who helped in the production of this manual.

About this manual

Its aim

The aim of this manual is to help you get the best value from your car. It can do so in several ways. It can help you decide what work must be done (even should you choose to get it done by a garage), provide information on routine maintenance and servicing, and give a logical course of action and diagnosis when random faults occur. However, it is hoped that you will use the manual by tackling the work yourself. On simpler jobs it may even be quicker than booking the car into a garage and going there twice to leave and collect it. Perhaps most important, a lot of money can be saved by avoiding the costs the garage must charge to cover its labour and overheads.

The manual has drawings and descriptions to show the function of the various components so that their layout can be understood. Then the tasks are described and photographed in a step-by-step sequence so that even a novice can do the work.

Its arrangement

The manual is divided into thirteen Chapters, each covering a logical sub-division of the vehicle. The Chapters are each divided into Sections, numbered with single figures, eg 5; and the Sections into paragraphs (or sub-sections), with decimal numbers following on from the Section they are in, eg 5.1, 5.2, 5.3 etc.

It is freely illustrated, especially in those parts where there is a detailed sequence of operations to be carried out. There are two forms of illustration: figures and photographs. The figures are numbered in sequence with decimal numbers, according to their position in the Chapter – eg Fig. 6.4 is the fourth drawing/illustration in Chapter 6. Photographs carry the same number (either individually or in related groups) as the Section or sub-section to which they relate.

There is an alphabetical index at the back of the manual as well as a contents list at the front. Each Chapter is also preceded by its own individual contents list.

References to the 'left' or 'right' of the vehicle are in the sense of a person in the driver's seat facing forwards.

Unless otherwise stated, nuts and bolts are removed by turning anti-clockwise, and tightened by turning clockwise.

Vehicle manufacturers continually make changes to specifications and recommendations, and these when notified are incorporated into our manuals at the earliest opportunity.

Whilst every care is taken to ensure that the information in this manual is correct, no liability can be accepted by the authors or publishers for loss, damage or injury caused by any errors in, or omissions from, the information given.

Introduction to the Alpine and Solara

Introduced in 1976 the Alpine received great acclaim from both the motoring press and the public alike and that same year won the Car of the Year Award.

The reason for the Alpine's success in both sales figures and proven performance is fairly straightforward; when planning the construction of the vehicle, Chrysler took all the best motor engineering design features and incorporated them into one car.

The four cylinder overhead valve engine is mounted transversely in the engine compartment and inclined rearwards to provide a low bonnet profile. This gives a smooth airflow over the body and reduces wind noise.

Power from the engine is transmitted through a four or five speed all-synchromesh gearbox (or three speed automatic transmission) and differential unit attached to the left-hand side of the engine, and then via short driveshafts to the front wheels.

Torsion bars and wishbone radius arms are used for the front suspension, while the independent rear suspension has trailing arms supported on coil springs. Telescopic dampers are fitted to both front and rear suspension systems.

The front wheel drive layout dispenses with the conventional propeller shaft, rear axle and associated floor bulges. This provides additional foot room in the passenger compartment and considerably increases the luggage area.

Although the Alpine is a fairly large car in terms of roominess, the performance, even with the smaller 1294 cc engine, is very lively. For the customer who requires increased performance with a minimal loss in fuel economy the larger 1442 cc and 1592 cc engines are available.

The Solara was introduced in April 1980 and is the saloon version of the Alpine. Mechanically it is identical to the Alpine although it is not available with a 1442 cc engine.

Various options and special edition models are available in both Alpine and Solara versions to provide a package to suit all preferences.

Chrysler/Talbot UK Ltd is now known as Peugeot Talbot Motor Company Ltd, and all references in the text to Chrysler should be regarded as references to Peugeot Talbot.

Contents

Three quarter view of a Chrysler Alpine

The Talbot Solara LS

Buying spare parts and vehicle identification numbers

Buying spare parts

Spare parts are available from many sources, for example: Chrysler/Talbot garages, other garages and accessory shops, and motor factors. Our advice regarding spare part sources is as follows:

Officially appointed Chrysler garages - This is the best source of parts which are peculiar to your car and are otherwise not generally available (eg; complete cylinder heads, internal gearbox components, badges, interior trim, etc). It is also the only place at which you should buy parts if your car is still under warranty - non-Chrysler components may invalidate the warranty. To be sure of obtaining the correct parts it will always be necessary to give the storeman your car's engine and chassis number, and if possible, to take the 'old' part along for positive identification. Remember that many parts are available on a factory exchange scheme - any parts returned should always be clean! It obviously makes good sense to go straight to the specialist on your car for this type of part for they are best equipped to supply you.

Other garages and accessory shops - These are often very good places to buy materials and components needed for the maintenance of your car (eg, oil filters, spark plugs, bulbs, fan belts, oils and greases, touch-up paint, filler paste etc). They also sell general accessories, usually have convenient opening hours, charge lower prices and can often be found not far from home.

Motor factors - Good factors will stock all of the more important components which wear out relatively quickly (eg, clutch components, pistons, valves, exhaust systems, brake cylinders/pipes/hoses/seals/shoes and pads etc). Motor factors will often provide new or reconditioned components on a part exchange basis - this can save a considerable amount of money.

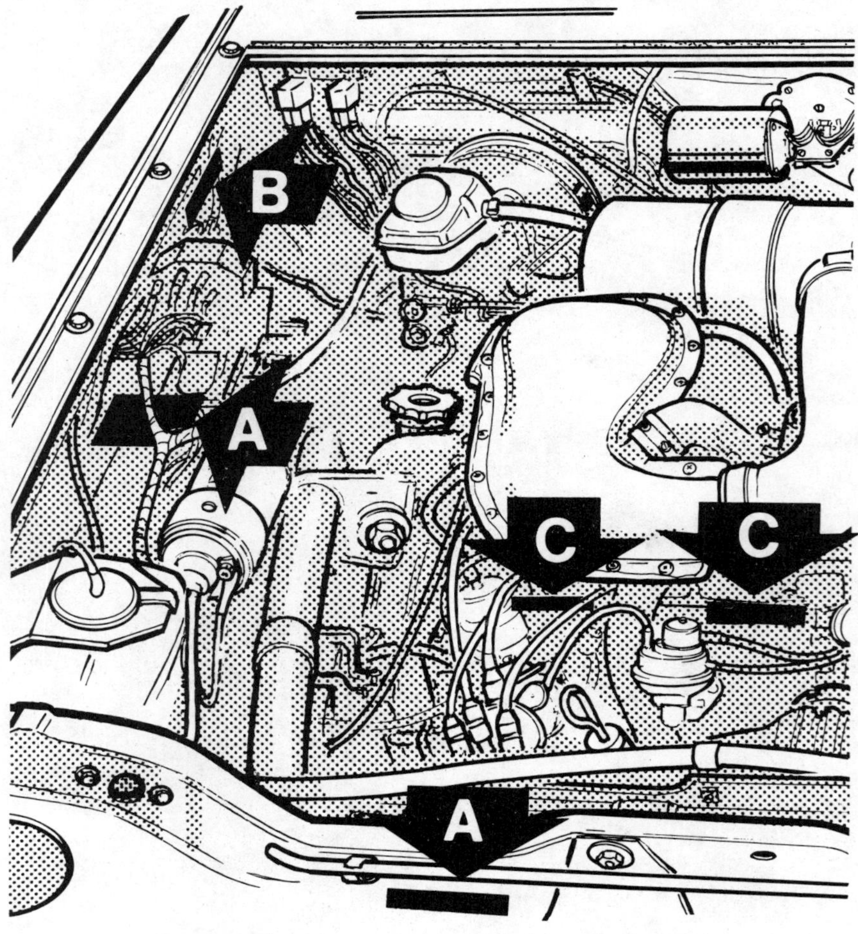

Location of vehicle identification numbers

A Vehicle identification plate B Body serial number C Engine number

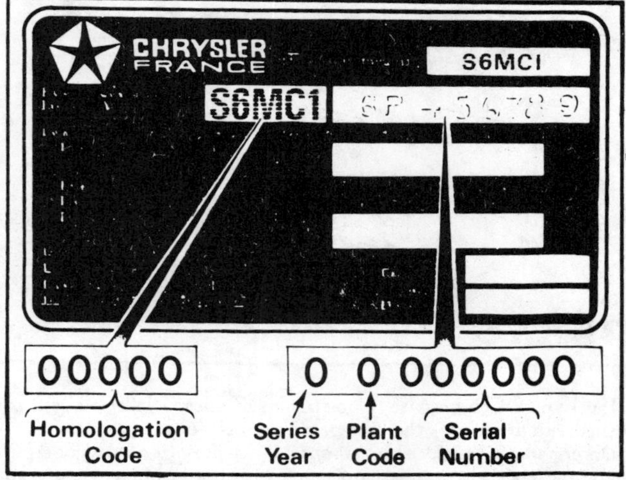

Example of vehicle identification plate

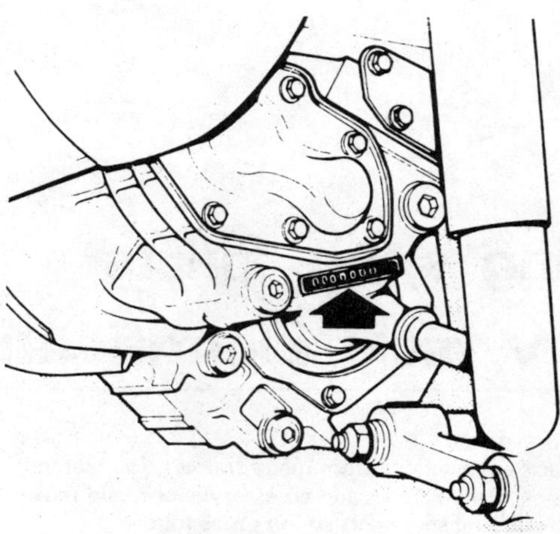

Location of gearbox number

Vehicle identification numbers

Always have details of the car, its serial and engine numbers available when ordering parts. If you can take along the part to be renewed, it is helpful. Modifications were and are being continually made and often are not generally publicised. A storeman in a parts department is quite justified in saying that he cannot guarantee the correctness of a part unless these relevant numbers are available.

The vehicle identification plate is attached to the right-hand wing valance in the engine compartment on some models. On other models it is located on the bonnet lock platform.

The body serial number is located on the right-hand wing valance.

The engine number is stamped on a plate secured to the front of the cylinder block. On some models it is adjacent to the distributor and on other models it is adjacent to the fuel pump. The engine number should also be quoted when ordering parts for the final drive assembly.

The gearbox number is located on the end of the casing.

Jacking and towing

Jacking

The jack and wheel brace are clipped to the wing valance inside the engine compartment. The spare wheel is secured beneath the rear end of the car. To remove it, lift up the luggage compartment carpet and slacken the carrier bolt using the wheelbrace until the wheel can be withdrawn from beneath the vehicle (photos).

Before jacking-up the car, ensure it is standing on level ground and set the handbrake on firmly. Chock the wheel diagonally opposite the one being raised.

Insert the jack arm into the square socket nearest to the wheel being removed. Ensure the jack is upright and standing on firm ground before raising the car.

If work is to be carried out beneath the vehicle while it is jacked up, axle stands or suitable wooden blocks **must** be placed beneath the car to take its weight in the eventuality of jack failure.

It is permissible to locate a workshop jack under the front crossmember in order to raise the front end provided a shaped wooden block is used to both insulate the crossmember and to prevent the jack slipping.

When raising the rear of the car, use only the rear under-sill jacking points.

Location of jacking equipment and spare wheel

Lowering the spare wheel

Spare wheel and carrier

Jacking up the car

Towing

There is no provision for towing other vehicles. When being towed, however, the tow-rope or chain should be attached to the front lower suspension wishbone. Always attach the tow-rope or chain to the same side of the towing vehicle to prevent any side scuffing movement.

Note that the eyes located on the front and rear bumber mounting brackets should not be used for towing. They are provided for lashing-down during transportation.

Remember that servo assistance will not be available after the first few applications of the foot pedal, so be prepared to use greater pedal pressures than normal.

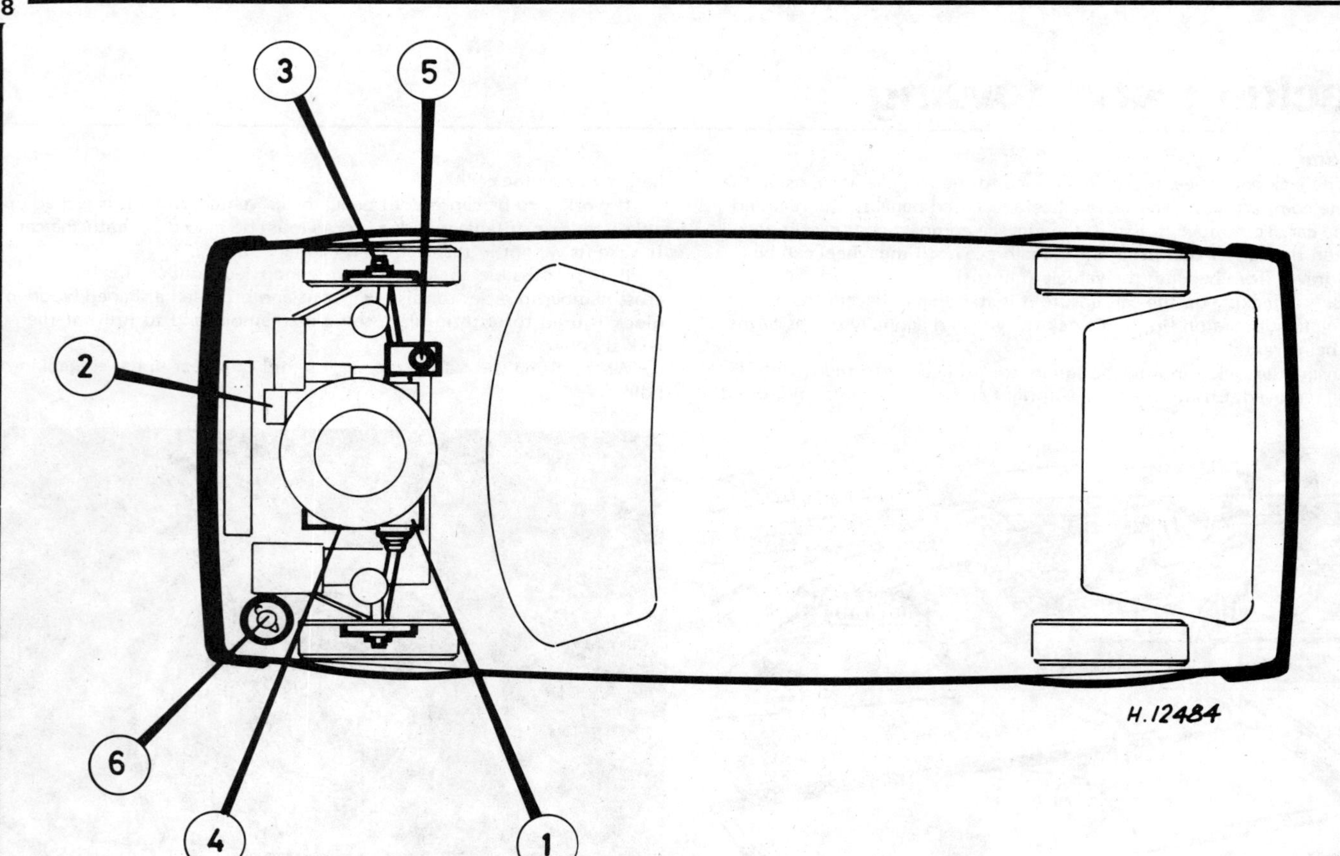

Recommended lubricants and fluids

Component or system		Lubricant type or specification
1	Engine	Multigrade engine oil
2	Manual gearbox and final drive	Gear oil SAE 90 (for later models - see Supplement)
	Automatic transmission and final drive	Dexron® automatic transmission fluid
3	Wheel bearings	NLG1 grade 2 EP grease
4	Manual steering rack	Graphite grease
	Power steering pinion and rack ends	Gear oil SAE 90
	Power steering pump and rack	Dexron® automatic transmission fluid
5	Brake and clutch master cylinders	Hydraulic fluid to DOT 3
6	Cooling system	Antifreeze to BS 3152

The above are general recommendations only. Different operating territories require different lubricants and if in doubt consult your nearest Agent.

Routine maintenance

For additional information and revised intervals, see Page 172

Maintenance is essential for ensuring safety and desirable for the purpose of getting the best in terms of performance and economy from the car. Over the years the need for periodic lubrication - oiling, greasing and so on - has been drastically reduced if not totally eliminated. This has unfortunately tended to lead some owners to think that because no such action is required the items either no longer exist or will last for ever. This is a serious delusion. It follows therefore that the largest initial element of maintenance is visual examination. This may lead to repairs or renewals.

In the summary given here the 'essential for safety' items are shown in **bold type.** These **must** be attended to at the regular frequencies shown in order to avoid the possibility of accidents and loss of life. Other neglect results in unreliability, increased running costs, more rapid wear and more rapid depreciation of the vehicle in general.

Checking the tyre pressures

Every 250 miles (400 km) travelled or weekly

Steering

Check the tyre pressures, including the spare wheel (photo).
Examine tyres for wear or damage.
Is steering smooth and accurate?

Brakes

Check reservoir fluid level.
Is there any fall off in braking efficiency?
Try an emergency stop. Is adjustment necessary?

Lights, wipers and horns

Do all bulbs work at the front and rear?
Are the headlamp beams aligned properly?
Do the wipers and horns work?
Check windscreen washer fluid level.

Engine

Check the sump oil level and top-up if required (photo).
Check the radiator coolant level and top-up if required.
Check the battery electrolyte level and top-up to the level of the plates with distilled water as needed.
Check the drivebelt tension.

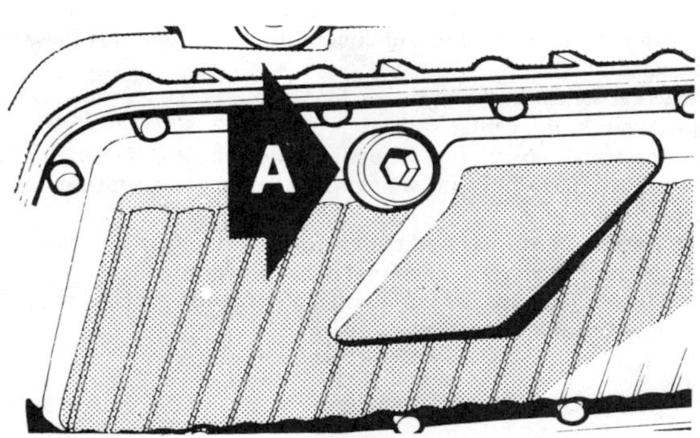

Engine oil drain plug 'A'

First 1000 miles (1600 km)

With new cars, or cars fitted with new engines the following checks should be carried out:

Change engine oil and gearbox oil.
Check tension of alternator drivebelt.
Check distributor points and ignition timing.
Check tightness of bolts on engine and manifolds.
Check valve clearances.
Check engine idle speed and fuel mixture.
Check cooling system hoses and fuel pipe connections for leaks and damage.
Check brake system pipes and hoses for leaks and damage.
Make a general check of all chassis and body components.

Service 'A'

Every 5000 miles (8000 km) or 5 months, whichever occurs first

Check electric cooling fan for correct operation.
Check strength of anti-freeze and top up if necessary.

Alternator drive belt and tensioner bolt

Change engine oil and fit a new oil filter element.
Check alternator belt tension and adjust if necessary.
Clean air filter element using compressed air if available.
Inspect the disc brake pads for wear and examine calipers for any signs of fluid leakage.
Check the brake and clutch hydraulic fluid levels (photo).
Check carburettor idling and mixture control settings and adjust if necessary.
Check and if necessary, top-up gearbox oil level.
Check and if necessary, top-up final drive oil level.
Check brake hydraulic system for leaks, damaged pipes etc.
Examine thickness of tyre treads.
Check and adjust, if necessary, front wheel alignment.
Examine exhaust system for corrosion and leakage.
Lubricate all controls and linkages.
Check for wear in steering gear and balljoints and condition of rubber bellows, dust excluders and flexible coupling.
Clean and adjust spark plugs.
Clean the ignition HT cables and examine for signs of chafing.
Lubricate all door locks and hinges and ensure door drain holes are clear.

Service 'B'

Every 10000 miles (16000 km) or 10 months, whichever occurs first

Carry out the maintenance tasks listed under Service 'A', plus the following:

Blow out any dirt from the engine side of the radiator using a compressed air line.
Check valve clearances and adjust if necessary.
Separate the halves of the flame trap, clean the filter discs in paraffin, wipe dry, and reassemble.
Renew air filter element.
Clean fuel pump filter and renew the in-line carburettor filter.
Remove rear brake drums and check linings for wear. Renew if necessary.
Check rear wheel brake cylinder for fluid leakage.
Adjust handbrake cable.
Examine shock absorbers for leakage and renew if necessary.
Check front driveshaft gaiters for damage or deterioration and renew if necessary.
Clean battery terminals and coat with petroleum jelly (Vaseline).
Drain and refill gearbox and final drive oil (photo).
Check fuel filler hose for leaks and tighten clips if necessary.
Fit a new set of spark plugs.

Topping up engine oil

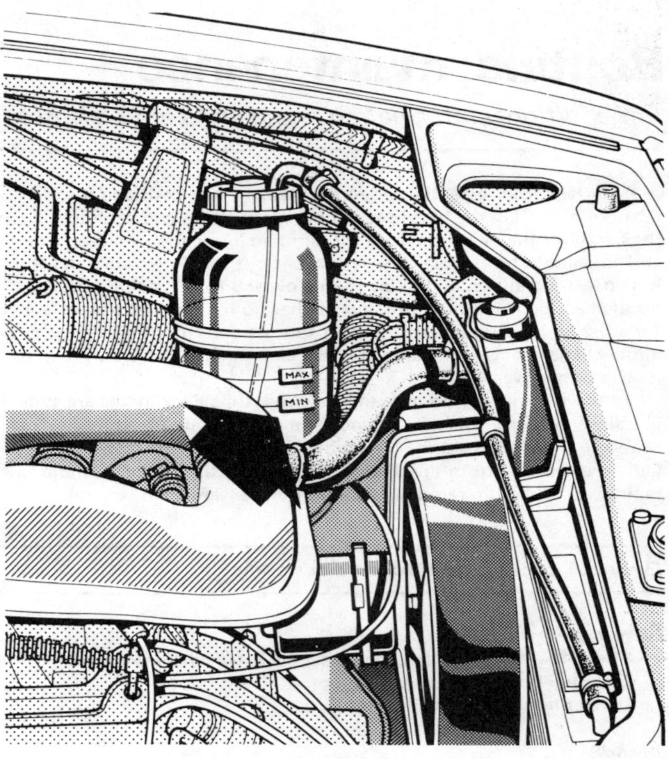

Cooling system expansion bottle

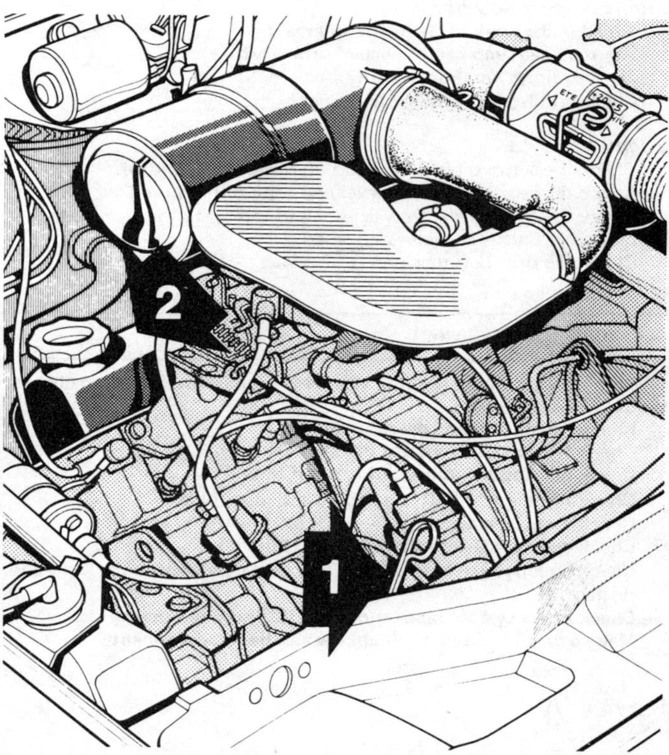

Engine oil checking points

1 Dipstick 2 Oil filler cap

Brake and clutch fluid reservoir

Refilling transmission with gear oil

Service 'C'

Every 15000 miles (24000 km) or 15 months, whichever occurs first

Carry out the maintenance tasks listed under Service 'A', plus the following:

Examine the handbrake ratchet for excessive wear and renew if necessary.

Check the handbrake cable for corrosion or fraying and renew if necessary.

Check the handbrake linkage pivots and pins for wear.

Clean and re-pack the rear wheel hubs with grease and reset the end float as described in Chapter 8.

Service 'D'

Every 30000 miles (48000 km) or 30 months, whichever occurs first

Carry out the maintenance tasks listed under Services 'B' and 'C', plus the following:

Renew the brake servo unit filter element (refer to Chapter 10).

Examine all the metal brake pipes for signs of corrosion and renew if necessary.

Check the shock absorber mountings for security and renew the rubber bushes if worn or perished.

Renew the air cleaner element.

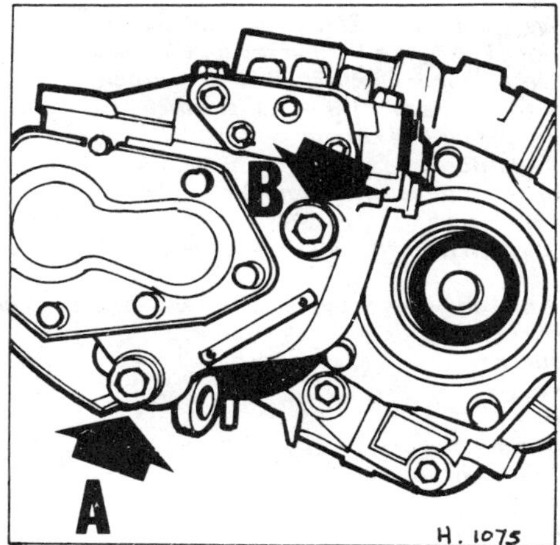

Gearbox (type AT) - 'A' drain plug, 'B' filler plug

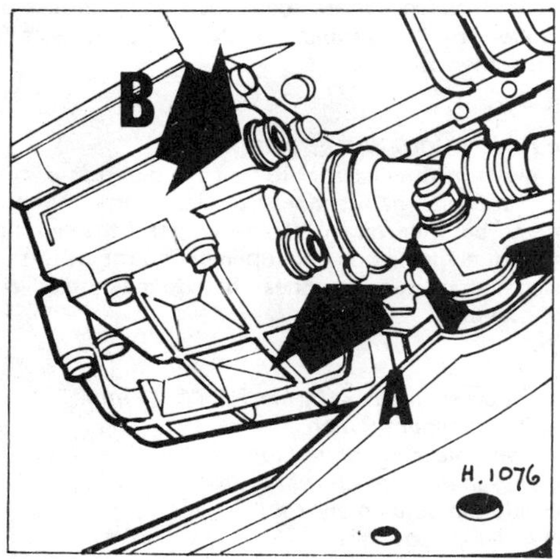

Final drive unit (type AT) - 'A' drain plug, 'B' filler plug

Tools and working facilities

Introduction

A selection of good tools is a fundamental requirement for anyone contemplating the maintenance and repair of a motor vehicle. For the owner who does not possess any, their purchase will prove a considerable expense, offsetting some of the savings made by doing-it-yourself. However, provided that the tools purchased are of good quality, they will last for many years and prove an extremely worthwhile investment.

To help the average owner to decide which tools are needed to carry out the various tasks detailed in this manual, we have compiled three lists of tools under the following headings: *Maintenance and minor repair, Repair and overhaul,* and *Special*. The newcomer to practical mechanics should start off with the *Maintenance and minor repair* tool kit and confine himself to the simpler jobs around the vehicle. Then, as his confidence and experience grow, he can undertake more difficult tasks, buying extra tools as, and when, they are needed. In this way, a *Maintenance and minor repair* tool kit can be built-up into a *Repair and overhaul* tool kit over a considerable period of time without any major cash outlays. The experienced do-it-yourselfer will have a tool kit good enough for most repair and overhaul procedures and will add tools from the *Special* category when he feels the expense is justified by the amount of use these tools will be put to.

It is obviously not possible to cover the subject of tools fully here. For those who wish to learn more about tools and their use there is a book entitled *How to Choose and Use Car Tools* available from the publishers of this manual.

Maintenance and minor repair tool kit

The tools given in this list should be considered as a minimum requirement if routine maintenance, servicing and minor repair operations are to be undertaken. We recommend the purchase of combination spanners (ring one end, open-ended the other); although more expensive than open-ended ones, they do give the advantages of both types of spanner.

Combination spanners - 10, 11, 13, 14, 17 mm
Adjustable spanner - 9 inch
Engine sump/gearbox drain plug key
Spark plug spanner (with rubber insert)
Spark plug gap adjustment tool
Set of feeler gauges
Brake bleed nipple spanner
Screwdriver - 4 in. long x 1/4 in. dia. (plain)
Screwdriver - 4 in. long x 1/4 in. dia. (crosshead)
Combination pliers - 6 inch
Hacksaw, junior
Tyre pump
Tyre pressure gauge
Oil can
Fine emery cloth (1 sheet)
Wire brush (small)
Funnel (medium size)

Repair and overhaul tool kit

These tools are virtually essential for anyone undertaking any major repairs to a motor vehicle, and are additional to those given in the *Maintenance and minor repair* list. Included in this list is a comprehensive set of sockets. Although these are expensive they will be found invaluable as they are so versatile - particularly if various drives are included in the set. We recommend the ½ in square-drive type, as this can be used with most proprietary torque spanners. If you cannot afford a socket set, even bought piecemeal, then inexpensive tubular box wrenches are a useful alternative.

The tools in this list will occasionally need to be supplemented by tools from the *Special* list.

Sockets (or box spanners) to cover range in previous list
Reversible ratchet drive (for use with sockets)
Extension piece, 10 inch (for use with sockets)
Universal joint (for use with sockets)
Torque wrench (for use with sockets)
Mole wrench - 8 inch
Ball pein hammer
Soft-faced hammer, plastic or rubber
Screwdriver - 6 in long x 5/16 in dia (flat blade)
Screwdriver - 2 in long x 5/16 in square (flat blade)
Screwdriver - 1½ in long x ¼ in dia (cross blade)
Screwdriver - 3 in long x 1/8 in dia (electricians)
Pliers - electricians side cutters
Pliers - needle nosed
Pliers - circlip (internal and external)
Cold chisel - ½ inch
Scriber (this can be made by grinding the end of a broken hacksaw blade)
Scraper (this can be made by flattening and sharpening one end of a piece of copper pipe)
Centre punch
Pin punch
Hacksaw
Valve grinding tool
Steel rule/straight edge
Allen keys
Selection of files
Wire brush (large)
Axle-stands
Jack (strong scissor or hydraulic type)

Special tools

The tools in this list are those which are not used regularly, are expensive to buy, or which need to be used in accordance with their manufacturers' instructions. Unless relatively difficult mechanical jobs are undertaken frequently, it will not be economic to buy many of these tools. Where this is the case, you could consider clubbing together with friends (or a motorists' club) to make a joint purchase, or borrowing the tools against a deposit from a local garage or tool hire specialist.

The following list contains only those tools and instruments freely available to the public, and not those special tools produced by the vehicle manufacturer specifically for its dealer network. You will find occasional references to these manufacturers' special tools in the text of this manual. Generally, an alternative method of doing the job without the vehicle manufacturers' special tool is given. However, sometimes, there is no alternative to using them. Where this is the case and the relevant tool cannot be bought or borrowed you will have to entrust the work to a franchised garage.

Valve spring compressor
Piston ring compressor
Balljoint separator
Universal hub/bearing puller
Impact screwdriver
Micrometer and/or vernier gauge
Dial gauge
Stroboscopic timing light
Dwell angle meter/tachometer
Universal electrical multi-meter
Cylinder compression gauge
Lifting tackle
Trolley jack
Light with extension lead

Buying tools

For practically all tools, a tool dealer is the best source since he will have a very comprehensive range compared with the average garage or accessory shop. Having said that, accessory shops often offer excellent quality tools at discount prices, so it pays to shop around.

Remember, you don't have to buy the most expensive items on the shelf, but it is always advisable to steer clear of the very cheap tools. There are plenty of good tools around at reasonable prices, so ask the proprietor or manager of the shop for advice before making a purchase.

Care and maintenance of tools

Having purchased a reasonable tool kit, it is necessary to keep the tools in a clean serviceable condition. After use, always wipe off any dirt, grease and metal particles using a clean, dry cloth, before putting the tools away. Never leave them lying around after they have been used. A simple tool rack on the garage or workshop wall, for items such as screwdrivers and pliers is a good idea. Store all normal spanners and sockets in a metal box. Any measuring instruments, gauges, meters, etc, must be carefully stored where they cannot be damaged or become rusty.

Take a little care when tools are used. Hammer heads inevitably become marked and screwdrivers lose the keen edge on their blades from time to time. A little timely attention with emery cloth or a file will soon restore items like this to a good serviceable finish.

Working facilities

Not to be forgotten when discussing tools, is the workshop itself. If anything more than routine maintenance is to be carried out, some form of suitable working area becomes essential.

It is appreciated that many an owner mechanic is forced by circumstances to remove an engine or similar item, without the benefit of a garage or workshop. Having done this, any repairs should always be done under the cover of a roof.

Wherever possible, any dismantling should be done on a clean flat workbench or table at a suitable working height.

Any workbench needs a vice: one with a jaw opening of 4 in (100 mm) is suitable for most jobs. As mentioned previously, some clean dry storage space is also required for tools, as well as the lubricants, cleaning fluids, touch-up paints and so on which become necessary.

Another item which may be required, and which has a much more general usage, is an electric drill with a chuck capacity of at least 5/16 in (8 mm). This, together with a good range of twist drills, is virtually essential for fitting accessories such as wing mirrors and reversing lights.

Last, but not least, always keep a supply of old newspapers and clean, lint-free rags available, and try to keep any working area as clean as possible.

Spanner jaw gap comparison table

Jaw gap (in)	Spanner size
0.250	1/4 in AF
0.276	7 mm
0.313	5/16 in AF
0.315	8 mm
0.344	11/32 in AF; 1/8 in Whitworth
0.354	9 mm
0.375	3/8 in AF
0.394	10 mm
0.433	11 mm
0.438	7/16 in AF
0.445	3/16 in Whitworth; 1/4 in BSF
0.472	12 mm
0.500	1/2 in AF
0.512	13 mm
0.525	1/4 in Whitworth; 5/16 in BSF
0.551	14 mm
0.563	9/16 in AF
0.591	15 mm
0.600	5/16 in Whitworth; 3/8 in BSF
0.625	5/8 in AF
0.630	16 mm
0.669	17 mm
0.686	11/16 in AF
0.709	18 mm
0.710	3/8 in Whitworth; 7/16 in BSF
0.748	19 mm
0.750	3/4 in AF
0.813	13/16 in AF
0.820	7/16 in Whitworth; 1/2 in BSF
0.866	22 mm
0.875	7/8 in AF
0.920	1/2 in Whitworth; 9/16 in BSF
0.938	15/16 in AF
0.945	24 mm
1.000	1 in AF
1.010	9/16 in Whitworth; 5/8 in BSF
1.024	26 mm
1.063	1 1/16 in AF; 27 mm
1.100	5/8 in Whitworth; 11/16 in BSF
1.125	1 1/8 in AF
1.181	30 mm
1.200	11/16 in Whitworth; 3/4 in BSF
1.250	1 1/4 in AF
1.260	32 mm
1.300	3/4 in Whitworth; 7/8 in BSF
1.313	1 5/16 in AF
1.390	13/16 in Whitworth; 15/16 in BSF
1.417	36 mm
1.438	1 7/16 in AF
1.480	7/8 in Whitworth; 1 in BSF
1.500	1 1/2 in AF
1.575	40 mm; 15/16 in Whitworth
1.614	41 mm
1.625	1 5/8 in AF
1.670	1 in Whitworth; 1 1/8 in BSF
1.688	1 11/16 in AF
1.811	46 mm
1.813	1 13/16 in AF
1.860	1 1/8 in Whitworth; 1 1/4 in BSF
1.875	1 7/8 in AF
1.969	50 mm
2.000	2 in AF
2.050	1 1/4 in Whitworth; 1 3/8 in BSF
2.165	55 mm
2.362	60 mm

General dimensions and weights

Dimensions			
Overall length	...	167 in (4245 mm)	
Overall width	...	66 in (1680 mm)	
Overall height (at kerb weight)	...	54.7 in (1390 mm)	
Wheelbase	...	102.5 in (2604 mm)	
Track (front)	...	55.7 in (1415 mm)	
Track (rear)	...	54.7 in (1390 mm)	
Ground clearance	...	5.0 in (130 mm)	
Weights			
Kerb weight (approximate):			
1294 cc	...	2315 lb (1050 kg)	
1442 cc (manual)	...	2370 lb (1075 kg)	
1592 cc (automatic)	...	2414 lb (1095 kg)	
Maximum trailer weight		**Braked**	**Unbraked**
1294 cc	...	1650 lb (750 kg)	1166 lb (530 kg)
1442 cc (manual)	...	1980 lb (900 kg)	1177 lb (535 kg)
1592 cc (automatic)	...	2426 lb (1100 kg)	1201 lb (545 kg)
Maximum roof rack load	...	110 lb (50 kg)	
Maximum gross vehicle weight	...	Refer to identification plate on car	

Chapter 1 Engine

For modifications, and information applicable to later models, see Supplement at end of manual

Contents

Specifications

General

Engine type ...	ohv, four cylinder in-line, crossflow head
Location ...	Transverse, inclined to rear 41°
Material ...	Crankcase - cast iron
	Cylinder head - aluminium alloy
Firing order ...	1 - 3 - 4 - 2 (No. 1 at flywheel end)
Direction of crankshaft rotation ...	Anticlockwise viewed from flywheel end

Engine data (1294 cc type)

Bore	3.02 in (76.7 mm)
Stroke	2.75 in (70 mm)
Compression ratio	9.5 : 1
Maximum rpm	5,700 in 4th gear
Idling	850

Engine data (1442 cc type)

Bore	3.02 in (76.7 mm)
Stroke	3.07 in (78 mm)
Compression ratio	9.5 : 1
Maximum rpm	6,100 in 4th gear
Idling rpm	900

Crankshaft and connecting rods

Main bearing journal diameter (red)	2.0478-2.0482 in (51.975-51.985 mm)
Main bearing journal diameter (blue)	2.0474-2.0478 in (51.966-51.976 mm)
Shell thickness (red)	0.075-0.076 in (1.915-1.924 mm)
Shell thickness (blue)	0.076-0.077 in (1.924-1.933 mm)
Running clearance	0.00158-0.00307 in (0.040-0.078 mm)
Big-end crankpin diameter (red)	1.6137-1.6140 in (40.957-40.965 mm)
Big-end crankpin diameter (blue)	1.6133-1.6137 in (40.949-40.957 mm)
Shell thickness (red)	0.059-0.060 in (1.492-1.501 mm)
Shell thickness (blue)	0.060-0.061 in (1.500-1.509 mm)
Running clearance	0.00118-0.00252 in (0.030-0.064 mm)
Endfloat of crankshaft	0.0035-0.011 in (0.09 to 0.27 mm)
Connecting rod side play	0.004-0.011 in (0.10 to 0.27 mm)
Crankshaft undersizes	0.004-0.008-0.020 in (0.1-0.2-0.5 mm)
Width of centre main bearing journal	1.238-1.239 in (31.43-31.47 mm)
Small-end bore diameter	0.8654-0.8660 in (21.965-21.980 mm)
Crankshaft thrust washer thickness	0.091-0.093 in (2.31-2.36 mm)

Pistons

Material	Aluminium alloy
Rings	2 compression, 1 oil control
Maximum weight difference allowable between any two pistons	3 grammes

Piston diameters

Class A	3.0203-3.0206 in (76.6575-76.6650 mm)
Class B	3.0206-3.0208 in (76.6650-76.6725 mm)
Class C	3.0208-3.0211 in (76.6725-76.6800 mm)
Class D	3.0211-3.0214 in (76.6800-76.6875 mm)
Oversize pistons available	0.0039-0.0157 in (0.1-0.4 mm)
Piston to bore clearance	0.0008-0.0015 in (0.022-0.037 mm)

Piston ring gaps

Top ring	0.010 - 0.018 in (0.25 - 0.45 mm)
Second ring	0.010 - 0.018 in (0.25 - 0.45 mm)
Oil control ring	0.008 - 0.016 in (0.20 - 0.40 mm)

Gudgeon pins

Material	Steel
Length	2.627 in (66.7 mm)
Outside diameter	0.8664-0.8666 in (21.991-21.995 mm)
Clearance in piston	0.0004-0.0007 in (0.010-0.019 mm)
Interference fit in connecting rod	0.0006-0.0012 in (0.016-0.030 mm)

Camshaft

Bearing journal diameter (numbered from flywheel end):	
No 1	1.3962-1.3970 in (35.439-35.459 mm)
No 2	1.6129-1.6137 in (40.939-40.959 mm)
No 3	1.6326-1.6334 in (41.439-41.459 mm)
Bearing clearance	0.00098-0.00299 in (0.025-0.076 mm)
Endfloat	0.004-0.008 in (0.10-0.20 mm)

Tappets (cam followers)

Outside diameter	0.9051-0.9062 in (22.974-23.000 mm)
Clearance in bore	Zero-0.018 in (zero-0.047 mm)
Length	1.556-1.595 in (39.5-40.5 mm)

Pushrod lengths

1294 cc engine	7.919 in (201.0 mm)
1442 cc engine	8.530 in (216.5 mm)

Valves

Inlet valves	
Head diameter	1.33858 in (34 mm)
Stem diameter	0.3138 to 0.3144 in (7.970 to 7.985 mm)
Stem to guide clearance	0.0015 to 0.0028 in (0.037 to 0.070 mm)
Face angle	46°
Exhaust valves	
Head diameter	1.10 in (28 mm)
Stem diameter	0.313 - 0.314 in (7.950 - 7.965 mm)
Stem to guide clearance	0.0022 - 0.0035 in (0.057 - 0.090 mm)
Face angle	46°

Valve guides

Inside diameter	0.3160-0.3167 in (8.022-8.040 mm)
Outside diameter (standard)	0.5517-0.5520 in (14.003-14.012 mm)
Length	2.047 in (52.0 mm)

Valve seats

Seat angle	46°
Seat width	0.059 in (1.5 mm)

Valve springs

Free length	1.905 in (48.4 mm)

Valve clearances

Inlet (cold)	0.010 in (0.25 mm)
Inlet (hot)	0.012 in (0.30 mm)
Exhaust (cold)	0.012 in (0.30 mm)
Exhaust (hot)	0.014 in (0.35 mm)

Lubrication system

Oil pump	Externally mounted, camshaft driven
Oil filter	Full flow, disposable cartridge type
Oil pump shaft side play	0.002 - 0.020 in (0.05 - 0.50 mm)
Sump capacity	5.28 Imp. pints (3.0 litres)

Torque wrench settings

	lb f ft	kg f m
Cylinder head bolts	52	7.18
Main bearing cap bolts	47	6.50
Connecting rod cap bolts	26	3.59
Oil relief valve cap	20	2.77
Flywheel to crankshaft bolts	40	5.53
Drive plate to crankshaft (auto. gearbox)	40	5.53
Drive plate to torque converter (Auto. gearbox)	26	3.59
Crankshaft pulley bolt	96	13.28
Oil pressure switch	22	3.04
Inlet manifold nuts	10	1.38
Exhaust manifold nuts	14.5	2.01
Spark plugs	20	2.77
Crankshaft oil seal housing to cylinder block	8.5	1.18
Camshaft retaining plate	10.5	1.45
Camshaft sprocket bolts	10.5	1.45
Timing gear cover bolts	14.5	2.01
Sump to block bolts	8.5	1.18
Sump pan bolts	6.5	0.89
Oil pump retaining bolts	8.5	1.18

1 General description

The models covered in this manual are available with either the 1294 cc or 1442 cc engine. Both engines have the same bore, but the larger capacity is achieved by fitting longer connecting rods in conjunction with a taller block, increasing the piston stroke.

The engines are identical in design comprising a four cylinder in-line overhead valve unit, with a cast iron block and aluminium cylinder head. The engine is located in a transverse position, (east-west) within the engine compartment and is inclined to the rear to provide clearance for the low profile bonnet.

The power unit transmits motion to the front road wheels through a unitary gearbox/differential (mounted ahead of and below left of the engine) and open drive shafts.

A forged steel crankshaft is fitted, supported on five bearings with renewable shells.

The valves are operated by push rods and the pistons are of aluminium alloy with two compression and one oil control ring. Piston assemblies are removed through the top of the block.

Lubrication is by means of an externally mounted, camshaft driven pump. A full flow oil filter is incorporated in the system and closed circuit type of crankcase breathing is employed.

Cooling is by sealed system and the radiator is cooled by an electrically driven, thermostatically controlled fan.

An electronic ignition system is fitted and in addition to being extremely reliable, requires no adjustment, unlike the earlier contact breaker type.

2 Major operations with engine in position in vehicle

The following operations may be carried out with the power unit in position in the vehicle.

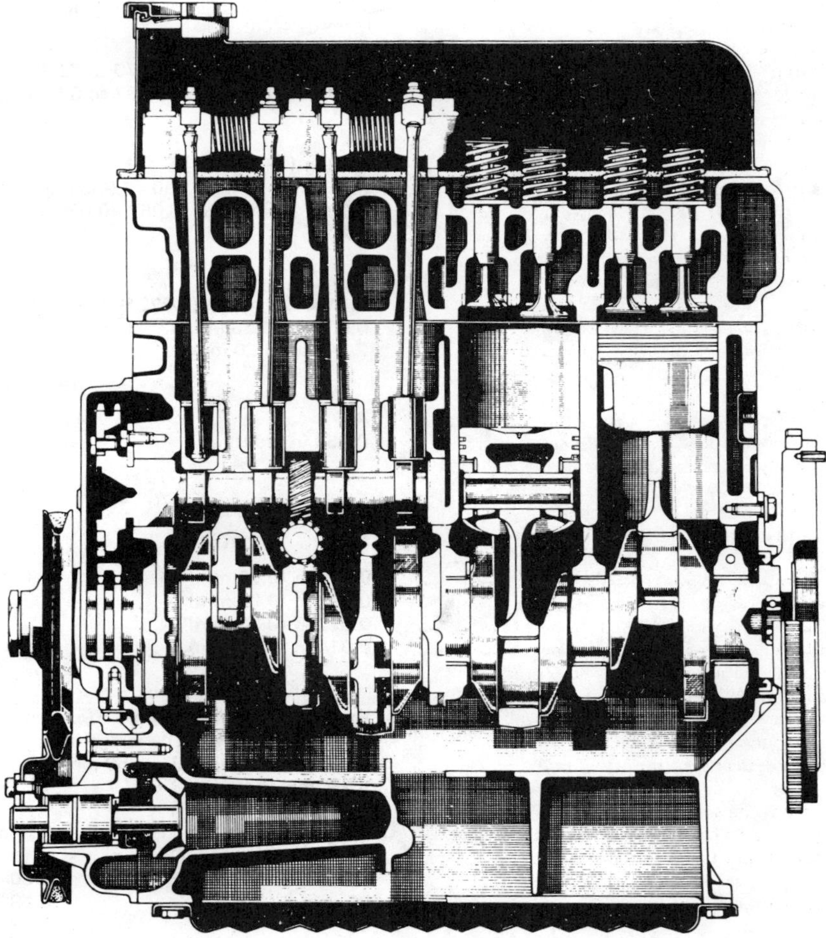

Fig. 1.1. Cross-sectional view of engine

a) Removal and refitting of the cylinder head assembly.
b) Removal and refitting of the oil pump.
c) Removal and refitting of sump.
d) Removal and refitting of pistons and connecting rods. (After removal of sump and cylinder head).
e) Removal and refitting of the engine and gearbox mountings.
f) Removal of the gearbox/differential for access to clutch and flywheel.
g) Removal and renewal of the major ancillaries, starter motor, water pump, distributor, alternator, oil filter.

3 Major operations with engine/gearbox removed from vehicle

a) Renewal of engine main bearings.
b) Removal of crankshaft.

4 Engine/gearbox/differential (single unit) removal

1 There is no point in removing the gearbox/differential unit separately from the engine when major servicing operations are to be carried out on the engine but the combined engine/gearbox/differential should be removed as a single unit for later separation.
2 Before commencing operations, hire or borrow a hoist or lifting tackle and a trolley onto which the engine unit can be lowered after removal from **below** the vehicle.
3 Jack up the car securely. A sufficient height is essential to enable the power unit to be withdrawn forwards from the engine compartment. Suitable jacks, stands and chocks should therefore be acquired. Alternatively, an inspection pit may be used but this method will present difficulties in moving the car after engine removal and hoisting from the pit floor.
4 Engine lifting eyes are provided in the front engine mountings and to ensure the engine is supported at the correct angle, a wire or terylene rope sling 44.5 in (1130 mm) in length and terminating in 'S' hooks should be used.
5 Begin by disconnecting both battery terminals.
6 Mark the position of the bonnet hinges using a soft-leaded pencil or tape. Get an assistant to support the weight of the bonnet, remove the two bolts from each hinge and lift away the bonnet assembly.
7 Remove the filler cap from the coolant expansion bottle, unscrew the drain plug from the bottom of the water pump housing (Fig. 1.3) and drain the coolant into a suitable container.
8 Apply the handbrake firmly, place chocks behind the rear wheels and slacken the front wheel securing bolts.
9 Using a trolley jack placed beneath the front suspension cross-member, raise the car as high as possible and support it on axle stands placed under the roll bar support brackets or front jacking pads. Use wooden packing pieces if necessary.
10 Remove both front road wheels.
11 Referring to Fig. 1.4 unscrew the alternator drive belt guard bolts and remove the guard.
12 Disconnect the electrical leads from the rear of the alternator.
13 Disconnect the cable and leads from the starter motor solenoid.
14 Remove the clutch slave cylinder from the bellhousing and tie it out of the way. Do not disconnect the hydraulic feed pipe from the cylinder otherwise it will be necessary to bleed the system.
15 Detach the gearchange linkage balljoints and brackets where necessary and lower it away from the engine.
16 Remove the front exhaust pipe from the engine manifold flange.
17 Drain the gearbox and final drive units into a suitable container.

18 Remove both front telescopic dampers and in their place fit the special retaining rods. These are obtainable from your Chrysler dealer. (Tool No. CF 0004). These should be fitted as shown in Fig. 1.5.
19 The purpose of these rods is to prevent the suspension arms springing apart when at a later stage the bottom balljoint is removed.
20 An alternative method is to place a jack beneath the outer end of the suspension arm and raise it sufficiently to counteract the downward thrust of the suspension torque rod. However, extreme care must be taken to ensure there is no possibility of the jack slipping off the end of the suspension arm. If the latter method is used it is not necessary to remove the telescopic dampers.
21 Remove the balljoint securing nuts from the lower suspension arms and using a suitable extractor disconnect the joints from the suspension arms (photo).
22 Remove the large nut from the centre of each wheel hub and carefully pivot the complete hub and brake assembly outwards and upward until the outer splined end of the drive shaft can be withdrawn from the centre of the hub (see Fig. 1.6 and photo).
23 The inner end of each drive shaft is retained in the differential unit by a spring ring and the shaft should be pulled firmly outwards to disengage it from the final drive unit. If possible try and avoid pulling the inner universal joint apart during this operation. If difficulty is experienced in extracting the drive shafts, try prising them out using a large screwdriver placed between the inner end of the drive shaft and the final drive casing.
24 When both shafts have been removed, lower the hub and brake assembly back into position and temporarily refit the balljoints to the suspension arms and refit the securing nuts. Fit special tool 80317M or two pieces of tubing into the differential side gears to prevent them being displaced while the driveshafts are removed (pre-1985 models - see Chapter 13, Section 7, paragraph 238).
25 Remove the complete engine rear mounting rubber and bracket assembly from the engine and crossmember (see Fig. 1.7).
26 Remove the air cleaner, intake hoses and carburettor intake adaptor.
27 Detach the brake servo hose from the servo unit.
28 Disconnect all the cooling system and heater hoses from the engine.
29 Disconnect the choke and throttle controls from the carburettor referring to Chapter 3 if necessary.
30 Disconnect the fuel feed hose from the fuel pump. Plug the hose and tie it back out of the way.
31 Disconnect the earthing leads from the engine block.
32 Disconnect all electrical leads from the engine. Tie them back out of the way and mark them for ease of reconnection.
33 Remove the distributor cover and rotor arm.
34 Fit the prepared lifting sling to the engine.
35 Take the weight of the engine on a hoist and then unscrew and remove the engine side mounting bolts to leave the bonded rubber blocks still secured to the vehicle frame.
36 Lower the engine carefully onto a trolley jack taking care not to damage the oil filter assembly.

4.21 Removing the lower arm balljoint

37 Remove the lifting sling and withdraw the engine/gearbox/transmission unit forward from beneath the car. Adjust the height of the supporting jacks if necessary to accomplish this.

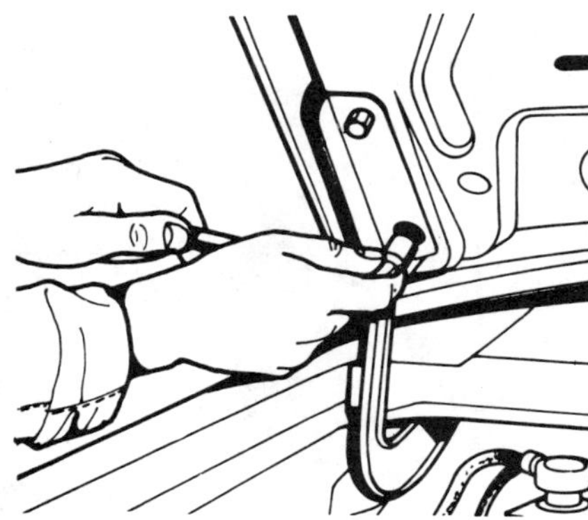

Fig. 1.2. Removing the bonnet hinge bolts

Fig. 1.3. Location of coolant drain plug

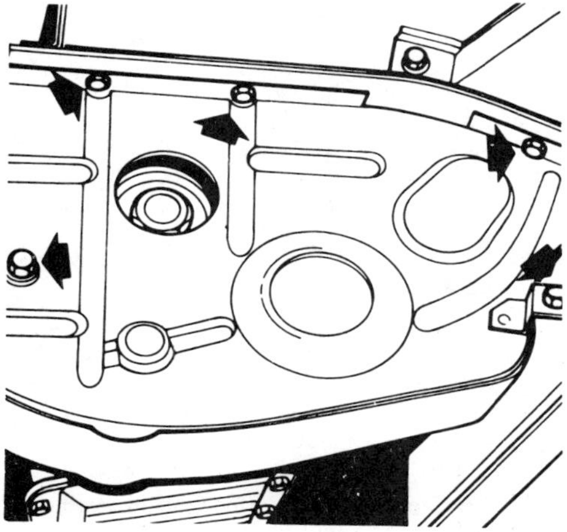

Fig. 1.4. Alternator drive belt cover securing bolts

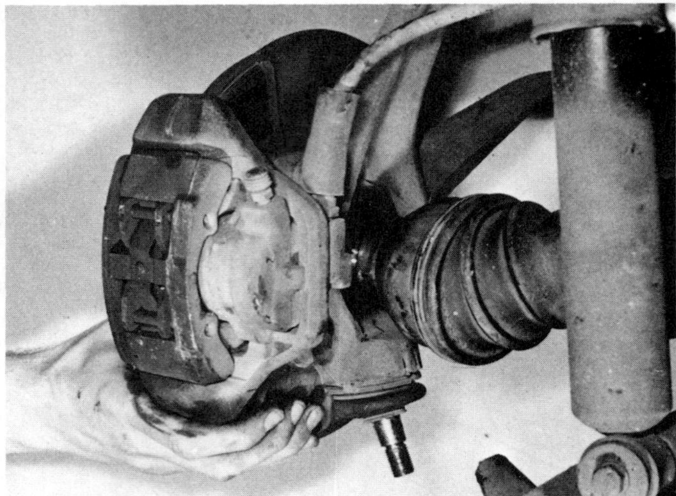

4.22 Lifting the brake and hub assembly clear of the suspension arm

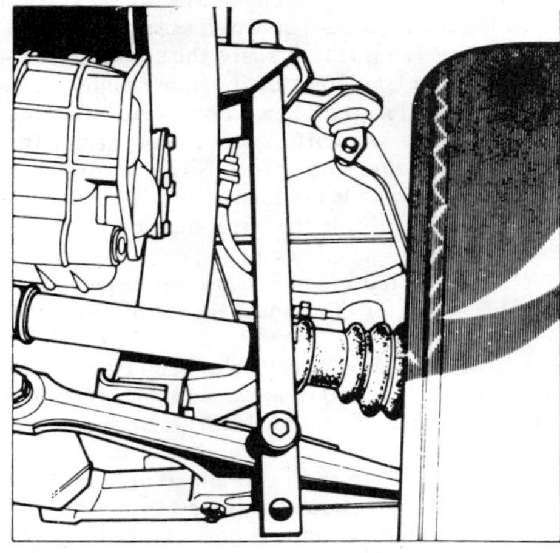

Fig. 1.5. Suspension tool No. CF0004 correctly positioned

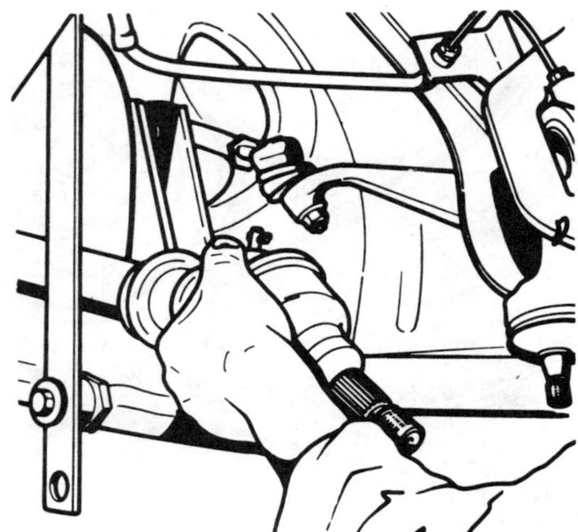

Fig. 1.6. Removing a drive shaft

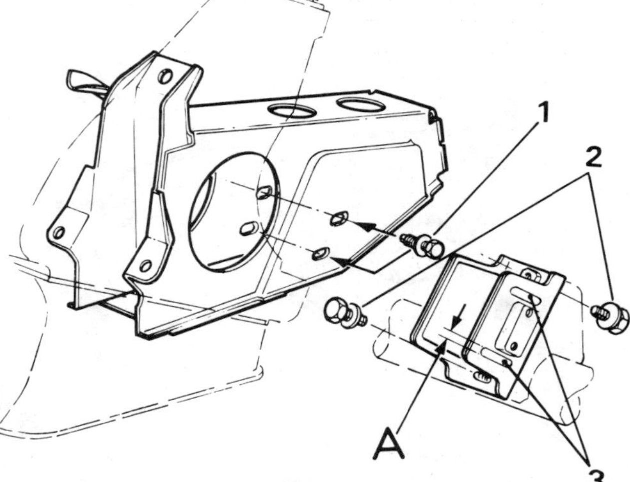

Fig. 1.7. Engine rear mounting bracket assembly

A Engine mounting rubber
1 Bolts securing mounting rubber to support bracket
2 Bolts securing mounting to crossmember
3 Slots in rubber mounting

5 Engine dismantling - general

1 It is best to mount the engine on a dismantling stand but if one is not available, then stand the engine on a strong bench so as to be at a comfortable working height. Failing this, the engine can be stripped down on the floor.

2 During the dismantling process the greatest care should be taken to keep the exposed parts free from dirt. As an aid to achieving this, it is a sound scheme to clean down the outside of the engine, removing all traces of oil and congealed dirt.

3 Use paraffin or a good grease solvent such as 'Gunk'. The latter compound will make the job much easier, as, after the solvent has been applied and allowed to stand for a time, a vigorous jet of water will wash off the solvent and all the grease and filth. If the dirt is thick and deeply embedded, work the solvent into it with a stiff paint brush.

4 Finally, wipe down the exterior of the engine with a rag and only then, when it is quite clean should the dismantling process begin. As the engine is stripped, clean each part in a bath of paraffin or petrol.

5 Never immerse parts with oilways in paraffin, eg, the crankshaft, but, to clean, wipe down with a petrol dampened rag. Oilways can be cleaned out with wire. If an air line is present all parts can be blown dry and the oilways blown through as an added precaution.

6 Re-use of old engine gaskets is a false economy and can give rise to oil and water leaks, if nothing worse. To avoid the possibility of trouble after the engine has been reassembled **ALWAYS** use new gaskets throughout.

7 Do not throw the old gaskets away as it sometimes happens that an immediate replacement cannot be found and the old gasket is then very useful as a template. Hang up the old gaskets as they are removed on a suitable hook or nail.

8 To strip the engine it is best to work from the top down. The sump provides a firm base on which the engine can be supported in an upright position. When the stage where the sump must be removed is reached, the engine can be turned on its side and all other work carried out with it in this position.

9 Wherever possible, refit nuts, bolts and washers finger-tight from wherever they were removed. This helps avoid later loss and muddle. If they cannot be refitted then lay them out in such a fashion that it is clear from where they came.

6 Engine/gearbox/differential - separation

1 The gearbox/differential unit should now be removed from the engine. First remove the starter motor.

2 Unscrew the securing bolts and remove the flywheel cover plates. There are two of these, one adjacent to the starter motor and the other near the differential.

3 Remove the bolts which secure the differential unit to the clutch

housing, noting their differing lengths.
4 Remove the bolts which secure the clutch housing to the cylinder block and withdraw the gearbox. During this operation do **NOT** allow the weight of the gearbox to hang upon the primary shaft while it is still engaged with the splined hub of the clutch friction disc or damage to the clutch components may result.

7 Ancillary engine components - removal

1 Before basic engine dismantling begins the engine should be stripped of all its ancillary components. These items should also be removed if a factory exchange reconditioned unit is being purchased. The items comprise:

Warm air intake
Rocker cover
Spark plugs
Distributor and attachment plate, vacuum pipe and HT cables
Fuel pump (retain the thick insulating washer)
Carburettor
Inlet and exhaust manifolds
Sump breather plug and pipe
Thermostat housing on cylinder head
Temperature sender unit
Water inlet pipe on the crankcase
Water manifold
Oil dipstick and bracket
Alternator
Water pump and pulley
Oil filter element
Oil pump and housing assembly

2 Remove the sump plug and drain the oil into a suitable container.
3 The ancillary components have now been removed and the engine can now be dismantled as described in the following Sections.

8 Cylinder head - removal

1 This operation may be carried out with the engine still in position in the car but of course the cooling system must first be drained.
2 Unscrew each cylinder head bolt, a turn at a time, in reverse to the order shown in Fig. 1.9.
3 When all the bolts have been withdrawn, lift off the rocker assembly.
4 Withdraw the pushrods from the tappet blocks (cam followers) by using a twisting motion to break the sealing effect of the oil and prevent displacement of the tappet blocks. Keep the pushrods in exact order for refitting.
5 Remove the cylinder head. Should it be stuck to the block, strike it in several places with a wooden or plastic faced mallet or refit the spark plugs and turn the engine over to enable compression to assist its removal. Never insert a chisel in the gasket joint or attempt to prise the head from the block.
6 Withdraw the tappet blocks and keep them in strict order for exact refitting.

9 Valves and rocker gear - dismantling

1 The valves should be removed from the cylinder head with the aid of a conventional valve spring compressor. Take great care to protect the alloy surfaces of the head from damage during these operations.
2 Compress the valve springs and remove the split cotters, cups and valve springs, seals and washers.
3 Place each valve and its components in strict order so that they may be refitted in their original positions.
4 Where necessary, the rocker shaft assembly may be dismantled by drifting the pins from their locations in the support brackets (Fig. 1.12).

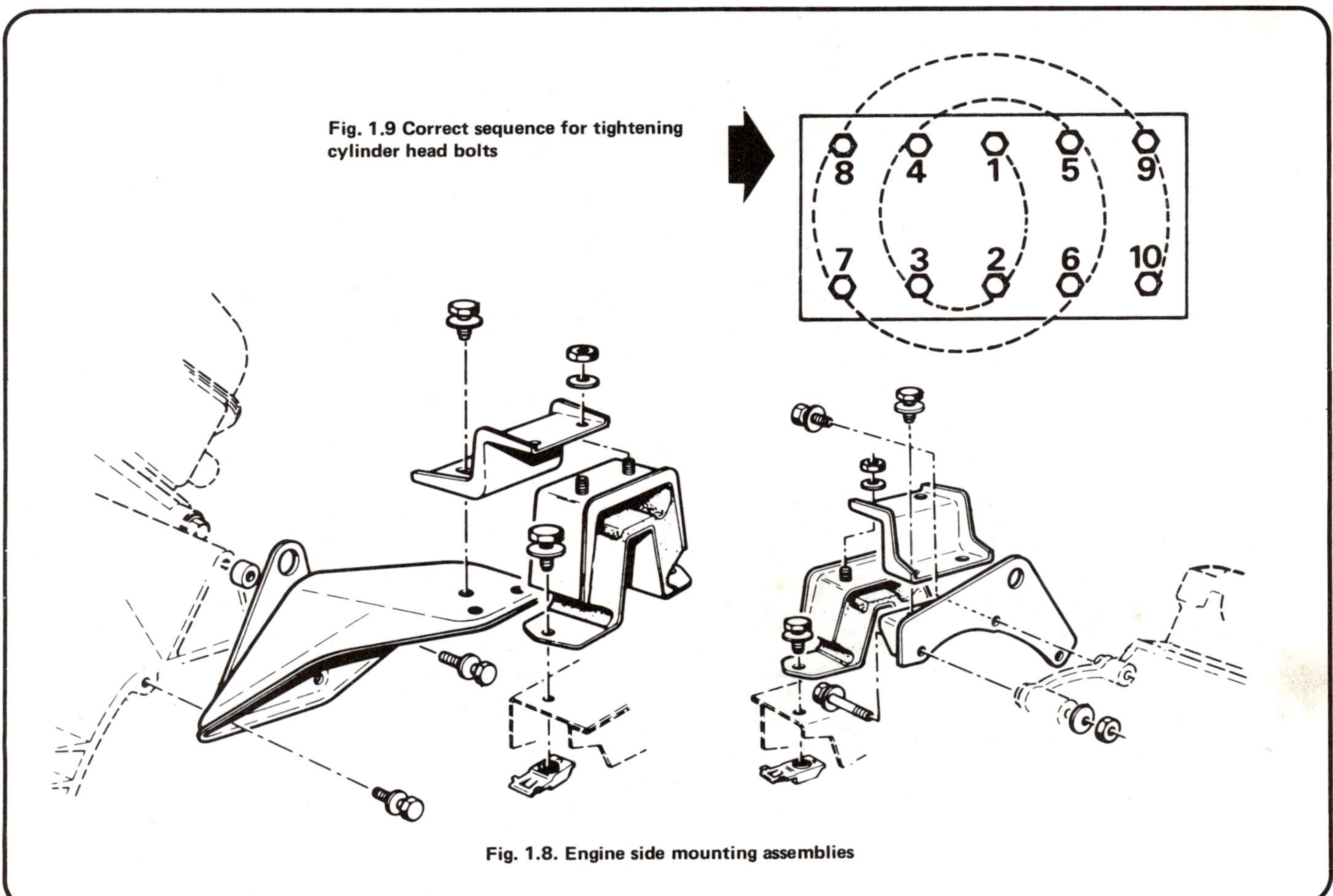

Fig. 1.9 Correct sequence for tightening cylinder head bolts

Fig. 1.8. Engine side mounting assemblies

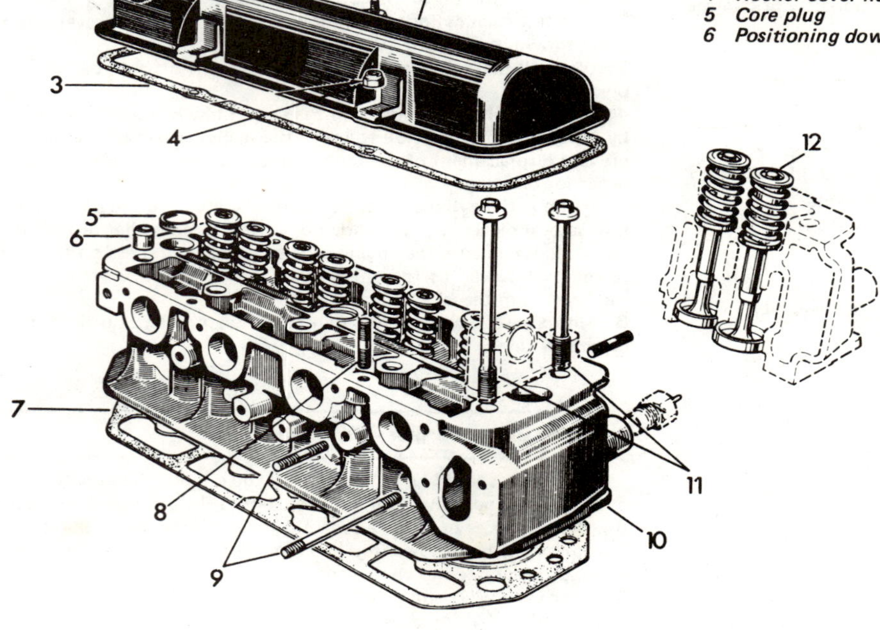

Fig. 1.10. Cylinder head assembly

1 *Rocker cover*
2 *Oil filler cap and washer*
3 *Gasket*
4 *Rocker cover nut*
5 *Core plug*
6 *Positioning dowel*
7 *Cylinder head gasket*
8 *Stud*
9 *Manifold stud*
10 *Cylinder head*
11 *Cylinder head bolts*
12 *Valve assemblies*

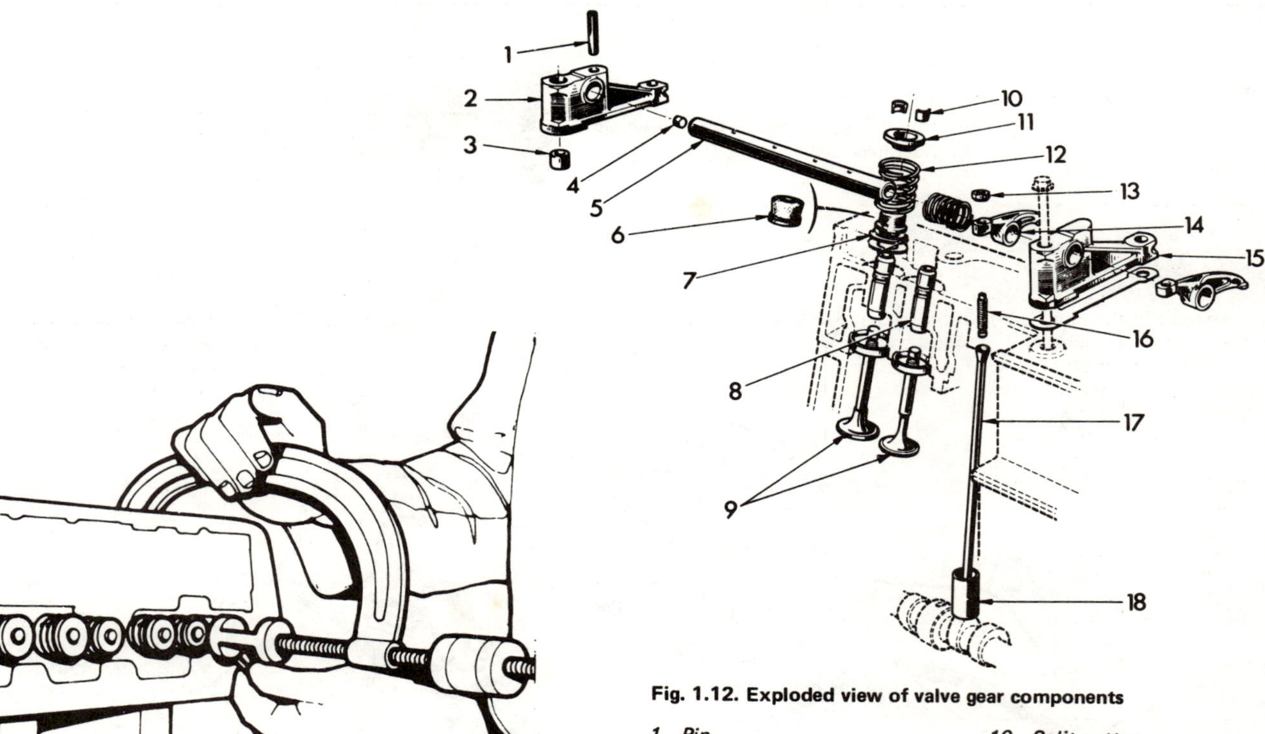

Fig. 1.11. Removing a valve

Fig. 1.12. Exploded view of valve gear components

1 *Pin*
2 *Rocker shaft pillar*
3 *Positioning dowel*
4 *Plug*
5 *Rocker shaft*
6 *Seal*
7 *Washers*
8 *Valve guide*
9 *Valve*
10 *Split cotters*
11 *Cup*
12 *Spring*
13 *Locknut*
14 *Rocker arm*
15 *Rocker shaft pillar*
16 *Adjuster screw*
17 *Pushrod*
18 *Tappet (cam follower)*

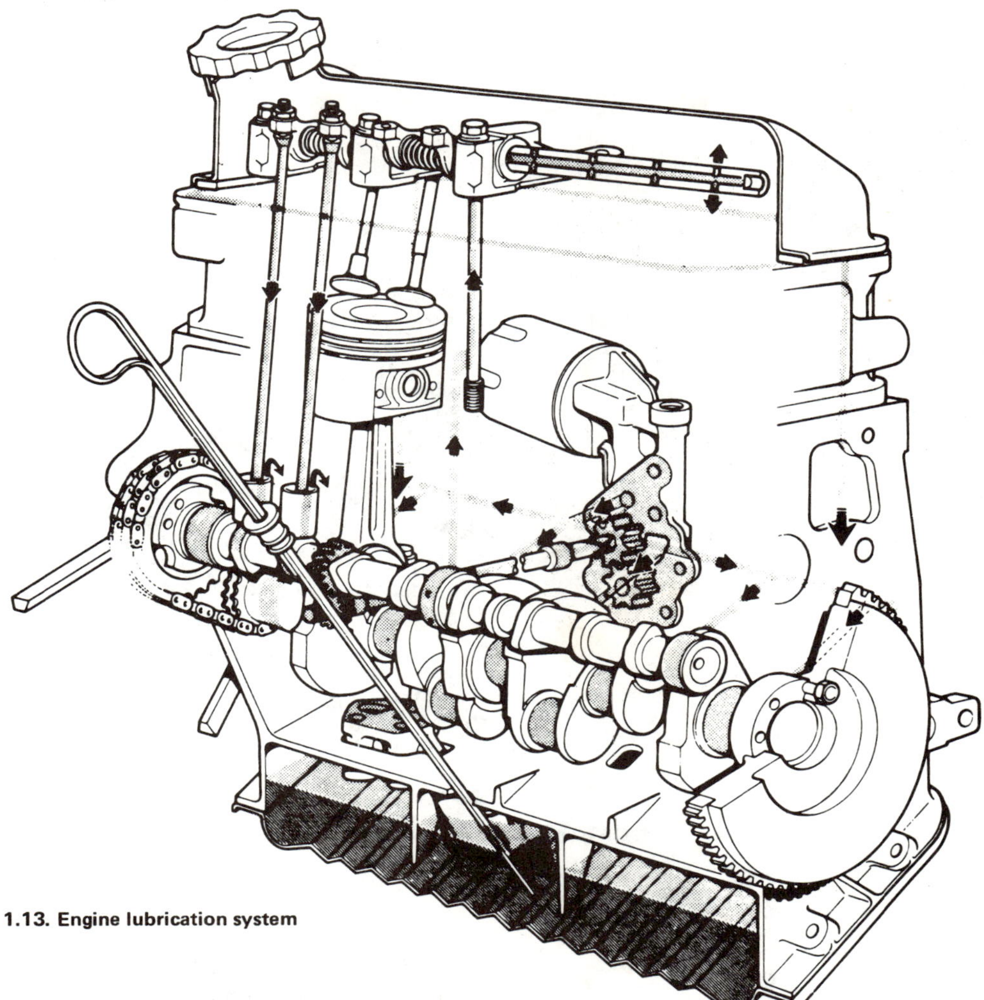

Fig. 1.13. Engine lubrication system

10 Lubrication system and oil sump - removal

1 The engine lubrication system is shown in diagrammatic form in Fig. 1.13.
2 Oil contained in the sump is drawn through a strainer and delivered under pressure by the action of a camshaft-driven pump mounted on the rear of the cylinder block.
3 The oil passes through a full-flow type filter and is fed to the connecting rod, camshaft and crankshaft bearings, the rocker assembly and the distributor and oil pump drive shafts. The cylinder bores, gudgeon pins and valve stems are splash lubricated.
4 To remove the sump assembly, turn the engine upside down, undo the securing screws and lift off the sump base plate and gasket.
5 Remove the oil pump strainer assembly.
6 Remove all the bolts securing the sump to the cylinder block not forgetting the bolts inside the sump casting. Carefully lift the sump assembly away from the cylinder block and remove the paper gasket.
7 If required, the sump can be removed with the engine still in the car by draining the cooling system and engine oil and then removing the starter motor and water pump assembly.

11 Clutch assembly - removal

1 Unscrew and remove the securing bolts which retain the clutch pressure plate cover to the flywheel. Unscrew them a few turns at a time in alternate sequence and mark the position of the cover relative to the flywheel for exact refitting.
2 Lift the cover away and catch the driven plate. Do not let it fall to the ground or it may fracture. Damage could also occur to the friction linings.

12 Crankshaft pulley and timing gear - removal

1 Lock the engine to prevent rotation by inserting a lever or large screwdriver blade in the flywheel ring gear.
2 Unscrew the crankshaft pulley. Several sharp blows on the arm of the spanner will loosen the retaining bolt where leverage will fail.
3 Remove the securing bolts from the timing cover.
4 Unscrew and remove the three bolts which secure the camshaft sprocket flange.
5 Remove the camshaft sprocket with chain which should be detached from the crankshaft sprocket.
6 Remove the crankshaft sprocket which may require the use of a puller or two levers placed behind it. Withdraw the key.

13 Camshaft and oil pump drive shaft - removal

1 From the distributor location hole in the side of the cylinder block, withdraw the distributor dog from the splined end of the drive shaft.
2 Using a small screwdriver, extract the circlip retaining the drive gear to the shaft, remove the gear and withdraw the shaft from the oil pump side of the cylinder block (see Fig. 1.15).
3 Unbolt and remove the camshaft thrust plate.
4 Withdraw the camshaft carefully from the cylinder block so that the bearings are not knocked or damaged by the passage of the cams through them.

14 Piston/connecting rod assemblies - removal

1 Rotate the crankshaft by turning the flywheel so that each piston

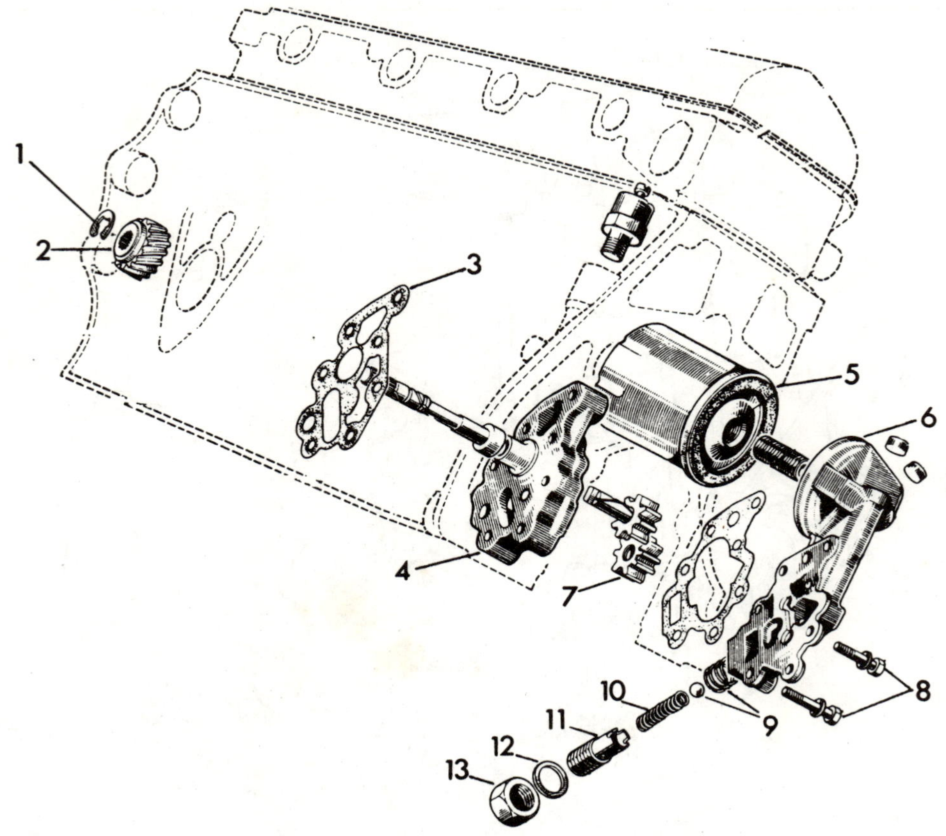

Fig. 1.14. Oil pump and driveshaft components

1	*Circlip*	*4*	*Body*	*7*	*Pump gears*	*10*	*Relief valve spring*
2	*Drive gear*	*5*	*Filter*	*8*	*Securing bolts*	*11*	*Relief valve body*
3	*Gasket*	*6*	*Pump cover*	*9*	*Relief valve ball*	*12*	*Washer*
						13	*Locknut*

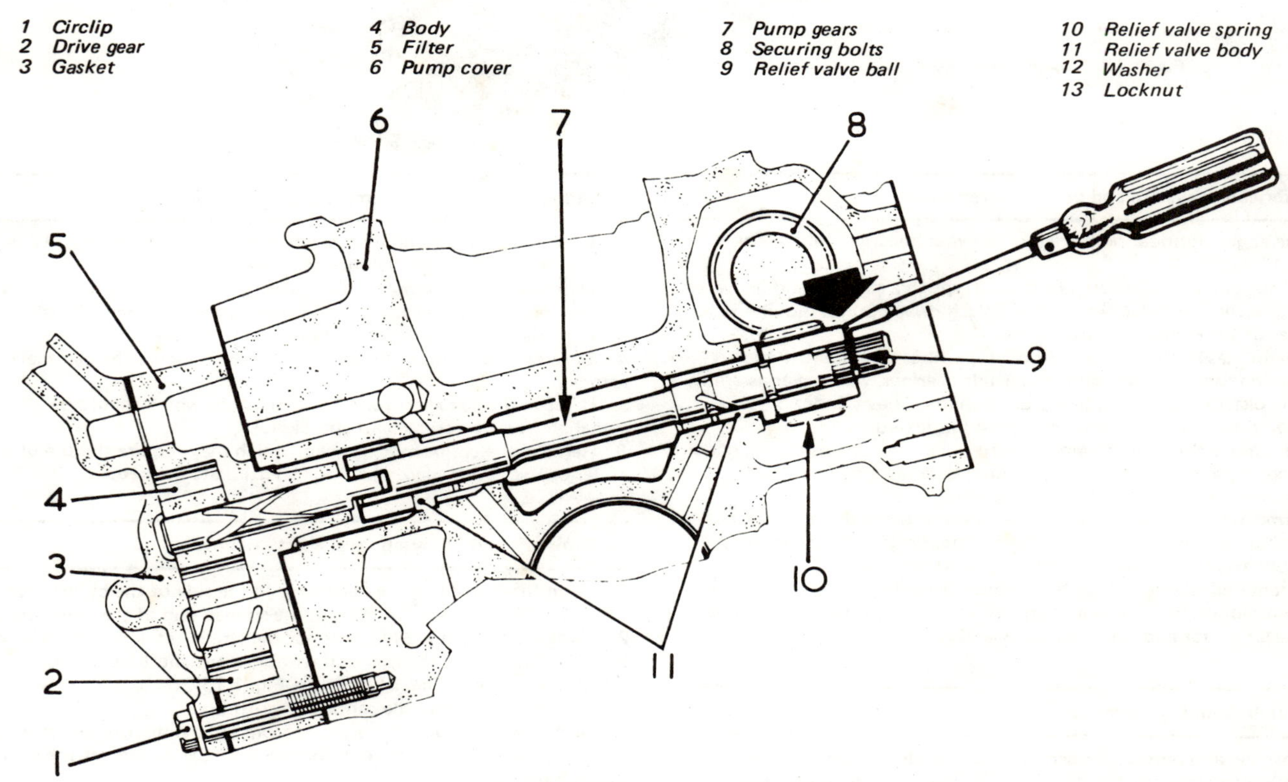

Fig. 1.15. Sectional view of oil pump and driveshaft

1	*Fixing bolt*	*4*	*Driven gear*	*7*	*Driveshaft*	*10*	*Drive gear*
2	*Idler gear*	*5*	*Body*	*8*	*Camshaft*	*11*	*Driveshaft bushes*
3	*Cover*	*6*	*Crankcase*	*9*	*Circlip*		

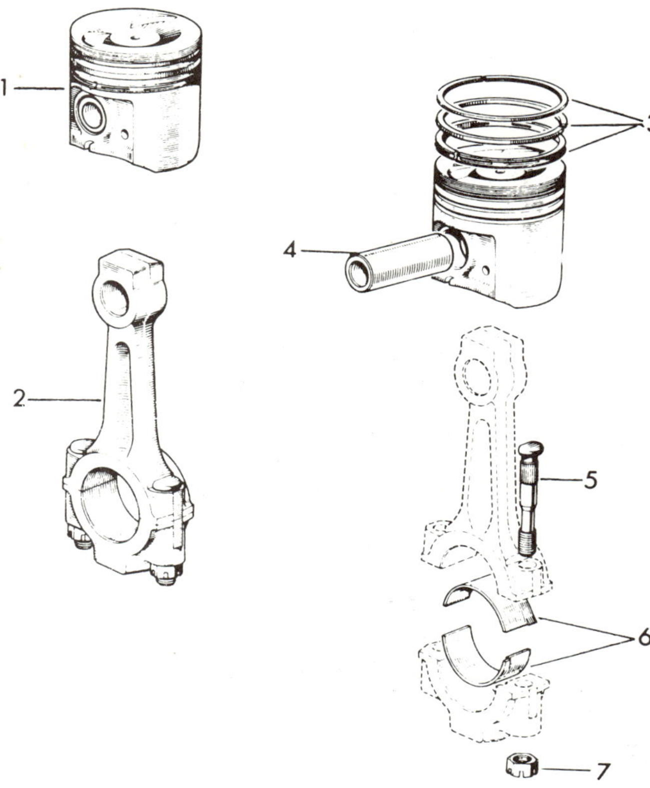

Fig. 1.16. Piston and connecting rod assembly

1 Piston
2 Connecting rod
3 Piston rings
4 Gudgeon pin
5 Big-end bolt
6 Big-end shell bearings
7 Big-end bolt nut

is (approximately) positioned half way down each bore.

2 Taking extreme care, scrape off the carbon ring which will be found to have formed at the top of each cylinder bore. Where the engine has seen considerable service, a 'wear' ring will also have been formed due to the lower portion of the cylinder bore having worn away. It is essential not to score the lower surfaces of the cylinder bore during the scraping operation. Removal of the carbon and 'wear' rings is necessary to permit the pistons and rings to pass out through the top of the block during withdrawal without fracturing the rings.

3 Invert the cylinder block and turn the flywheel until number one piston is either at TDC or alternatively the big-end is at its lowest point.

4 Unscrew and remove the big-end bearing cap nuts.

5 Remove the big-end bearing cap and extract the shells.

6 Push the piston/rod assembly out through the top of the cylinder block but restraining the outward expansion of the piston rings as they emerge from the bore to prevent them breaking.

7 Ensure that the big-end bolt threads do not score the inside of the bore as they travel upwards.

8 Refit the bearing cap to the connecting rod by screwing on the securing nuts a few turns. Mark the bearing cap with the bore location number and note particularly the correct orientation of the connecting rod for refitting.

9 Repeat the operations on the remaining three piston/connecting rod assemblies.

15 Flywheel, crankshaft and main bearings - removal

1 Unscrew and remove the securing bolts from the flywheel.

2 Pull off the flywheel and inset plate.

3 Unscrew and remove the cover plate and gasket and extract the oil seal.

4 Unbolt the five main bearing caps and remove them, noting carefully their numbered sequence.

5 Lift the crankshaft from the crankcase bearings.

6 Remove the shells from the crankshaft main bearing caps and from the crankcase locations. Temporarily refit the main bearing caps.

7 Retain the thrust washers (two) noting their location.

8 The engine is now completely dismantled and checking, examination and renovation of the components should commence as described in the following Sections.

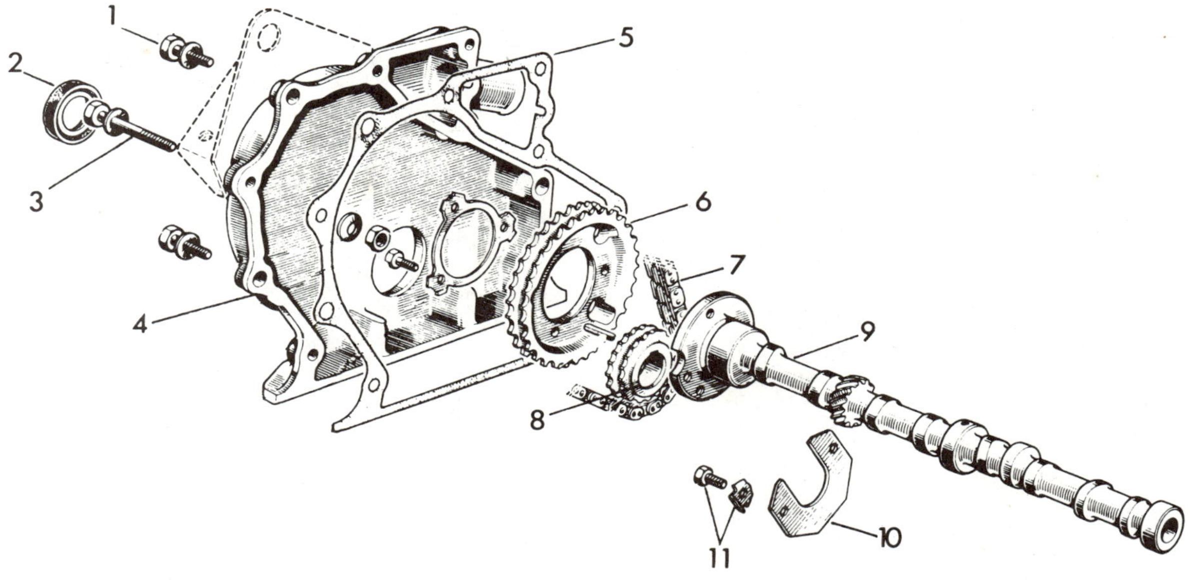

Fig. 1.17. Camshaft and front timing cover

1 Bolt
2 Crankshaft pulley oil seal
3 Timing cover bolt
4 Timing cover
5 Gasket
6 Camshaft sprocket
7 Timing chain
8 Crankshaft sprocket
9 Camshaft
10 Camshaft thrust plate
11 Thrust plate bolt

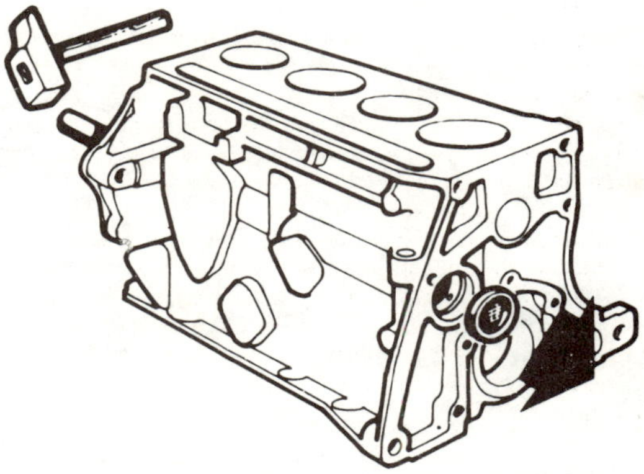

Fig. 1.19. Removing the camshaft end plug

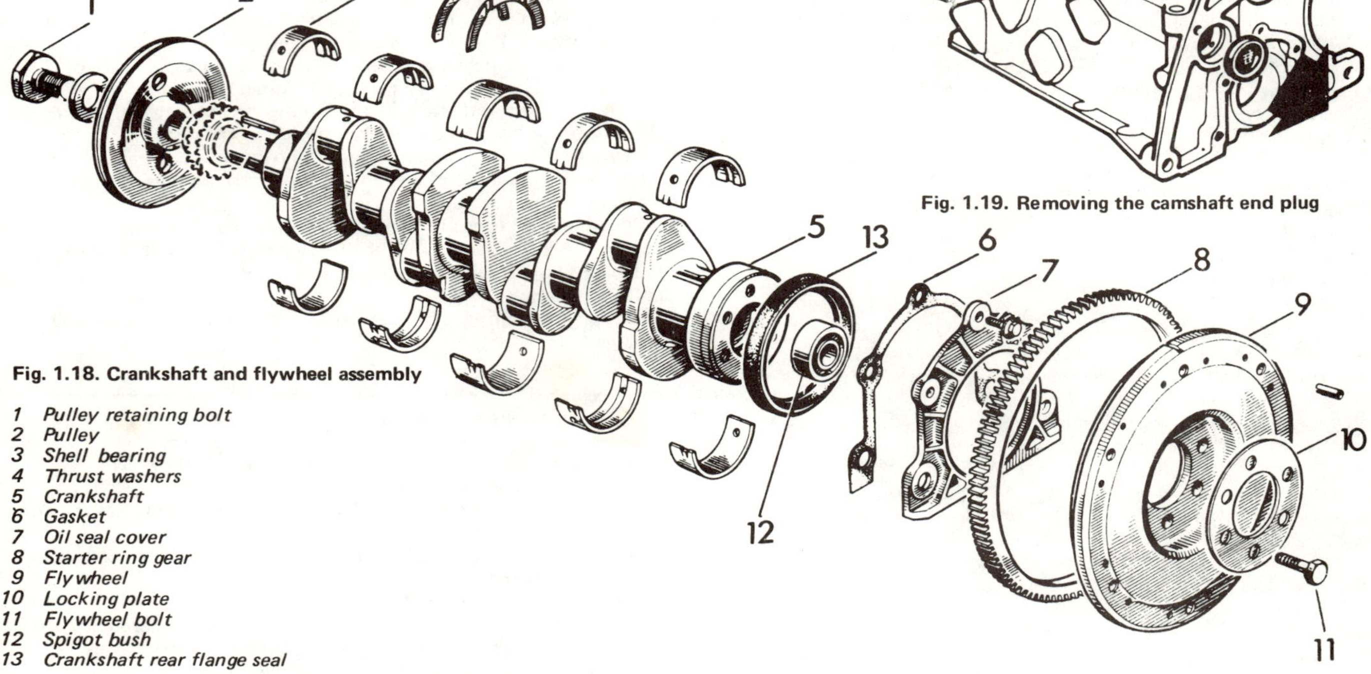

Fig. 1.18. Crankshaft and flywheel assembly

1 *Pulley retaining bolt*
2 *Pulley*
3 *Shell bearing*
4 *Thrust washers*
5 *Crankshaft*
6 *Gasket*
7 *Oil seal cover*
8 *Starter ring gear*
9 *Flywheel*
10 *Locking plate*
11 *Flywheel bolt*
12 *Spigot bush*
13 *Crankshaft rear flange seal*

16 Engine components - examination for wear

Before any detailed examination of the dismantled components can be carried out, they must be thoroughly cleaned. A paraffin bath and stiff brush is an ideal method for the removal of grease, oil and grit. Where heavy deposits of grease and dirt have to be cleaned from the exterior of the engine casting, the use of a high pressure air hose is recommended, but this should have been done before dismantling the engine.

Fig. 1.20. Removing a core plug

17 Crankcase and cylinder block - examination and renovation

1 Thoroughly clean the interior and exterior surfaces of the casting and inspect for cracks. Should any be apparent then the cost of repairing by a specialist welder must be compared with the purchase of a new or secondhand unit.

2 Check the security of studs and for stripped threads in all tapped holes. If necessary have thread inserts fitted such as 'Helicoil' or examine the possibility of fitting oversize bolts and studs.

3 Where there is evidence of water or oil leakage from the various core plugs these should be renewed. Remove the cap or plug by either drilling a central hole and levering out or tapping a thread and using a bolt and bridge piece as an extractor. With the smaller plugs or caps, they may be tapped inwards on one side and levered out. Protect the edge of the hole with a piece of wood as shown in Fig. 1.20. When driving a new plug or cap, do not knock it harder than is required to effect a good seating.

4 Examine the camshaft and if the bearing bushes are worn, scored or chipped, renew them as described in Section 21.

5 Examine the oil pump drive shaft bearings and renew them if necessary as described in Section 26.

6 Examine the cylinder bores for scoring and wear. This operation is undertaken in conjunction with the examination of the pistons and by consideration of the previous history of oil consumption and smoke emission. Reference should be made to Section 24.

18 Crankshaft - examination and renovation

1 A rough visual check for wear in the crankshaft main bearings and big-end bearings may be carried out before removal of the crankshaft from the crankcase. If movement can be felt by pushing and pulling it and also slackness observed in the big-ends, then almost certainly the crankshaft will have to be reground.
2 With the crankshaft removed, all journals should be measured at two or three different points for ovality with a micrometer. If the measurements taken at one journal differ by more than 0.011 inch (0.03 mm) then the crankshaft must be reground.
3 Regrinding must be undertaken by a specialist firm who will regrind to the permitted tolerances and supply matching oversize shell bearings (see Specifications).

19 Main and big-end shell bearings - examination and renewal

1 When connecting rods are removed from the crankshaft the bearing shells will be released and even though the crankshaft journals are in good condition the bearings may need renewal. Certainly if their bearing surfaces are anything other than an even, matt grey colour they should be renewed. Any scores, pitting or discolouration is an indication of damage by metal particles or the top bearing surface wearing away. If there are any doubts it is always a good idea to renew them anyway, unless there is a definite record that they have only been fitted for a small mileage. The backs of the shells are marked with serial numbers and an indication of whether or not they are undersized due to the crankshaft having been reground previously. If in doubt take them to your supplier who will be able to ensure that you are sold new ones of the correct type. If the crankshaft is being reground new bearings will be required anyhow and these are always available from the firm which does the regrinding.

20 Tappets - examination and renovation

1 The tappets should be checked in their respective bores in the crankcase and no excessive side-play should be apparent. The faces of the tappets which bear against the camshaft lobes should also have a clear, smooth shiny surface. If they show signs of pitting or serious wear they should be renewed. Re-facing is possible with proper grinding facilities but the economics of this need investigating first.

21 Camshaft and bearings - examination and renovation

1 The lobes of the camshaft should be examined for any indications of flat spots, pitting or extreme wear on the bearing surfaces. If in

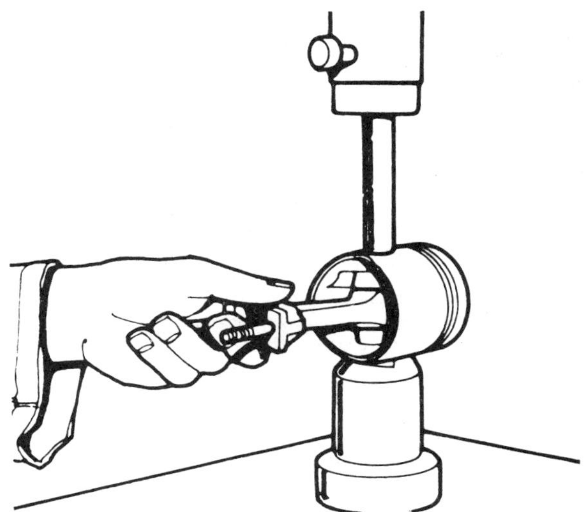

Fig. 1.22. Removing a gudgeon pin using a press

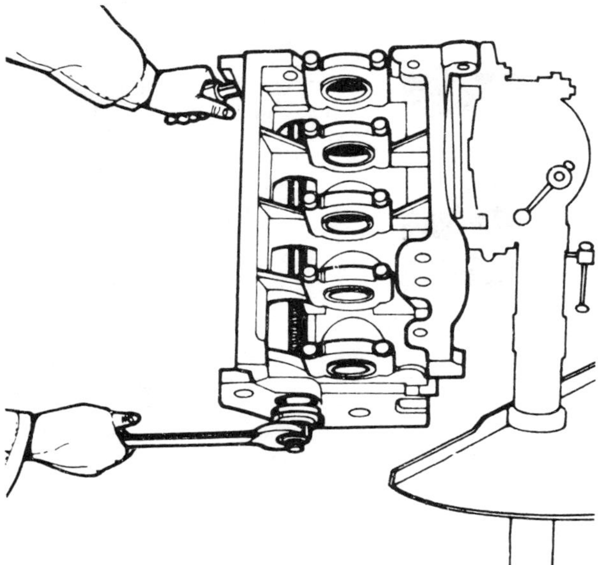

Fig. 1.21. Method of fitting new camshaft bearings

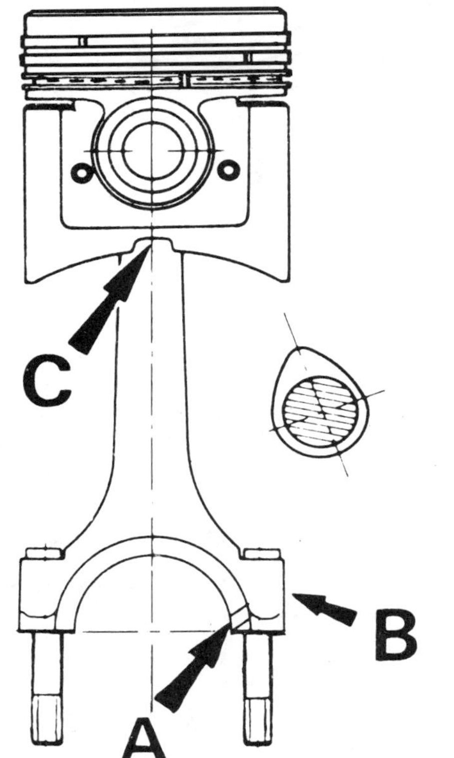

Fig. 1.23. Correct position of piston and connecting rod in engine

A Lubrication groove
B Production marking
C Notch in piston skirt faces towards timing cover

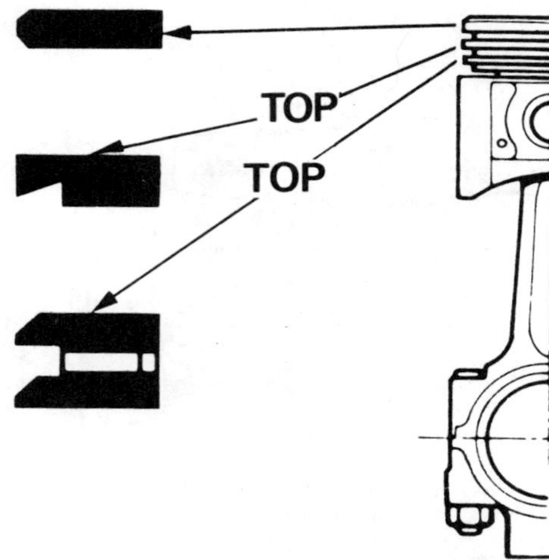

Fig. 1.24. Identification of piston rings

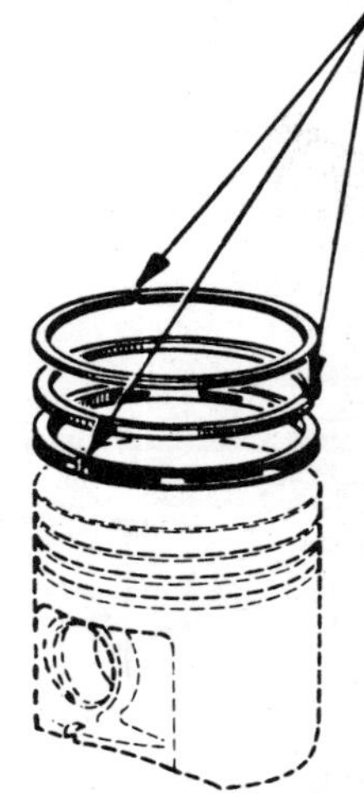

Fig. 1.25. Correct positioning of piston ring gaps

Fig. 1.26. Checking the piston ring gaps in cylinder bore

doubt get the profiles checked against the specification dimensions with a micrometer. Minor blemishes may be smoothed down with a 120 grain oil stone and polished with one of 300 grain. The bearing journals also should be checked in the same way as those on the crankshaft. The camshaft bearings are renewable.

2 Drive out the camshaft rear bearing (flywheel end) sealing cap. To do this, insert a rod or tube from the other end of the crankcase (Fig. 1.19).

3 Removal of the front, rear and centre camshaft bushes is best accomplished by the use of a length of threaded rod and nuts with suitable tubular distance pieces (see Fig. 1.21).

4 Note the precise positioning of each bearing bush before removal and ensure that the bearing seats are not damaged during the removal operation.

5 Fit the new bearing bushes using the same method as for removal, starting with the centre one. It is essential that the bearing oil hole is in exact alignment with the one drilled in the bearing seat and marks should be made on the edge of the bearing bush and seat before pulling into position.

6 Fit a new camshaft front bearing sealing cap.

22 Connecting rods and small end bushes - examination and renovation

1 It is unlikely that a connecting rod will be bent except in cases of severe piston damage and seizure. It is not normally within the scope of the owner to check the alignment of a connecting rod with the necessary accuracy, so if in doubt have it checked by someone with the proper facilities. It is in order to have slightly bent connecting rods straightened - the manufacturers provide special jigs for the purpose.

2 If a connecting rod is bent or damaged beyond repair all four rods must be renewed as they are only supplied in weight matched sets of four.

3 The connecting rod small end is a shrink fit on the gudgeon pin and therefore is not subject to wear. The gudgeon pin pivots within the piston, and excessive wear in the piston bores will necessitate the renewal of the piston and gudgeon pin assembly as described in the following Section.

23 Pistons and piston rings - removal and refitting

1 A piston ring should be removed by prising the open ends of the ring just sufficiently far to enable three feeler blades or strips of tin to be slid round behind it. Position the strips at equidistant points to provide guides so that the piston ring may be slid over the hands and other grooves and removed.

2 The gudgeon pin is a press fit in the connecting rod small end bearing and it must be removed on a press using suitable distance pieces to prevent damage to the soft alloy of the piston body. (It is too difficult to do at home).

3 Refitting of the gudgeon pin is carried out on a press but the small end bearing must be expanded by placing the connecting rod in an oven (220 - 250°C) or immersing it in hot oil.

4 When assembling the piston to the connecting rod ensure that the notch in the bottom of the piston skirt faces towards the front, (timing cover end) of the engine, and the oil groove in the face of the big-end is towards the camshaft (see Fig. 1.23).

5 Check that when the small end is located centrally between the bosses inside the piston, the ends of the gudgeon pin are inset an equal distance from the outside of the piston bosses. If necessary adjust the position of the gudgeon pin by pressing it in the appropriate direction.

6 Refit the piston rings using the reverse of the removal procedure, but first check the ring gap in the cylinder bore as described in the next Section. Three rings are fitted; the top compression ring can be fitted either way up but the second scraper ring and bottom oil control ring must be fitted with the word 'TOP' uppermost (see Fig. 1.24).

7 Set the ring gaps at an angle of 120° to each other around the perimeter of the piston to reduce gas blow-by, (see Fig. 1.25).

24 Piston, ring and cylinder bore wear - examination and renovation

1 Piston and cylinder bore wear are contributory factors to excessive

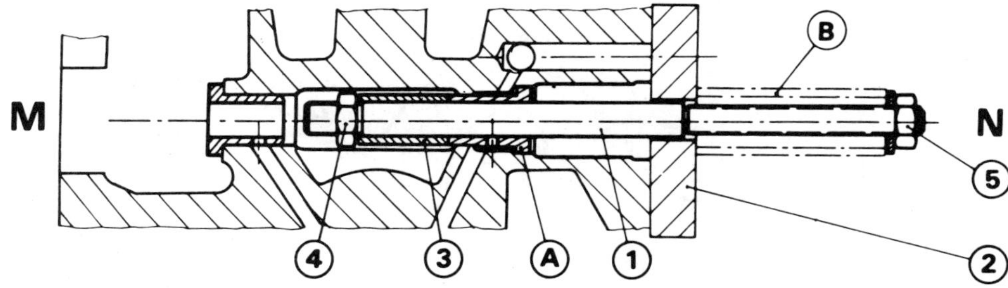

Fig. 1.27. Extracting the oil pump/distributor driveshaft bushes

1 Rod
2 Plate
3 Sleeve
4 Nut
5 Nut
A Shaft bush
B Old gudgeon pin
M Side towards distributor
N Side towards oil pump

oil consumption (over 1 pint to 300 miles) and general engine noise. They also affect engine power output due to loss of compression. If you have been able to check the individual pressures before dismantling so much the better. They will indicate whether one or more is losing compression which may be due to cylinders and pistons if the valves are satisfactory.

2 Determining the degree of wear on pistons and cylinders is complementary. In some circumstances the pistons alone may need renewal, the cylinders not needing reboring. If the cylinders need reboring then new pistons must be fitted. First check the cylinders. A preliminary check can be done simply by feeling the walls about ½ inch down from the top edge. If a ridge can be felt at any point then the bores should be measured with an inside micrometer or calipers to see how far they vary from standard. The measurement should be taken across the bore of the cylinder about 15 mm (0.6 in) down from the top edge at right angles to the axis of the gudgeon pin. Then measure the piston, also at right angles to the gudgeon pin across the skirt at the bottom. The two measurements should not differ by more than 0.008 to 0.0015 in (0.022 to 0.037 mm).

3 Further measurement of the cylinder across the bore will indicate whether or not the wear is mostly on the piston. If the cylinder bore is uniform in size fitting new pistons in the original bore size is possible. However, it is a very short sighted policy. If new pistons are needed anyway the cost of reboring will add 20 - 25% to the cost of the pistons so it would be as well to get it done whilst the engine is out of the vehicle.

4 Another feature of the pistons to check is the piston ring side clearance in the grooves. This should not exceed 0.12 mm (0.0047 in) for the top ring and 0.10 mm (0.004 in) for the other two. Usually however, this wear is proportionate to the rest of the piston which is otherwise apparently little worn. If you think that only a new set of rings is required it would be a good idea to take your pistons to the supplier of the new rings and check the new rings in the gaps. You may change your mind about how worn the pistons really are! Once a cylinder has been rebored, twice (to + 0.0157 in (0.4 mm) diameter) it must not be rebored again.

5 Remove the piston rings from each piston in turn and keep them identified for their respective bore. Similarly, if new rings are being fitted, keep them identified in respect of piston location and bore.

6 Press each ring squarely an inch or two (25 to 50 mm) down the cylinder bore and check the ring gap with feeler gauges. The gaps should be 0.010 to 0.018 in (0.25 to 0.45 mm) for the compression rings and 0.008 to 0.016 in (0.20 to 0.40 mm) for the oil control rings.

25 Flywheel - examination and renovation

1 The clutch friction disc mating surface of the flywheel should be examined for scoring. If this is apparent then it should either be exchanged for a new unit or if the scoring is very light it may be skimmed.

2 The starter ring gear should be examined and if the teeth are worn or chipped it must be renewed.

3 To remove the ring cut or drill at the root of two teeth, then support the flywheel and drive off the ring gear using a bronze or steel bar.

4 Take care not to damage the flywheel locating dowels during this operation or they will have to be renewed.

5 Place the new ring gear in position on the flywheel ensuring that the teeth have their lead-ins (chamfer) facing the correct way as originally fitted.

6 Using a blow lamp or torch, heat the ring gear evenly all round until it just starts to drop into position.

7 Drive the ring gear squarely onto the flywheel shoulder using a drift.

26 Oil pump and drive shaft bushes - examination and renovation

1 The oil pump gears and drive shaft should be checked for wear and play compared with the tolerances given in Specifications. Renew components as required or exchange the complete unit.

2 Check the pressure relief valve components and renew as appropriate if the ball or body are scored or seats appear pitted.

3 Check the wear of the oil pump/distributor drive shaft bushes. Where these have to be renewed, either extract them using a threaded rod and nuts and distance pieces similar to the method used in Fig. 1.27.

4 Alternatively, tap a thread in each bush and use a bolt as an extractor.

5 Fit the shorter bush to the distributor side (M) pulling it in tight to the machined surface of the cylinder block. It is vital that the oil hole in the bush aligns with the crankshaft bearing oil passage.

6 Fit the bush to the oil pump side (N) again using the threaded rod and nut method and avoiding damage to the bush just fitted. There is no need to align the oil hole on this bush as it opens into a circular oil chamber in the cylinder block.

7 Test the drive shaft in the bushes for ease or rotation. A hard spot will indicate mis-alignment or distortion.

27 Cylinder head, rocker gear, valves - examination and renovation

1 As previously described, the alloy cylinder head must be handled very carefully to avoid scoring or damage. Do not stand it on a bench which is covered with filings or they may become embedded in the surface of the head.

2 Remove carbon using a blunt scraper taking great care not to damage the machined surfaces.

3 If there are any visible cracks the head should be scrapped. Cracks are most likely to occur round the valve seats or spark plug holes. Bearing in mind that a head will cost (new) nearly 20% of the cost of a complete, new engine, economies should be considered as well as the likelihood of obtaining a used head from a breaker's yard. If the latter, make sure that the head you get is the same type as the old one - and in better condition!

4 The valve seats should be examined for signs of pitting or ridging. Slight pitting can be ground away using carborundum paste and an **old** valve. New valves are specially plated and must not be used to grind in the seats. If the valve faces are burnt or cracked, new valves must be obtained. If the valve seats require re-cutting, ensure that the width is maintained at 0.060 in (1.5 mm) and at an angle of 46°.

5 The rocker gear should be dismantled and thoroughly cleaned of the sludge deposits which normally tend to accumulate on it. The

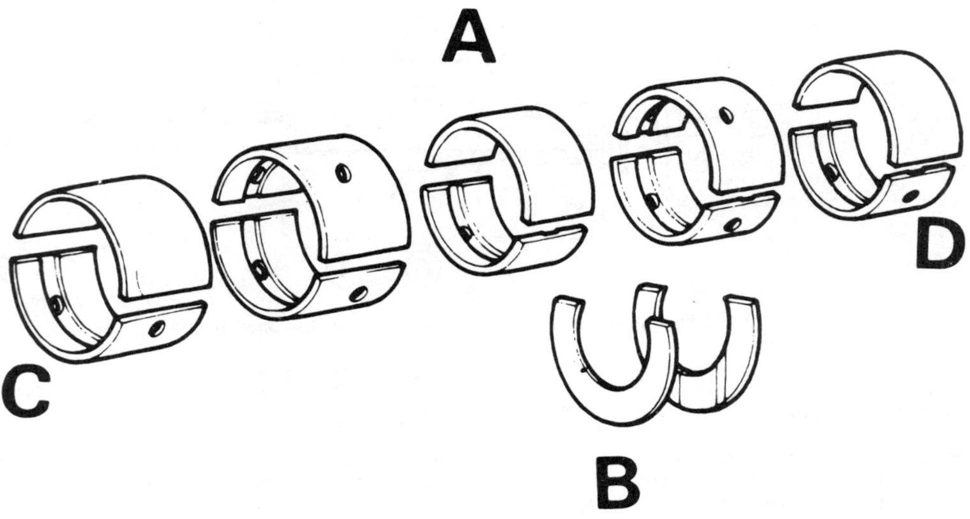

Fig. 1.28. Identification of main bearing shells and thrust bearings

A Lower (cap) half-shells *B Upper (cyl. block) thrust bearings* *C Timing cover end* *D Flywheel end*

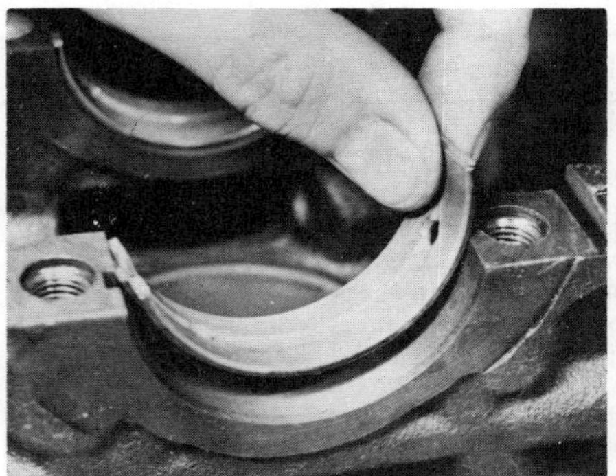

29.1 Fitting the main bearings in the crankcase

29.3 Correct locations of crankshaft thrust washers

29.5 Lowering the crankshaft into position

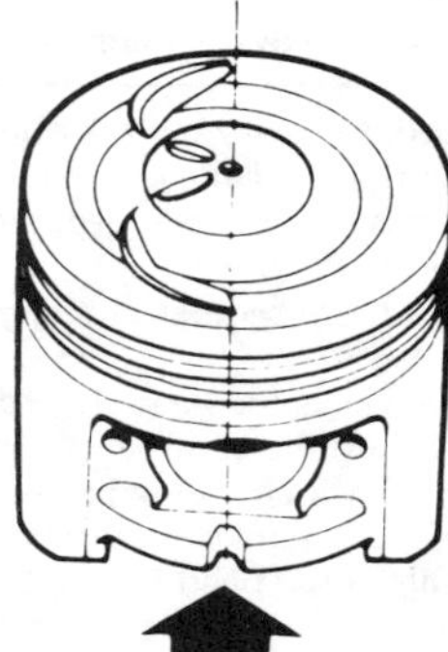

Fig. 1.29. Correct position of valve cut-outs in piston as viewed from timing cover end of engine

rocker arms should be a smooth fit on the shaft with no play. If there is any play it is up to the owner to decide whether it is worth the cost of renewal. The effects on engine performance and noise may not be serious although wear tends to accelerate once it is started. The valve clearance adjusting screws should also be examined. The domed ends that bear on the valve stems tend to get hammered out of shape. If bad, renewal is relatively cheap and easy.

6 The valves themselves must be thoroughly cleaned of carbon. The head should be completely free of cracks or pitting and must be perfectly circular. The edge which seats into the cylinder head should also be unpitted and unridged and if any are evident, the valve must be renewed.

7 Refit the valve into its guide in the head and note if there is any sideways movement which denotes wear between the stem and guide. Here again the degree of wear can vary. If excessive, the performance of the engine can be noticeably affected and oil consumption increased. The maximum tolerable sideways rock, measured at the valve head with the end of the valve stem flush with the end of the guide, is 0.8 mm (0.031 in). Wear is normally in the guide rather than on the valve stem but check a new valve in the guide if possible first. Valve guide renewal is a tricky operation and should be left to a Chrysler Dealer.

8 Check that the end face of the valve stem is not 'hammered' or ridged by the action of the rocker arm. If it is, dress it square and smooth with an oilstone.

9 The cylinder head surface can be ground to a maximum of 0.024 in (0.6 mm) providing a cylinder head gasket of 0.070 in (1.8 mm) is used on reifting to maintain the correct compression ratio. The correct type of gasket is obtainable from a Chrysler Dealer.

28 Engine reassembly - general

Ensure absolute cleanliness during assembly operations and lubricate each component with clean engine oil before fitting.

Renew all lockwashers, gaskets and seals as a matter of course.

Take care to tighten nuts and bolts to the torque specified in Specifications and watch out for any differences in bolt lengths which might mean attempting to screw a long bolt into a short hole with subsequent fracturing of a casting.

Follow the sequence of reassembly given in the following Sections and do not skip any adjustment procedure essential at each fitting stage.

29 Crankshaft and main bearings - reassembly

1 Fit the new main bearing shells to the crankcase locations (photo).
2 Fit the two semi-circular thrust washers.
3 The thrust washers should be retained in position with grease at each side of the main bearing centre web of the crankcase and must have their oil grooves facing outwards (photo).
4 Oil the crankcase bearing shells liberally.
5 Lower the crankshaft carefully into position in the crankcase (photo).
6 Fit the shell bearings to the main bearing caps (photo).
7 Place the main bearing caps in order, ready for fitting. The caps are numbered 1 to 5, number 1 being positioned nearest the flywheel. The shells in caps 2 and 4 have oil grooves, 1, 3 and 5 do not. Fit the main bearing caps in their correct order (photo).
8 Fit the main bearing cap bolts and tighten them to the correct torque of 47 lb f ft (6.4 kg f m) (photo).
9 The crankshaft endfloat must now be checked. To do this, lever the crankshaft in one direction and then the other and, using feeler gauges, measure the gap between the thrust washers and the face of each side of the centre main bearing web. The total endfloat should be between 0.0035 and 0.011 in (0.09 and 0.27 mm), photo. Where the clearance is outside the permitted tolerance then the thrust washers must be changed for ones of different thickness. These are available in a variety of thicknesses.

30 Flywheel, spigot bearing and oil seal - refitting

1 It is an advantage to fit the flywheel at this stage as it will provide a means of rotating the crankshaft when installing the connecting rod/piston assemblies.
2 Check the condition of the spigot bearing which is located in the centre of the flywheel mounting flange. If it is worn, extract it and drift in a new one (photo).
3 Stick a new paper gasket in position on the cylinder block.
4 Fit a new oil seal to the inner rim of the crankshaft/flywheel flange oil seal cover.
5 Locate the oil seal cover so that the lip of the oil seal is not damaged and then fit and tighten the five securing bolts, each fitted with a lockwasher (photo).
6 If the paper gasket stands proud of the lower joint face, trim it.
7 Locate the flywheel on the crankshaft mounting flange (photo).
8 Locate the circular plate and insert the securing bolts and tighten them to 40 lb f ft (5.4 kg f m).
9 The flywheel can only be fitted one way as the bolt holes are not equidistant.

Note: Loctite should be smeared on the threads of the flywheel bolts.

31 Connecting rod/piston assemblies - refitting to crankshaft

1 Fit the piston rings to the piston as described in Section 23. Oil them liberally and ensure that the gaps are staggered.
2 Fit a piston ring clamp and then locate the shell bearings in the connecting rod big-end.
3 Oil the cylinder bores and insert the piston/connecting rod assembly into its cylinder. Take care not to damage the bore surfaces with the big-end bolts (photo).
4 Note that with the piston correctly positioned in the cylinder bore the valve recesses in the top of the piston must be nearest the left-hand side of the cylinder block when viewing the engine from the timing cover end (see Fig. 1.29). Gently tap the pistons into the bore using a piece of wood or hammer handle.

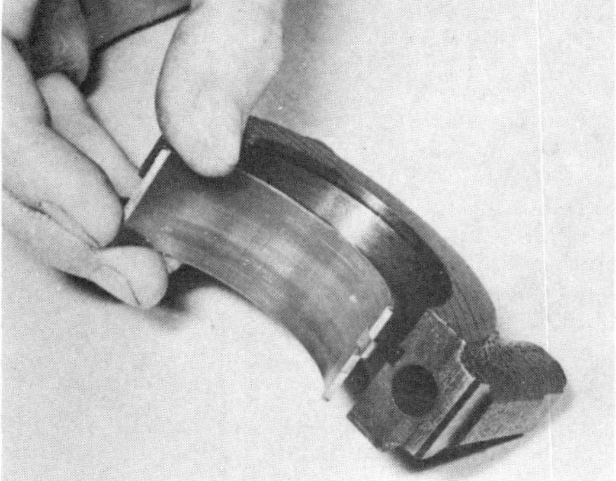

29.6 Fitting the shell into the main bearing cap

29.7 Installing the main bearing caps

29.8 Tightening the main bearing cap bolts

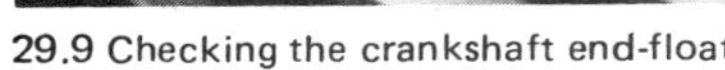

29.9 Checking the crankshaft end-float

30.2 Spigot bearing in crankshaft flange

30.5 Crankshaft oil seal cover in position

30.7 Flywheel in position

30.8 Tightening the flywheel bolts

31.3 Inserting a piston into the block

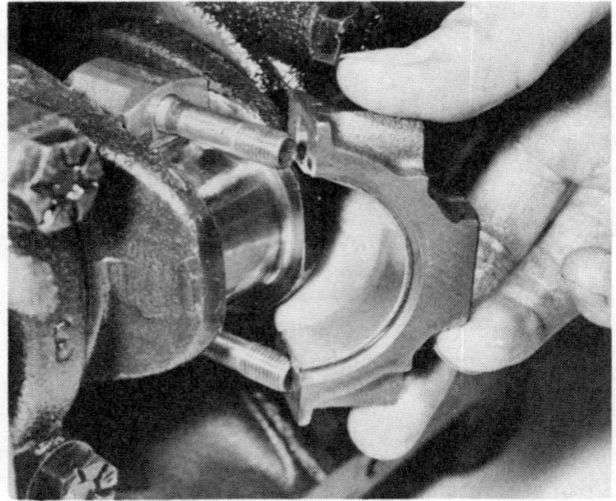
31.6 Fitting a big-end bearing cap

31.7 Tightening the big-end cap nuts

31.9 Crankshaft sprocket in position

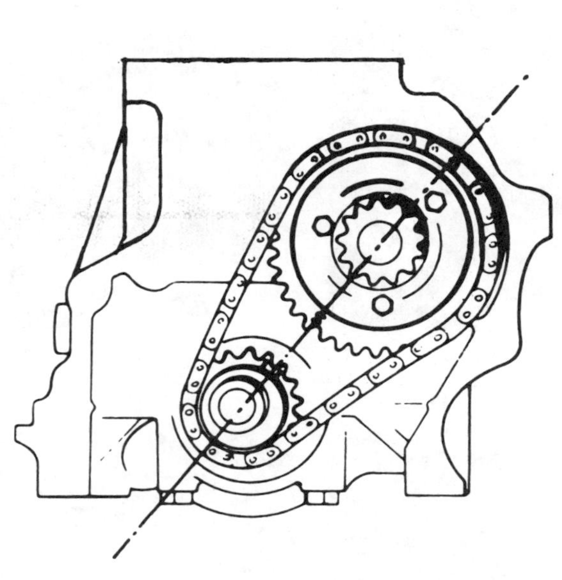
Fig. 1.30. Correct alignment of timing sprockets (No. 1 piston at TDC)

32.1 Installing the camshaft

32.3 Camshaft thrust plate correctly fitted

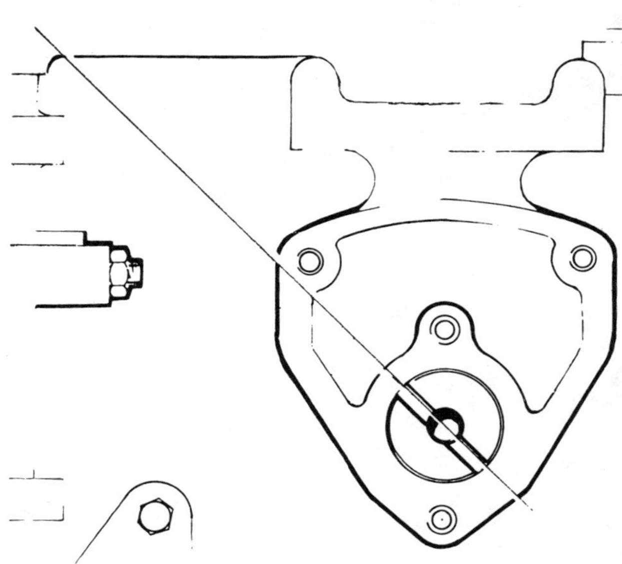

Fig. 1.31. Correct position of distributor drive dog with No. 1 piston 10° BTDC (Note smaller segment is towards the bottom of cylinder block)

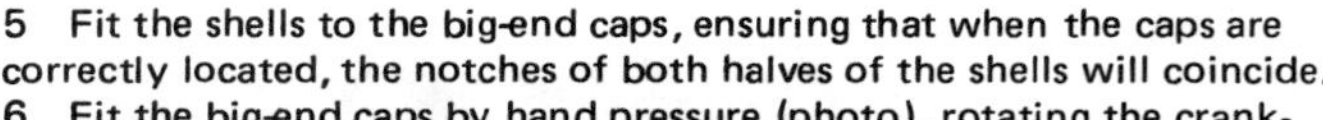

5 Fit the shells to the big-end caps, ensuring that when the caps are correctly located, the notches of both halves of the shells will coincide.
6 Fit the big-end caps by hand pressure (photo), rotating the crankshaft by means of the flywheel to tdc or bdc to facilitate fitting.
7 Fit the big-end nuts and tighten them to a torque of 26 lb f ft (3.5 kg f m) (photo).
8 Fit the Woodruff key to the crankshaft.
9 Fit the crankshaft sprocket, using a tubular drift if necessary and ensuring that the countersunk side of the sprocket faces the engine (photo).

32 Camshaft, timing gear, distributor/oil pump drive - reassembly

1 Oil the camshaft bearings liberally and insert the camshaft into the cylinder block taking care not to damage the bearings as the cam lobes pass through them (photo).
2 Fit the camshaft thrust plate so that it engages in the groove of the journal.
3 Fit and tighten the thrust plate retaining screws using new locking tabs (photo).
4 At this stage check the camshaft endfloat, using either a dial gauge or feelers. The tolerance is between 0.005 and 0.008 in (0.10 and 0.20 mm) and endfloat outside these recommendations should be rectified by the substitution of a new thrust plate.
5 Bend over the thrust plate bolt locking tabs.
6 Turn the crankshaft by means of the flywheel until the No. 1 piston (nearest the flywheel) is in the tdc position.
7 Temporarily fit the sprocket onto the camshaft ensuring the bolt holes are aligned and then rotate the camshaft until the single mark on the sprocket is in line with the two marks on the crankshaft sprocket as shown in Fig. 1.30.
8 Remove the camshaft sprocket without rotating the camshaft and lift the timing chain over both sprockets. Refit the sprocket onto the camshaft (photo), and check that the timing marks are in alignment using a straight edge, if necessary, and refit the three camshaft sprocket securing bolts.
9 Tighten the bolts to the specified torque wrench setting and bend over the locking tabs.

32.8 Fitting the camshaft sprocket and timing chain (Note position of timing marks)

32.10 Inserting the oil pump driveshaft

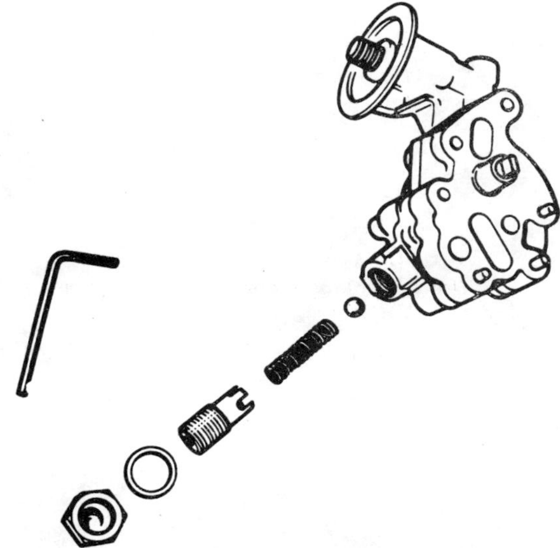

Fig. 1.32. Oil pump relief valve assembly

10 Smear the oil pump drive shaft with engine oil and insert it in position (photo).
11 Insert the shaft drive gear, engaging it with the camshaft worm. The gear must be fitted the correct way round so that the splined part of the bore is to the outside. Secure the gear with the circlip (photo).
12 From the opposite side to the oil pump insert the distributor drive so that the large and small segments are in the position shown in Fig. 1.31.
13 It is imperative that during the foregoing operations No. 1 piston is on the compression stroke and the flywheel timing mark is aligned with the 10° BTDC mark in the clutch housing 'window' (photo).
14 Locate the distributor base plate so that its boss is towards the timing cover and tighten the securing bolts (photo).

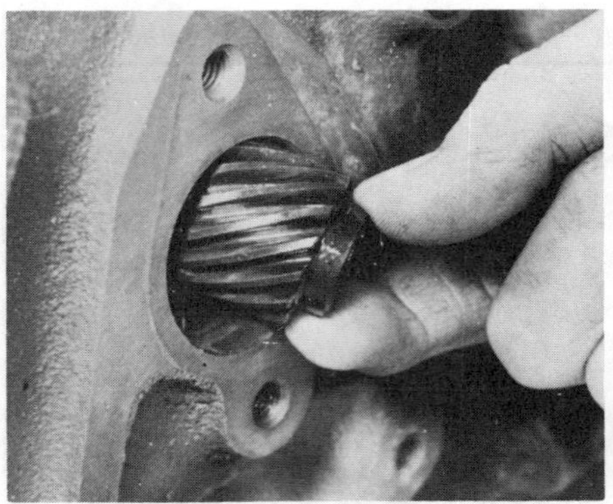
32.11A Inserting the drive shaft gear ...

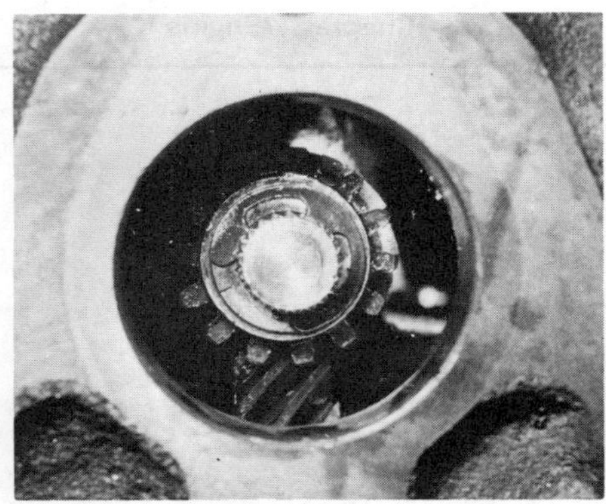
32.11B ... and securing it with the circlip

32.13 Flywheel timing marks

32.14 Fitting the distributor base plate

33.3 Fitting the oil pump

34.3 Correct location of timing cover oil seal

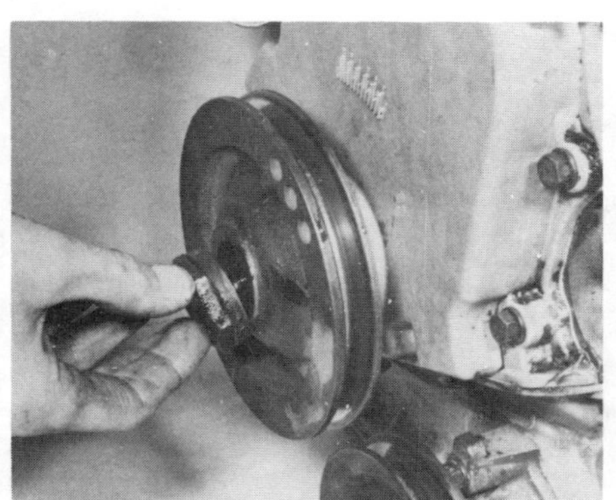
34.6 Fitting the crankshaft pulley

35.2 Installing the sump case

35.3 Fitting the oil pump strainer

35.5 Fitting the sump pan

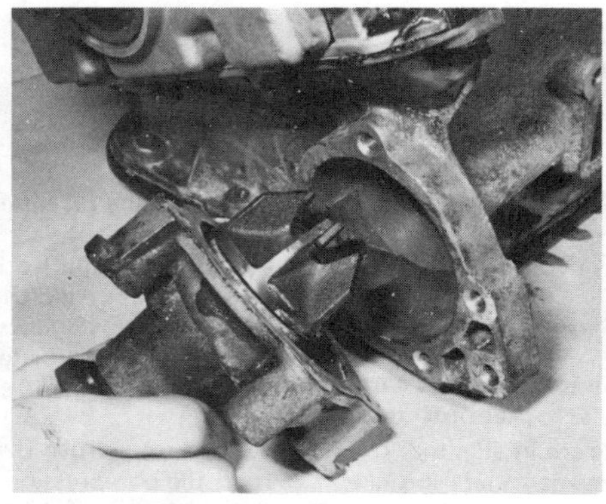
36.2 Fitting the water pump

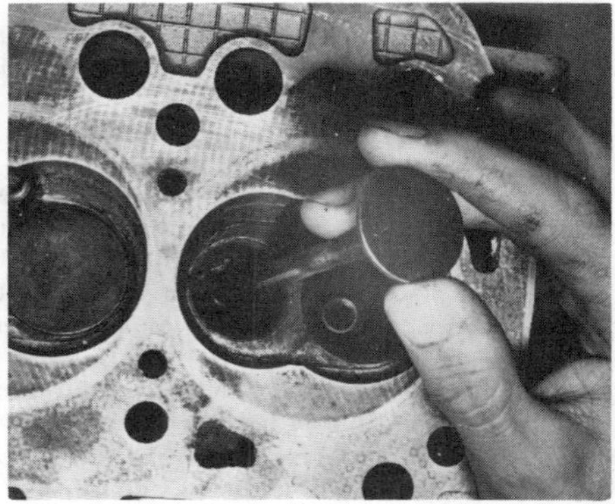
37.2 Inserting a valve

33 Oil pump - refitting

1 Locate the oil pump body and a new gasket on the cylinder block.
2 Insert the oil pump gears.
3 Locate the oil pump cover and fit the securing bolts (photo).
4 Tighten the bolts evenly to the specified torque wrench setting.
5 If the relief valve has been dismantled, reassemble it and screw it securely into position (see Fig. 1.32).

34 Timing cover oil seal and crankshaft pulley - refitting

1 Examine the timing cover oil seal for wear or damage and renew it, also if there has been previous evidence of oil leakage.
2 Extract the old seal and drive in the new one squarely using a suitable drift.
3 Before fitting the seal, check that the directional arrow on the seal follows the direction of rotation of the engine and that the seal spring faces inwards (photo).
4 Using a new gasket, fit the timing cover to the cylinder block.
5 Tighten the retaining bolts noting that two different sizes of setscrews and one nut and bolt are used.
6 Fit the crankshaft pulley and retaining bolt (photo).

35 Sump and oil strainer - refitting

1 Locate a new sump case gasket to the cylinder block, using grease to hold it in position.
2 Lower the sump case into position on the block and secure with the internal and external bolts (photo).
3 Fit the oil pump strainer into position (photo).
4 Locate the cork sump pan gasket in position.
5 Locate the sump pan (photo) and tighten the bolts evenly.

36 Water pump - refitting

1 Locate the water pump gasket to the sump case using grease to hold it in position.
2 Fit the water pump and tighten the securing bolts to the correct torque (photo).
3 Connect the water hose between the water pump and the crankcase outlet.
4 Re-fit the water pump drain plug and sealing washer.

37 Cylinder head, valves and rocker gear - reassembly and refitting

1 The components of the cylinder head should have been examined for wear, carbon removed and the valves renewed, if necessary, as described in Section 27.
2 Insert the first valve into its guide in the cylinder head, and, if they have been kept in strict order, the valve will be fitted to its original location (photo).
3 Place the lower valve spring seat in position (photo).
4 Place a new oil seal in position (photo).
5 Place the valve spring and collar in position on the valve stem (photo). If the original valve springs have been in service for 15000 miles (24000 km) or more they should be renewed.
6 Using a suitable compressor, compress the valve spring and insert the split collets (photo).
7 Release the valve spring compressor and check that the components of the valve are correctly fitted. Place a small block of wood on the end face of the valve stem and give it a blow with a hammer in order to settle the collets and to ensure that later valve clearance adjustment is precise. Repeat the fitting operations on the other seven valves.
8 Insert the tappet blocks (cam followers) into their original sequence in the cylinder block (photo).
9 Clean both mating surfaces of block and head, ensuring that no

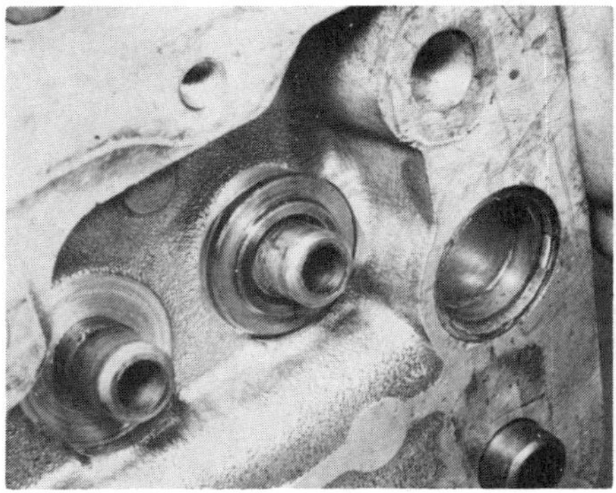

37.3 Valve spring seat in position

37.4 Fitting the valve oil seal

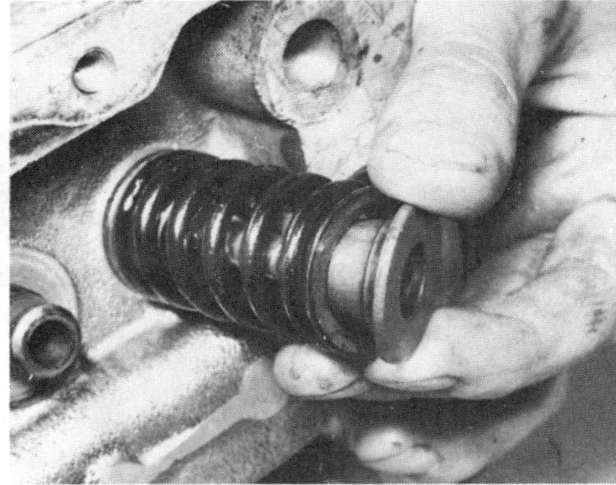

37.5 Fitting the valve spring and collar

37.6 Compressing a valve spring

37.8 Inserting a cam follower into the block

37.10 Cylinder head gasket in position

trace of carbon, scale or grit remains.
10 Locate a new cylinder head gasket in position on the block, ensuring that the word **'DESSUS'** is visible on the upper surface (photo)
11 Check that the two alignment dowels are correctly positioned through the gasket. The gasket should normally be fitted dry but if there has been evidence of oil or water leaks from the head gasket or if the surfaces of the block or head are scratched, then a thin film of non-setting compound should be applied to both sides of the gasket. This compound also provides a protection against corrosion of the alloy head.
12 Lower the cylinder head gently into position on the cylinder block (photo).
13 At this stage, check that the rocker pillar alignment dowels are securely in position on the top face of the cylinder head.
14 Fit the pushrods, carefully engaging their lower ends in the tappet blocks.
15 If the rocker shaft assembly has been dismantled for the renewal of worn components, it should be reassembled with the parts fitted in the sequence in Fig. 1.33.
16 Fit a new pin to the centre rocker shaft pillar, ensuring that the plugged ends of the hollow shafts face outwards.
17 Lower the assembled rocker shaft onto the locating dowels on the cylinder head.
18 Drop the cylinder head bolts into their holes, noting that the longer bolts locate in the thicker side of the rocker pillars (photo).
19 Check that the pushrods locate correctly with each rocker arm. It will be essential to have slackened the rocker arm adjuster screws right off to ensure that there is no valve spring pressure applied when tightening down the cylinder head bolts.
20 Using a torque wrench, tighten the cylinder head bolts to the specified torque in the sequence shown in Fig. 1.9 (photo).
21 Adjustment of the valve clearances may be carried out now or at a later stage when the engine has been fitted but if the operation is undertaken now, it will permit the rocker cover and other ancillaries to be fitted and obviate their removal again later.
22 The valve clearances must be adjusted when the engine is cold. The inlet valves should be set to 0.010 in (0.25 mm) and the exhaust valves to 0.012 in (0.30 mm). Counting from the flywheel end of the engine, the inlet valves are numbers 1, 3, 6, and 8, and exhaust valves are 2, 4, 5, and 7,
23 Turn the engine by means of the flywheel and refer to the following table, adjusting each valve clearance in the sequence given. Counting from the flywheel end of the block.

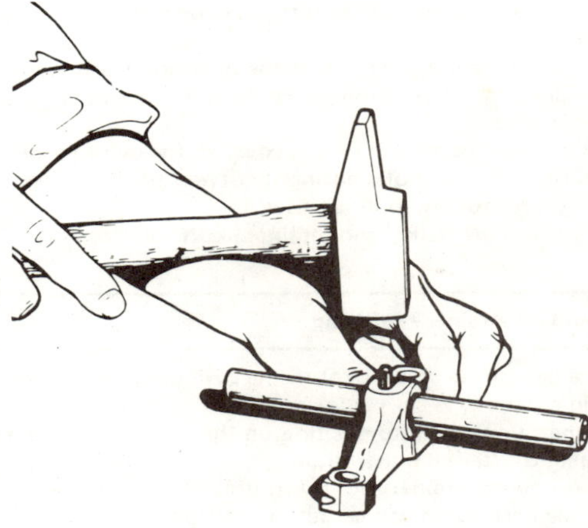

Fig. 1.34. Fitting the rocker shaft bracket retaining pin

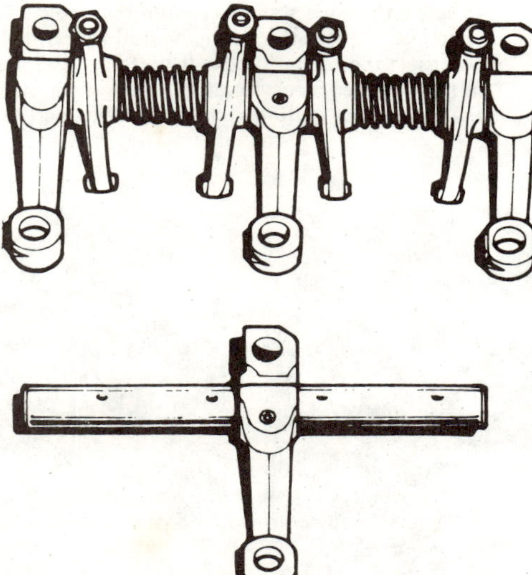

Fig. 1.33. Correct assembly of rocker shaft

37.12 Lowering the cylinder head into position

37.18 Correct installation of rocker shaft assembly

37.19 Tightening the cylinder head bolts

37.24 Setting the valve clearances

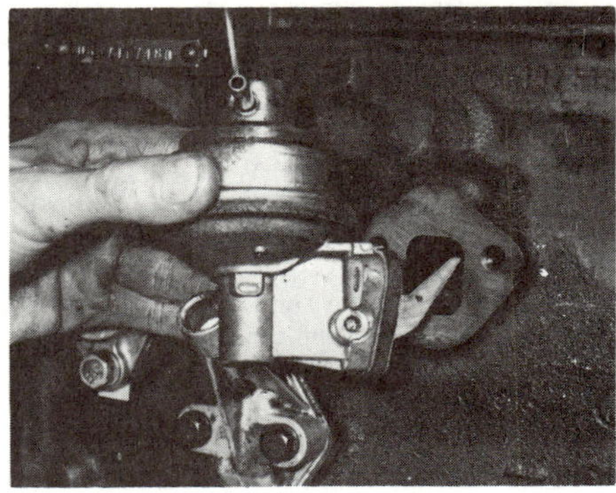
38.2 Fitting the fuel pump

38.3 Dipstick retainer installed

38.8 Fitting the inlet manifold

38.19 Correct position of carburettor heater box

39.2 Refitting the gearbox to engine

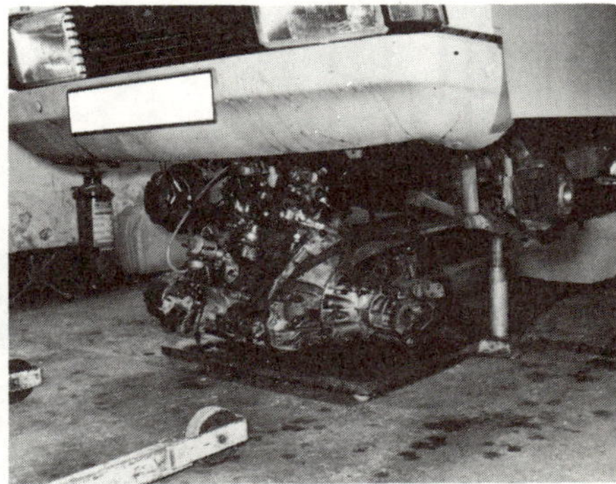
40.1 Positioning the engine beneath the car

Valve clearances to adjust	*Valves rocking (exhaust closing, inlet opening)*
1 and 2	7 and 8
3 and 4	5 and 6
5 and 6	3 and 4
7 and 8	1 and 2

24 Insert the feeler blade between the heel of the rocker arm and the valve stem end face. Loosen the locknut, then turn the adjuster screw until the feeler is a stiff sliding fit, then tighten the locknut (photo).
25 Repeat the adjustment on the other valves, then recheck all clearances and fit the rocker cover.

38 Engine ancillaries - refitting

1 Place a new gasket either side of the insulating washer which locates between the fuel pump and the crankcase.
2 Fit the fuel pump to the crankcase ensuring that the pump actuating arm fits between the crankcase wall and the camshaft (photo).
3 Fit the dipstick retainer to the crankcase, using a new gasket (photo).
4 Rotate the crankshaft until the segments on the distributor drive are in the position shown in Fig. 1.31 and insert the distributor into the crankcase aperture, ensuring that the drive dogs are correctly engaged.
5 Check that the distributor rotor is in line with the timing mark on the distributor body as described in Chapter 4.
6 Fit the distributor retaining plate by sliding it behind the distributor mounting bolt head.
7 Fit the thermostat housing to the cylinder head (two bolts).
8 Fit the inlet manifold to the cylinder head using a new gasket (photo). Before completing this operation, fit the pre-heat pipe from the thermostat elbow, pull the inlet manifold downwards and slide the pre-heat pipe under the manifold securing studs.
9 Connect the crankcase breather tube to the carburettor.
10 Fit the fuel pipe to the carburettor.
11 Fit the pre-heat pipe which runs between the water pump and the inlet manifold noting carefully its correct positioning against the crankcase.
12 Fit a new exhaust manifold gasket and bolt the manifold to the cylinder head, using only the specified brass nuts.
13 Fit the clutch assembly to the flywheel. Full details of the fitting procedure and centralising the driven plate are given in Chapter 5.
14 Position the thermostat in its housing, noting the tab for correct location.
15 Fit the thermostat housing cover, using a new gasket.
16 Fit the water pump pulley.
17 Fit the water pump pulley spacer and locking tab ring. Tighten the securing bolts and bend up the tabs.
18 Fit the alternator and drive belt. Adjust the belt tension as described in Chapter 2 and tighten the alternator securing bolts.
19 Lower the heater box (which provides warmed air to the carburettor) onto the rocker cover. Ensure that the nut underneath the heater box is tight before fitting it (photo).

39 Gearbox to engine - reconnection

1 If previously removed, reassemble the clutch operating mechanism, release bearing and fork as described in Chapter 5.
2 Position the engine and gearbox units as shown (photo).
3 Having aligned the driven plate (Chapter 5) the gearbox should slide easily as the gearbox primary shaft passes through the splined hub of the driven plate and engages with the spigot bearing which is located in the centre of the flywheel to crankshaft mounting flange.
4 Insert the gearbox to engine securing bolts and tighten to a torque of 32 lb f ft (4.4 kg f m).
5 Fit the cover plate on the exhaust side of the engine and secure with the three retaining bolts.
6 Fit the starter motor in position and secure it with the three retaining bolts.

40 Combined engine/gearbox/differential unit - refitting to vehicle

1 Check that the vehicle is raised sufficiently high at the front to

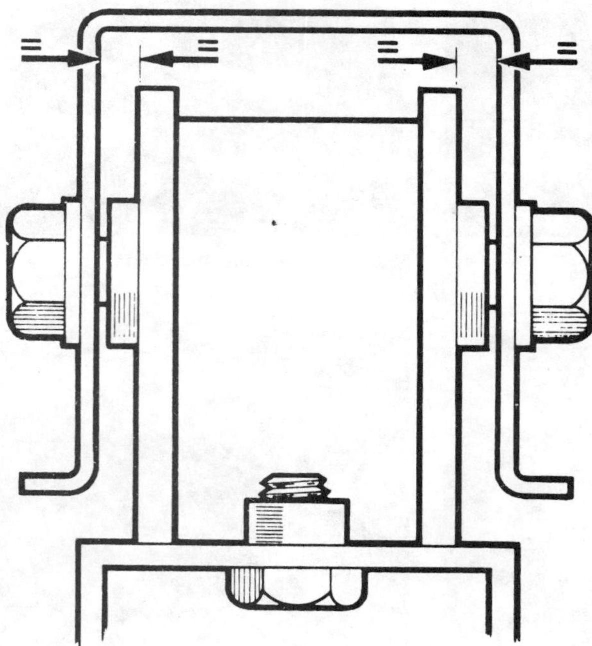

Fig. 1.35. Positioning the engine rear mounting bracket

permit the engine/gearbox unit to pass beneath it (photo).
2 Attach suitable lifting chains or slings.
3 Slide the power unit under the vehicle and then lift it upwards into the engine compartment using a suitable hoist.
4 Enter the three engine mounting bolts into the left-hand mounting (photo).
5 Enter the two engine mounting bolts on the right-hand side (photo).
6 Fit the engine rear mounting bracket to the crankcase but do not tighten the rubber mounting bolts at this stage.
7 Refit the drive shafts into the final drive unit and if necessary tap the outer end of the shafts with a soft faced mallet to ensure the inner splines are fully engaged in the differential.
8 Engage the outer splined end of each shaft through the front wheel hub assemblies, fit the lower balljoints in each bottom suspension arm and secure with the retaining nuts. Tighten both nuts to the specified torque wrench setting.
9 Now rock the engine from side to side to centralize it and then position the engine rear rubber mounting in the slotted holes, so that there is equal clearance on each side of the bracket as shown in Fig. 1.35. Check that the slots are parallel and tighten the securing bolts.
10 Remove the clamping rods that were fitted to retain the front suspension in position and refit the dampers.
11 Refit the driveshaft securing nuts, get an assistant to apply the footbrake and tighten the nuts. The nuts can be finally tightened to the specified torque wrench setting after the wheels have been fitted and the vehicle is lowered to the ground.
12 Attach the exhaust pipe flange to the manifold and secure the clamping bracket.
13 Refit the accelerator brackets and balljoints and adjust the brackets to obtain a clearance of 1/8 to 3/16 in (3 to 5 mm) between the peg and the slot in the gearbox selector shaft (see Chapter 3).
14 Refit the alternator drivebelt cover beneath the front right-hand wheel arch.
15 Jack-up the front of the car, remove the axle stands and lower the vehicle to the ground. Tighten the wheel nuts and check that the driveshaft securing nuts are tightened to the specified torque wrench setting (photo). Lock the nuts by peening the flange of each nut into the driveshaft groove.

40.4 LH side engine mounting bolts

40.5 RH side engine mounting bolts

40.15 Tightening the drive shaft nuts

41.7 One of the two engine earthing leads

41.8A Water temperature transmitter unit

41.8B Oil pressure transmitter unit

41 Engine - final reassembly, connections and replenishment of lubricants

1 Refit the clutch slave cylinder ensuring that the pushrod is correctly engaged in the operating lever.
2 Fit the rotor arm and distributor cap. Refit the spark plugs and HT leads and connect the LT leads to the coil.
3 Reconnect the fuel pipe to the pump and the pipe from the pump to the carburettor.
4 Refit the heater hoses and top and bottom radiator hoses.
5 Refit the coolant expansion bottle and attach the hoses to the radiator.
6 Refit the throttle cable and choke cable on the carburettor, referring to Chapter 3 if necessary.
7 Refit the two earth leads to the engine block (photo).
8 Reconnect the leads to the water temperature and oil pressure transmitter units (photo).
9 Fit the air cleaner assembly, referring to Chapter 3 if necessary.
10 Reconnect the leads to the rear of the alternator and the starter motor solenoid.
11 Fit a new oil filter, check that the sump drain plug is tight and fill the engine with the correct grade and quantity of oil.
12 Fill the radiator with coolant and top the expansion bottle up to the level mark on the side.
13 Fill the gearbox and final drive unit with the correct grade and quantity of oil.
14 With the help of an assistant, refit the bonnet ensuring the hinges line up with the marks made before removal.
15 Carry out a careful check to make sure all hoses, electrical leads and controls have been correctly refitted and then connect the battery terminals.

42 Engine - initial start up after overhaul or major repair

Make sure that the battery is fully charged and that all lubricants, coolant and fuel are replenished.

If the fuel system has been dismantled, it will require several revolutions of the engine on the starter motor to pump petrol to the carburettor. It will help to remove the spark plugs, which will enable the engine to turn over much easier, and enable oil to be pumped around the engine before starting it.

Refit the spark plugs and as soon as the engine fires and runs, keep it going at a fast tickover only (no faster) and bring it up to normal working temperature.

As the engine warms up, there will be odd smells and some smoke from parts getting hot and burning off oil deposits. Look for water or oil leaks which will be obvious if serious. Check also the clamp connection of the exhaust pipe to the manifold as these do not always 'find' their exact gas tight position until the warmth and vibration have acted on them, and it is almost certain that they will need tightening further. This should be done, of course, with the engine stationary.

When the engine running temperature has been reached, adjust the idling speed as described in Chapter 3.

Stop the engine and wait a few minutes to see if any lubricant or coolant leaks.

Road test the car to check that the timing is correct and giving the necessary smoothness and power. Do not race the engine. If new bearings and or pistons and rings have been fitted, it should be treated as a new engine and run in at reduced revolutions for 500 miles (800 km).

43 Modifications

The operations described in this Chapter apply to both engine types but it is essential when ordering spare parts that exact particulars of the vehicle (model, engine number, etc), are quoted. See 'Ordering spare Parts'.

44 Fault finding

Symptom	Reason/s	Remedy
Engine will not turn over when starter switch is operated	Flat battery Bad battery connections Bad connections at solenoid switch and/or starter motor	Check that battery is fully charged and that all connections are clean and tight.
	Starter motor jammed	Rock car back and forth with a gear engaged. If ineffective remove starter.
	Defective solenoid	Remove starter and check solenoid.
	Starter motor defective	Remove starter and overhaul.
Engine turns over normally but fails to fire and run	No spark at plugs	Check ignition system according to procedures given in Chapter 4.
	No fuel reaching engine	Check fuel system according to procedures given in Chapter 3.
	Too much fuel reaching the engine (flooding)	Slowly depress accelerator pedal to floor and keep it there while operating starter motor until engine fires. Check fuel system if necessary as described in Chapter 3.
Engine starts but runs unevenly and misfires	Ignition and/or fuel system faults	Check the ignition and fuel system as though the engine had failed to start.
	Incorrect valve clearances	Check and reset clearances.
	Burnt out valves	Remove cylinder heads and examine the overhaul as necessary.
Lack of power	Ignition and/or fuel system faults	Check the ignition and fuel system for correct ignition timing and carburettor settings.
	Incorrect valve clearances	Check and reset the clearances.
	Burnt out valves	Remove cylinder head and examine and overhaul as necessary.
	Worn out piston or cylinder bores	Remove cylinder head and examine pistons and cylinder bores. Overhaul as necessary.

Symptom	Reason/s	Remedy
Excessive oil consumption	Oil leaks from crankshft oil seal, rocker cover gasket, oil pump, drain plug gasket, sump plug washer, oil cooler	Identify source of leak and repair as appropriate.
	Worn piston rings or cylinder bores resulting in oil being burnt by engine Smoky exhaust is an indication	Fit new rings or rebore cylinders and fit new pistons depending on degree of wear.
	Worn valve guides and/or defective valve stem seals	Remove cylinder heads and recondition valve stem bores and valves and seals as necessary.
Excessive mechanical noise from engine	Wrong valve to rocker clearances	Adjust valve clearances.
	Worn crankshaft bearings Worn cylinders (piston slap)	Inspect and overhaul where necessary.
Unusual vibration	Misfiring on one or more cylinders	Check ignition system.
	Loose mounting bolts	Check tightness of bolts and condition of flexible mountings.

Note: *When investigating starting and uneven running faults do not be tempted into snap diagnosis. Start from the beginning of the check procedure and follow it through. It will take less time in the long run. Poor performance from an engine in terms of power and economy is not normally diagnosed quickly. In any event the ignition and fuel systems must be checked first before assuming any further investigation needs to be made.*

Chapter 2 Cooling system

For modifications, and information applicable to later models, see Supplement at end of manual

Contents

Specifications

System type ... Semi-sealed, thermo-syphon, water pump assisted with electric cooling fan

Radiator material ... Steel

Cooling area ... 194 in^2 (1250 cm^2)

Thermostat fully open ... 96°C (205°F) begins to open 83°C (181°F)

Electric fan operation

Actuates at water temperature of ...	95°C (± 1.5°C) 203°F (± 2.7°F)
Switches off at water temperature of ...	86°C (± 2°C) 187°F (± 3.6°F)

Capacities

System including vehicle heater ...	10.75 pint (6.1 litres)
Difference between 'MAX' and 'MIN' marks on expansion bottle ...	4/5 pint (0.45 litre)

Water pump belt tension ... ½ in (12 mm) maximum free movement of longest run between pulleys.

Torque wrench settings	**lb f ft**	**kg f m**
Thermoswitch to radiator ...	25	3.46
Radiator mounting bolts ...	12	1.68
Thermostat housing to cylinder head ...	8.5	1.18

Thermostat housing cover	8.5	1.18
Water temperature transmitter to cylinder head	10.5	1.45
Water pump to sump casing	8.5	1.18
Water pump inlet housing to sump casing	8.5	1.18
Water pump drain plug	11.5	1.59
Water pump pulley bolt	10.5	1.45
Water inlet housing to cylinder block	14.5	2.01
Water return manifold to cylinder block	12.0	1.66

1 General description

The cooling system is of conventional type and operates by means of thermo-syphon action with the assistance of a belt driven water pump. A diagrammatic presentation of the circulation pattern and the location of the major components is shown in Fig. 2.1.

The coolant in the radiator flows from left to right instead of in the more common downward direction and the 'header' tanks are situated on either side of the radiator matrix.

Coolant heated in the cylinder jackets is cooled by the ram effect of air passing through the radiator matrix when the car is in motion and assisted by a thermostatically, electrically operated fan which operates within a pre-determined temperature range.

Coolant from the system also circulates through the vehicle heater and is used to pre-heat the inlet manifold.

The system is semi-pressurised and incorporates an expansion bottle to accept coolant displaced when the engine is hot and to act as a reservoir when the system cools down or in the event of minor leakage.

A thermostat is fitted in the system to restrict coolant circulation until the normal engine operating temperature has been reached.

The original coolant is effective indefinitely but where a loss has occurred or the strength of the coolant (antifreeze mixture) is suspect, then the system should be drained, flushed and refilled as described in later Sections of this Chapter.

All flexible hose joints are secured by Corbin type clips and they should only be prised open sufficiently far to enable them to slide over the hose. Ordinary pliers may be used for their removal but special types are more satisfactory. However, the cost of these need not be incurred if the Corbin type are replaced by reliable worm drive clips.

2 Cooling system - draining

1 If it is wished to retain the coolant for further use, place a clean receptacle beneath the engine and then set the car interior heater to the full on position.

2 Unscrew and remove the plug and sealing washer from the base of the water pump and remove the lid of the expansion bottle.

3 To expedite the draining operation, the radiator filler cap may be removed.

4 Do not allow the coolant to come into contact with the paintwork of the car as its antifreeze content will damage the surface of the finish.

5 The coolant should be retained in a covered vessel pending return to the system and any sediment which precipitates should be discarded.

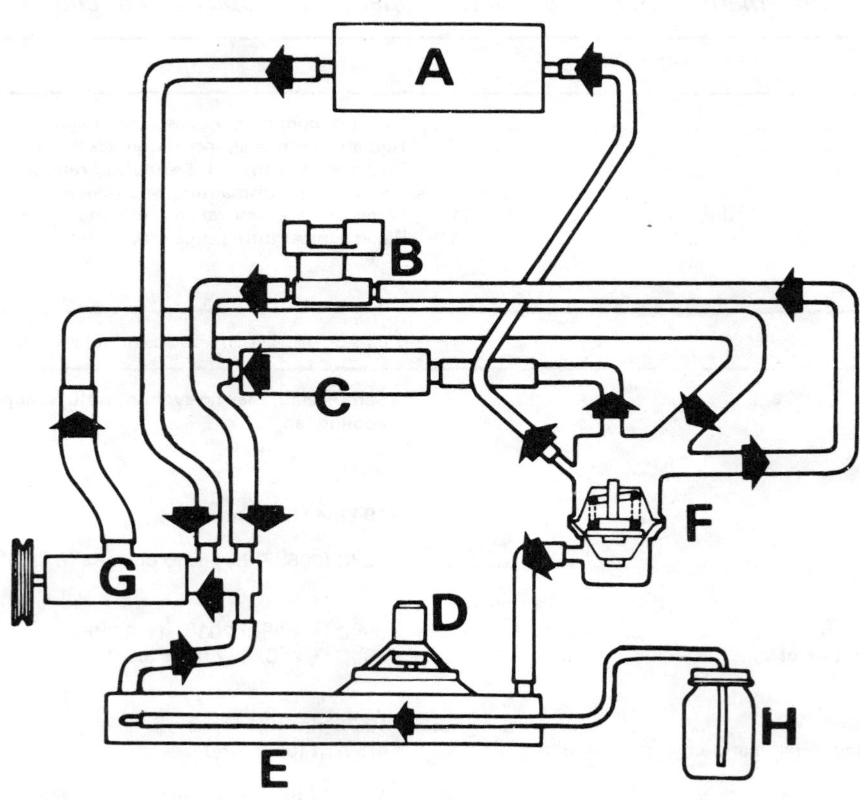

Fig. 2.1. Layout of cooling and interior heating system

A Car heater matrix
B Heated carburettor flange (1294 cc single carb)
C Heated inlet manifold
D Electric cooling fan
E Radiator
F Thermostat
G Water pump
H Expansion bottle

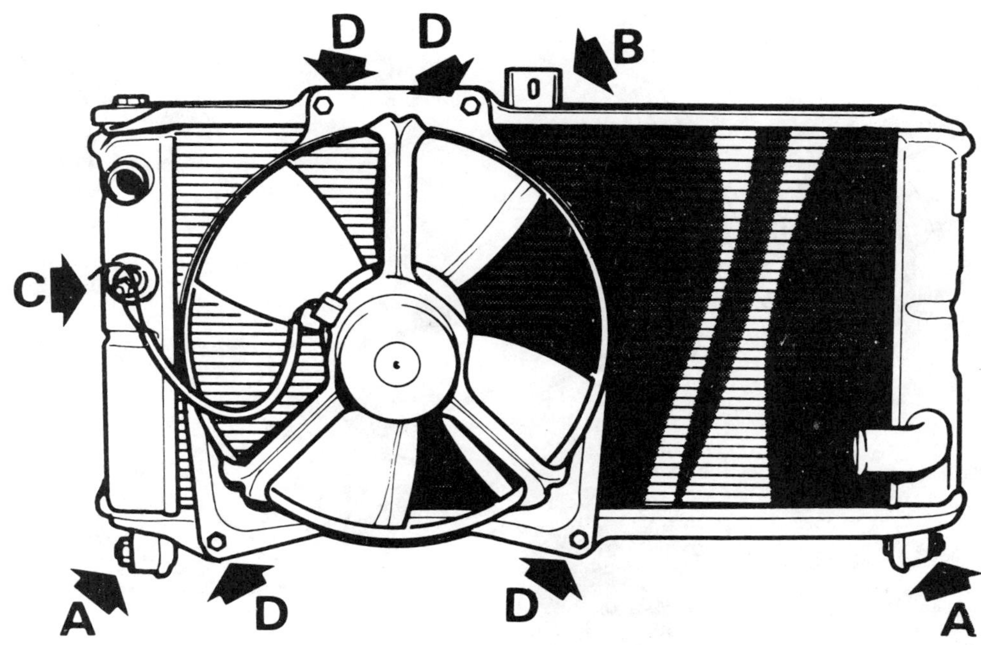

Fig. 2.2. Radiator and cooling fan assembly

A Lower radiator attachment points *B Top radiator attachment point* *C Temperature sensing unit* *D Cooling fan securing bolts*

3 Cooling system - flushing

1 Provided the cooling system is maintained in good order then periodic flushing should not be necessary. Where the coolant has discoloured or become contaminated with oil due to gasket failure, then the system should be thoroughly cleansed.
2 To do this, remove the radiator filler cap and insert a hose in the filler neck, then, with the water pump drain plug removed and the heater control full on, allow the water to flow until it is quite clear when emerging from the water pump drain plug.
3 If the radiator appears blocked then it should be removed as described in Section 6 and reverse flushed. This is carried out by placing the hose in the right-hand radiator outlet so that the water flow is in the opposite direction to normal.
4 The removal of scale from the system should not normally be a problem as, with a semi-sealed circuit, only initial scaling occurs unless, due to leaks, continual topping up is required.
5 The use of chemical de-scalers and cleansers is not recommended as, unless specifically formulated, damage to the aluminium cylinder head, water pump and thermostat housings may occur.
6 Never flush a hot engine cooling system with cold water or cracks or distortion of the block or head may be caused.
7 In the event of blockages in the heater matrix then this should be removed as described in Chapter 12 and serviced as previously described for the radiator.

4 Cooling system - filling

1 Place the heater control to the full on position.
2 Remove the expansion bottle cap.
3 Remove the radiator cap.
4 Fill the radiator slowly right to the top of the filler neck with coolant of the correct antifreeze mixture strength (see next Section). If a new solution is being made up it is preferable to use soft or de-mineralised water.
5 Fill the expansion bottle to the 'maximum' mark.
6 Refit the radiator cap but do not overtighten it.
7 Refit the expansion bottle cap.
8 Start the engine and run it at an even speed until bubbles cease to be visible in the coolant contained in the expansion bottle.
9 Top up the level of the coolant in the expansion bottle until it reaches the 'maximum' mark.
10 Do not allow coolant solution to contaminate the valve in the expansion bottle cap due to overfilling or gas pressure which could be caused by a blown gasket. The valve should be renewed in either event.

5 Antifreeze mixture

1 The use of antifreeze mixture in the cooling system fulfills two purposes. To protect the engine and heater components against fracture during periods of low ambient temperature and to utilise the effects of the rust and corrosion inhibitors incorporated in the antifreeze product.
2 A 'long-life' type of antifreeze mixture may be used but where a normal commercial type is used it is wise to renew it or at least check its strength with a hydrometer every year.
3 Ensure that the mixture is of a type compatible with aluminium components and refer to the following table for strength recommendations.

Quantity of antifreeze	Gives protection to
1.7 pints (1 litre)	*−17.8°C (0°F)*
2.0 pints (1.3 litres)	*−28.9°C (−20°F)*
2.3 pints (1.43 litres)	*−34.5°C (−30°F)*
3.0 pints (1.7 litres)	*−40°C (−40°F)*

4 Due to the searching action of antifreeze mixture, always check the security of hose clips and gasket joints before filling.

6 Radiator - removal, inspection, cleaning and refitting

1 Drain the cooling system as described in Section 2 of this Chapter.
2 Disconnect the battery earth terminal.
Note: It is possible to remove the radiator and fan unit as one assembly, however, the task is easier if the fan unit is removed first.

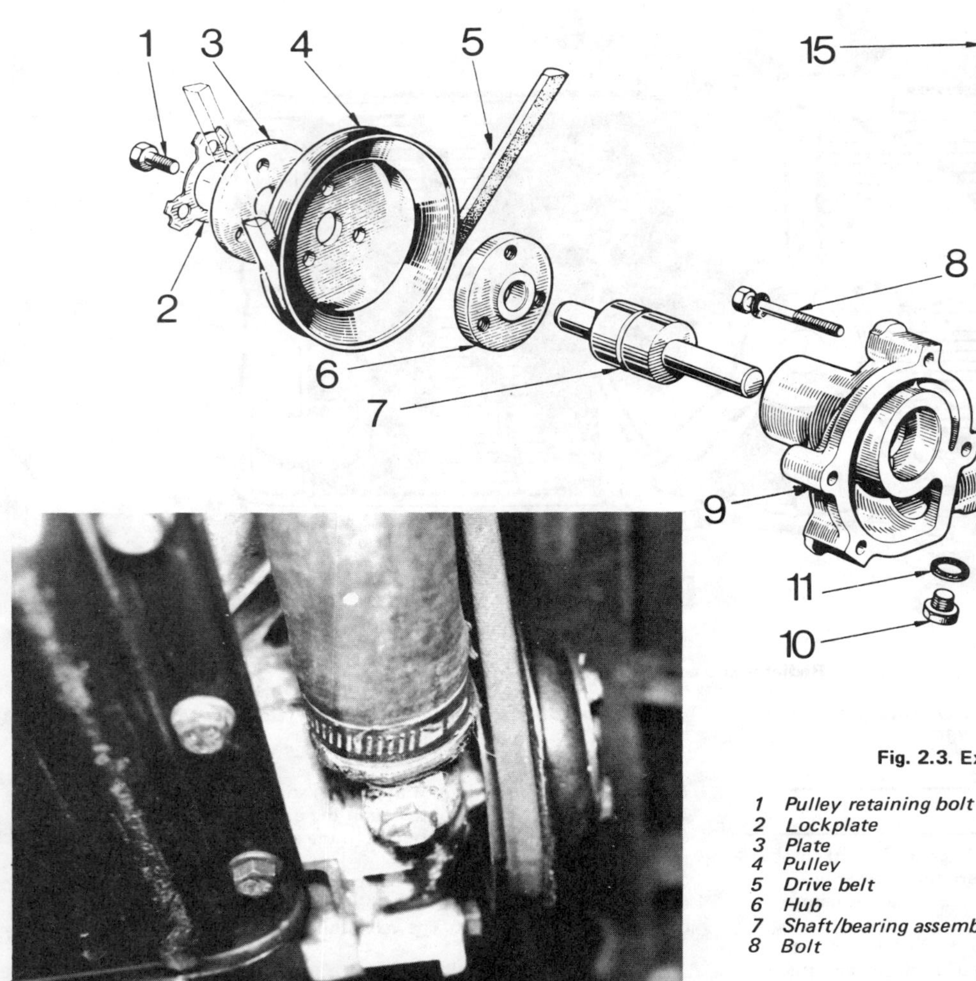

8.5 Water pump to engine hose (note the drain plug)

Fig. 2.3. Exploded view of water pump

1 *Pulley retaining bolt*
2 *Lockplate*
3 *Plate*
4 *Pulley*
5 *Drive belt*
6 *Hub*
7 *Shaft/bearing assembly*
8 *Bolt*
10 *Drain plug*
11 *Drain plug seal*
12 *Seal*
13 *Impeller*
14 *Gasket*
15 *Assembled water pump*

Fig. 2.4. Fitting the water pump seal

3 Disconnect the two wires from the temperature sensing unit located at the rear left-hand side of the radiator.
4 Referring to Fig. 2.2, remove the bolts securing the fan unit to the radiator and lift away the fan and cowl.
5 Disconnect the top and bottom radiator hoses.
6 Disconnect the expansion bottle hose from the top of the radiator and slide it out of the retaining clip.
7 Remove the nuts and washers from both the lower radiator mountings and the top mounting.
8 Tilt the radiator assembly towards the engine and then lift it clear of the lower mounting plates.
9 The radiator matrix should be cleaned internally as described in Section 3. Any accumulation of flies on the radiator fins should be removed by lightly brushing or blowing out with compressed air.
10 If the radiator is leaking, do not attempt to repair it yourself as the heat used for soldering must be carefully localised if further leaks are not to be created. Take the unit to a specialist repairer or exchange the unit for a factory reconditioned one. The use of any type of leak sealant is at best, a temporary cure and its use may clog the fine tubes of the heater matrix and damage the water pump seals.
11 Refitting the radiator is a reversal of removal. Refill the system as described in Section 4.

7 Thermostat - removal, testing and refitting

1 Drain the cooling system as previously described.
2 Unscrew and remove the two bolts which secure the thermostat housing cover in position.
3 Remove the cover and gasket and withdraw the thermostat. If it is stuck in its seat do not try to lever it out but cut round its seat joint with a pointed blade to break the seal.

Fig. 2.5. Fitting the water pump shaft/bearing assembly

4 To test whether the thermostat is serviceable, suspend it in a pan of water into which a thermometer has been placed. Heat the water and check that the thermostat begins to open when the water temperature reaches that at which the thermostat is rated (see Specifications).
5 Similarly, when the thermostat is fully open, place it into cooler water and observe its closure. Any failure in the opening or closing actions of the unit will necessitate renewal. Fit one with the specified temperature marked on it, nothing is to be gained by fitting one having a different operating temperature range and could cause cool running or overheating of the engine and heater inefficiency.
6 Refitting is a reversal of removal but ensure that the locating tab on the thermostat is correctly aligned and use a new cover gasket. Do not overtighten the cover securing bolts.
7 Refill the system as described in Section 4.

8 Water pump - removal and refitting

1 Disconnect the battery negative terminal.
2 Set heater control to full on, remove the expansion bottle cap and unscrew the drain plug at the base of the water pump. Retain the coolant if required.
3 From under the wing remove the pulley driving belt cover as described in Chapter 1.
4 Slacken the adjuster bracket bolt on the alternator and having pushed the alternator in towards the engine, slip off the driving belt.
5 Disconnect the water pump to crankcase hose (photo).
6 Unscrew and remove the four water pump cover retaining bolts and lift the pump from its crankcase location.
7 Peel off the old gasket and clean the mating faces of both the pump and crankcase.
8 Refitting is a reversal of removal but always use a new gasket.
9 Refill the cooling system and reconnect the battery.

9 Water pump - dismantling and reassembly

1 The water pump is designed for long trouble-free service and if it has been in operation for a considerable mileage then it will probably be more realistic to exchange the complete unit for one which has been factory reconditioned, rather than attempt to repair a pump without the necessary experience and tools.
2 Check the condition of the impeller for corrosion. Other than this, any fault will lie with a worn shaft seal (Fig. 2.3).
3 Press off the hub from the shaft. To achieve this, support the rear face of the hub and exert pressure on the end of the shaft. (If only the seal is to be renewed, do not remove the hub).
4 Turn the pump over and again, adequately supporting the rear face of the impeller, press the end of the shaft to expel it from the impeller bore.
5 Now immerse the pump body in boiling water for two or three minutes. Remove the pump quickly and drive out the shaft/bearing assembly by driving it out from the seal end with a copper or plastic faced hammer.
6 Drive the seal from the water pump body by means of a drift inserted from the front end of the pump.
7 The shaft/bearing assembly cannot be separated, and if necessary,

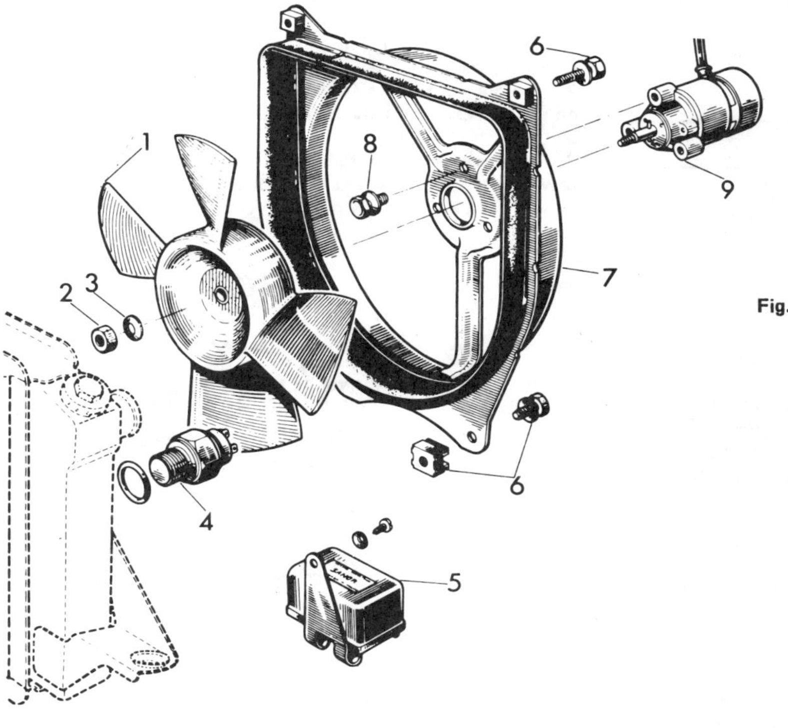

Fig. 2.6. Cooling fan and motor assembly

1 Fan
2 Securing nut
3 Lockwasher
4 Thermostatic contactor
5 Relay
6 Mounting bolt
7 Frame
8 Motor securing bolt
9 Electric motor

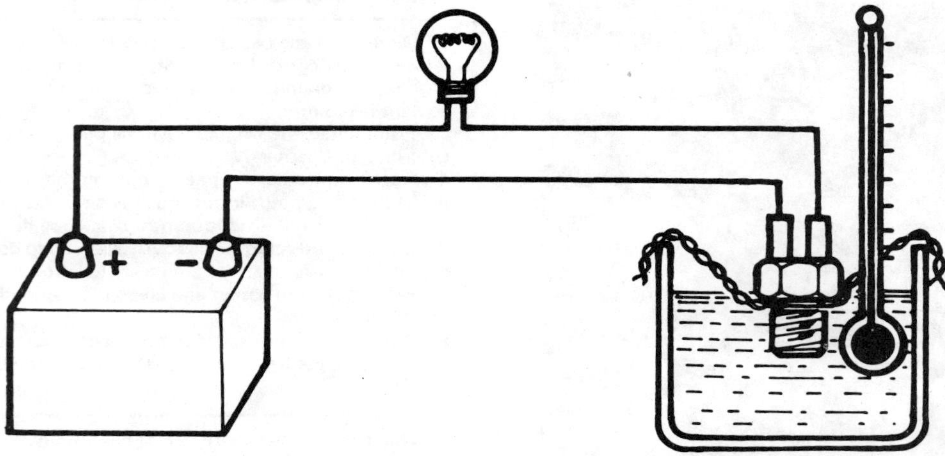

Fig. 2.7. Method of testing the temperature sensor unit

must be renewed as a single component.

8 Check the condition of the seal and shaft bearing mating surfaces of the pump body. Where necessary, these may be improved with grade '600' paper. If the seal is to be renewed, make sure the new seal is of the correct type. The seal with a long brass outer skirt must only be fitted to the later short impeller, otherwise the impeller clearance will be incorrect.

9 Press the water pump seal into its seating in the body (Fig. 2.4).

10 Again heat the pump body in boiling water and then press the shaft bearing assembly into the body ensuring that the bearing seats securely against the bore inner shoulder. Note that the longer end of the shaft is at the seal end (Fig. 2.5).

11 Supporting the end face of the shaft at the seal end, press on the flanged hub so that its boss is towards the shaft bearing. The hub is correctly positioned on the shaft when there is a clearance of 0.050 in (1.270 mm) between the pump and hub faces measured with feeler gauges.

12 Support the front end face of the shaft and press on the impeller (vanes outwards) until there is again a clearance of 0.050 in (1.270 mm) between the pump and impeller faces.

13 Check for free rotation by turning the vanes of the impeller.

10 Radiator cooling fan - description and servicing

1 The radiator cooling fan is shown in exploded form in Fig. 2.6.

2 The assembly comprises a four bladed fan attached to the driving spindle of an electric motor.

3 The fan motor is controlled by a temperature sensing unit that is screwed into the rear left-hand side of the radiator.

4 When the coolant temperature reaches 95°C (203°F) the temperature sensor contacts close and energise the fan. When the water temperature falls to between 86 and 88°C (187 and 191°F) the contacts open and the fan is de-energised.

Warning: *It must be realised that during adjustments within the engine compartment, with the engine at operating temperature, the fan blades may turn unexpectedly if the ignition is switched on. From the point of view of safety, disconnect the fan motor leads before carrying out adjustments in close proximity to the radiator, but watch the coolant temperature if the engine is running!*

5 If the operation of the temperature sensing unit is suspect it should be removed for testing.

6 Disconnect the battery earth terminal, drain the cooling system as described in Section 2 and unscrew the sensor unit from the radiator.

7 Suspend the sensor unit in a water filled container with a thermometer of a suitable temperature range and connect up a 12V battery and test lamp as shown in Fig. 2.7.

8 Heat the water on a stove and check that the lamp lights at a temperature of 95°C (203°C) and extinguishes when the water is allowed to cool to 86°C (187°C).

9 If the lamp does not light, or lights and extinguishes at a different temperature than specified, the sensor unit is faulty and must be renewed.

10 Should the sensor unit function correctly then the fault must be in the fan motor itself.

11 Remove the fan motor and blade assembly after disconnecting the electrical leads and withdrawing the four securing bolts.

12 Unscrew and remove the nut and spring washer which secures the fan blades to the motor shaft. This has a left-hand thread and the nut must therefore be unscrewed in a clockwise direction.

13 Unscrew and remove the three bolts which secure the motor to the fan assembly outer frame. Renew the motor on an exchange basis.

14 Reassembly and refitting of the fan cooling unit is a reversal of removal and dismantling.

11 Drivebelt - adjustment, removal and refitting

1 To adjust the alternator drivebelt, slacken the pivot bolt and locking bolt shown in Fig. 2.8 and move the alternator in the necessary direction.

2 The correct tension for the belt is ½ in (12 mm) of free movement on the longest run of the belt. When correctly adjusted tighten the pivot and locking bolts.

3 To remove the belt, slacken the bolts and push the alternator in towards the engine as far as possible. Ease the belt off the pulleys rotating the alternator if necessary.

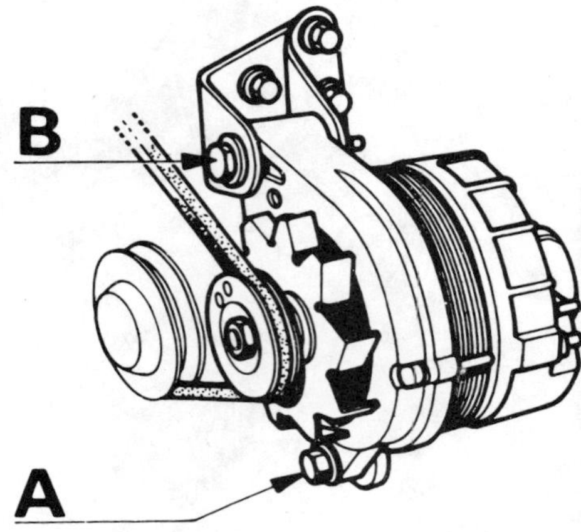

Fig. 2.8. Drive belt adjusting bolts

A Locking bolt *B Pivot bolt*

4 Fit the new belt using the reverse procedure to removal and adjust the tension as described previously.

12 Water temperature gauge - fault finding

1 Correct operation of the water temperature gauge is important as the engine could attain a considerable degree of overheating, unnoticed, if giving false readings.

2 To check the correct operation of the gauge, first disconnect the 'Lucar' connector from the sender unit plug screwed into the side of the inlet manifold. With the ignition 'on' the gauge should be at the cold mark. Then earth the lead to the engine block when the needle should indicate hot, at the opposite end of the scale. This test shows that the gauge on the dash is functioning properly. If it is not then it will need renewal. If there is still a fault in the system with this check completed satisfactorily, there will be a fault in the sender unit or the wire leading from it to the gauge. Renew these as necessary.

13 Fault finding - cooling system

Symptom	Reason/s	Remedy
Overheating		
Heat generated in cylinder not being successfully disposed of by radiator	Drivebelt slipping	Check tension.
	Insufficient water in cooling system	Top up radiator.
	Cooling fan inoperative	Overhaul or renew fan.
	Radiator core blocked or radiator grille restricted	Reverse flush radiator, remove obstruction.
	Bottom water hose collapsed, impeding flow	Remove and fit new hose.
	Thermostat not opening properly	Remove and fit new thermostat.
	Ignition advance and retard incorrectly set (accompanied by loss of power, and perhaps, misfiring)	Check and reset ignition timing.
	Carburettor(s) incorrectly adjusted (mixture too weak)	Tune carburettor(s).
	Exhaust system partially blocked	Check exhaust pipe for constrictive dents and blockages.
	Oil level in sump too low	Top up sump to full mark on dipstick.
	Blown cylinder head gasket (water/steam being forced down the radiator overflow pipe under pressure)	Remove cylinder head, fit new gasket.
	Engine not yet run-in	Run-in slowly and carefully.
	Brakes binding	Check and adjust brakes if necessary.
Underheating		
Too much heat being dispersed by radiator	Thermostat jammed open	Remove and renew thermostat.
	Incorrect grade of thermostat fitted allowing premature opening of valve	Remove and replace with new thermostat which opens at a higher temperature.
	Thermostat missing	Check and fit correct thermostat.
Loss of cooling water		
Leaks in system	Loose clips on water hoses	Check and tighten clips if necessary.
	Top, bottom, or by-pass water hoses perished or leaking	Check and renew any faulty hoses.
	Radiator core leaking	Remove radiator and repair.
	Thermostat gasket leaking	Inspect and renew gasket.
	Radiator cap seal ineffective	Renew radiator pressure cap seal.
	Blown cylinder head gasket (pressure in system forcing water/steam into expansion bottle)	Remove cylinder head and fit new gasket.
	Cylinder wall or head cracked	Dismantle engine, despatch to engineering works for repair.

Chapter 3 Carburation; fuel and exhaust systems

For modifications, and information applicable to later models, see Supplement at end of manual

Contents

Specifications

Fuel pump

Type	Mechanical, driven from camshaft
Make	SEV
Pressure at zero output	200-300 m/bars (2.9-4.4 lb in^2)
Mean operating pressure	133 m/bars (1.9 lb in^2)

Carburettor type and data (Solex)

Engine type	**1294 cc single carb. (except Germany)**	**1294 cc single carb. (Germany only)**
Engine code	6 G 1	6 G 1D
From engine no.	67 400 021	67 530 021
Compression ratio	9.5 : 1	8.2 : 1
Carburettor make	Solex	Solex
Type	32 BISA 5	32 BISA 5
Ref no.	81	87
Choke tube	27	23
Main jet	140 ± 2.5	112 ± 2.5
Air correction jet	165 ± 5	185 ± 5
Emulsion tube	E7	E7

Econostat fuel jet	45	40
Idling fuel jet	39 to 45	39 to 45 with damper
Idling air jet	180	180
Progression holes	4 x 0.6	4 x 0.6
Pump injector	—	45
Pump stroke	3 mm	3 mm
Float needle	1.5	1.5
Float weight	5.7 g	5.7 g
Starting system	Strangler	Strangler
Fast idle gap (at throttle plate)	1.00 mm ± 0.05	1.05 mm ± 0.05
Constant CO circuit:		
Air intake aperture	500	500
Fuel calibration	30	30
Air calibration	100	130
CO content at idling	1 to 2%	1 to 2%

Carburettor type and data (Weber)

Engine type	1294 cc 2 double choke carbs		1442 cc 1 double choke carb	
Engine code	6 G 4		6 Y 2	
From engine no.	67 600 021		68 000 021	
Compression ratio	9.5 : 1		9.5 : 1	
Carburettor make	Weber		Weber	
Type	36 DCNF		36 DCNV	
Ref no.	49 - 50		2	
Choke tube	29	29	28	28
Secondary venturi	3.5 long	3.5 long	4.5 short	4.5 short
Main jet	125	125	132 ± 2.5	132 ± 2.5
Air correction jet	200 ± 15	200 ± 5	175 ± 5	175 ± 5
Emulsion tube	F 36	F 36	F 36	F 36
Idling fuel jet	45	45	40 - 42	40 - 42
Idling air jet	135 ± 15	135 ± 15	145 ± 5	145 ± 5
Progression holes	80-90-90-105	80-90-90-105	—	—
Pump injector	40	40	50	50
Pump bypass	40		—	
Float needle	1.75		1.75	
Float weight	20 g		20 g	
Float level	52 ± 0.25 mm (54 ± 0.25 mm on DCNF-A)		52 ± 0.25 mm	
Starting system	Subsidiary jets		Strangler	
Air jet	250 (220*)	250	—	—
Fuel jet	70 (80*)	70	—	—
Emulsion tube	F10 (F4*)	F10	—	—
Fast idle gap	—	—	0.35 - 0.40 mm	
Vacuum kick opening	—	—	6.0 - 6.5 mm	
Mechanical opening	—	—	8.0 - 8.5 mm	
CO content at idling	1.5 to 3.0%		1 to 2.0%	

* (from January 1977)

Idling speeds

1294 cc engine (single carb)	850 rpm
1294 cc engine (twin carbs)	950 rpm
1442 cc engine (single carb)	900 rpm

Torque wrench settings

	lb f ft	kg f m
Inlet manifold	10.0	1.38
Exhaust manifold nuts	14.5	2.01
Carburettor mounting flange nuts	14.5	2.01
Air cleaner securing bolts	9.5	1.31
Fuel pump to crankcase	10.5	1.45

1 General description

The basic components of the fuel system comprise, a rear mounted fuel tank, a mechanically driven fuel pump, and either a single or twin downdraught type carburettor with all the associated fuel pipes and controls.

Several types of fuel pumps have been fitted but they are all similar in operation and the removal and servicing procedure is basically the same.

The 1294 cc engine is fitted with either the single Solex carburettor or two twin choke Weber types. The 1442 cc engine is equipped with one twin choke Weber.

Each carburettor is carefully tested during production and the most suitable sized jets are selected for that carburettor and, in the case of the Weber type, the jets fitted to one barrel may differ in size from those of the other. Therefore when removing the jets a careful note must be made of which barrel they belong to.

The inlet manifold is water heated on all models and the Solex carburettors have a water heated flange to improve fuel atomization.

2 Air cleaner - description and servicing

1 All models are fitted with a renewable paper element air filter housed in a plastic casing on top of the engine which is retained by a quick-release strap.

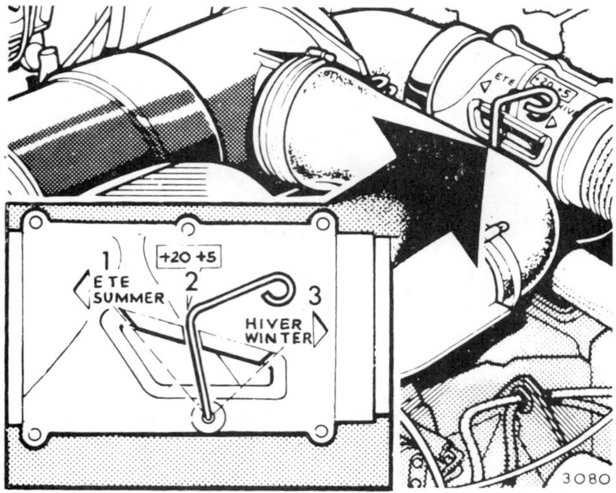

Fig. 3.1. Air inlet temperature control lever

1 Summer position
2 Mid position
3 Winter position

2.6 Air cleaner assembly complete with Solex adaptor

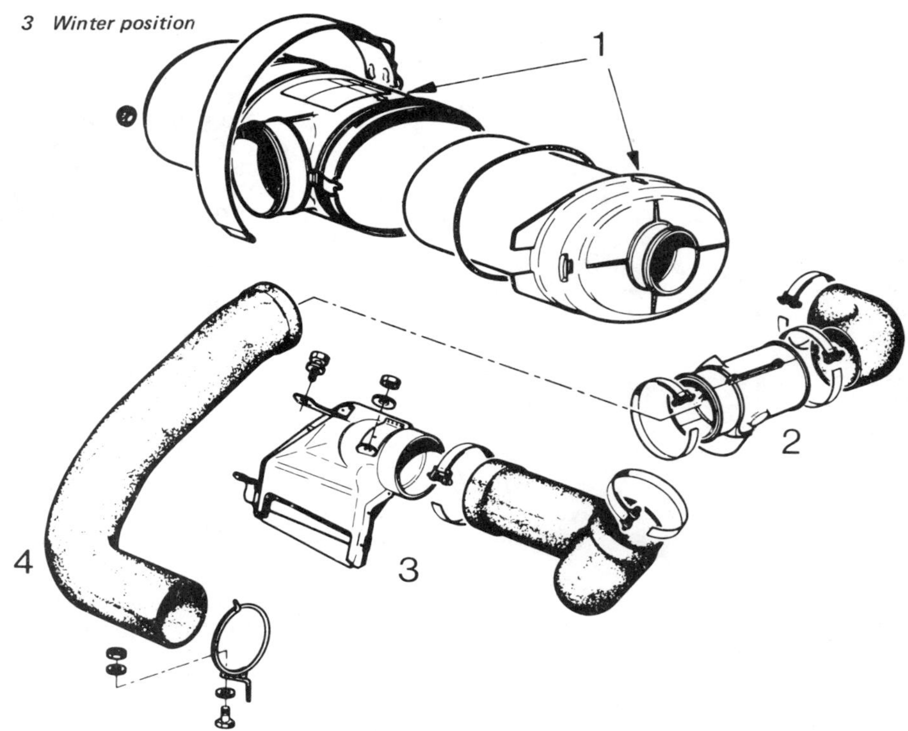

Fig. 3.2. Air cleaner components

1 Alignment marks *2 Temperature control duct* *3 Hot air pick-up duct* *4 Cool air hose*

2 The temperature of the air entering the air cleaner can be controlled by means of a SUMMER(ETE)/WINTER(HIVER) selector lever on the intake duct (see Fig. 3.1). In the 'summer' position cold air is drawn into the air cleaner assembly via a hose leading in from the front wing, while the 'winter' position allows air to be drawn in through a hot air duct fitted over the exhaust manifold.
3 To renew the filter element, disconnect the inlet hose from the end cover, release the spring clips and withdraw the end cover complete with filter element (See Fig. 3.2).
4 The end cover is integral with the filter element and the complete assembly must be discarded.
5 When fitting the new filter and end cover, ensure that the arrows on top of the cover and filter housing are in alignment before fastening the clips and refitting the intake hose.
6 To remove the complete filter housing assembly, remove the intake and outlet hoses, release the large retaining strap and lift the filter assembly away from the engine (photo).
7 Three different carburettor-to-air cleaner adaptors are used depending on the carburettor(s) fitted.
8 To remove the adaptor on all models, first remove the complete air cleaner as previously described. Disconnect the inlet hose from the adaptor and remove the fume extractor hose if fitted.
9 On the 1294 cc engine remove the nuts securing the adaptor to the carburettor(s), lift the adaptor off the studs and retrieve the sealing ring (single carburettor) or two gaskets (twin carburettors).
10 In the case of the 1442 engines, release the seven clips securing the two halves of the adaptor together and prise off the top half. Remove the retaining nuts, lift off the bottom half and retrieve the gasket and stiffener plate (Fig. 3.3).).
11 Refit the adaptor assemblies using the reverse procedure to that of removal. On the 1442 cc engine ensure that the seal is correctly fitted between the two halves of the adaptor.

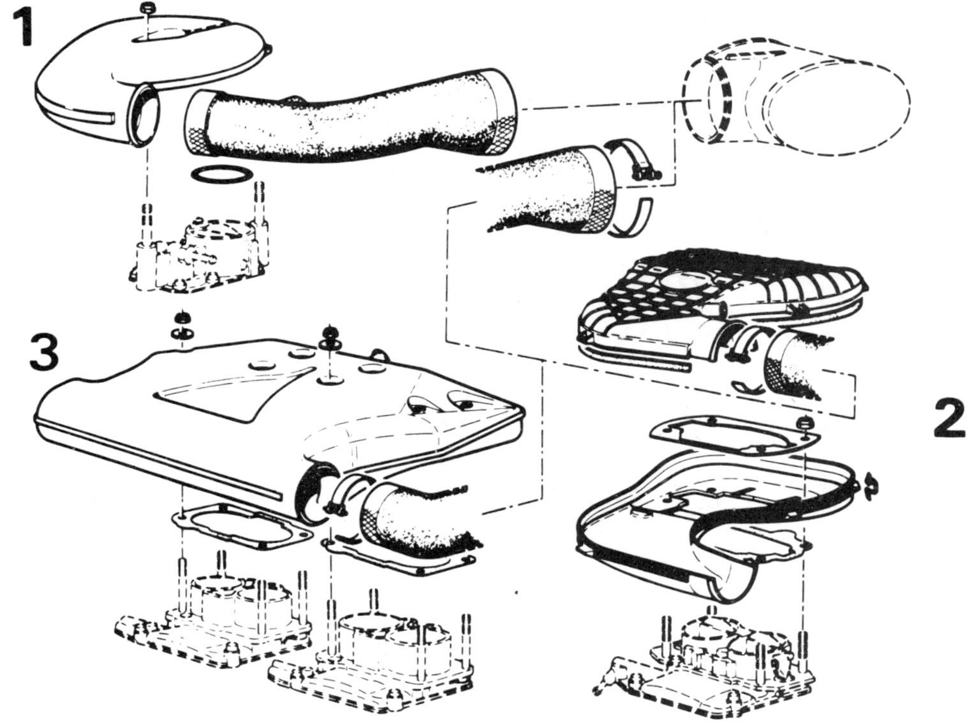

Fig. 3.3. Carburettor adaptor assemblies

1 Solex *2 Single Weber* *3 Twin Weber*

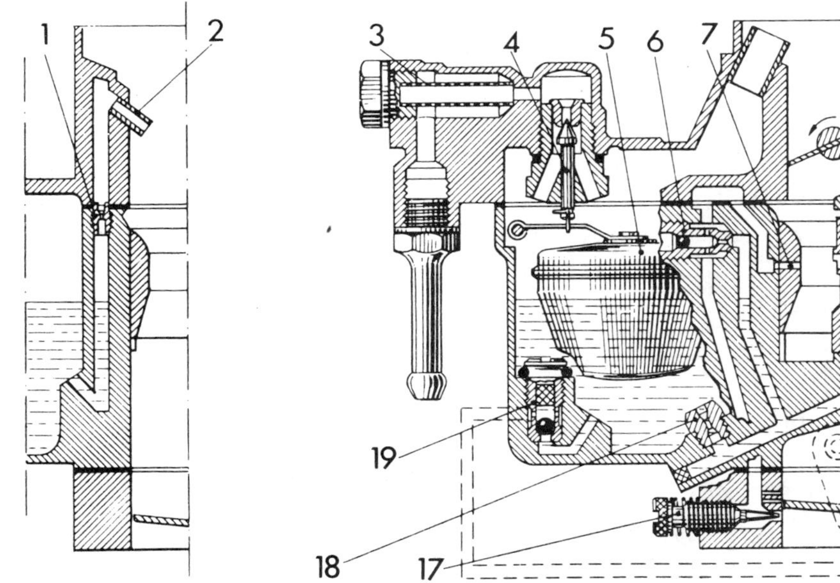

Fig. 3.4. Sectional view of Solex carburettor

1 Econostat jet
2 Econostat discharge nozzle
3 Fuel filter
4 Fuel needle valve
5 Fuel float
6 Idling jet
7 Idling air bleed (calibrated drilling)
8 Choke plate
9 Choke plate by-pass valve
10 Emulsion tube and air correction jet
11 Accelerator pump discharge nozzle
12 Venturi (choke tube)
13 Accelerator pump diaphragm
14 Accelerator pump lever
15 Pre-heater water connections
16 Throttle plate
17 Idling mixture adjustment screw
18 Main jet
19 Accelerator pump non-return valve

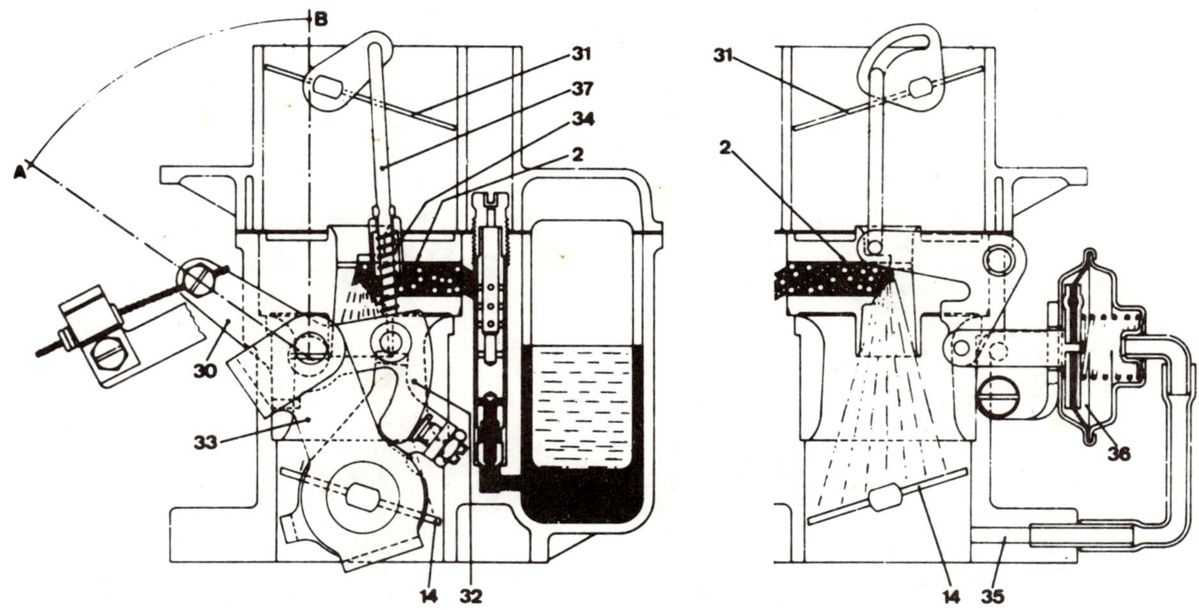

Fig. 3.5. Choke operation of Weber single carburettor (type 36DCNV). Refer to Section 3 for key details.

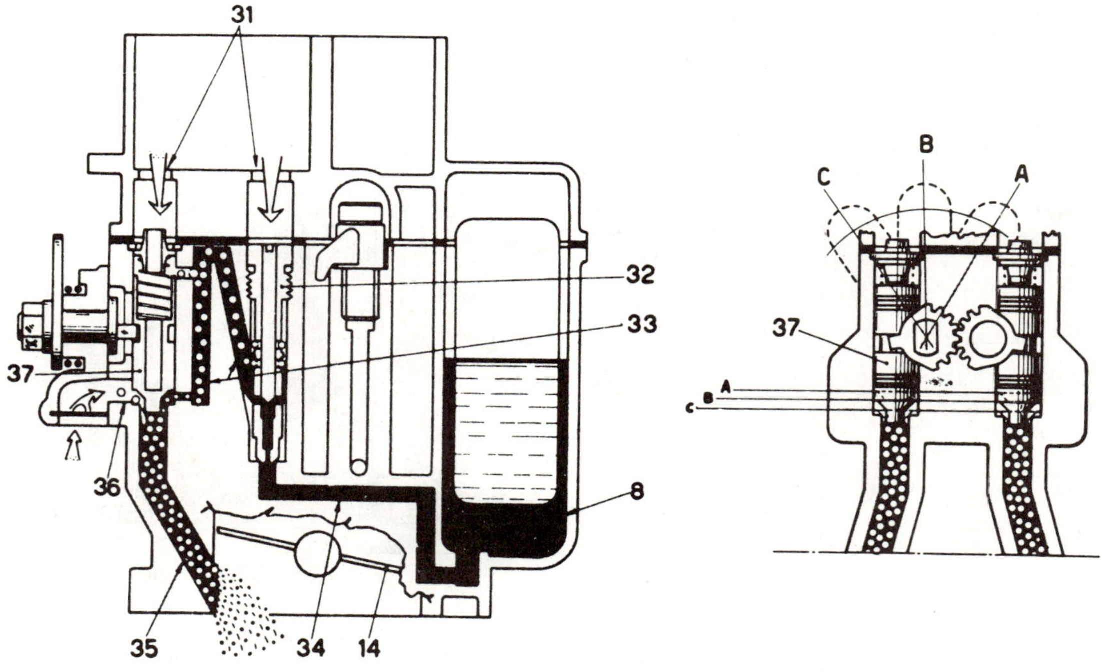

Fig. 3.6. Choke operation of Weber twin carburettor (type 36DCNF). Refer to Section 3 for key details.

3 Carburettors - description and operation

1 Three types of carburettor are used with the engines covered by this manual (see Specifications for engine/carburettor type). These comprise, the Solex 32 BISA 5, the Weber 36 DCNV and the Weber 36 DCNF. The 36 DCNV type is used on single carburettor models and the 36 DCNF is used on the twin carburettor versions. Both the Weber types are very similar in construction, the main difference being in the type of choke mechanism used.
2 The Solex carburettor is shown in sectional form in Fig. 3.4.
3 It comprises three main units, the throttle block, the main body and the top cover. The throttle block embodies the throttle plate and control assembly and incorporates a water heated jacket connected to the engine cooling system to provide pre-heating of the carburettor.
4 The main body incorporates the choke tube, float chamber, accelerator pump and the jets, also the distributor vacuum and crankcase breather connections.
5 The top cover comprises the starting plate assembly, needle valve, inlet connection and filter and the econostat discharge nozzle.
6 The starting device comprises a choke plate fitted with a bypass valve and a fast idle connecting link to the throttle control arm.
7 The accelerator pedal is connected to the throttle spindle which in turn actuates the accelerator pump to discharge neat fuel into the choke tube when accelerating hard.
8 The econostat device operates from the air intake flow and comes into action at high engine speeds only, to provide a rich mixture at full load conditions.
9 The twin type carburettors (Weber DCNF) fitted to the 1294 cc engine are of twin choke downdraught design with manually operated choke. Each choke tube supplies one engine cylinder and the twin carburettors therefore provide the effect of four individual units.
10 The carburettors are identical in construction but the throttle control layout varies between left- and right-hand units.
11 The throttle and cold start controls are both connected to one carburettor but the design of the control arms permits synchronising of the two carburettors and the individual cold-start devices are interconnected.
12 As stated previously the single Weber carburettor is virtually identical to the type used on the twin carburettor installation with the exception of the choke mechanism. The following paragraphs describe the various operating cycles of the Weber type carburettors.
13 *Choke in operation (single carburettor Fig. 3.5).* The choke valves (31) are offset on the spindle. A lever on one end of the spindle is connected to another lever (30) operated by the choke control knob and a lever on the other end of the spindle is connected to the vacuum kick diaphragm (36). When the choke valves are closed by the spring loaded rod (37), the cam (32) on the operating lever opens the throttle valves (14) a pre-set amount. When the starter motor is operated a rich mixture is drawn from the main discharge passage (2) in the secondary venturi. When the manifold depression increases as the engine starts, the vacuum kick valve assists the operation of the choke valves via the passage (35) and control linkage.
14 *Choke in operation (twin carburettors Fig. 3.6).* Fuel from the bowl (8) passes through channels (34) and jets (32) and is emulsified by air from the holes (31). The fuel reaches the valve seat (37) through the channels (33) and again emulsified by air from the holes (36). It flows through channels (35) into the area below the butterfly valve (14).
15 *Idling and progression (Fig. 3.7).* The fuel flows to the idling jets (19) through the channels (18) and emulsified with air supplied through the jets (20). The volume of air admitted is regulated by the size of the jets (20). The fuel then passes through the channels (17) and holes (15) to the tubes located below the butterfly valve (14). From idling, as the butterfly valves (14) are opened progressively, the mixture reaches the tubes and flows through the progression holes (13) so enabling the engine speed to be increased progressively.
16 *Normal running at constant throttle opening (Fig. 3.8).* The fuel flows through the needle valve (12) towards the bowl (8) where the float (9) pivoted on the spindle (10) regulates the position of the inlet needle valve (11) and so maintains a pre-determined level of fuel in the bowl. The fuel then passes through jets (7) to the wells (6). It then mixes with air from the emulsion tubes (5) and correction jets (1) and as a vapour passes through holes (2) to reach the carburation area formed by the venturi components (3) and (4).
17 *Accelerator pump (Fig. 3.9).* The action of the pump is twofold.

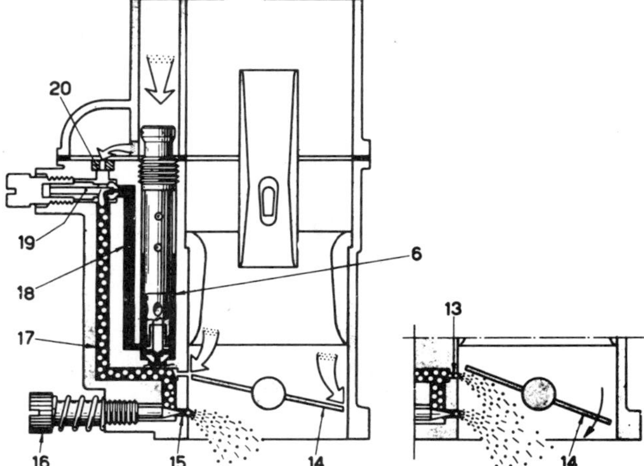

Fig. 3.7. Idling and progression cycle of Weber carburettors (see Section 3 for key details)

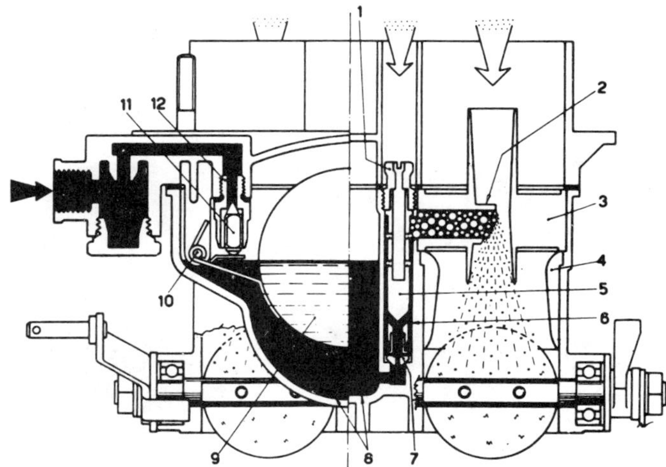

Fig. 3.8. Normal running cycle of Weber carburettors (see Section 3 for key details)

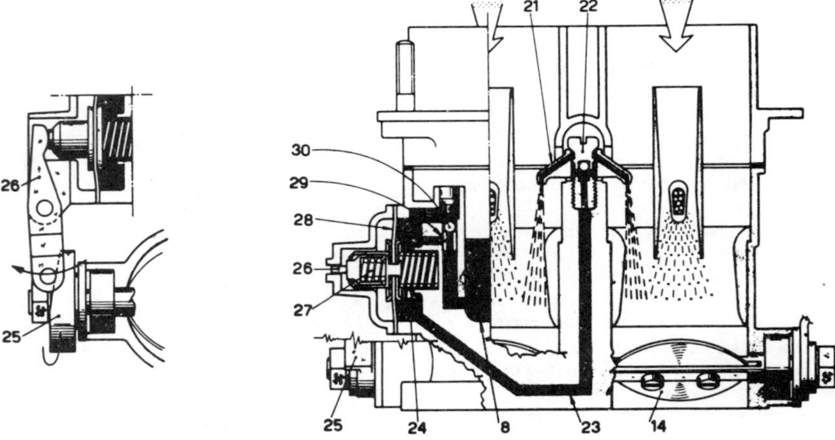

Fig. 3.9. Function of accelerator pump in Weber carburettors (See Section 3 for key details)

When the butterfly valve (14) is being closed, the lever (26) releases the diaphragm (28) which, due to the effect of the spring (24) draws fuel by suction from the bowl (8) through the ball valve (30). As the butterfly valve is being opened by means of the cam (25) and lever (26) the diaphragm (28) ejects fuel into the carburettor tubes (23) through the valve (22) and the jets (21). The spring (27) dampens the action of the butterfly valve and so extends the period of fuel injection. Excess fuel from the accelerator pump is returned to the carburettor bowl together with the fumes from the pump chamber through the calibrated hole (29).

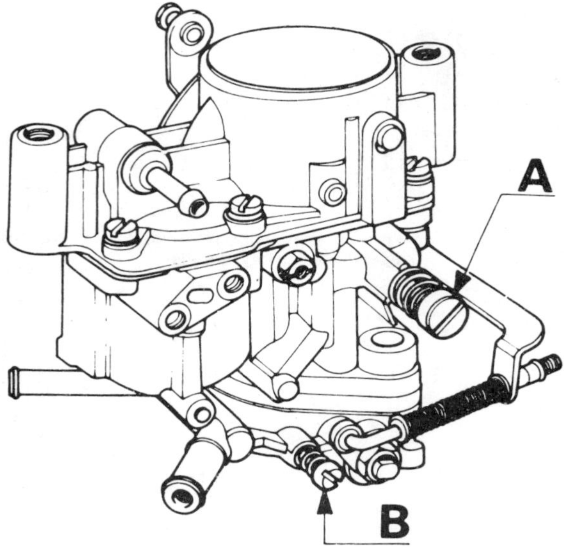

Fig. 3.10. Solex carburettor adjustment screws

A Volume screw *B Mixture screw*

4 Solex carburettor - idling adjustment

1 Before altering any carburettor settings, check the following for condition and adjustment. The ignition timing, valve clearances, spark plugs, air cleaner, fuel pump screen and the ease of operation of the accelerator control mechanism. Ensure that the engine is at normal operating temperature before proceeding.
2 If a tachometer is available, connect it to the engine following the manufacturer's instructions, start the engine and allow it to idle.
3 Turn the volume screw as necessary until the engine is idling at the speed given in the Specifications.
4 This is normally the only adjustment necessary on these carburettors. However, if the idling is unsatisfactory and all other engine variables as described above are in order, then adjustment of the mixture screw may be carried out as follows.
5 Start the engine and if necessary reset the idling speed by means of the volume screw.
6 Turn the mixture screw in or out to obtain the highest possible idling speed, consistent with even running.
7 Return the idling speed to the correct setting again by means of the volume screw. Switch off the engine and disconnect the tachometer.

5.2 Accelerator and choke cable connections on the Solex carburettor

5 Solex carburettors - removal and refitting

1 Remove the air cleaner and adaptor assembly as described in Section 2 and retain the rubber sealing ring.
2 Disconnect the choke and accelerator cables (photo).
3 Disconnect the vacuum advance pipe and fuel supply pipe.
4 Remove the filler cap from the expansion bottle and partially drain the cooling system.
5 Disconnect the two water hoses from their throttle block connections and plug both hoses to prevent loss of coolant.
6 Unscrew and remove the two carburettor securing nuts and lock-washers, lift off the carburettor and peel off the gasket.
7 Refitting is a reversal of removal but use a new flange gasket and check the satisfactory operation of the choke and accelerator cables.
8 Top up the cooling system expansion bottle if required.

6 Solex carburettors - setting and adjustment of components

1 *Float level.* Incorrect setting will be indicated by flooding or fuel leakage at the carburettor or evidence of fuel starvation causing stalling or loss of power of the engine.
2 Remove the air cleaner and disconnect the fuel pipe and choke cable from the carburettor.
3 Disconnect the fast idle link from the throttle control arm.
4 Remove the retaining screws and lift off the carburettor top cover.
5 Invert the top cover so that the float arm rests against the fuel inlet needle valve.
6 Use a small rule and measure the distance between the cover mating face of the carburettor and the upper edge of the float collar (Fig. 3.11).
7 The correct distance is 0.5315 to 0.6496 in (13.5 to 16.5 mm) and should be adjusted, if necessary, by gently bending the float operating arm.
8 Flooding of fuel from the float chamber may sometimes be caused by the needle valve housing not being fully tightened in its seat or a damaged sealing washer. Fuel would, in these circumstances, enter the float chamber without passing through the needle valve orifice.
9 *Accelerator pump stroke.* The carburettor must be removed in order to make this adjustment. Invert the carburettor and fully open the throttle butterfly valve.
10 Close the throttle butterfly valve slowly until the adjuster nut just contacts the accelerator pump lever, then check that a 3 mm (0.125 in) diameter twist drill will just slide between the edge of the throttle butterfly valve and the barrel wall (see Fig. 3.12).
11 If adjustment is necessary, turn the adjuster nut in the required direction, but note that the rod is crimped into the nut and if difficulty is encountered it may be necessary to renew those items.
12 *Throttle butterfly valve plate, fast idle setting.* This adjustment will again require removal of the carburettor. Invert the carburettor.
13 Pull the choke plate to the fully closed position and check the clearance between the edge of the throttle valve and the barrel wall.
14 Again using twist drills as gauges, a 1 mm diameter drill should pass through but a 1.1 mm drill should not.
15 Where the clearance is incorrect, adjust the fast idle screw as necessary (Fig. 3.13).

7 Solex carburettors - dismantling, servicing and reassembly

1 Disconnect the fast idle link (9) (Fig. 3.14).
2 Remove the top cover securing screws and remove the cover and gasket.
3 Remove the fuel filter retaining plug (15) and extract the filter (14).

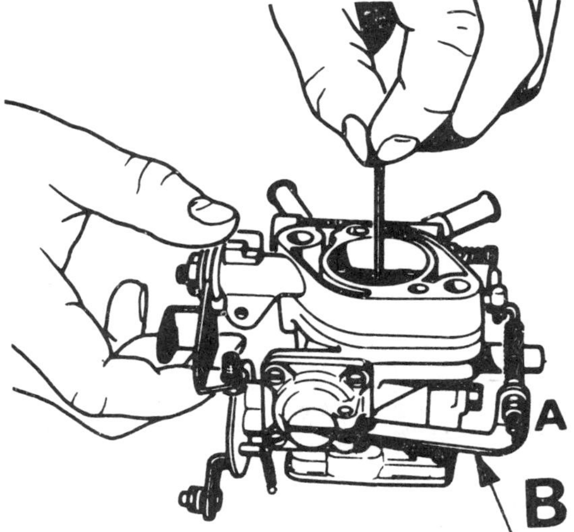

Fig. 3.12. Setting the accelerator pump stroke on the Solex carburettor

A Adjusting nut *B Pump lever*

4 Unscrew and remove the fuel inlet valve (5) and retain the sealing washer.
5 If absolutely necessary, the choke plate (3) and spindle (4) may be removed after removal of the two retaining screws and the spindle circlip.
6 Disconnect the accelerator pump operating rod (13).
7 Remove the two screws which secure the throttle block to the carburettor body.
8 Withdraw the throttle block and gasket, (and spacer if fitted).
9 If absolutely necessary, the throttle spindle (17) and throttle valve plate (10) may be removed after unscrewing the spindle nut and the two plate retaining screws.
10 Remove the float (6) and unscrew the various jets.
11 Remove the four securing screws from the accelerator pump cover assembly (12) and withdraw the diaphragm and return spring.
12 With the carburettor now dismantled, clean the components in clean fuel. Blow out the jets using air from a tyre pump; never probe with wire.
13 Examine the spindles and their housing bores for wear and slackness in operation. If this is evident then it will be realistic to exchange the complete unit for a reconditioned one on an exchange basis.
14 Check the jet sizes and other calibrated components against the

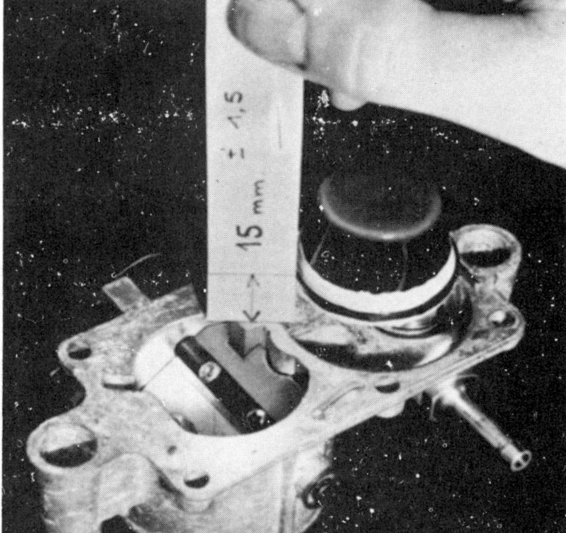

Fig. 3.11. Checking the float position on the Solex carburettor

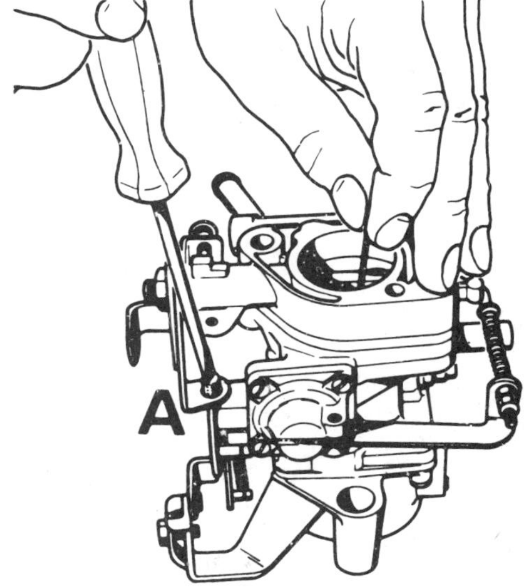

Fig. 3.13. Setting the fast idle adjusting screw 'A' (Solex)

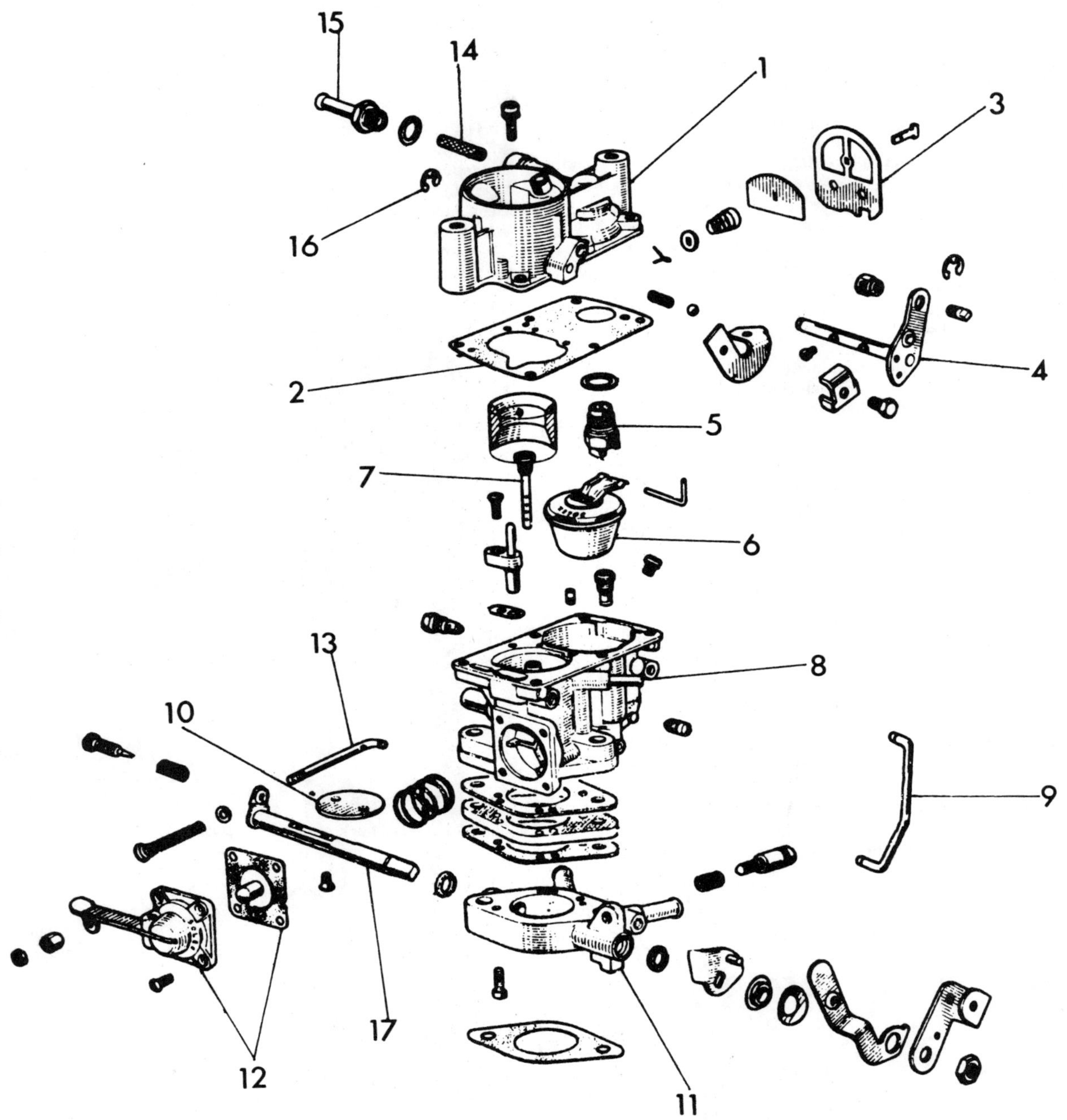

Fig. 3.14. Exploded view of Solex carburettor

1 Top cover
2 Cover gasket
3 Choke plate
4 Choke spindle
5 Inlet valve
6 Float
7 Air correction jet
8 Body
9 Fast idle link
10 Throttle plate
11 Water heated throttle block
12 Accelerator pump assembly
13 Accelerator pump operating rod
14 Filter
15 Inlet tube and filter retaining plug
16 Choke spindle circlip
17 Throttle spindle

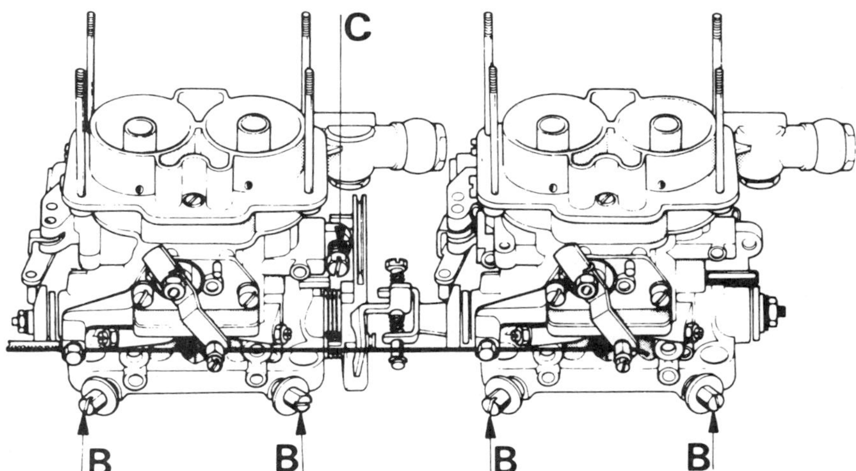

Fig. 3.15. Idling adjustment control screws (Twin Weber installation)

B Idling mixture screw
C Throttle stop screw

specification listed at the beginning of this Chapter and renew any that have previously been incorrectly substituted.
15 Obtain a repair kit which will contain a complete set of new gaskets and washers and other essential items.
16 Reassembly is a reversal of dismantling but remember to check the various settings and adjustments as described in the preceding Section. Use all the items found in the repair kit and refit the original sealing washer beneath the inlet valve housing. Any alteration in the washer will affect the fuel level in the carburettor bowl.

8 Weber (twin) carburettors - idling, adjustment and synchronisation

1 Before adjusting the carburettors, run through the check list given in Section 4.
2 The flow rates through the balancing channels are set in production and the screws 10, 11, 12 and 13 should not be altered, Fig. 3.16.
3 With the engine at normal operating temperature, set the idling speed adjustment screw (C) in Fig. 3.15, so that the engine idling speed is set to the figure given in the Specifications.
4 Screw the first mixture control screw (B) in Fig. 3.15 in or out until the engine speed is the highest obtainable. Adjust the idling speed screw so that the engine again runs at approximately 900 rpm.
5 Repeat the sequence of operations with each of the other mixture control screws (B) in turn.
6 Test check the four screws again to obtain the best balance at the required engine speed and note that screw number (9) will require unscrewing two turns more than the others to achieve the best idling adjustment as there is an additional air flow connection to its internal passage.
7 With the mixture control screws now correctly set, the units must be synchronised. To do this it is preferable to obtain one of the several proprietary balancing devices which indicate (visually) correct synchronisation of twin carburettor installations. This is achieved simply by balancing the intake suction of each carburettor by locating the device in the intake of each carburettor in rotation and turning the screw (5) (Fig. 3.16) until the correct readings are obtained. An alternative method of synchronising is to place a length of hose in each of the carburettor intakes in turn and attempt to match the 'hiss' from each carburettor. At best this is a temporary expedient until more precise tuning can be carried out.

9 Weber (single) carburettor - idling adjustment

1 Adjust the idling mixture control screws (two only) in a similar manner to that already described for the twin carburettor installation.
2 No synchronisation will of course, be required.

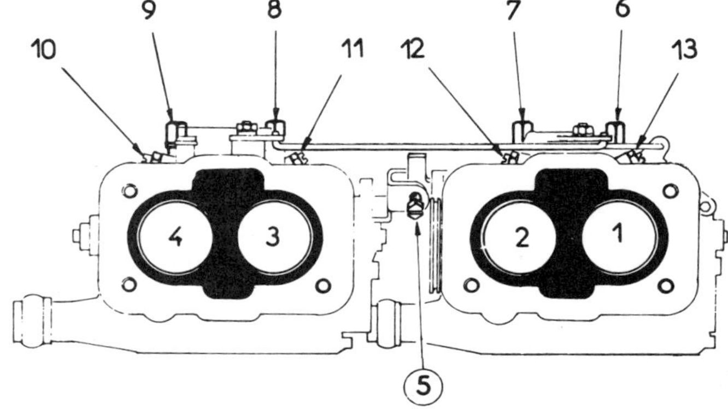

Fig. 3.16. Adjustments screws (Twin Weber installation)

1 to 4 Carburettor throats
5 Synchronising screw
6 to 9 Idling mixture adjustment screws
10 to 13 Fuel flow rate control screws (not to be altered)

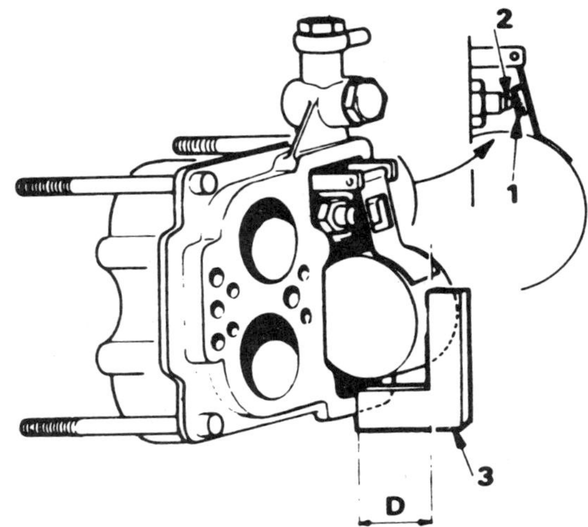

Fig. 3.17. Checking the float position (Weber carburettors)

1 Float arm tab
2 Needle valve
3 Measuring tool
D Float level dimension

10 Weber carburettors - removal and refitting

1 Remove the air cleaner and adaptor assembly as described in Section 2 and retain the gasket(s).
2 Disconnect the choke and accelerator control cables.
3 Disconnect the vacuum advance pipe and fuel supply pipe.
4 Remove the carburettor retaining nuts and lift the carburettor(s) away from the manifold studs. On the twin carburettor installation, disengage the throttle control linkage by pulling the two carburettors away from each other.
5 Remove the heat insulators and gaskets and place a piece of clean cloth in the manifold aperture(s) to prevent any dirt getting in.
6 Refitting is a reversal of removal but use new flange gaskets and check that the choke and accelerator controls operate correctly.

11 Weber carburettors - setting and adjustment of components

1 *Float level adjustment.* This may be required where there is evidence of flooding or fuel starvation.
2 Remove the carburettor top cover and hold it vertically with the float hanging down (see Fig. 3.17). With the float arm in light contact with the spring tensioned ball in the end of the needle, measure the dimension from the cover face (without gasket) to the outside face of the float.
3 If the dimension is not as given in the Specifications, bend the tab on the float arm as necessary.

12 Weber carburettors - dismantling, servicing and reassembling

1 Refer to Fig. 3.19.
2 Unscrew and remove the cover securing screws and withdraw the lid and gasket.
3 Remove the float arm pivot pin and withdraw the float (6).
4 Unscrew and remove the inlet needle valve and its housing retaining the sealing washer located beneath it.
5 Remove the filter plug (14) and withdraw the filter (15).
6 Remove the venturis (19) from the carburettor body.
7 Unscrew the jets and adjusting screws (not flow rate screws see Fig. 3.15).
8 Disconnect the link from the choke control arm (20) and the starting device operating lever. Remove the retaining screw and detach the choke control arm.
9 On the type 36 DCNF (twin carburettors), remove the retaining screws and the starting device cover (21).
10 The starting device piston assembly (13) may be dismantled after removal of the retaining circlip (22).
11 Remove the four retaining screws from the accelerator pump assembly cover (7).
12 Withdraw the diaphragm (23) and return spring (24).

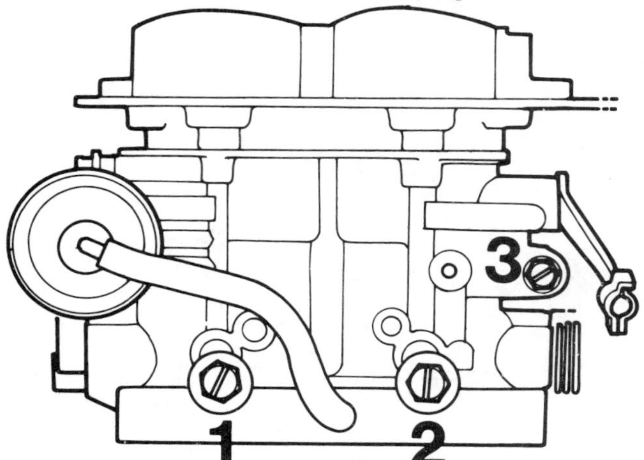

Fig. 3.18. Idling adjustment control screws (Single Weber carburettor)

1 Mixture screw, left-hand barrel
2 Mixture screw, right-hand barrel
3 Throttle stop screw

13 Dismantle the throttle plate and spindle only if essential. The throttle plate may be slid from its locating slot after the two retaining screws have been removed.
14 Having dismantled the carburettor into its components, carefully inspect them for wear, damage or distortion and renew as appropriate. If there are any cracks in the carburettor body exchange the complete unit for a reconditioned one.
15 Blow all jets clear with air from a tyre pump, never probe with wire.
16 Obtain a repair kit, specifying carefully the exact carburettor type number.
17 Reassembly is a reversal of dismantling, use the new components and gaskets supplied in the repair kit and carry out the settings and adjustments described in the preceding Section.
18 When the carburettor is refitted to the engine, carry out the idling (and synchronising if applicable) adjustments as described in Sections 8 and 9.
19 Do not screw the mixture control screws fully home or damage to their seating may result. Do not alter the thickness of the inlet needle valve seating washer and check that the chamfer on the throttle plate is fitted so that it seals against the choke tube when closed.

13 Fuel pump - routine maintenance

1 Detach the fuel supply hose from the fuel pump and plug the hose to prevent loss of fuel.
2 Unscrew the cover retaining screw and remove the cover from the pump.
3 Withdraw the filter screen and wash it in petrol.
4 Remove all traces of dirt and sediment from the interior of the pump chamber.
5 Refit the filter screen, check that the cover gasket is in good condition. If it is not, renew it. Fit the cover but do not use excessive force on the retaining screw.
6 Reconnect the fuel line, start the engine and check for leaks.

14 Fuel pump - removal and refitting

1 Disconnect the fuel supply line from the pump and plug the line to prevent loss of fuel.
2 Disconnect the fuel lines between the pump and carburettor.
3 Unscrew and remove the two pump securing bolts and remove the pump, noting the insulating and sealing washers, fitted between the pump flange and the cylinder block.
4 Withdraw the pump in a downward direction so that the pump actuating arm can be extracted from between the crankcase wall and the camshaft.
5 Refitting is a reversal of removal but check the location and condition of the flange washers.
Note: The fuel pump cannot be dismantled for repair, and if faulty must be renewed.

15 Fuel system - fault finding

There are three main types of fault to which the fuel system is prone. They may be summarised as follows:

a) Lack of fuel at engine
b) Weak mixture
c) Rich mixture.

16 Lack of fuel at engine

1 If it is not possible to start the engine, first check that there is fuel in the tank, and then check the ignition system as detailed in Chapter 4. If the fault is not in the ignition system then disconnect the fuel inlet pipe from the carburettor and turn the engine over by the starter relay switch.
2 If petrol squirts from the end of the inlet pipe, reconnect the pipe and check that the fuel is getting to the float chamber. This is done by unscrewing the bolts from the top of the float chamber, and lifting the

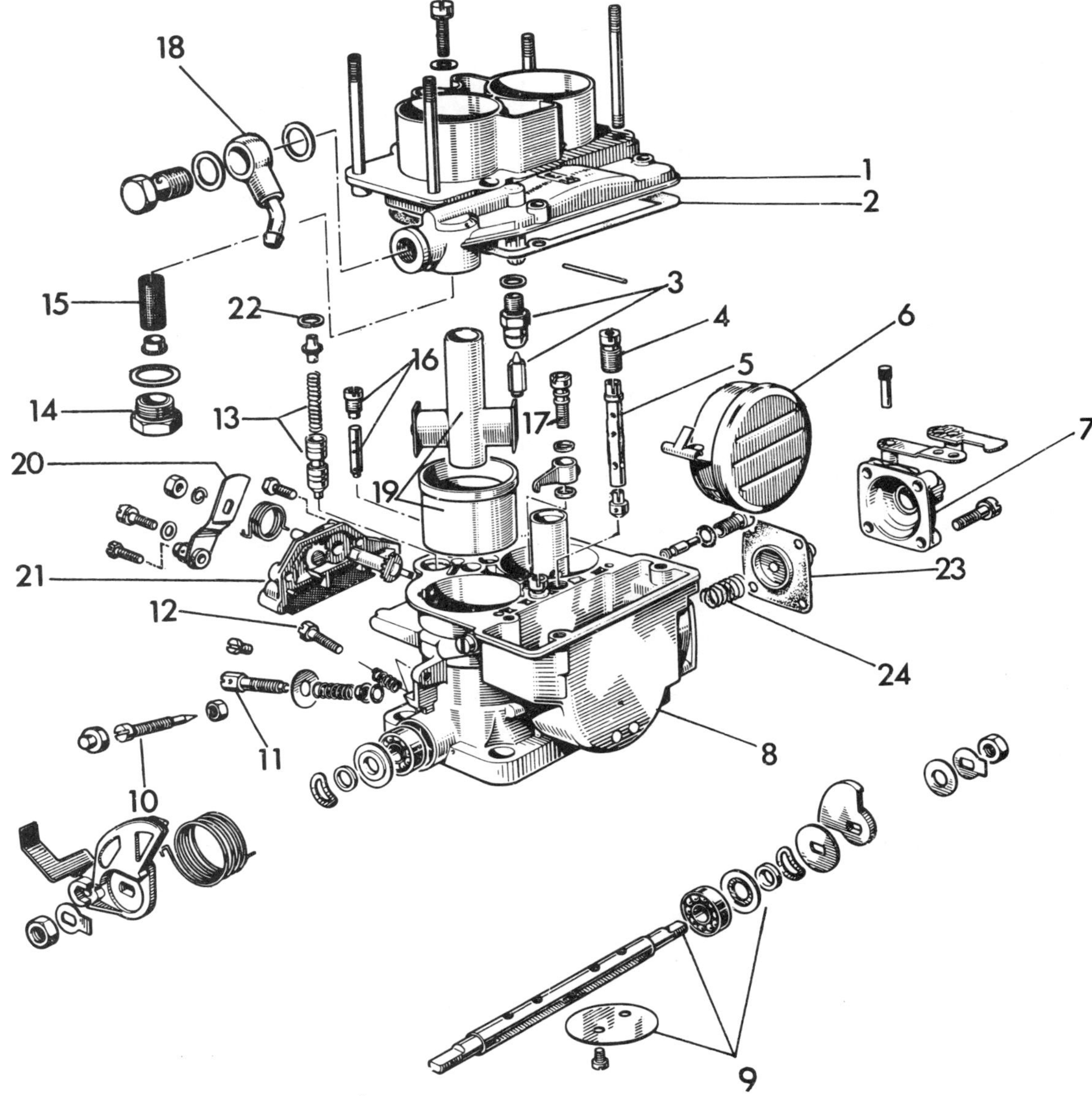

Fig. 3.19. Exploded view of Weber type 36DCNF carburettor. (Weber type 36DCNV is similar except for choke details)

1 Cover
2 Gasket
3 Inlet needle valve
4 Main air correction jet
5 Emulsion tube
6 Float
7 Accelerator pump
8 Body
9 Throttle spindle butterfly
10 Flow rate screw
11 Idling mixture screw
12 Idling speed screw
13 Starting device piston
14 Filter plug
15 Filter
16 Starting fuel jet
17 Pump discharge valve
18 Fuel inlet coupling
19 Venturi
20 Choke control arm
21 Starting device cover
22 Circlip (starting device piston
23 Accelerator pump diaphragm
24 Accelerator pump return spring

cover just enough to see inside.

3 If fuel is there then it is likely that there is a blockage in the starting jet, which should be removed and cleaned.

4 No fuel in the float chamber, is caused either by a blockage in the pipe between the pump and float chamber or a sticking float chamber valve. Alternatively on the Weber carburettor the gauze filter at the top of the float chamber may be blocked. Remove the securing nut and check that the filter is clean. Washing in petrol will clean it.

5 If it is decided that it is the float chamber valve that is sticking, remove the fuel inlet pipe, and undo the screws and lift the cover, complete with valve and floats, away.

6 Remove the valve needle and valve and thoroughly wash them in petrol. Petrol gum may be present on the valve or valve needle and this is usually the cause of a sticking valve. Refit the valve in the needle valve assembly, ensure that it is moving freely, and then reassemble the float chamber. It is important that the same washer be placed under the needle valve assembly as this determines the height of the floats and therefore the level of petrol in the chamber.

7 Reconnect the fuel pipe and refit the air cleaner.

8 If petrol does not squirt from the end of the pipe leading to the carburettor then disconnect the pipe leading to the inlet side of the fuel pump. If fuel runs out of the pipe then there is a fault in the fuel pump, and the pump should be checked as detailed in Section 14.

9 No fuel flowing from the tank when it is known that there is fuel in there indicates a blocked pipe line. The line to the tank should be blown out. It is unlikely that the fuel tank vent would become blocked, but this could be a reason for the reluctance of the fuel to flow. To test for this, blow into the tank down the filler orifice. There should be no build up of pressure in the fuel tank, as the excess pressure should be carried away down the vent pipe.

17 Weak mixture

1 If the fuel/air mixture is weak there are six main clues to this condition:

a) The engine will be difficult to start and will need much use of the choke, stalling easily if the choke is pushed in.
b) The engine will overheat easily.
c) If the spark plugs are examined (as detailed in Chapter 4), they will have a light grey/white deposit on the insulator nose.
d) The fuel consumption may be light.
e) There will be a noticeable lack of power.
f) During acceleration and on the overrun there will be a certain amount of spitting back through the carburettor.

2 As the carburettors are of the fixed jet type, these faults are invariably due to circumstances outside the carburettor. The only usual fault likely in the carburettor is that one or more of the jets may be partially blocked. If the car will not start easily but runs well at speed, then it is likely that the starting jet is blocked, whereas if the engine starts easily but will not rev, then it is likely that the main jets are blocked.
3 If the level of petrol in the float chamber is low this is usually due to a sticking valve or incorrectly set floats.
4 Air leaks either in the fuel lines, or in the induction system, should also be checked for. Also check the distributor vacuum pipe connection as a leak in this is directly felt in the inlet manifold.
5 The fuel pump may be at fault as has already been detailed.

18 Rich mixture

1 If the fuel/air mixture is rich there are five main clues to this condition:

a) If the spark plugs are examined they will be found to have a black sooty deposit on the insulator nose.
b) The fuel consumption will be heavy.
c) The exhaust will give off black smoke, especially when accelerating.
d) The interior deposits on the exhaust pipe will be dry, black and sooty (if they are wet, black and sooty this indicates worn bores, and oil is being burnt).
e) There will be a noticeable lack of power.

2 The faults in this case are usually in the carburettor and most usual is that the level of petrol in the float chamber is too high. This is due either to dirt behind the needle valve, or a leaking float which will not close the valve properly, or a sticking needle.
3 With very high mileages (or because someone has tried to clean the jets out with wire), it may be that the jets have become enlarged.
4 If the air correction jets are restricted in any way the mixture will tend to become very rich.
5 Occasionally it is found that the choke control is sticking or has been maladjusted.
6 Again, occasionally, the fuel pump pressure may be excessive so forcing the needle valve open slightly until a higher level of petrol is reached in the float chamber.

19 Fuel tank and transmitter unit - removal, servicing and refitting

1 The pressed steel fuel tank is located beneath the rear floor section of the vehicle (photo).
2 The air vent pipe is fitted with an expansion device to reduce the possibility of fuel blowback if the tank is filled too quickly.
3 The fuel gauge transmitter unit is fitted to the top of the tank and can be detached without having to remove the fuel tank from the car.
4 To remove the complete fuel tank assembly, first disconnect the battery earth terminal and remove the spare wheel from beneath the vehicle.
5 Pull back the luggage compartment mat, remove the fuel tank access cover securing screws and lift away the cover (photo).
6 Make a note of the colour and position of the two wires and disconnect them from the transmitter unit.
7 From beneath the car remove the clip securing the wiring harness to the rear of the fuel tank.
8 Place a clean container beneath the fuel tank outlet pipe, disconnect the pipe and drain the tank. Note that on some models there is a fuel

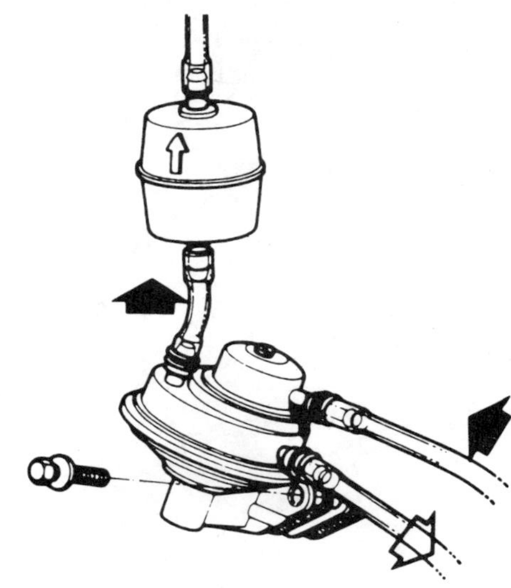

Fig. 3.20. Fuel pump assembly (arrows indicate fuel flow)

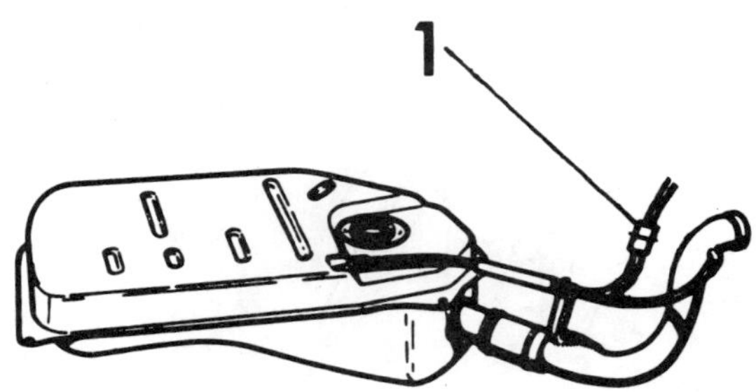

Fig. 3.21. Fuel tank and vent pipe (1)

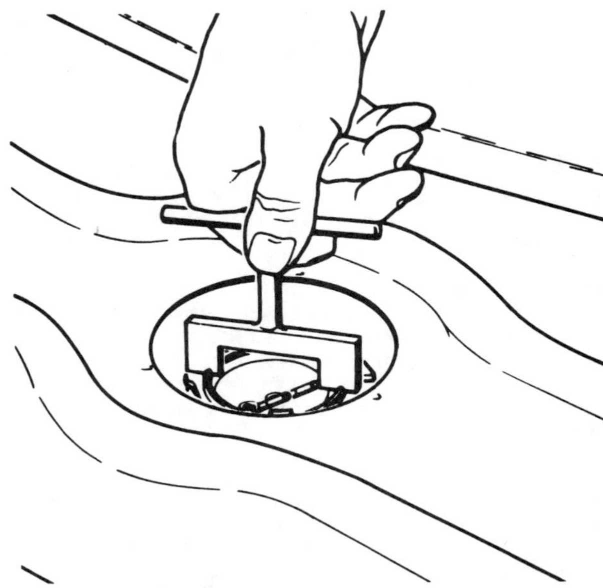

Fig. 3.22. Removing the fuel tank transmitter unit

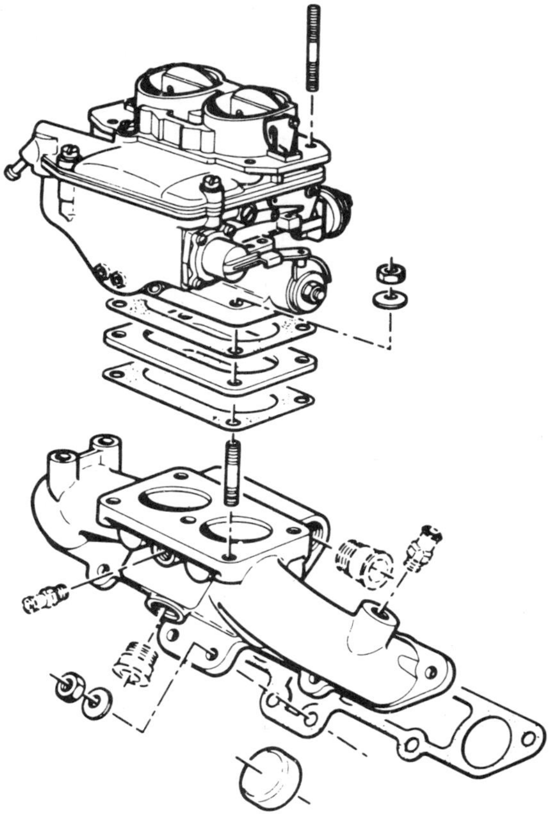

Fig. 3.23. Inlet manifold for single Weber carburettor

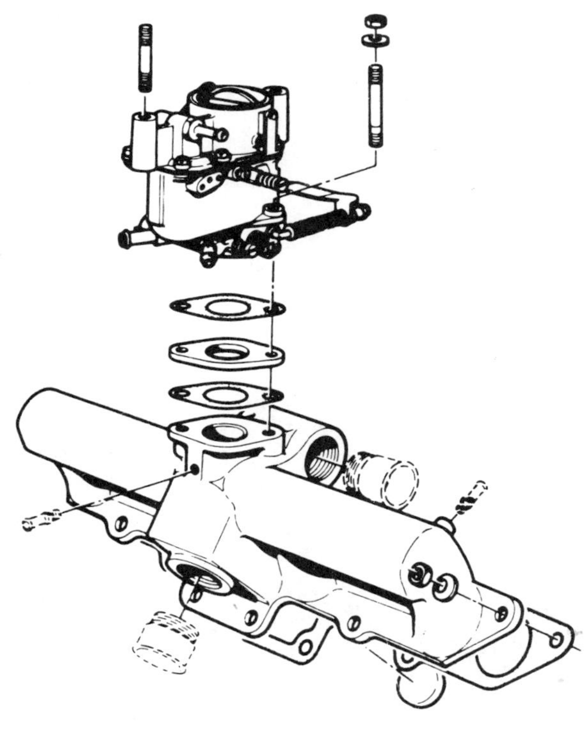

Fig. 3.24. Inlet manifold used for Solex carburettor

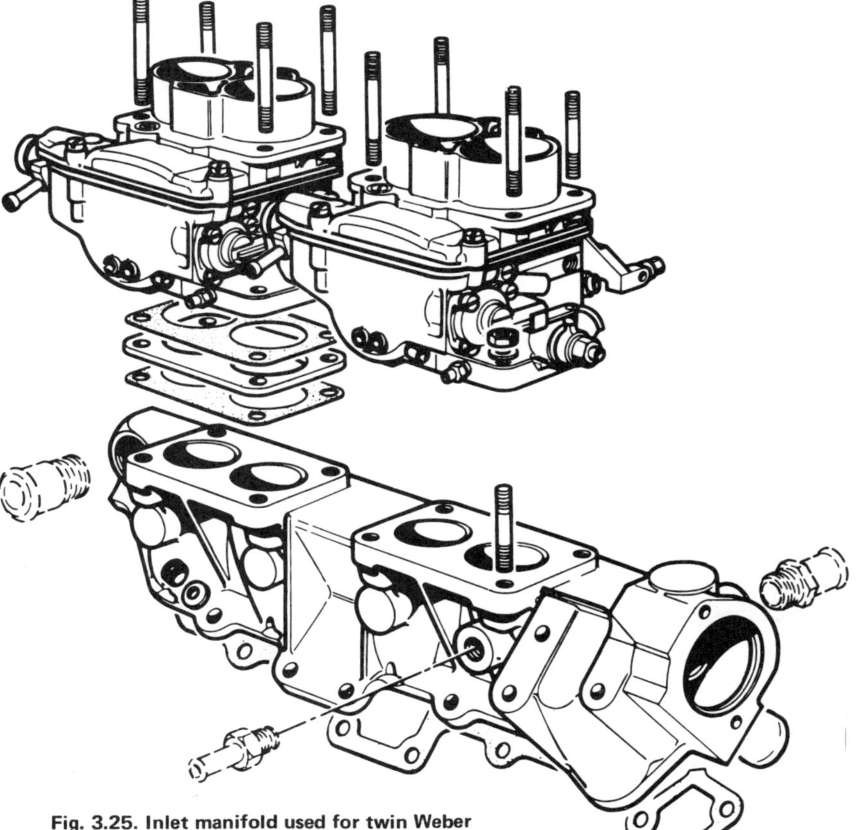

Fig. 3.25. Inlet manifold used for twin Weber carburettors

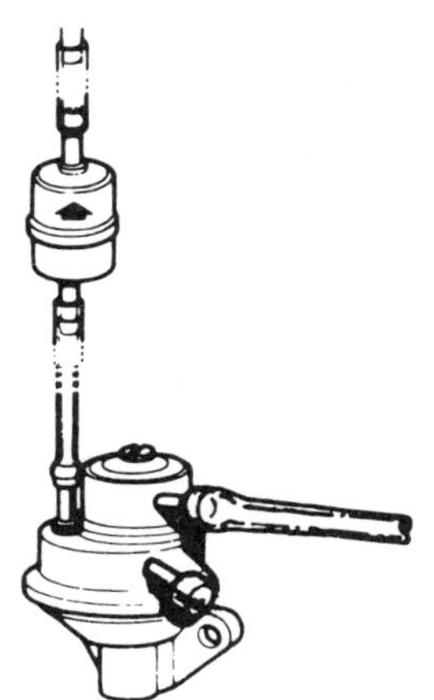

Fig. 3.26. Fuel in-line filter assembly

19.1 Fuel tank location

19.5 Fuel tank transmitter unit access hole

19.9 Fuel tank filler pipe

20.6A Rear flexible strap holding the centre pipe and tailpipe in position

20.6B Exhaust system rear section

20.6C Rear exhaust mounting

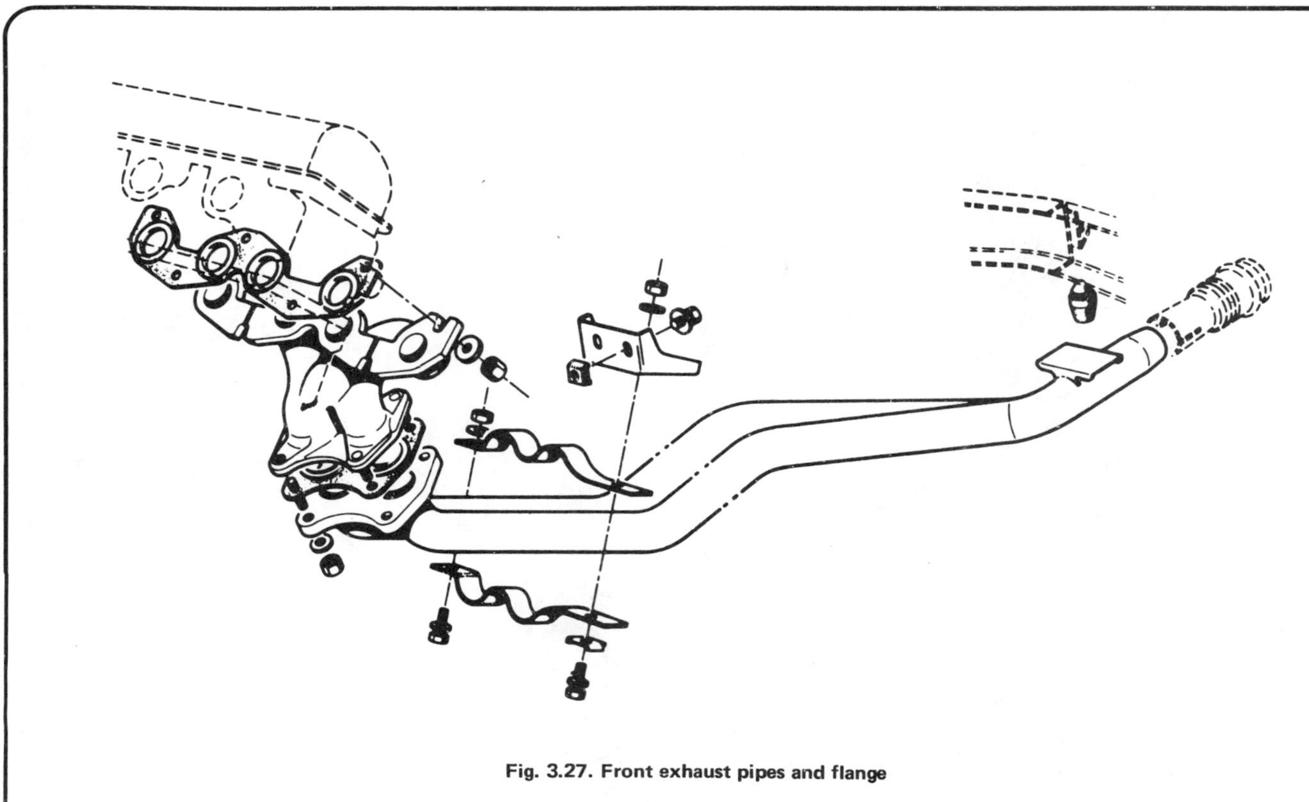
Fig. 3.27. Front exhaust pipes and flange

return pipe from the pump and this must also be disconnected.

9 Remove the clip and disconnect the tank filler hose from beneath the rear right-hand wing (photo).

10 Support the weight of the tank and remove the four securing bolts. Lower the tank sufficiently to disconnect the top vent pipe and then withdraw the tank from beneath the vehicle.

11 If the tank is dirty or contains sediment, use two or three lots of clean paraffin to wash it out and then let it drain thoroughly. Do not shake the tank too vigorously or use a stick to probe the interior or damage may be caused to the transmitter float and arm.

12 Do not try to solder a leak in a fuel tank. This is a specialists job. If the leak cannot be repaired with a cold setting compound then the tank should be renewed.

13 The fuel gauge transmitter unit can be unscrewed from the tank using a tool similar to that shown in Fig. 3.22. It is not necessary to remove the tank from the car. The unit is not repairable and if faulty must be renewed.

14 Refit the fuel tank and transmitter unit using the reverse procedure to that of removal. Ensure that the yellow wire is fitted to the outer terminal on the transmitter unit and the red wire to the inner (centre) terminal.

20 Inlet manifolds and exhaust system - general description

1 Three different types of inlet manifold are used on the models covered in this manual. One type is fitted to the 1442 cc engine in conjunction with the single Weber carburettor (see Fig. 3.23) and one of two types is fitted to the 1294 cc engine to cater for either the single Solex carburettor or the twin Weber installation (see Figs. 3.24 and .3.25).

2 All three types of manifold are water heated and it is therefore

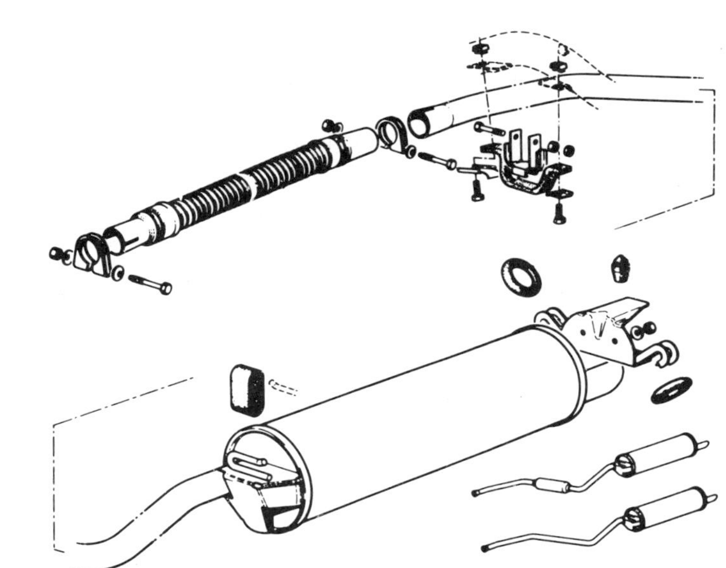

Fig. 3.28. Centre exhaust pipe section and silencer

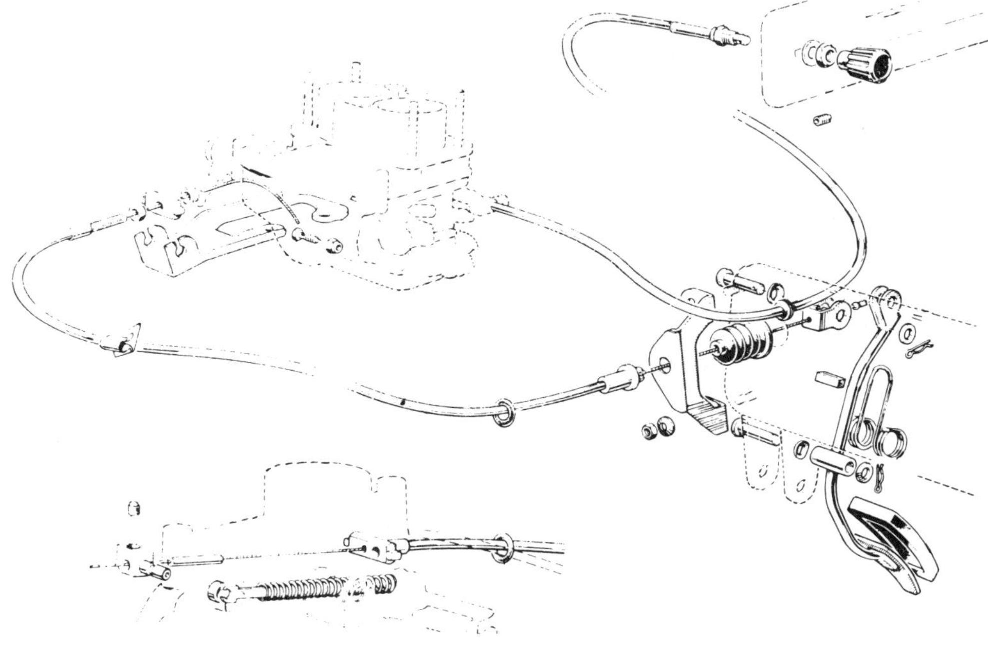

Fig. 3.29. Typical layout of throttle and choke control cables

necessary to drain the cooling system before removal.
3 Always use new gaskets when refitting the manifold and carburettor.
4 The exhaust system comprises three main sections, twin pipes are connected to the engine manifold flange and converge into a single flexible pipe section. This in turn is attached to the rear silencer and tailpipe.
5 The front pipes are attached to the engine manifold by means of a flange and gasket, and secured by four bolts. Whenever the joint is dismantled a new gasket should always be used (Fig. 3.27).
6 The centre pipe and rear silencer/tailpipe assembly is secured to the underfloor by means of rubber rings hooked onto brackets (see Fig. 3.28) and a rear flexible strap (photos).
7 It is wise to use only properly made exhaust system parts and exhaust pipe shackles. Remember that the exhaust pipes and silencers corrode internally as well as externally and therefore when any one section of pipe or a silencer demands renewal, it often follows that the whole system is best renewed.
8 It is most important when fitting exhaust systems that the twists and contours are carefully followed and that each connecting joint overlaps the correct distance (1.5 in/32.5 mm). Any stresses and strains imparted in order to force the system to fit the vehicle will result in premature system failure.
Note: The front twin pipe assembly fitted to the 1442 cc engine is slightly longer than that used on the 1294 cc engine and is identified by a paint mark.

21 In-line fuel filter - general

On some models a filter is fitted in the pipeline between the fuel pump and carburettors (Fig. 3.26).

It cannot be cleaned and should be renewed at the recommended service intervals.

22 Accelerator and choke controls

1 The layout of the carburettor controls is shown in Fig. 3.29.
2 Ensure that the accelerator cable has enough slack in the idling position for correct closure of the butterfly valve to occur when the accelerator pedal is released.
3 On the Solex and single Weber carburettors check that the choke butterfly valve is fully open when the control knob is pushed in and closes completely when the knob is pulled right out. Adjust the cable if necessary.
4 Lightly oil the moving levers and swivels of the controls at regular intervals.

23 Fault finding chart - carburation; fuel and exhaust systems

Symptom	Reason/s	Remedy
Fuel consumption excessive		
Carburation and ignition faults	Air cleaner choked and dirty causing a rich mixture	Remove, clean and refit air cleaner.
	Fuel leaking from carburettor(s), fuel pumps or fuel lines	Check for and eliminate all fuel leaks. Tighten fuel line union nuts.
	Float chamber flooding	Check and adjust float level.
	Generally worn carburettor(s)	Remove, overhaul and refit.
	Distributor condenser faulty	Remove, and fit new unit.
	Balance weights or vacuum advance mechanism in distributor faulty	Remove, and overhaul distributor.
Incorrect adjustment	Carburettor(s) incorrectly adjusted, mixture too rich	Tune and adjust carburettor(s).
	Idling speed too high	Adjust idling speed.
	Contact breaker gap incorrect	Check and reset gap.
	Valve clearances incorrect	Check rocker arm to valve stem clearances and adjust as necessary.
	Incorrectly set sparking plugs	Remove, clean and regap.
	Tyres under-inflated	Check tyre pressures and inflate if necessary.
	Wrong sparking plugs fitted	Remove and fit correct units.
	Brakes dragging	Check and adjust brakes.
Insufficient fuel delivery or weak mixture due to air leaks		
Dirt in system	Petrol tank air vent restricted	Remove petrol cap and clean out air vent.
	Partially clogged filters in pump and carburettor(s)	Remove and clean filters.
	Dirt lodged in float chamber needle housing	Remove and clean out float chamber and needle valve assembly.
Fuel pump faults	Incorrectly seating valves in fuel pump	Renew fuel pump.
	Fuel pump diaphragm leaking or damaged	Renew fuel pump.
	Gasket in fuel pump damaged	Renew fuel pump.
	Fuel pump valves sticking due to petrol gumming	Renew fuel pump.
Air leaks	Too little fuel in fuel tank (prevalent when climbing steep hills)	Refill fuel tank.
	Union joints on pipe connections loose	Tighten joints and check for air leaks.
	Split in fuel pipe on suction side of fuel pump	Examine, locate, and repair.
	Inlet manifold to block or inlet manifold to carburettor(s) gasket leaking	Test by pouring oil along joints - bubbles indicate leak. Renew gasket as appropriate.

Chapter 4 Ignition system

For modifications, and information applicable to later models, see Supplement at end of manual

Contents

Specifications

Type ... ... Electronic

Distributor

Make	SEV or Bosch
Direction of rotation	Clockwise
Firing order	1 - 3 - 4 - 2 (No. 1 piston at flywheel end of engine)

Ignition coil

Primary resistance	1.4 - 1.6 ohms
Secondary resistance	8000 - 10,000 ohms
Ballast resistor	0.50 to 0.60 ohms

Spark plugs

Type	
All models	Champion N9YC or equivalent
Gap	0.024 - 0.028 in (0.6 - 0.7 mm)

Ignition timing (dynamic)

Engine	btdc	Idling speed
1294 cc (single carb)	10°	850
1294 cc (twin carb)	10°	950
1442 cc (engine code Y2)	12°	900
1442 cc (engine code Y2D)	10°	900

Torque wrench settings

	lb f ft	kg f m
Spark plugs	22	3.04
Distributor clamp bolt	7	0.97
Distributor mounting bolts	15	2.07

1 General description

In order that the engine can run correctly it is necessary for an electrical spark to ignite the fuel/air mixture in the combustion chamber and at exactly the right moment in relation to engine speed and load. The ignition system is based on feeding low tension voltage from the battery to the coil where it is converted to high tension voltage. The high tension voltage is powerful enough to jump the spark plug gap in the cylinders many times a second under high compression pressures providing that the system is in good condition and that all adjustments are correct.

The Alpine models are fitted with an electronic ignition system which provides a high degree of reliability with virtually no servicing requirements apart from periodically checking the HT (high tension) and LT (low tension) lead connections.

This new system retains the normal ignition coil, distributor advance mechanism and distributor cap, but replaces the conventional contact breaker points and condenser with a reluctor and pick-up unit operating in conjunction with a control unit.

The reluctor unit is a four-toothed wheel, (one for each cylinder) that is fitted to the distributor shaft in place of the conventional contact breaker operating cam.

The pick-up unit is also located in the distributor and comprises basically of a coil and permanent magnet.

The control unit is a transistorised amplifier that is used to boost the voltage induced by the pick-up coil.

A simplified circuit diagram is shown in Fig. 4.2 while the complete wiring diagram with colour coding is given in Fig. 4.1.

When the ignition switch is ON, the ignition primary circuit is energised. When the distributor reluctor 'teeth' or 'spokes' approach the magnetic coil assembly, a voltage is induced which signals the amplifier to turn off the coil primary current. A timing circuit in the amplifier module turns on the coil current after the coil field has collapsed.

When switched on, current flows from the battery through the ignition switch, through the coil primary winding, through the amplifier module and then to ground. When the current is off, the magnetic field in the ignition coil collapses, inducing a high voltage in the coil secondary winding. This is conducted to the distributor cap where the rotor directs it to the appropriate spark plug. This process is repeated for each power stroke of the car engine.

The distributor is fitted with devices to control the actual point of ignition according to the engine speed and load. As the engine speed increases two centrifugal weights move outwards and alter the position of the armature in relation to the distributor shaft to advance the spark slightly. As engine load increases (for example when climbing hills or accelerating), a reduction in intake manifold depression causes the base plate assembly to move slightly in the opposite direction (clockwise) under the action of the spring in the vacuum unit, thus retarding the spark slightly and tending to counteract the centrifugal advance. Under light loading conditions (for example at moderate steady speeds) the comparatively high intake manifold depression on the vacuum advance diaphragm causes the baseplate assembly to move in a counterclockwise direction to give a larger amount of spark advance.

2 Distributor - servicing

1 Periodically the condition of the distributor cap and rotor should be checked.

2 Remove the protective boot from around the distributor cap, release the two spring clips and lift off the cap.

3 Clean the inside and outside of the distributor cap with a clean cloth and check the centre contact and four segments on the inside of the cap for excessive wear or burning. If evident the cap should be renewed.

4 Remove the rotor and check the end of the brass segment for burning. Renew if necessary.

5 The air gap between the reluctor rotor tips and the magnetic pick-up point is pre-set. It cannot be adjusted and requires no servicing.

6 Refit the rotor and cap and secure with the spring clips. Refit the protective boot.

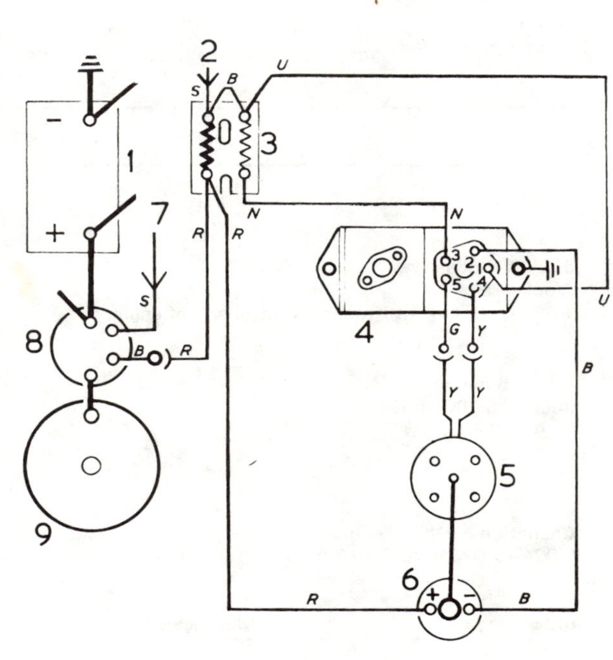

Fig. 4.1. Ignition system wiring circuit and colour code

1 Battery
2 Supply from Ign. sw.
3 Resistor block
4 Control unit
5 Distributor
6 Ignition coil
7 Supply from key start
8 Starter solenoid
9 Starter motor

Colour code
B = Black
G = Green
N = Brown
R = Red
S = Slate
U = Blue
Y = Yellow

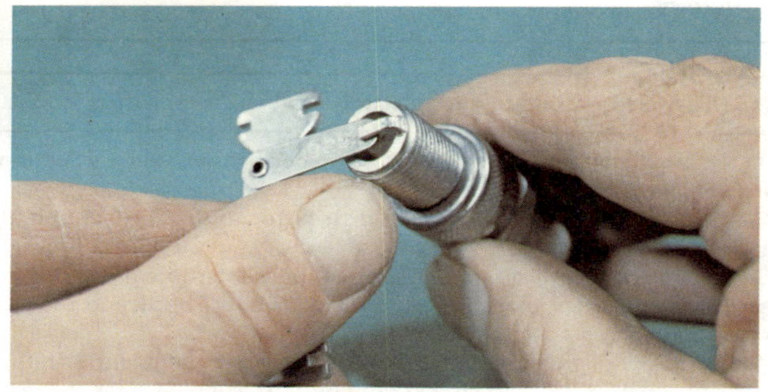

Measuring plug gap. A feeler gauge of the correct size (see ignition system specifications) should have a slight 'drag' when slid between the electrodes. Adjust gap if necessary

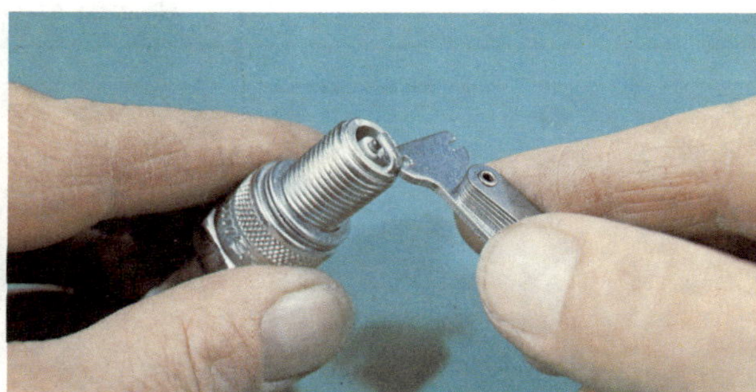

Adjusting plug gap. The plug gap is adjusted by bending the earth electrode inwards, or outwards, as necessary until the correct clearance is obtained. Note the use of the correct tool

Normal. Grey-brown deposits, lightly coated core nose. Gap increasing by around 0.001 in (0.025 mm) per 1000 miles (1600 km). Plugs ideally suited to engine, and engine in good condition

Carbon fouling. Dry, black, sooty deposits. Will cause weak spark and eventually misfire. Fault: over-rich fuel mixture. Check: carburettor mixture settings, float level and jet sizes; choke operation and cleanliness of air filter. Plugs can be re-used after cleaning

Oil fouling. Wet, oily deposits. Will cause weak spark and eventually misfire. Fault: worn bores/piston rings or valve guides; sometimes occurs (temporarily) during running-in period. Plugs can be re-used after thorough cleaning

Overheating. Electrodes have glazed appearance, core nose very white – few deposits. Fault: plug overheating. Check: plug value, ignition timing, fuel octane rating (too low) and fuel mixture (too weak). Discard plugs and cure fault immediately

Electrode damage. Electrodes burned away; core nose has burned, glazed appearance. Fault: pre-ignition. Check: as for 'Overheating' but may be more severe. Discard plugs and remedy fault before piston or valve damage occurs

Split core nose (may appear initially as a crack). Damage is self-evident, but cracks will only show after cleaning. Fault: pre-ignition or wrong gap-setting technique. Check: ignition timing, cooling system, fuel octane rating (too low) and fuel mixture (too weak). Discard plugs, rectify fault immediately

3.6A Distributor aperture and drive shaft

3.6B Fitting the distributor into the engine

3.7 Distributor correctly fitted

3 Distributor - removal and refitting

1 Remove the distributor cap as described in the previous Section and disconnect the vacuum pipe.
2 Disconnect the wires from the distributor to the control unit at the twin plug.
3 Using a small sharp screwdriver or similar instrument carefully scribe an alignment mark between the distributor body and the crankcase mounting bracket.
4 Undo the distributor clamp bolt, remove the clamp and lift out the distributor from the engine. Do **not** rotate the engine after the distributor has been removed unless absolutely necessary.
5 The distributor is not repairable and if it is known to be faulty (see Section 8) it must be renewed.
6 To refit the distributor, enter it into the crankcase aperture and ensure the tongue on the end of the distributor drive shaft engages with the offset slot in the engine drive shaft (photo).
7 Rotate the distributor body to line up the scribe marks, refit the clamp and tighten the securing bolt (photo).
8 Refit the remaining distributor components using the reverse procedure to that of removal.
9 Providing the engine was not rotated while the distributor was removed and the scribe marks are correctly aligned it should not be necessary to re-time the ignition. However, if any doubt exists, time the ignition using the procedure described in Section 5.
10 If the engine was rotated while the distributor was removed it will be necessary to re-time it to the engine as described in the following Section.

4 Distributor - timing to the engine

1 To time the distributor to the engine, first remove No. 1 spark plug (nearest the flywheel end of the engine).
2 Place a finger over the spark plug hole and rotate the engine until pressure is felt, this indicates the piston is on the compression stroke.
3 Continue turning the engine until the mark on the flywheel (seen through the clutch housing aperture) is opposite the degree mark given in the 'Specifications' (see Fig. 4.3).
4 With the distributor removed from the engine, turn the rotor until it is aligned with the groove in the body on Bosch type distributors (see Fig. 4.4) or, in the case of the SEV distributor, the mark on the base of the rotor is between the two lines scribed on the arch shield (see Fig. 4.5).
5 Refit the distributor to the engine as described in Section 3, ensuring that the driveshaft is correctly engaged. Check that the timing marks are still in alignment (rotate the distributor body if necessary) and lightly tighten the clamp bolt.
6 Finally, check the ignition timing using a strobe lamp as described in the following Section.

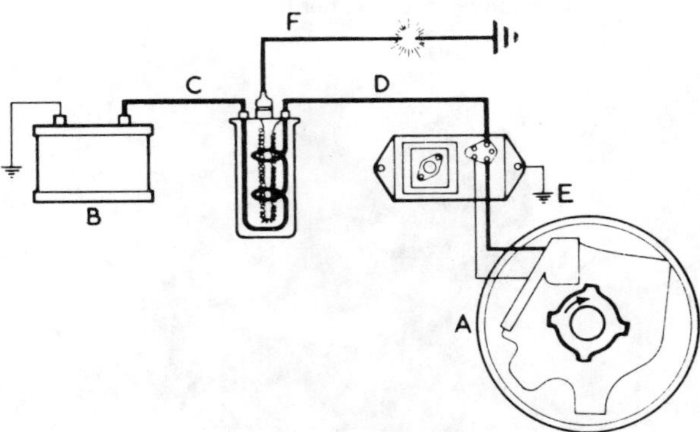

Fig. 4.2. Simplified diagram of ignition system

A Pick-up coil inducing negative voltage
B Battery
C Supply to ignition coil
D Coil L.T. to control unit
E L.T. open circuit in control unit
F H.T. induced in coil secondary

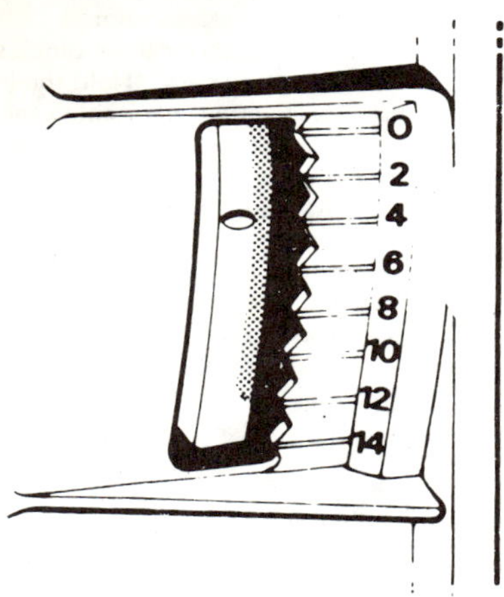

Fig. 4.3. Timing aperture in clutch housing

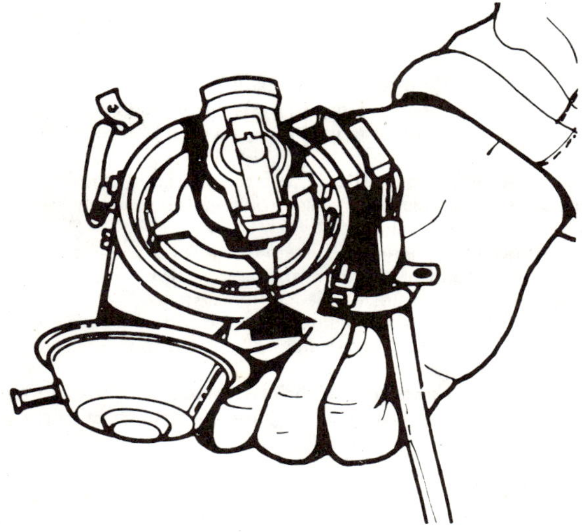

Fig. 4.4. Timing marks on Bosch distributor

5 Ignition timing

1 Because the distributor only gives a timing signal when the shaft is rotating, a strobe lamp must be used with the engine running at idling speed.
2 First disconnect the vacuum advance pipe from the distributor and connect the strobe lamp between No. 1 spark plug and its associated HT lead.
3 Remove the rubber plug from the clutch housing aperture. Clean the flywheel and housing timing marks with a piece of rag and mark them with a spot of white paint. For the correct degree mark, refer to the Specifications at the beginning of this Chapter.
4 Start the engine and set the idling speed using a tachometer to the correct idling speed given in the Specifications.
5 Aim the strobe light at the clutch housing aperture and check that the timing marks line up as the light flashes.
6 Rotate the distributor body clockwise to retard the ignition and anticlockwise to advance it. When the timing marks are in perfect alignment, tighten the distributor clamp.
7 Re-check the timing and then disconnect the strobe light and reconnect the vacuum advance pipe.
8 The centrifugal advance mechanism can also be adjusted, but it requires specialised bench testing equipment and the task is best left to your Chrysler Dealer.

6 Coil - general

The two LT wire connections on top of the coil should be periodically checked for security. Remove the rubber sleeve from the centre HT lead and make sure that the end of the wire is clean and making good contact with the coil.

A possible source of arcing is between the top of the coil or sleeve and the LT terminals (See Fig. 4.6). To avoid this, keep the top of the coil clean and renew the rubber sleeve if cracked or perished.

The coil cannot be repaired and if it is thought to be faulty, it should be either taken to an electrical dealer for testing or renewal.

As the coil operates in conjunction with two ballast resistors it **must** be replaced with one of the same specifications.

7 Spark plugs and leads

1 The correct functioning of the spark plugs is vital for the correct running and efficiency of the engine.
2 At intervals of 6,000 miles (9,500 km) the plugs should be removed, examined, cleaned, and if worn excessively, renewed. The condition of the spark plugs will also tell much about the overall condition of the engine.
3 If the insulator nose of the spark plug is clean and white, with no deposits, this is indicative of a weak mixture, or too hot a plug. (a hot plug transfers heat away from the electrode slowly - a cold plug transfers it away quickly).
4 The plugs fitted as standard are listed in the Specifications at the beginning of this Chapter. If the tip and insulator nose is covered with hard, black looking deposits, then this is indicative that the mixture is too rich. Should the plug be black and oily, then it is likely that the engine is fairly worn, as well as the mixture being too rich.
5 If the insulator nose is covered with light tan to greyish brown deposits, then the mixture is correct and it is likely that the engine is in good condition.
6 If there are any traces of long brown tapering stains on the outside of the white portion of the plug, then the plug will have to be renewed, as this shows that there is a faulty joint between the plug body and the insulator, and compression is being allowed to leak away.
7 Plugs should be cleaned by a sand blasting machine, which will free them from carbon more thoroughly than cleaning by hand. The machine will also test the condition of the plugs under compression. Any plug that fails to spark at the recommended pressure should be renewed.
8 The spark plug gap is of considerable importance, as, if it is too large or too small, the size of the spark and its efficiency will be seriously impaired. The spark plug gap should be set to the figure given in the Specifications at the beginning of this Chapter.
9 To set it, measure the gap with a feeler gauge, and then bend open, or close, the outer plug electrode until the correct gap is achieved. The centre electrode should never be bent as this may crack the insulation and cause plug failure if nothing worse.
10 When refitting the plugs, remember to use new plug washers, and refit the leads from the distributor in the correct firing order, which is 1 - 3 - 4 - 2, No. 1 cylinder being the one nearest the flywheel.
11 The plug leads require no routine attention other than being wiped over regularly and kept clean. At intervals of 6,000 miles (9,500 km) however, pull the leads off the plugs and distributor one at a time and make sure no water has found its way onto the connections. Remove any corrosion from the brass ends, wipe the collars on top of the distributor, and refit the leads.

8 Fault diagnosis - ignition system

By far the majority of breakdown and running troubles are caused by faults in the ignition system either in the low tension or high tension circuits.

There are two main symptoms indicating faults. Either the engine will not start or fire, or the engine is difficult to start and misfires. If it is a regular misfire (ie, the engine is running on only two or three cylinders), the fault is almost sure to be in the secondary or high tension circuit. If the misfiring is intermittent the fault could be in either the high or low tension circuits. If the car stops suddenly, or will not start at all, it is likely that the fault is in the low tension circuit. Loss of power and overheating, apart from faulty carburation settings, are normally due to faults in the distributor or to incorrect ignition timing.

By eliminating the conventional (and troublesome) contact breaker points and condenser the electronic ignition system fitted to the Alpine is extremely reliable.

If a fault does develop in the system it will usually result in complete engine stoppage or failure to start. Misfiring is unlikely to be caused by the electronic side of the ignition system.

Specialised electrical equipment is required to test the electronic ignition components and this task should be entrusted with your nearest Chrysler Dealer.

Before assuming however, that the electronic side of the ignition system has failed, the following basic ignition system checks should be carried out.

Engine fails to start

1 If the engine fails to start and the car was running normally when it was last used, first check there is fuel in the petrol tank. If the engine turns over normally on the starter motor and the battery is evidently well charged, then the fault may be in either the high or low tension circuits. First check the HT circuit. **Note:** If the battery is known to be fully charged, the ignition light comes on, and the starter motor fails to turn the engine **check the tightness of the leads on the battery terminals and also the secureness of the earth lead to its connection to the body.** It is quite common for the leads to have worked loose, even if they look and feel secure. If one of the battery terminal posts gets very hot when trying to work the starter motor this is a sure indication of a faulty connection to that terminal.

2 One of the commonest reasons for bad starting is wet or damp spark plug leads and distributor. Remove the distributor cap. If condensation is visible internally dry the cap with a rag and also wipe over the leads. Refit the cap.

3 If the engine still fails to start, check that voltage is reaching the plugs by disconnecting each plug lead in turn at the spark plug end, and holding the end of the cable about ¼ in (6 mm) away from the cylinder block. Spin the engine on the starter motor.

4 Sparking between the end of the cable and the block should be fairly strong with a strong regular blue spark. (Hold the lead with rubber to avoid electric shocks). If voltage is reaching the plugs, then remove them and clean and regap them. The engine should now start.

5 If there is no spark at the plug leads, take off the HT lead from the centre of the distributor cap and hold it to the block as before. Spin the engine on the starter once more. A rapid succession of blue sparks between the end of the lead and the block indicates that the coil is in order and that the distributor cap is cracked, and the rotor arm is faulty, or the carbon brush in the top of the distributor cap is not making good contact with the spring on the rotor arm.

6 If there are no sparks from the end of the lead from the coil, check the connection at the coil end of the lead. If it is in order start checking the low tension circuit.

7 Before checking the low tension circuit reference should be made to Fig. 4.2 which shows the ignition wiring diagram and the colour coding of each wire.

8 Using a voltmeter or test bulb, first check that current is reaching the coil. Switch on the ignition and connect the test wires between the (+) terminal on the coil and earth. No reading indicates either a break in the supply from the ignition switch or a fault in the ballast resistor. Check the ignition switch connections and test the input wire (coloured slate) on the resistor block located on the wing valance.

9 With the ignition switch still on check that a voltage reading can be obtained at the (—) terminal on the coil and the No. 2 terminal on the electronic control box.

10 If no voltage reading can be obtained, check the wires for breaks and the terminal connections for security. If a positive voltage reading is obtained but the engine still refuses to start it is possible that some part of the electronic circuitry is at fault and the assistance of a Chrysler Dealer will be required to check it out.

Engine misfires

1 If the engine misfires regularly run it at a fast idling speed. Pull off each of the plug caps in turn and listen to the note of the engine. Hold the plug cap in a dry cloth or with a rubber glove as additional

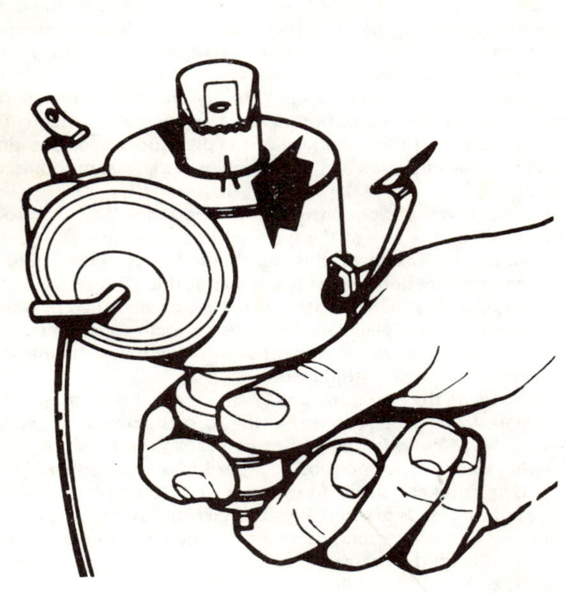

Fig. 4.5. Timing marks on SEV distributor

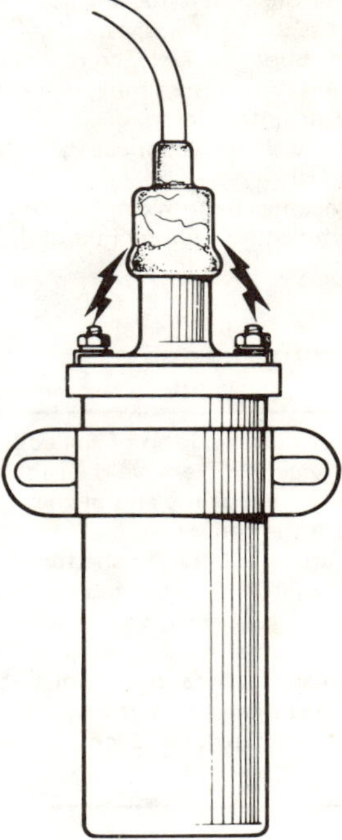

Fig. 4.6. Possible source of arcing on coil

protection against a shock from the HT supply.

2 No difference in engine running will be noticed when the lead from the defective circuit is removed. Removing the lead from one of the good cylinders will accentuate the misfire.

3 Remove the plug lead from the end of the defective plug and hold it about 3/16th in (5 mm) away from the block. Restart the engine. If the sparking is fairly strong and regular the fault must lie in the spark plug.

4 The plug may be loose, the insulation may be cracked, or the points may have burnt away giving too wide a gap for the spark to jump. Worse still, one of the points may have broken off. Either renew the plug, or clean it, reset the gap, and then test it.

5 If there is no spark at the end of the plug lead, or, if it is weak and intermittent, check the ignition lead from the distributor to the plug. If the insulation is cracked or perished, renew the lead. Check the connections at the distributor cap.

6 If there is still no spark, examine the distributor cap carefully for tracking. This can be recognised by a very thin black line running between two or more electrodes, or between an electrode and some other part of the distributor. These lines are paths which now conduct electricity across the cap thus letting it run to earth. The only answer is a new distributor cap.

7 Apart from the ignition timing being incorrect, other causes of misfiring have already been dealt with under the section dealing with the failure of the engine to start. To recap - these are that:

a) The coil may be faulty giving an intermittent misfire.
b) There may be a damaged wire or loose connection in the low tension circuit.
c) There may be a mechanical fault in the distributor (broken driving spindle)

8 If the ignition timing is too far retarded, it should be noted that the engine will tend to overheat, and there will be a quite noticeable drop in power. If the engine is overheating and the power is down, and the ignition timing is correct then the carburettor should be checked, as it is likely that this is where the fault lies.

Chapter 5 Clutch

Contents

Specifications

Type	Ferodo 180DBR, 190DBR or 200DBR, or 190 Borg and Beck, single dry plate	
Pressure plate		
Type	Diaphragm spring	
Outer diameter	9.17 in (233 mm)	
Total depth of assembly	1.77 in (45 mm)	
Driven plate (friction disc)		
Friction lining:		
Outer diameter	7.15 in (181.5 mm)	
Inner diameter	4.88 in (124 mm)	
Thickness (under load)	0.31 to 0.33 in (7.8 to 8.4 mm)	
Master cylinder		
Type	Lockheed	
Bore	0.75 in (19 mm)	
Stroke	0.875 in (22.5 mm)	
Slave cylinder		
Type	Lockheed, self-adjusting	
Bore	1.0 in (25.4 mm)	
Stroke	0.625 in (16.2 mm)	
Torque wrench settings	**lb f ft**	**kg f m**
Clutch assembly to flywheel	10.5	1.45
Slave cylinder to clutch housing	16.0	2.21
Release fork pivot pin to gearbox face	32.0	4.42
Release bearing support tube to gearbox	10.5	1.45

1 General description

The clutch assembly comprises a single (dry) driven plate, a pressure plate, release bearing and mechanism. The driven plate (friction disc) is free to slide along the gearbox primary shaft and is held in position between the flywheel and pressure plate faces by the pressure exerted by the diaphragm spring of the pressure plate. The friction linings are riveted to the driven plate which incorporates a spring cushioned hub to absorb transmission shocks and to assist smooth take up of the drive.

The diaphragm spring is mounted on shouldered pins and is held in place in the cover by fulcrum rings. The clutch is actuated by a pendant type foot pedal which operates a hydraulic master and slave cylinder. Depressing the clutch pedal pushes the release bearing, mounted on its hub, forward to bear against the spring fingers of the diaphragm. This action causes the diaphragm spring outer edge to deflect and move the pressure plate rearwards to disengage the pressure plate face from the driven plate linings. When the clutch pedal is released, the diaphragm spring forces the pressure plate into contact with the friction linings and sandwiches the driven plate between it and the flywheel so taking up the drive.

As the friction linings wear, the pressure plate automatically moves closer to the driven plate to compensate.

The release bearing is a ball race type, grease packed and sealed for life. The slave cylinder is the hydrostatic type that automatically compensates for clutch lining wear. No clutch adjustment is required or provided.

2 Clutch hydraulic system - bleeding

1 Gather together a clean jam jar, a length of rubber tubing which fits tightly over the bleed nipple in the slave cylinder, a tin of hydraulic brake fluid, and the help of an assistant.

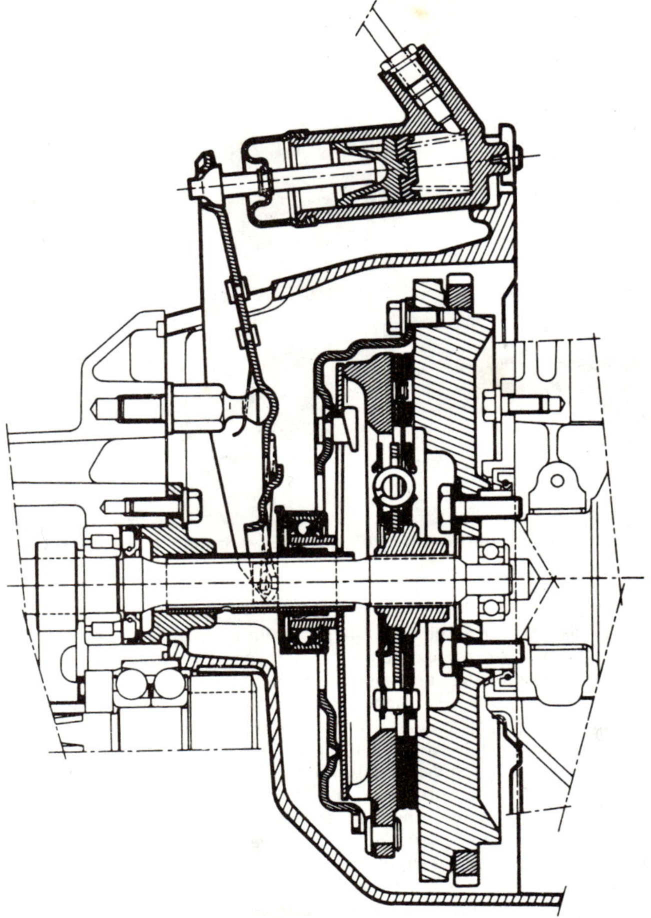

Fig. 5.1. Cross-sectional view of clutch assembly

2 Check that the master cylinder is full and if not, fill it, and cover the bottom inch of the jar with hydraulic fluid.
3 Remove the rubber dust cap from the bleed nipple on the slave cylinder and, with a suitable spanner, open the bleed nipple one turn.
4 Place one end of the tube securely over the nipple and insert the other end in the jam jar so that the tube orifice is below the level of the fluid.
5 The assistant should now pump the clutch pedal up and down slowly until air bubbles cease to emerge from the end of the tubing. He should also check the reservoir frequently to ensure that the hydraulic fluid does not get so low as to let air into the system.
6 When no more air bubbles appear, tighten the bleed nipple on the downstroke.
7 Refit the rubber dust cap over the bleed nipple. Allow the hydraulic fluid in the tin to stand for at least 24 hours before using it, to allow all the minute air bubbles to escape.
8 Never re-use old hydraulic fluid.

3 Clutch pedal - removal and refitting

1 Disconnect the clutch master cylinder pushrod from the clutch pedal by removing the pin and cotter.
2 Remove the pin from the pedal cross shaft (Fig. 5.2).
3 Detach the pedal return spring (if fitted).
4 Withdraw the cross shaft sufficiently far to enable the clutch pedal to be removed downwards. Take care to note the sequence of the various cross shaft washers, spacers and springs.
5 Refitting is a reversal of removal but grease the cross shaft and use new pins.

4 Clutch - removal and refitting

1 Access to the clutch may be gained in one of two ways. Either the complete engine/transmission unit should be removed as described in Chapter 1 and the gearbox/transmission separated from the engine or the engine left in the car and the gearbox/transmission unit removed independently. The latter procedure is fully described in the next Chapter, Section 2. Refer to Chapter 13 Supplement, Section 7 for details of removing Types AC and BE transmission for access to the clutch.
2 Note the alignment of the paint spots on both the flywheel and the pressure plate assembly for exact refitting.
3 Unscrew the six pressure plate retaining bolts from the flywheel, working in a diametrically opposite removal sequence and slackening the bolts only a few turns at a time (Fig. 5.3).
4 Withdraw the pressure plate assembly and driven plate.
5 It is important that no oil or grease gets on the clutch disc friction linings, or the pressure plate and flywheel faces. It is advisable to refit the clutch with clean hands and to wipe down the pressure plate and flywheel faces with a clean dry rag before assembly begins.
6 Place the clutch disc against the flywheel with the longer end of the hub, which is the end with the chamfered splines, facing the flywheel. On no account should the clutch disc be refitted with the shorter end of the centre hub facing the flywheel as on reassembly it will be found quite impossible to operate the clutch in this position.
7 To clarify the position some disc's are marked 'flywheel side' on the centre and it should be noted that the grooves in the friction lining must face towards the pressure plate (photo).
8 Refit the clutch cover assembly loosely on the two dowels. Refit the six bolts and spring washers and tighten them finger-tight so that the clutch disc is gripped but can still be moved.
9 The clutch disc must now be centralised so that when the engine and gearbox are mated the gearbox input shaft splines will pass through the splines in the centre of the driven plate hub.
10 Centralisation can be carried out quite easily by inserting a round bar or long screwdriver through the hole in the centre of the clutch, so that the end of the bar rests in the small hole in the end of the crankshaft containing the input shaft bearing bush.
11 Using the input shaft bearing bush as a fulcrum, moving the bar sideways or up and down will move the clutch disc in whichever direction is necessary to achieve centralisation.
12 Centralisation is easily judged by removing the bar and viewing the driven plate hub in relation to the hole in the release bearing. When the hub appears exactly in the centre of the release bearing hole all is correct.

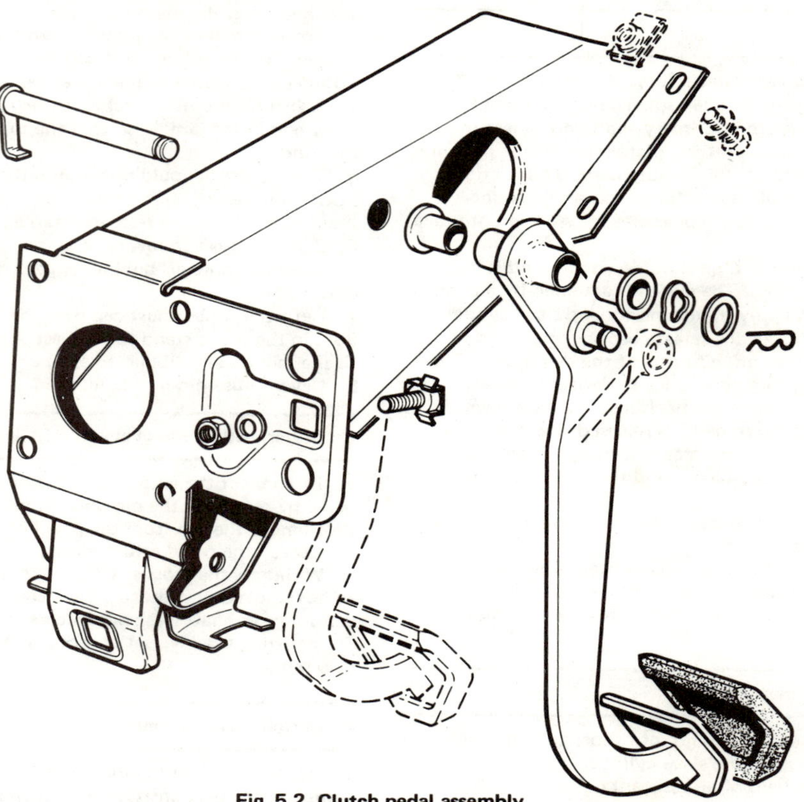

Fig. 5.2. Clutch pedal assembly

4.7 Refitting the clutch disc and cover assembly

4.13 Centralising the clutch disc using the gearbox primary shaft

13 An alternative and more accurate method of centralisation is to use the gearbox primary shaft if the gearbox has been dismantled or if there is an old shaft available (photo).
14 Tighten the clutch bolts in a diagonal sequence to ensure that the cover plate is pulled down evenly and without distortion of the flange.
15 Grease the splines of the gearbox primary shaft sparingly with molybdenum type grease.
16 Refit the gearbox as described in the next Chapter.

5 Clutch assembly - inspection

1 In the normal course of events, clutch dismantling and reassembly, is the term used for simply fitting a new clutch pressure plate and friction disc. Under no circumstances should the diaphragm spring clutch unit be dismantled. If a fault develops in the pressure plate assembly, an exchange unit must be fitted.
2 If a new clutch disc is being fitted it is false economy not to renew the release bearing at the same time. This will preclude having to renew it at a later date when wear on the clutch linings is very small.
3 Examine the clutch disc friction linings for wear or loose rivets, the disc for rim distortion, cracks and worn splines. If any of these faults are evident the disc must be renewed.
4 Check the machined faces of the flywheel and the pressure plate. If either is badly grooved it should be machined until smooth, or renewed. If the pressure plate is cracked or split it must be renewed.
5 Examine the hub splines for wear and also make sure that the centre hub is not loose.

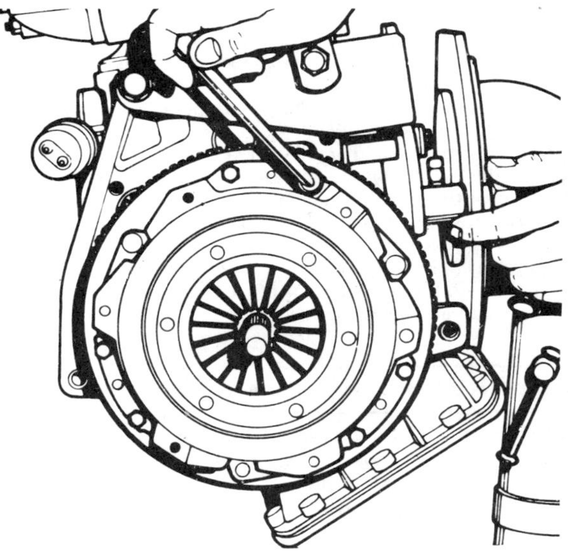

Fig. 5.3. Removing the clutch cover securing bolts

6.1 Clutch release bearing

6 Clutch release bearing - removal and refitting

1 With the gearbox and engine separated to provide access to the clutch, attention can be given to the release bearing located in the bell-housing, over the input shaft (photo).
2 The release bearing is a relatively inexpensive but important component and unless it is nearly new it is a mistake not to renew it during an overhaul of the clutch.
3 To remove the release bearing, first pull off the release arm rubber gaiter.
4 The release arm and bearing assembly can then be withdrawn from the clutch housing.
5 To free the bearing from the release arm simply unhook it, and then with the aid of two blocks of wood and a vice press off the release bearing from its hub.
6 Refit the release arm and bearing using the reverse procedure to that of removal. Make sure the spring clip on the release arm (photo) is correctly engaged on the pivot ball. Note that the small lug on the release bearing must be positioned between the bosses on the housing as shown in Fig. 5.4.
7 Refit the retaining spring on the release arm and engage the ends in the release bearing (see Fig. 5.4).
8 On some later models the release bearing is retained by lugs which engage with the release fork, rather than by a spring clip.
9 To remove a bearing retained by lugs, turn the lugs anti-clockwise to disengage them from the fork. Refitting is the reverse of the removal procedure.
10 Release bearings retained by spring clips should be lubricated with molybdenum grease. Bearings retained by lugs should be lubricated with silicone grease. With both types of bearing, make sure that no grease finds its way onto the friction surfaces of the clutch disc, pressure plate or flywheel.

7 Clutch master cylinder - removal, servicing and refitting

1 Remove the brake and clutch fluid reservoir cap, place a clean piece of polythene sheet over the aperture and refit the cap. This will reduce the loss of fluid when the hydraulic pipes are disconnected.
2 From the rear engine compartment bulkhead, disconnect the clutch master cylinder outlet pipe (Fig. 5.5).
3 From inside the car, place a container below the master cylinder to catch any fluid and disconnect the hydraulic supply pipe from the master cylinder.
4 Disconnect the master cylinder operating pushrod from the top of the clutch pedal.

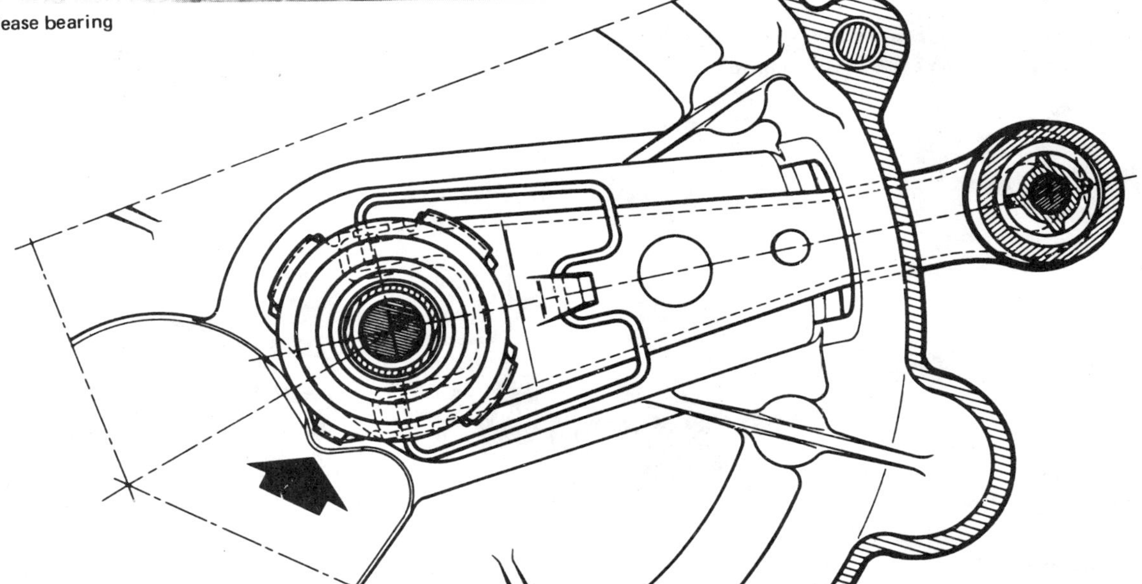

Fig. 5.4. Correct location of release bearing and retaining spring

5 Unscrew and remove the master cylinder retaining nuts and lift the unit from the bulkhead.
6 Expel the fluid from the master cylinder by depressing the pushrod two or three times.
7 Obtain a master cylinder repair kit which will contain the essential seals and components most likely to require renewing.
8 Pull off the rubber dust excluder.
9 Withdraw the pushrod.
10 Referring to Fig. 5.6, push the piston in slightly and remove the circlip and washer from the end of the cylinder.
11 Withdraw the piston complete with seals and return spring. If the piston sticks in the bore, gently tap the body of the master cylinder on the bench to release it.
12 With the master cylinder now completely dismantled, wash all components and the interior of the cylinder body with methylated spirit or clean hydraulic fluid.
13 Discard the old seals, first noting (and sketching if necessary) the way they are fitted to the piston in respect of chamfers and lips.
14 Examine the master cylinder bore and the piston surfaces for scoring and 'bright' spots. If these are evident, then the complete unit should be reassembled and exchanged for a factory reconditioned one.
15 Commence reassembly by dipping the new seals, supplied in the repair kit, in clean brake fluid and fitting them to the piston assembly, using the fingers only to manipulate them into position. Take particular care that the lips and chamfered edges have not been deformed or cut during storage in the repair kit packet.
16 Locate the return spring in the cylinder bore and then lubricate the bore liberally with clean hydraulic fluid. Insert the piston assembly using a twisting motion and ensuring that the lips of the seals are not trapped or pinched during the operation.
17 Refitting of the remaining components is a reversal of dismantling.
18 Refit the master cylinder to the bulkhead and reconnect the inlet and outlet hydraulic pipes and operating pushrod.
19 Remove the polythene sheeting from the fluid reservoir and top up the reservoir with the correct grade of clutch fluid.
20 Refit the cap and bleed the system as described in Section 2.

8 Clutch slave cylinder - removal, servicing and refitting

1 Remove the fluid reservoir cap, place a thin sheet of polythene over the aperture and screw the cap back on. This will seal the system and reduce the loss of fluid.
2 Disconnect the supply pipe from the slave cylinder and plug the

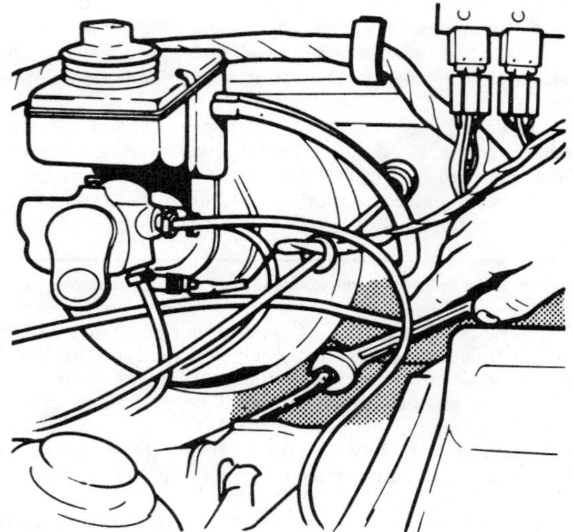

Fig. 5.5. Disconnecting the clutch master cylinder outlet pipe

6.6 Spring retaining clip on the release arm

Fig. 5.6. Clutch master cylinder components

1 *Circlip*
2 *Stop washer*
3 *Secondary cup*
4 *Piston*
5 *Primary cup*
6 *Spring*

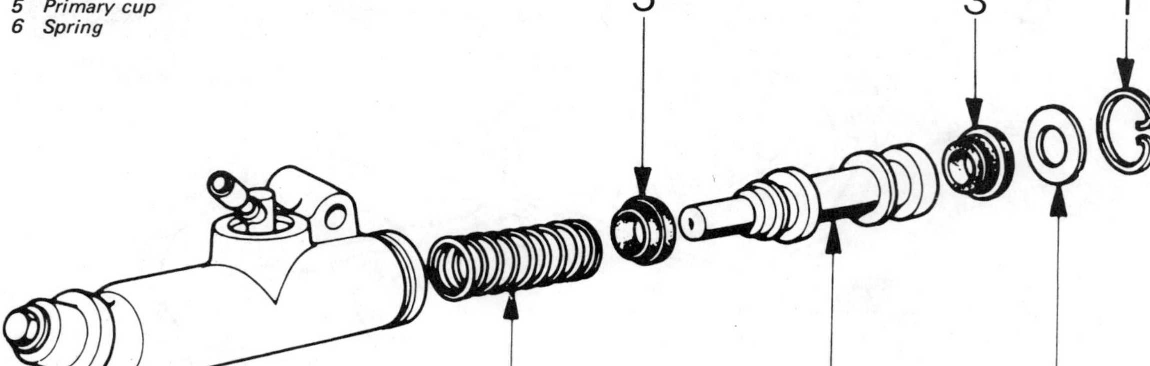

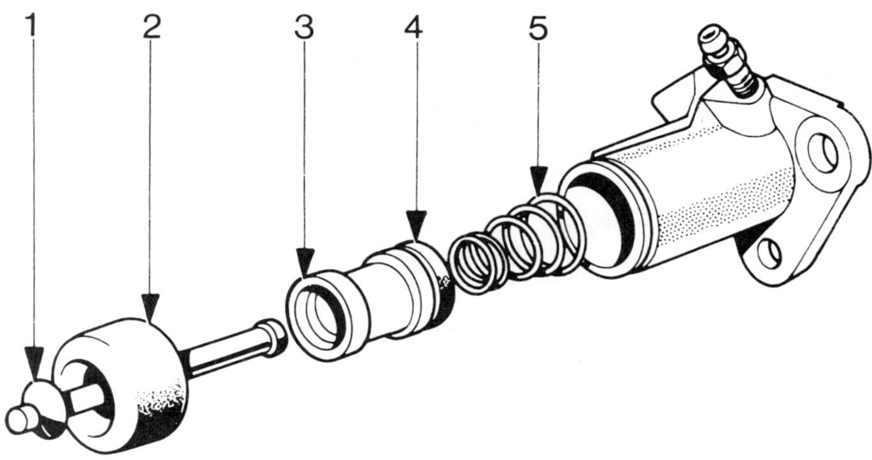

Fig. 5.7. Clutch slave cylinder components

1 Pushrod
2 Rubber boot
3 Piston
4 Seal
5 Spring

end of the pipe to prevent dirt ingress.
3 Remove the bolts securing the slave cylinder to the clutch housing, push the outer end of the clutch release fork towards the front (timing cover) of the engine and withdraw the complete slave cylinder assembly.
4 Obtain a slave cylinder repair kit of the correct type.
5 Referring to Fig. 5.7, remove the pushrod and rubber dust excluder.
6 Withdraw the piston complete with seal followed by the return spring.
7 Inspect the components for wear and fit a new piston seal using the procedures described for the master cylinder in the preceding Section.
8 Reassemble the slave cylinder and refit it to the engine using the reverse procedure to that of removal.
9 Remove the polythene sheet from the reservoir, top up the system with fluid and bleed it as described in Section 2.
10 The slave cylinder is the hydrostatic type and no adjustment is necessary.

9 Fault diagnosis - clutch

There are four main faults to which the clutch and release mechanism are prone. They may occur by themselves or in conjunction with any of the other faults. They are squeal, slip, spin and judder.

Clutch squeal - diagnosis and cure

1 If on taking up the drive or when changing gear, the clutch squeals, this is a sure indication of a badly worn clutch release bearing.
2 As well as regular wear due to normal use, wear of the clutch release bearing is much accentuated if the clutch is ridden, or held down for long periods in gear, with the engine running. To minimise wear of this component the car should always be taken out of gear at traffic lights and for similar holdups.

Clutch slip - diagnosis and cure

3 Clutch slip is a self evident condition which occurs when the clutch friction plate is badly worn, when oil or grease have got onto the flywheel or pressure plate faces, or when the pressure plate itself is faulty.
4 The reason for clutch slip is that, due to one of the faults listed above, there is either insufficient pressure from the pressure plate, or insufficient friction from the friction plate to ensure solid drive.
5 If small amounts of oil get into the clutch, they will be burnt off under the heat of clutch engagement, and in the process, gradually darken the linings. Excessive oil on the clutch will burn off leaving a carbon deposit which can cause quite bad slip, or fierceness, spin and judder.
6 If clutch slip is suspected, and confirmation of this condition is required, there are several tests which can be made.
7 With the engine in top gear and pulling lightly up a moderate incline sudden depression of the accelerator pedal may cause the engine to increase its speed without any increase in road speed.
8 In extreme cases of clutch slip the engine will race under normal acceleration conditions.
9 The permanent cure is, of course, to renew the clutch driven plate and trace and rectify the oil leak.

Clutch spin - diagnosis and cure

10 Clutch spin is a condition which occurs when the release arm travel is excessive, there is an obstruction in the clutch either on the primary gear splines or in the operating lever itself, or the oil may have partially burnt off the clutch linings and have left a resinuous deposit which is causing the clutch to stick to the pressure plate or flywheel.
11 The reason for clutch spin is that due to any, or a combination of the faults just listed, the clutch pressure plate is not completely freeing from the centre plate even with the clutch pedal fully depressed.
12 If clutch spin is suspected, the condition can be confirmed by extreme difficulty in engaging first gear from rest, difficulty in changing gear, and very sudden take up of the clutch drive at the fully depressed end of the clutch pedal travel as the clutch is released.
13 Check the clutch master and slave cylinders and the connecting hydraulic pipe for leaks. Fluid in one of the rubber boots fitted over the end of cylinders is a sure sign of a leaking piston seal.
14 If these points are checked and found to be in order then the fault lies internally in the clutch, and it will be necessary to remove it for examination.

Clutch judder - diagnosis

15 Clutch judder is a self evident condition which occurs when the gearbox or engine mountings are loose or too flexible, when there is oil on the faces of the clutch friction plate, or when the clutch pressure plate has been incorrectly adjusted during assembly.
16 The reason for clutch judder is that due to one of the faults just listed, the clutch pressure plate is not freeing smoothly from the friction disc, and is snatching.
17 Clutch judder normally occurs when the clutch pedal is released in first or reverse gears, and the whole car shudders as it moves backwards or forwards.

Chapter 6 Gearbox and final drive

For modifications, and information applicable to later models, see Supplement at end of manual

Contents

Specifications

Gearbox

Type	Four forward speeds (all with synchromesh) and reverse
Gear ratios	
1st	3.9 : 1
2nd	2.31 : 1
3rd	1.52 : 1
4th	1.08 : 1
Reverse	3.77 : 1

Final drive

Type	Helical gears
Gear ratios	
1294 cc (Germany only)	3.937 : 1
1294 cc (except Germany)	3.706 : 1
1442 cc (Germany only)	3.706 : 1
1442 cc (except Germany)	3.588 : 1

Lubricants and oil capacities

Gearbox	1 pint (0.60 litre)
Final drive	0.9 pint (0.50 litre)
Lubricant type	SAE 90EP gear oil

Torque wrench settings

	lb f ft	kg fm
Clutch bellhousing to cylinder block bolts	32.0	4.43
Gearbox to clutch bellhousing bolts	16.0	2.21
Gearbox selector fork cover to gearbox bolts	8.5	1.18
Gearbox and final drive drain and filler plugs	25.0	3.46
Clutch release fork pivot pin to gearbox	10.5	1.45
Clutch thrust bearing support tube to gearbox bolts	10.5	1.45
Output shaft locknut (LH thread)	107	14.79
Differential half housing bolts	16.0	2.21
Large bearing thrust plate bolts	16.0	2.21

1 General description

The integrated manual gearbox and final drive unit is located on the right-hand end of the engine when viewed from the front of the car.

The front roadwheels are driven through driveshafts of unequal lengths.

The clutch bellhousing incorporates the final drive unit and the gearbox is mounted on the outer face of the clutch housing (Fig. 6.1).

The differential assembly is of conventional design having the inner ends of the driveshafts splines to the differential gears.

2 Gearbox - removal and refitting

1 The removal of the gearbox with the engine is described in Chapter 1. The gearbox may be removed without removing the engine using the following method.

2 Disconnect the leads from the battery terminals.

3 Raise the front of the vehicle and support it on stands.

4 Disconnect the gearshift rod assembly from the gearbox, see Section 10.

5 Unbolt the clutch slave cylinder from the clutch bellhousing and release the pushrod from the clutch release arm. The slave cylinder may now be tied up out of the way without disconnecting the hydraulic fluid line.

6 Make up an engine support which will take the weight of the unit when the gearbox mounting is removed.

7 Remove the left-hand front roadwheel.

8 Remove the left-hand damper and then drain the oil from the gearbox.

9 Disconnect the left-hand front engine mounting as described in Chapter 1.

10 Obtain two guide studs each 2.17 inch (55 mm) in length and threaded (8 x 125) over 0.6 inch (15 mm) of their length. Insert them

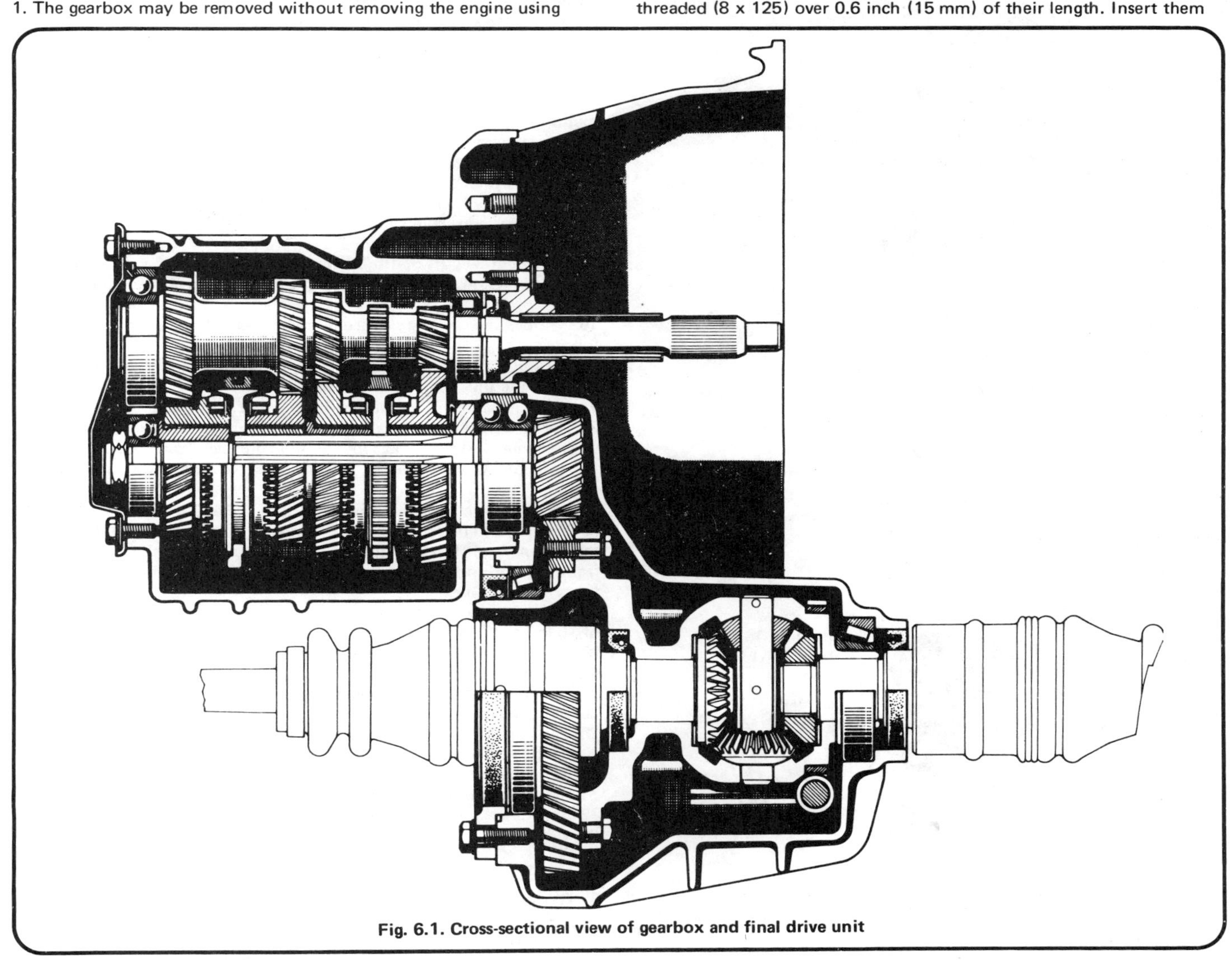

Fig. 6.1. Cross-sectional view of gearbox and final drive unit

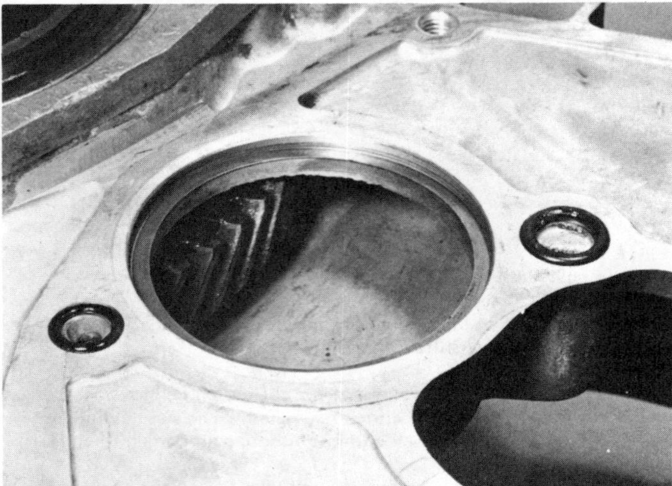

2.15 Location of 'O' ring seals in transmission casing

2.17 Shims fitted in pinion aperture

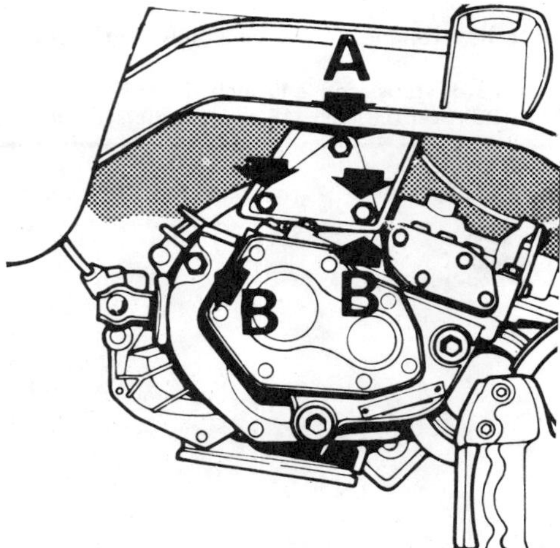

Fig. 6.2. Gearbox securing bolts

A Gearbox mounting bolts *B Guide bolt holes*

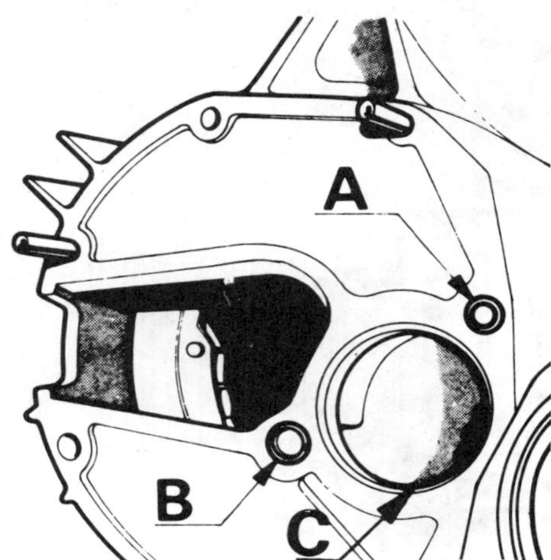

Fig. 6.3. Rear face of clutch/final drive housing

A 'O' ring seal
B 'O' ring seal
C Gearbox pinion aperture

in the holes, (B) (Fig. 6.2).

11 Take out the bolts which secure the gearbox to the clutch bell-housing.

12 Withdraw the gearbox, carefully retaining the shims fitted at the gearbox pinion aperture (Fig. 6.3).

13 Before refitting the gearbox, secure the clutch release arm in position using a piece of wire or string to ensure it does not become dislodged during refitting.

14 Select a gear and then refit the shims removed at (C) Fig. 6.3, using grease to retain the shims in position.

15 Check that the two small sealing 'O' rings are correctly located round the gearbox to differential connecting oil passages (photo).

16 Where major overhaul or renewal of internal components has been carried out, then the correct sealing of the gearbox to the clutch bellhousing/final drive unit must be checked. To do this, first drive the pinion fully home using a soft faced mallet.

17 Fit the original shims into the recess in the clutch/final drive housing (photo).

18 Place four lengths of 1 mm diameter soft soldering wire at equidistant points on the bearing mating surface of the clutch/final drive recess.

19 Remove the 'O' ring seal from the gearbox pinion bearing recess (photo).

20 Carefully locate the gearbox onto the clutch/final drive housing. To facilitate this, remove the clutch assembly.

21 Tighten the securing bolts to 16 lb f ft (2.21 kg fm).

22 Remove the bolts and gearbox and measure the thickness of the crushed wire and then add shims to the equivalent thickness.

23 Refit the 'O' ring seal into the groove surrounding the output shaft pinion.

24 Pass the clutch release bearing through the aperture in the clutch bellhousing and insert the gearbox input shaft into the splined hub of the driven plate. The driving gear may require moving slightly to facilitate meshing of the splines and to engage the gearbox housing on the two guide studs.

25 Screw in and tighten the gearbox retaining bolts evenly to a torque of 16 lb f ft (2.21 kg fm).

26 Remove the guide bolts and fit the engine mounting assembly and spacer.

27 Remove the temporary engine support bar.

28 Refit the clutch slave cylinder.

29 Reconnect the gearshift control mechanism, and adjust if necessary, see Section 10.

30 Refit the damper and roadwheel.

31 Fill the gearbox and check the oil level in the final drive unit.

32 Reconnect the battery leads and remove the jacks from the front of the vehicle.

3 Gearbox - dismantling

1 Clean the external surfaces of the gearbox and dry thoroughly.

Place the unit in a servicing frame or secure it to a bench. Remove the mounting assemblies from the outside of the gearbox housing.
2 Remove the gearshift selector cover, using a swivelling action to clear the forks (photo).
3 Remove the rear cover (Fig. 6.4) and the clutch release bearing guide tube.
4 Slide two gears into engagement to lock the gearbox.
5 Remove the nut (20) which locks the rear bearing (19) to the output shaft, Fig. 6.5. Note that it has a left-hand thread.
6 Make up a plate to the dimensions shown in Fig. 6.6.
7 Fit this between the face of the first gear wheel and the inside face of the gearbox housing. Using a strengthening plate on the outside of the gearbox and a bridge piece and bolt, press the shaft through the gears and rear bearing until it can be withdrawn completely from the gear case.
8 Withdraw the spacer and special roller thrust washer from the front of the gearbox.
9 Lift out the 1st/2nd gear cluster complete with synchromesh unit (Fig. 6.7).
10 Lift out the 3rd/4th gear cluster complete with synchromesh unit.
11 Remove the circlip locating pin from the reverse gear shaft (Fig. 6.8).
12 Using a suitable sized drift, carefully tap the shaft out through the front of the casing. If difficulty is experienced a special tool (No. 39963) is available from your Chrysler dealer.
13 Remove the single securing screw and withdraw the clutch release bearing guide tube from the front of the casing.
14 Using a suitable sized piece of tubing or special tool No. 20888V, drive out the input shaft complete with bearings from the front to rear of the casing (Fig. 6.10).
15 Discard the oil seal from the front of the shaft and retrieve the shim(s).
16 Do not attempt to remove the primary shaft bearings or the output shaft bearings unless you have a press and extractors suitable for the job. It is better to take them to a Chrysler agent.
17 If necessary, the gearshift control and selector fork assembly may be dismantled. Commence by setting the selector shafts and forks in neutral and removing the end cover (Fig. 6.11).
18 Unscrew and remove the 3rd/4th selector fork securing screw, withdraw the selector shaft slowly and retain the detent ball as it is removed.
19 Remove the 3rd/4th selector fork.
20 Remove the 1st/2nd selector fork screw, withdraw the shaft just sufficiently far to permit removal of the 1st/2nd selector fork.
21 Rotate the 1st/2nd selector shaft through 90° to facilitate removal of the detent ball.
22 Withdraw the 1st/2nd selector shaft completely but retaining the detent balls and springs during the operation.
23 Remove the reverse selector fork securing screw, drift out the reverse shaft sufficiently far to enable the fork to be removed. Withdraw the reverse shaft completely and retain the detent and interlocking balls and springs.
24 Remove the securing screw from the fork control lever.
25 Withdraw the spindle and fork control lever and the three springs which impinge upon the detent balls.
26 The gearbox is now completely dismantled and both the inside of the gearbox case and all components should be thoroughly washed in paraffin so that examination may be carried out as described in the following Section.

4 Gearbox - examination

1 Examine each component for wear, distortion, chipping or scoring and renew if apparent.
2 Examine the gear wheels, particularly for wear and chipping of teeth and renew as necessary.
3 Check all ball and roller bearings for play. If even the slightest wear is evident then they must be withdrawn from their shafts and renewed.
4 The condition of the synchromesh units will be known from previous driving experience. The units can be dismantled for inspection and renewal of components using the following method.

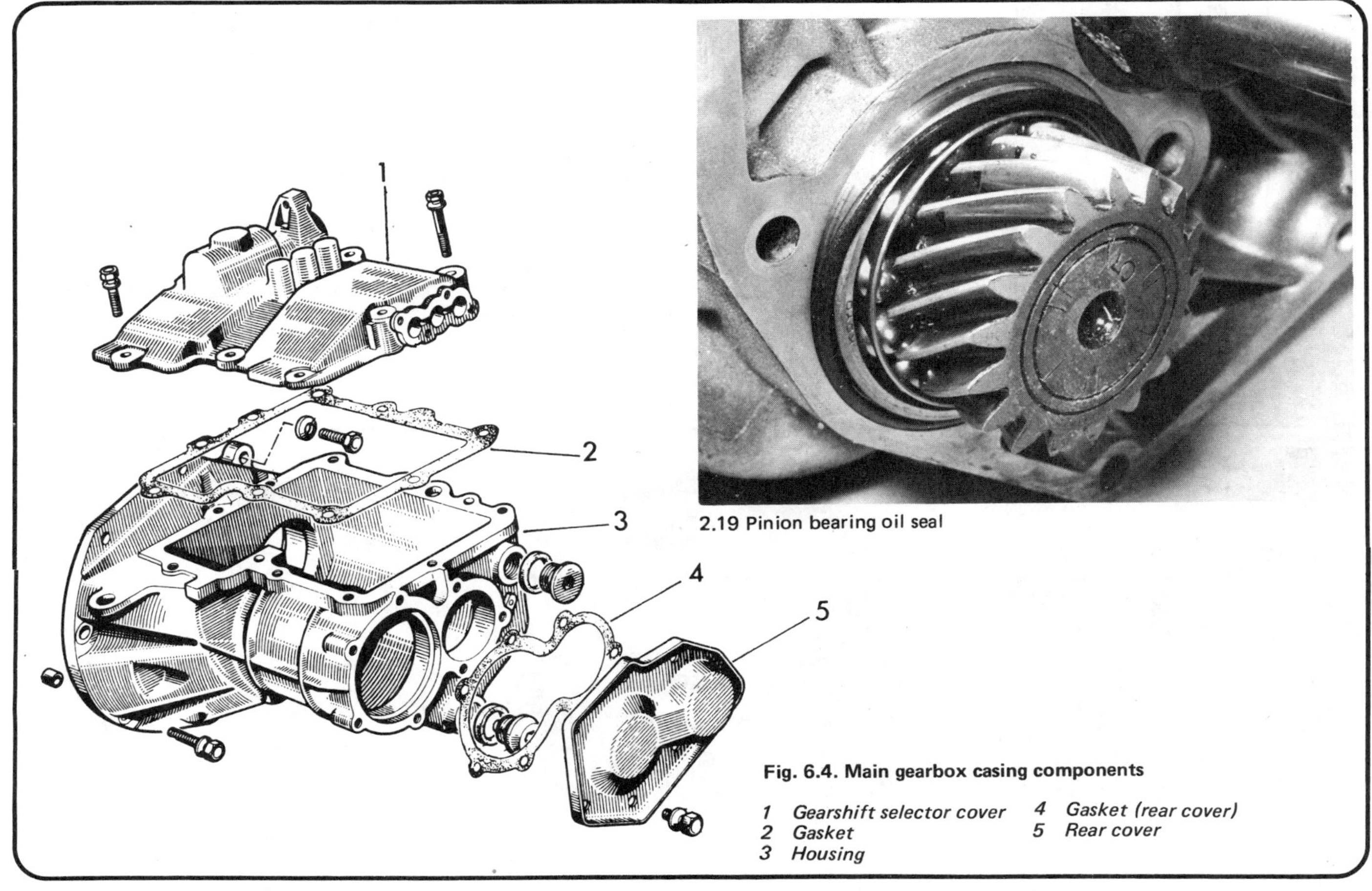

2.19 Pinion bearing oil seal

Fig. 6.4. Main gearbox casing components

1 Gearshift selector cover
2 Gasket
3 Housing
4 Gasket (rear cover)
5 Rear cover

OUTPUT SHAFT FRONT SECTION

REAR SECTION OUTPUT SHAFT

SEE FRONT SECTION

Fig. 6.5. Exploded view of gearbox output shaft (mainshaft)

1 *Spacer*
2 *O-ring seal*
3 *Double ball race*
4 *Oil seal*
5 *Bush*
6 *1st gear*
7 *Synchroniser ring*
8 *Synchro locking piece*
9 *Synchro circlip*
10 *Synchro hub*
11 *1st/2nd synchro sleeve (reverse on periphery)*
12 *Retainer stop*
13 *Spring*
14 *1st/2nd synchro assembly (part)*
15 *2nd gear*
16 *3rd gear*
17 *3rd/4th synchro unit*
18 *4th gear*
19 *Bearing*
20 *Locking nut*

Fig. 6.6. Mainshaft packing plate (dimensions in mm) (thickness 4 mm)

3.2 Removing the gearbox top cover

4.6 Synchro unit with circlip removed

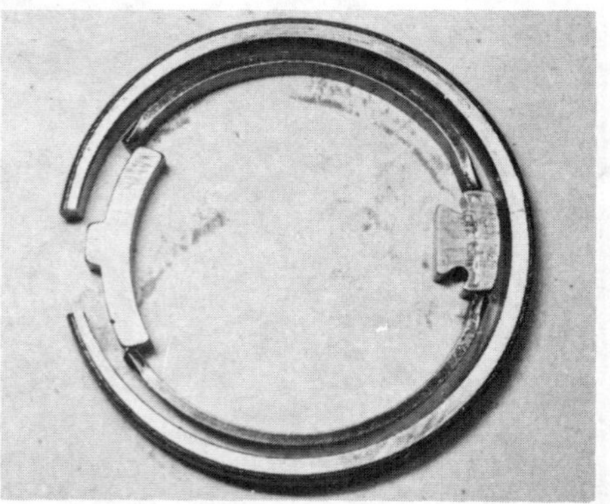
4.10A Correct positioning of synchro unit components prior to assembly

5 Using a strong pair of circlip pliers, carefully remove the large circlip from the synchro unit (Fig. 6.13).
6 With the circlip removed, make a careful note of the position of the baulk ring, the two curved springs and pawls (see photo).
7 Lift out the components and check the ends of the springs for burrs and ensure the edges of the spring stop and locking pawl are smooth and square.
8 Check the small internal teeth for wear or breakage. The sliding sleeve and hub are meant to be a fairly loose fit, and this should not be mistaken for wear.
9 If difficult gearchanging has been experienced or any doubts exist about the synchro units, the best policy is to obtain complete new units.
10 If fitting new components, reassemble them on the bench first (photo) before fitting them into the gear. Carefully secure them with the circlip and refit the sleeve and hub (photo).
11 Examine the ends of the selector forks at their points of contact. Comparison of their profiles with new components will give a guide to renewal requirements.
12 Check the gearbox casing for cracks, particularly around the shaft bearing and bolt holes.

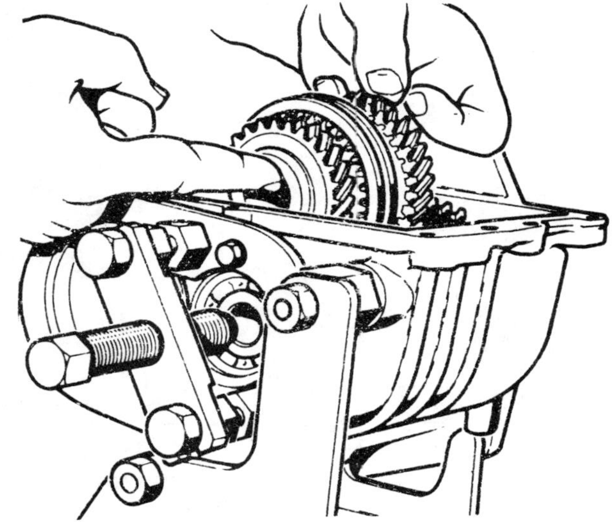

Fig. 6.7. Lifting out the 1st/2nd gear cluster

5 Gearbox - reassembly

1 From the rear end of the gearbox casing, carefully drive in the primary shaft complete with bearings using a soft faced hammer.
2 Continue to drive the shaft in until the rear bearing circlip is firmly seated against the casing (photo).
3 Assemble the 1st/2nd and 3rd/4th gears and synchro assemblies (photos).
4 Fit the front mainshaft ball bearing into its outer race in the casing (photo).
5 Insert the thick spacer into the front bearing aperture (photo). **Note:** If there is a groove in the spacer it must face towards the outside of the gear casing.
6 Fit the roller bearing thrust washer onto the front face of the 1st gear assembly (photo). Use grease to hold it in place if necessary.
7 Lower the 3rd/4th gear assembly into the casing (photo), followed by the 1st/2nd gear assembly and thrust washer.
8 Holding the gear assemblies in place, carefully insert the mainshaft through the front of the casing. Rotate the gears as necessary to align them with the shaft (photo).
9 Fit the mainshaft rear bearing (photo) and gently tap it into the casing.
10 Fit the mainshaft rear nut but do not attempt to tighten it fully at this stage.

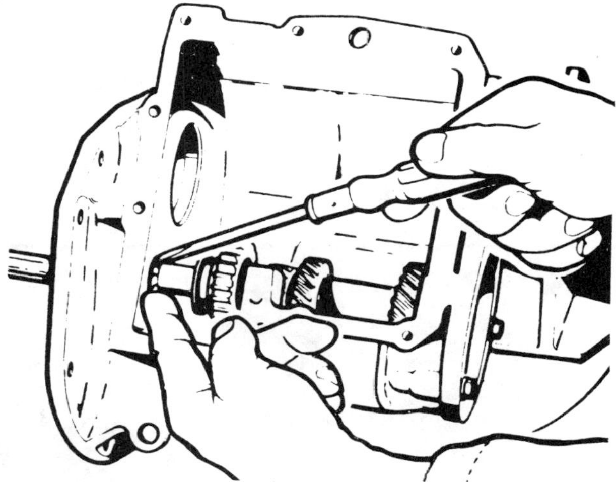

Fig. 6.8. Removing the reverse gear shaft circlips

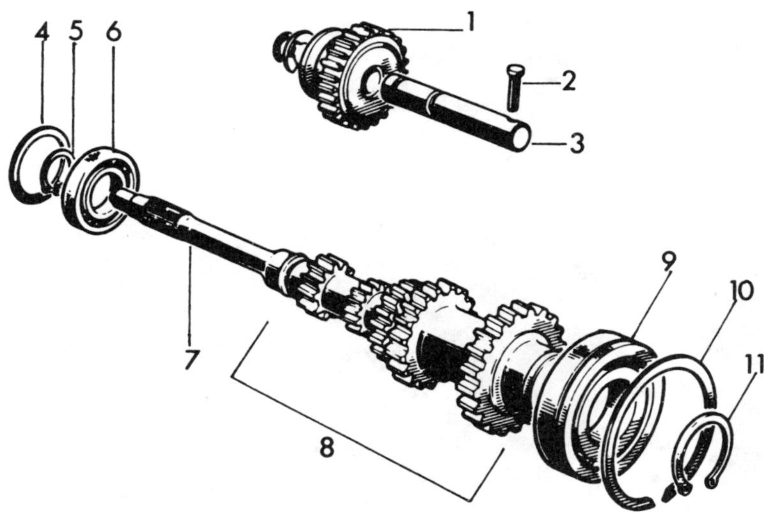

Fig. 6.9. Exploded view of reverse gear assembly and input shaft

1 Reverse gear
2 Pin
3 Reverse idler shaft
4 Spacer
5 Circlip
6 Bearing
7 Primary shaft
8 Gear assembly
9 Bearing
10 Bearing snap ring
11 Circlip

Fig. 6.10. Driving out the input shaft and bearings

Fig. 6.11. Exploded view of top cover and selector forks

1	Cover	5	1st/2nd selector fork
2	Fork control lever	6	1st/2nd selector shaft
3	Reverse selector fork	7	3rd/5th selector fork
4	Reverse selector shaft	8	3rd/4th selector shaft

4.10B Assembling the synchro sleeve and hub

5.2 Fitting the primary (input) shaft complete with bearings

5.3A 1st/2nd gear cluster assembly

5.3B 3rd/4th gear cluster assembly

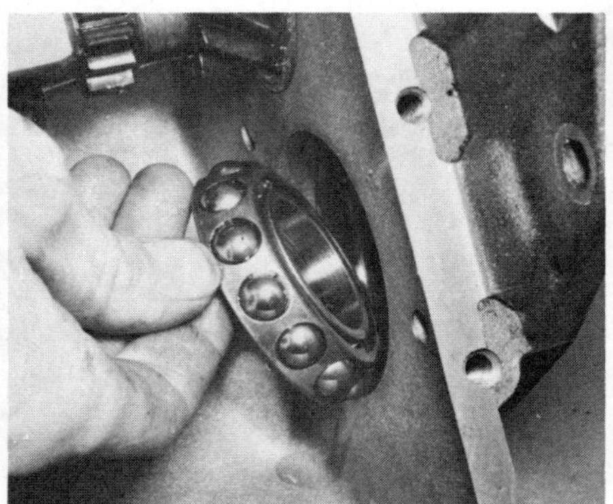

5.4 Inserting the front mainshaft bearing ...

5.5 ... followed by the spacer

5.6 Fitting the mainshaft thrust washer

5.7 Positioning the 3rd/4th gear cluster in the casing

5.8 Sliding the mainshaft through the gears

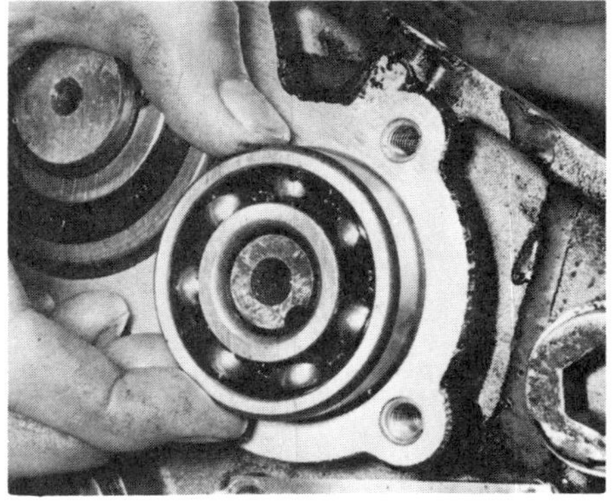
5.9 Fitting the mainshaft rear bearing

5.11 Fitting the reverse roller gear and shaft

5.12 Inserting the reverse gear shaft pin

5.14 Mainshaft rear retaining nut peened over

5.16 Fitting the primary shaft shims ...

5.17 ... followed by the oil seal

5.18 Location of clutch release bearing guide

5.22 Fitting the gearbox rear cover

5.23 Clutch release arm in position

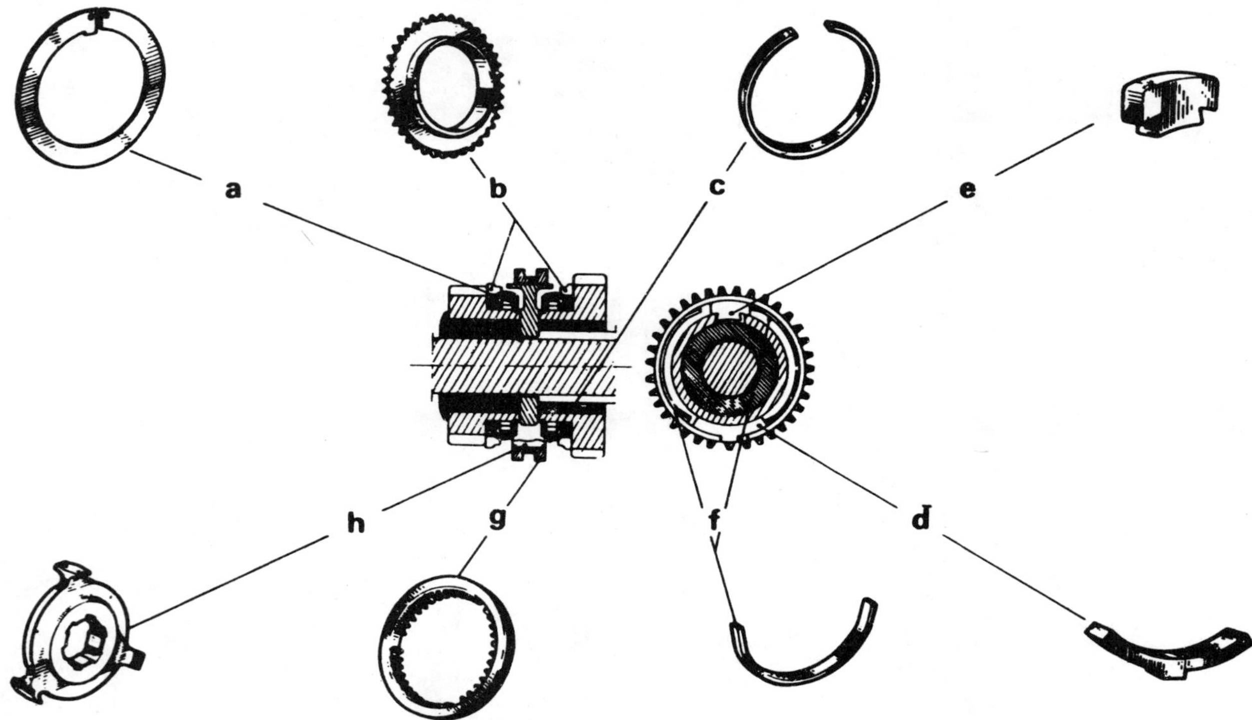

Fig. 6.12. Exploded view of synchromesh assembly

A Circlip
B Dog tooth ring, driven gear
C Baulk ring
D Locking pawl
E Spring stop
F Drive springs
G Sliding sleeve
H Driving hub

11 Fit a new 'O' ring on the reverse idler gear shaft, hold the reverse gear in position and slide the shaft through the casing and gear (photo).
12 Fit the circlip on the front of the reverse gear shaft and insert the retaining pin into the top of the casing (photo).
13 Engage the reverse gear and one forward gear to lock the shafts, and tighten the mainshaft rear nut to the specified torque wrench setting. Do not forget that the nut has a left-hand thread.
14 Lock the nut by peening it into the groove in the mainshaft (photo).
15 Disengage the gears and check that both shafts rotate freely and the gears mesh correctly.
16 Fit the original shim(s) into the primary shaft front bearing recess (photo).
17 Coat the outer diameter of a new oil seal with oil and fit it into the front bearing recess, ensuring the internal lip faces towards the bearing (photo).
18 Slide the clutch release bearing guide tube over the primary shaft and tighten the single securing bolt (photo).
19 Fit a new gasket to the gearshift fork cover.
20 Engage the selector forks and locate and bolt the cover into position.
21 Fit a new rear cover gasket.
22 Fit the cover and gasket to the gearbox (photo).
23 If the clutch release arm pivot has been removed then it should be refitted (photo).
24 Refit the mounting plates.

6 Final drive (differential) unit - removal and refitting

1 Remove the gearbox as described in Section 2 of this Chapter.
2 Drain the oil from the differential unit by removing the drain plug.
3 Unbolt the speedometer cable from the lower differential unit.
4 Remove the left-hand driveshaft/stub axle assembly as described for engine removal (Chapter 1).
5 Unscrew and remove the gearbox guide bolts which were fitted to facilitate removal of the gearbox.
6 Remove the starter motor.
7 Remove the bolts securing the gearchange relay bracket to the final drive unit.

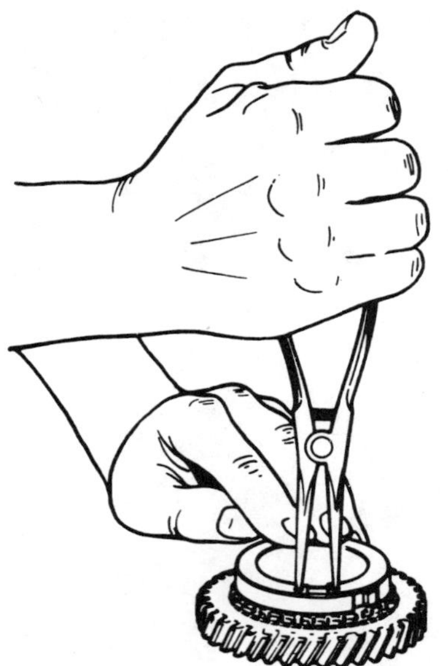

Fig. 6.13. Removing a synchromesh unit circlip

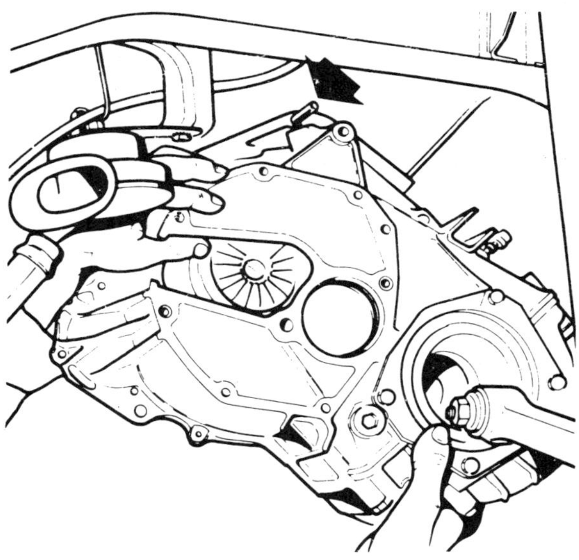

Fig. 6.14. Location of guide stud used when replacing transmission

Fig. 6.15. Lifting out the differential assembly

Fig. 6.16. Removing the crownwheel securing bolts

8 Remove the dirt shield from the clutch housing.

9 Remove the bolts which secure the clutch bellhousing to the engine block.

10 Pull the combined clutch bellhousing, final drive unit outwards sufficiently far to disengage the right-hand driveshaft. Do not allow the driveshaft to drop but have an assistant support it and lower it gently.

11 To refit the clutch bellhousing/final drive unit, first screw in a guide stud as shown in Fig. 6.14.

12 Offer up the unit and engage it on the guide stud.

13 Engage the right-hand driveshaft in the differential.

14 Slowly push the clutch bellhousing/final drive unit into its correct position, turning the driveshaft at the same time.

15 Fit the securing bolts and tighten them to a torque of 32 lb f ft (4.43 kg fm). Remove the guide stud.

16 Refit the left-hand driveshaft, stub axle and brake disc as described in Chapter 1.

17 Reconnect the speedometer drive cable to the final drive unit.

18 The gearbox may now be refitted as described in Section 2 of this Chapter.

7 Final drive (differential) unit - dismantling

1 Place the clutch bellhousing face downwards on the bench and remove the securing bolts from the bearing thrust plate.

2 Remove the thrust plate, retain the shims fitted below it and remove the 'O' ring seal.

3 Remove the securing bolts from the differential half housing and then remove the half housing complete with the seal located on the smaller bearing side.

4 Lift the differential assembly from the casing (Fig. 6.15).

5 Wash all components and the interior of the differential housing in paraffin.

8 Final drive (differential) unit - inspection and servicing

1 Inspect each component for wear, scoring or damage. Examine the teeth of the crownwheel for chipping and also the gear wheels in the differential cage.

2 Examine the roller bearings for wear and cracks in the inner and outer tracks.

3 To remove the crownwheel, (if this component is renewed, the gearbox output shaft with matched pinion will also have to be renewed) support the unit in a vice and unscrew the securing bolts (Fig. 6.16).

4 The differential bearings may be renewed if a suitable extractor is available (Fig. 6.17).

Fig. 6.17. Removing a differential roller bearing using a puller

8.6 Speedometer driven gear

8.9 Differential inner oil seal

8.10 Differential housing oil seal (small end)

9.1 Differential unit positioned in main housing

9.2 Final drive lower housing in position

9.7 Fitting the final drive bearing shims

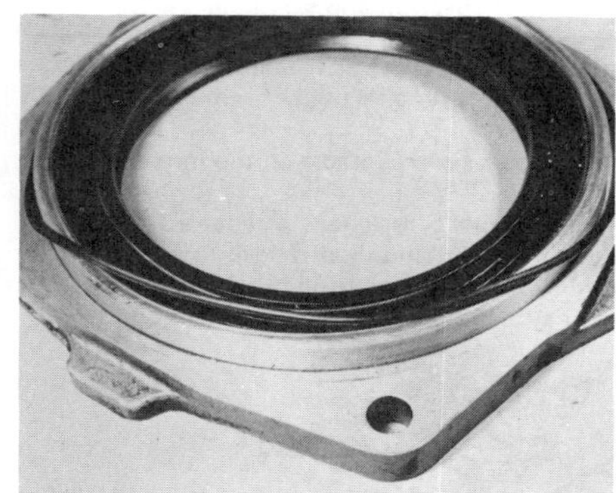
9.12 Thrust plate 'O' ring and oil seal

10.1 Lower end of gearshift lever

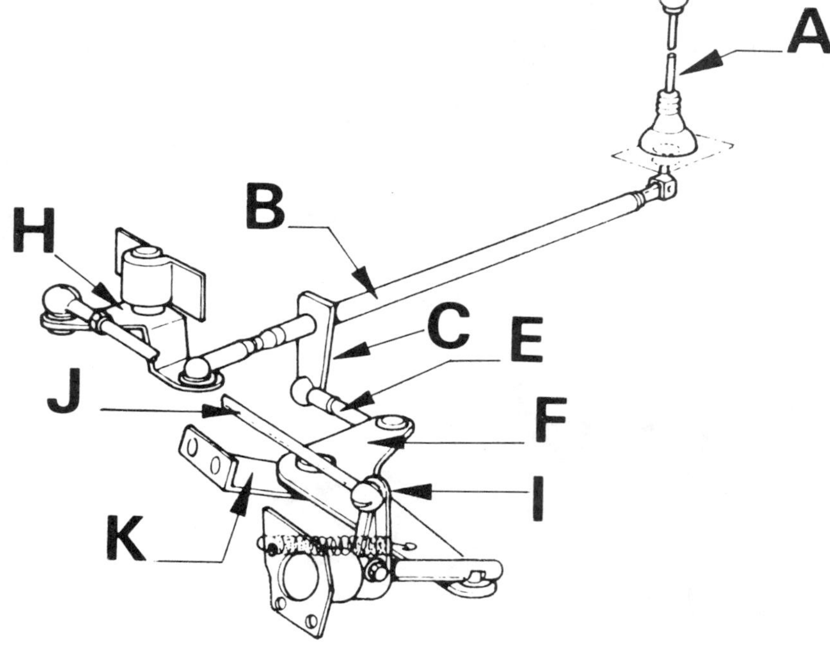

Fig. 6.18. Layout of gearshift linkage

A Gear lever
B Control tube
C Operating arm, gear selection
E Adjustable rod, gear selection
F Relay lever, gear selection
H Relay lever, gear engagement
I Relay lever
J Adjustable rod, gear engagement
K Relay support, final drive

5 The speedometer drive gear is a press fit on the outside of the differential cage. An extractor will be needed to remove it.
6 Refitting of the bearings and the speedometer drive gear will require the use of a press and it will be better to let a Chrysler dealer carry out this work if the correct tools are not available. Where the speedometer drive gear is to be renewed, then the meshing gear in the lower differential half housing must also be renewed as a matched pair (photo).
7 Dismantling of the sun and planet bevel gears is not recommended. It is better to obtain a reconditioned factory exchange differential cage complete.
8 The differential oil seals should be renewed as a matter of course. These may be renewed without removing the unit from the vehicle but the oil must first be drained and the gearbox may need to be pulled forward.
9 Extract the inner oil seal and carefully drive in a new one ensuring that its lip is towards the differential cage (photo).
10 Drive out the oil seal from the small bearing end of the housing and fit a new one squarely with its lip towards the roller bearing (photo).
11 Tap out the oil seal from the large bearing flange and fit a new one using a soft faced hammer ensuring that the lip will be towards the large bearing when it is bolted up (Fig. 6.19).
12 The differential and final drive assembly is now ready for refitting to its housing but certain adjustments are necessary. They are described in the following Section.

9 Final drive (differential) unit - reassembly and adjustment

1 Lower the assembled final drive unit into the housing and at the same time locate the large roller bearing track ring (photo).
2 Fit the lower half housing into position but do not tighten the bolts more than finger tight (photo).
3 A thrust plate dummy tool should now be borrowed or hired from your Chrysler dealer (part No. 20886K) or a substitute made up. Its purpose is to align the surfaces of the upper and lower differential housings and to locate the differential bearings correctly in their seats.
4 Bolt the tool first to the main housing face and then to the lower housing (Fig. 6.20).
5 Tighten the central threaded plate of the tool until the roller bearings are seated, rotating the differential housing at the same time. Repeat this operation several times and then release the plate of the tool and tighten the half housing bolts to 16 lb ft (2.21 kg fm).
6 Remove the dummy thrust plate tool and check that the differential housing will turn easily without binding or tight spots.
7 Fit the original bearing shims (photo).
8 If a new crownwheel carrier, carrier bearing flange, carrier bearing, or casing is being fitted, the bearing pre-load must be set up using the following method.

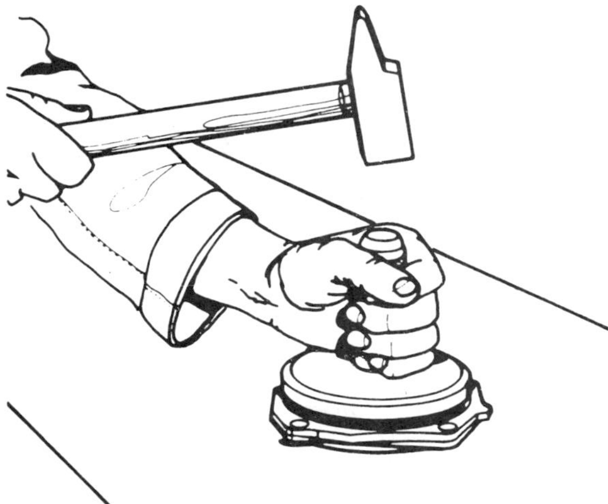

Fig. 6.19. Fitting a new seal into the thrust plate flange

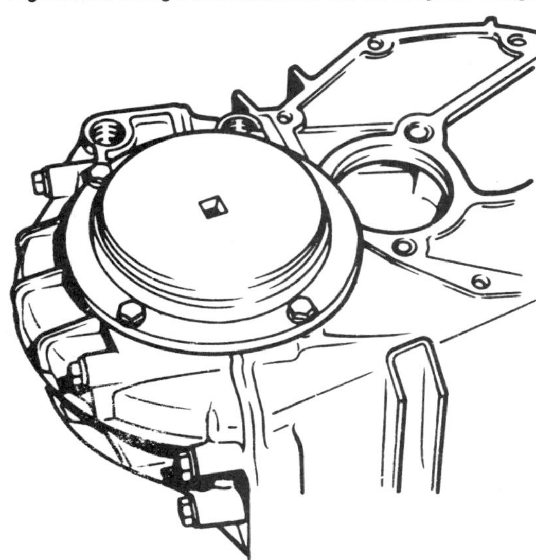

Fig. 6.20. Special tool fitted in place of the thrust plate

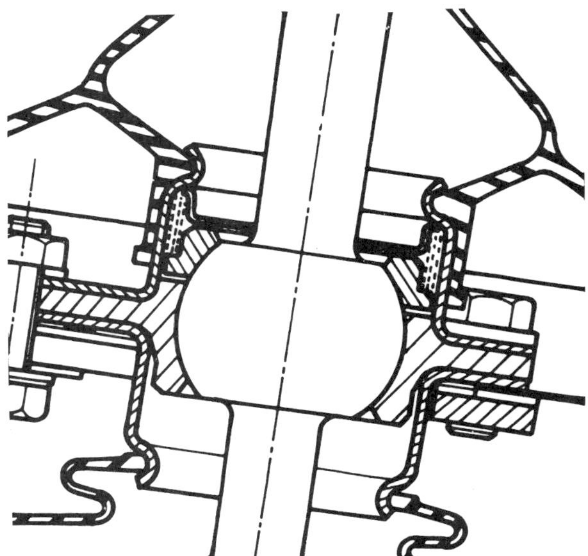

Fig. 6.22. Cross-sectional view of gear lever housing

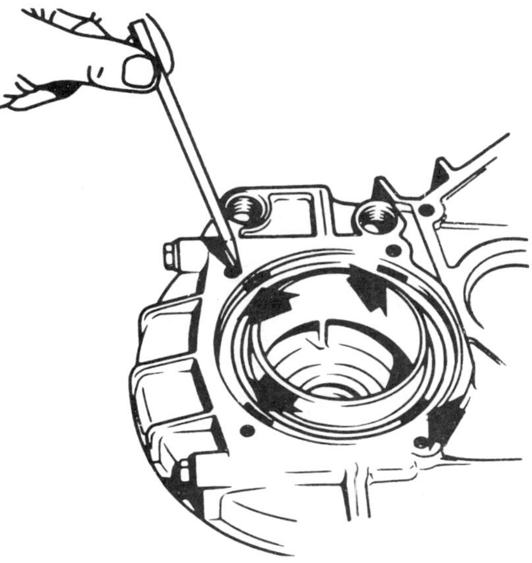

Fig. 6.21. Location of pieces of soft solder used for measuring bearing pre-load

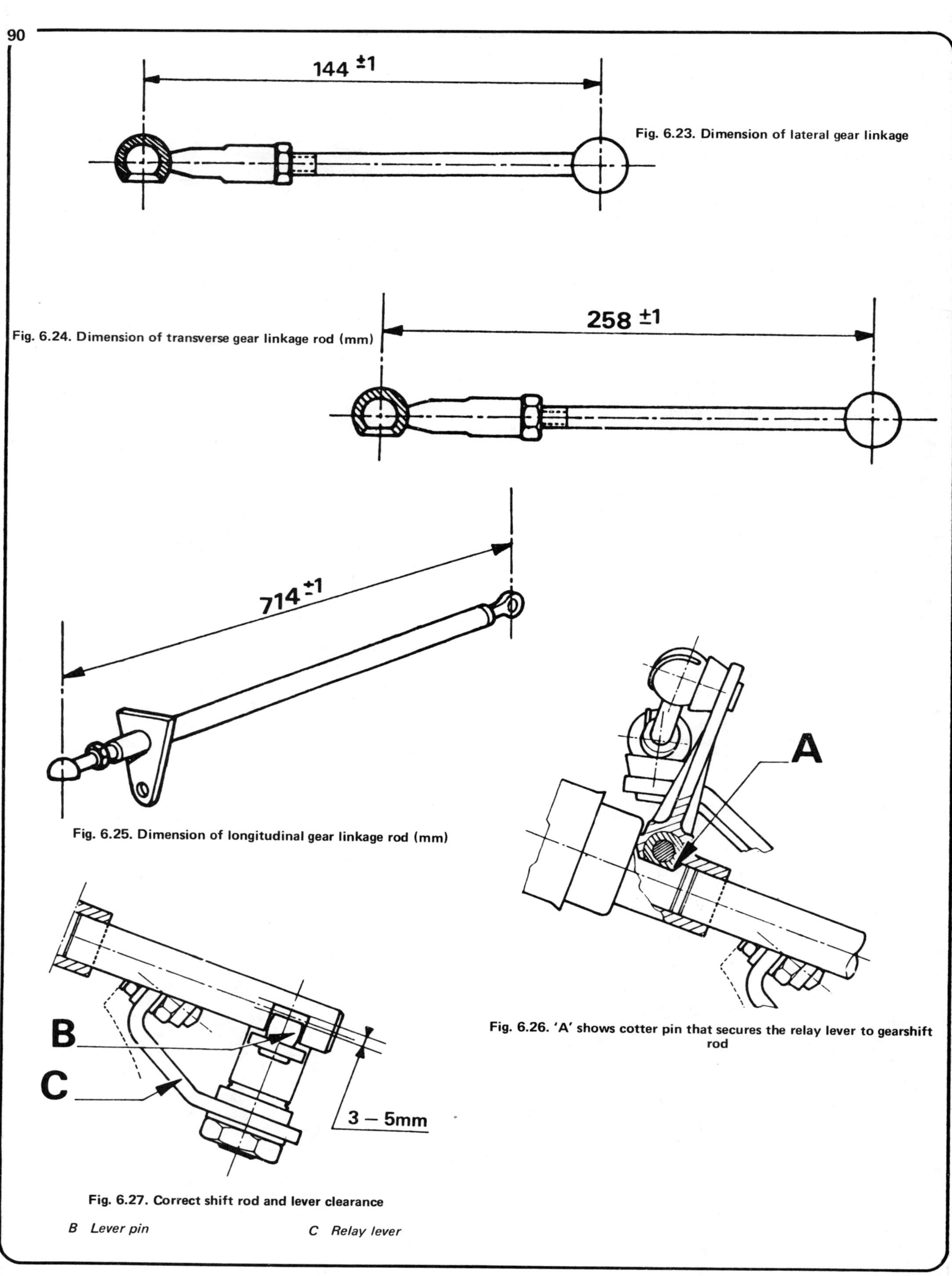

Fig. 6.23. Dimension of lateral gear linkage

Fig. 6.24. Dimension of transverse gear linkage rod (mm)

Fig. 6.25. Dimension of longitudinal gear linkage rod (mm)

Fig. 6.26. 'A' shows cotter pin that secures the relay lever to gearshift rod

Fig. 6.27. Correct shift rod and lever clearance

B Lever pin *C Relay lever*

9 Place four lengths of 1 mm diameter soft soldering wire (or Plastigage) at equal points on top of the original shims, bolt the thrust plate (without 'O' ring seal) into position, tightening the bolts to 16 lb f ft (2.21 kg fm).
10 Remove the thrust plate, measure the thickness of the crushed wire with a micrometer and this thickness plus 0.04 inch (0.1 mm) will be the required **additional** shim pack thickness to provide the necessary bearing pre-load. Shims are available in a range of thicknesses to suit most requirements. The next higher thickness should be used however where an exact tolerance cannot be matched precisely by a combination of shim thicknesses.
11 Locate the shim pack in position.
12 Fit a new 'O' ring to the thrust plate (photo).
13 Bolt the thrust plate to the housing, tightening the bolts to a torque of 16 lb f ft (2.21 kg fm).
14 The final drive (differential) unit is now ready for fitting to the vehicle as described in Section 6, followed by the gearbox, Section 2.

10 Gearshift lever and mechanism - adjustment

1 The gearshift lever is mounted in a ball housing bolted to the car floor (see Fig. 6.22 and photo).
2 Movement of the lever is transmitted to the gearbox through a longitudinal control rod and a series of levers and rods as shown in Fig. 6.18.
3 The balljoints have a nylon insert and can be prised apart after first removing the wire clip where fitted.
4 If gearchanging is difficult, and the angle of the lever appears to be incorrect or new linkage components are being fitted, check the length of the lateral rods 'E' and 'J' (Fig. 6.18).
5 If necessary remove the rod(s) and adjust the length by slackening the locknut and turning the balljoint end in the required direction. Tighten the locknut (Fig. 6.23, 6.24 and 6.25).
6 Check that the cotter pin securing the lever to the shift rod protruding from the gearbox (Fig. 6.26) is tight and there is no free movement between rod and lever.
7 Set the lever in the neutral position and check that there is a clearance of 1/8 to 3/16 in. (3 to 5 mm) between the slot in the end of the shift rod and the pin on the relay lever (Fig. 6.27).
8 If necessary obtain the correct clearance by slackening the nuts securing the relay bracket to the final drive casing and moving the bracket up or down as necessary. Tighten the nuts when the adjustment is correct.

11 Fault diagnosis - gearbox and final drive

Symptom	Reason/s	Remedy
Weak or ineffective synchromesh	Synchronising springs worn, split or damaged	Dismantle and overhaul transmission unit. Fit new gear wheels and synchronising cones.
	Synchromesh dogs worn, or damaged	Dismantle and overhaul transmission unit. Fit new synchromesh unit.
Jumps out of gear	Broken gearchange fork rod spring	Dismantle and renew spring.
	Transmission unit coupling dogs badly worn	Dismantle transmission unit. Fit new coupling dogs.
	Selector fork rod groove badly worn	Fit new selector fork rod.
	Selector fork rod securing screw and locknut loose	Tighten securing screw and locknut.
Excessive noise	Incorrect grade of oil in transmission unit or oil level too low	Drain, refill, or top up transmission unit with correct grade of oil.
	Transmission bearings worn or damaged	Dismantle and overhaul transmission unit. Renew bearings.
	Gearteeth excessively worn or damaged	Dismantle, overhaul transmission unit. Renew gear wheels.
	Mainshaft thrust washer worn allowing excessive end play	Dismantle and overhaul transmission unit. Renew thrust washers.
Excessive difficulty in engaging gear	Clutch pedal adjustment incorrect	Adjust clutch pedal correctly.
	Gearshift relay rod incorrectly adjusted	Adjust length of relay rods to the correct dimensions.

Chapter 7 Front suspension; driveshafts and hubs

For modifications, and information applicable to later models, see Supplement at end of manual

Contents

Specifications

Suspension type ... Torsion bar, anti-roll bar and telescopic hydraulic shock absorbers

Driveshafts

Type ... Three piece solid shaft comprising inner, intermediate and stub axle sections.

Driveshafts joints

Inner joint ... Constant velocity type
Outer joint ... Universal type

Lubricant capacity

CV joint (inner) ... 5 oz (145 g)
Universal joint (outer) ... 10 oz (280 g)

Torsion bar identification

Right-hand bar ... Orange paint mark
Left-hand bar ... White paint mark

Torque wrench settings

	lb f ft	kg f m
Shock absorber, lower mounting	18	2.49
Shock absorber, upper mounting	11	1.52
Hub nut, driveshaft	144	19.91
Anti-roll bar to tie-rod	7	0.97
Upper suspension arm to crossmember	37	5.12
Bump stops to lower arm	11	1.52
Brake disc to hub	37	5.12
Stiffening bracket to body	37	5.12

Stiffening bracket to crossmember	15	2.07
Brake caliper to stub axle	55	7.61
Spindle nuts, lower suspension arms	55	7.61
Spindle nuts, upper suspension arms	41	5.67
Torsion bar adjusting lever to crossmember	55	7.61
Anti-roll bar bush to side member	18	2.49
Lower balljoint nut to arm	55	7.59
Lower balljoint to stub axle	250	34.58
Upper balljoint to stub axle	26	3.59
Slotted bearing nut in stub axle	214	29.59
Lower crossmember to body	80	11.04
Upper crossmember to body	37	5.11
Sub-frame to body	33	4.56
Torsion bar anchor crossmember to body	15	2.07
Torsion bar hub fork to crossmember	11	1.52

1 General description

The front suspension on the Alpine is of the independent double wishbone type with longitudinal torsion bars and telescopic dampers.

The upper suspension arms pivot on a pressed steel crossmember which is bolted to the longitudinal frame members.

The lower suspension arms pivot on spindles attached to a sub-frame comprising two tubular crossmembers and two pressed steel sidemembers.

The stub axle carriers swivel on upper and lower balljoints which are lubricated and sealed for life.

The front end of each torsion bar fits into a hexagonal socket on the inner ends of the lower suspension arms while the rear ends are located in the centre crossmember. Adjustable arms on the ends of each torsion bar sleeve enable the loading on the torsion bars to be altered to reset the vehicle ground clearance.

An anti-roll bar is attached to the frame by rubber bushes and the outer ends are connected to each lower suspension arm by tie rods.

Each driveshaft has a constant velocity joint at the inner end and a universal joint at the outer end. The joints are lubricated for life and providing the rubber gaiters are maintained in good condition the joints should last a considerable time before renewal is necessary.

2 Driveshaft - removal and refitting

1 The method of removing and refitting the driveshafts is given in Chapter 1, Sections 4 and 40 respectively.

2 It must be emphasised that care must be taken to prevent the lower suspension arm dropping right down due to the twisting effort of the pre-loaded torsion bars. Although the lower arm can be supported using a jack as described in Chapter 1, it is recommended that the telescopic damper is removed and the special restraining tool, (Chrysler Part No. CF 0004) is fitted in its place (Fig. 7.2).

Pre-1985 models

3 If both driveshafts are being removed at the same time, insert special tool 80317M or two pieces of tubing into the differential side gears to prevent these gears being displaced.

3 Driveshaft joints - servicing

(a) Inner constant velocity joint

1 Secure the driveshaft in a vice and remove the circlip which retains the wider mouth of the gaiter (Fig. 7.3).

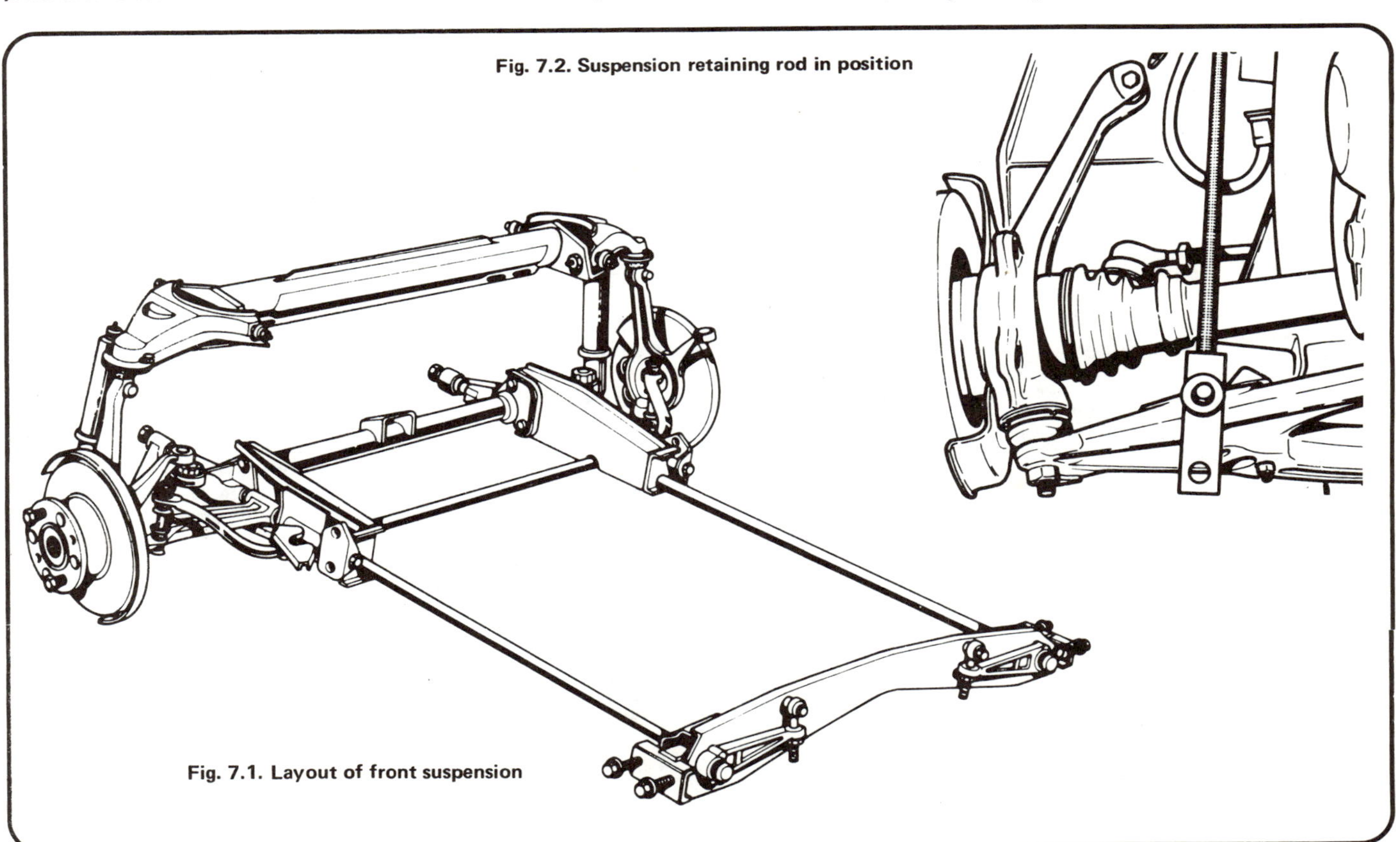

Fig. 7.2. Suspension retaining rod in position

Fig. 7.1. Layout of front suspension

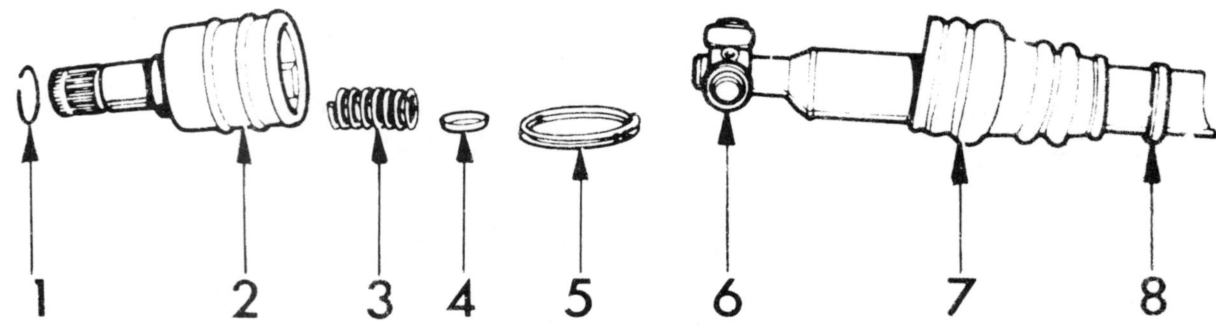

Fig. 7.3. Constant Velocity joint components

1 Spring
2 Metal cap

3 Spring
4 Spring seat

5 Gaiter retaining ring
6 Spider and roller assembly
7 Gaiter
8 Gaiter retaining ring

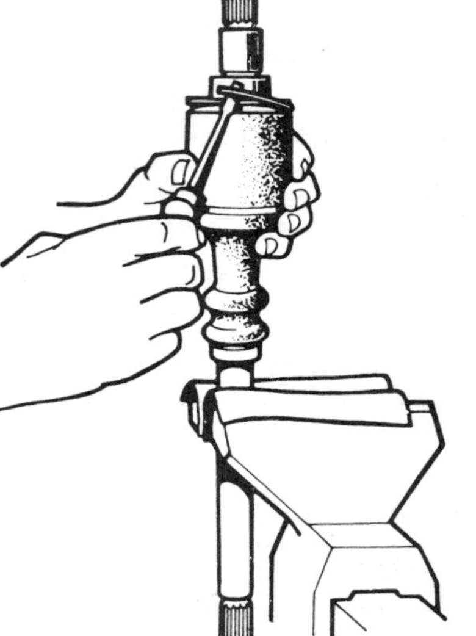

Fig. 7.4. Removing the spring ring securing the driveshaft rubber gaiter

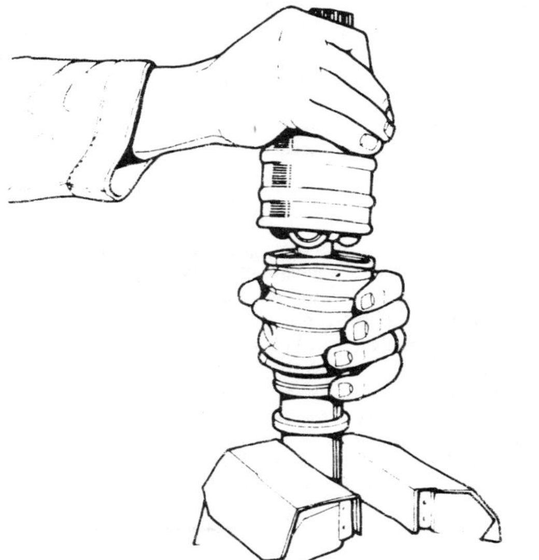

Fig. 7.5. Withdrawing the yoke from the rubber gaiter

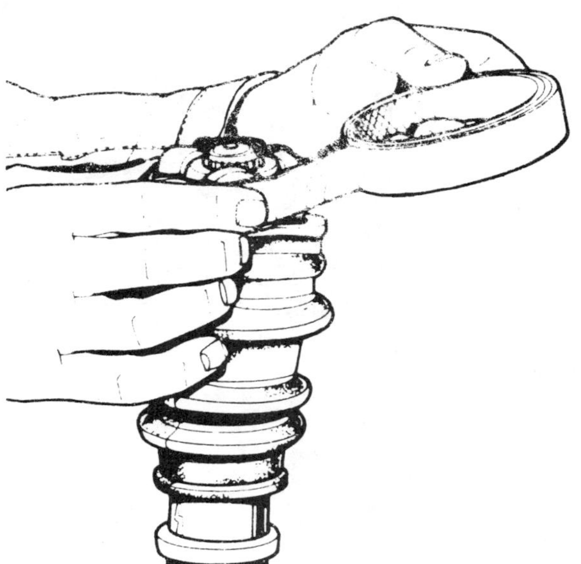

Fig. 7.6. Taping the driveshaft joint bearings

2 Draw the tulip shaped yoke upwards and remove it (Fig. 7.5).
3 Remove the spring and cap and then wipe out as much grease as possible from the joint assembly.
4 Fit masking or insulating tape round the ends of the three joint trunnions to prevent the bearings becoming dislodged (Fig. 7.6). **Note:** The inner constant velocity joint spider and the outer universal joint spider must be offset from each other by $57^\circ \pm 3^\circ$ as shown in Fig. 7.7. The position of the spider in relation to the shaft should therefore be marked as shown in Fig. 7.8 before removing it.
5 Using a press and adaptor tool No. 20937M or similar, press the splined shaft through the spider assembly (Fig. 7.9).
6 Obtain a repair kit which will contain all renewable components except the tulip shaped yoke and intermediate shaft. Check all renewable parts for wear, damage, or corrosion, and renew if necessary.
7 Hold the intermediate shaft vertically in a vice. During the assembly it is necessary to make sure that every part is clinically clean.
8 Fit the new rubber retaining ring and sealing gaiter to the shaft.
9 Fill the gaiter and the tulip shaped yoke, complete with its metal cover, with grease.
10 Secure the new needle roller bearings on the new joint tripod and retain them with tape as previously described.
11 Fit the joint spider complete with bearings to the intermediate splined shaft, using a tubular drive, Fig. 7.10.
12 Peen the end of the intermediate shaft at three equidistant points to secure the joint tripod in position.
13 Remove the temporary retaining tape from the needle bearings.
14 Fit the cap and spring onto the curved end of the intermediate shaft.
15 Insert the tulip shaped yoke and metal cover into position in the bellows (Fig. 7.11).

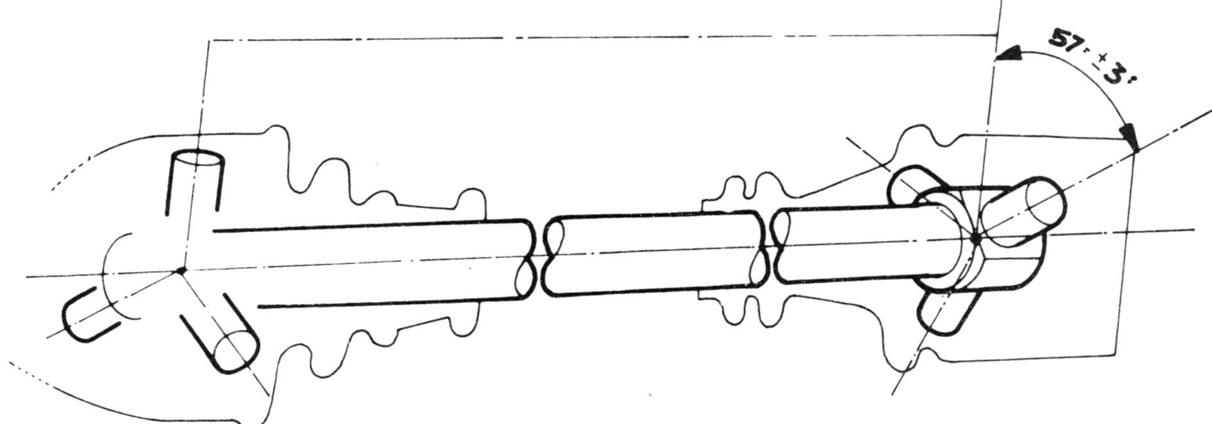

Fig. 7.7. Correct degree of offset for driveshaft inner and outer splines

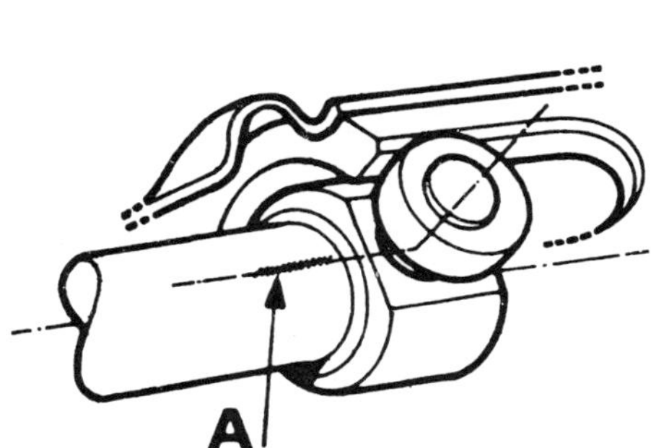

Fig. 7.8. Spider alignment mark

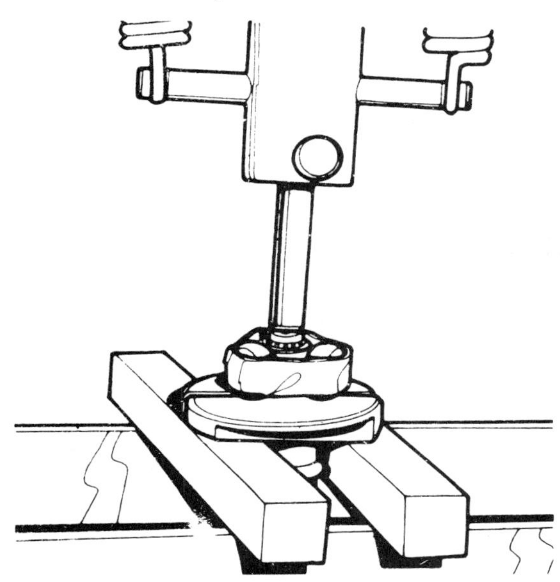

Fig. 7.9. Pressing the driveshaft from the spider

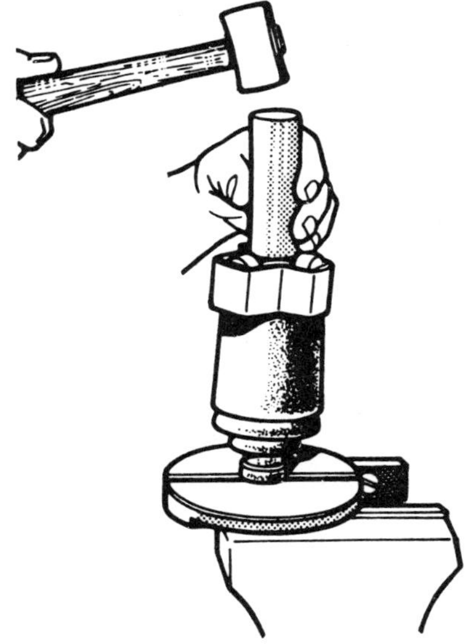

Fig. 7.10. Drifting a new spider onto the shaft

Fig. 7.11. Refitting the driveshaft yoke and cover into the rubber gaiter

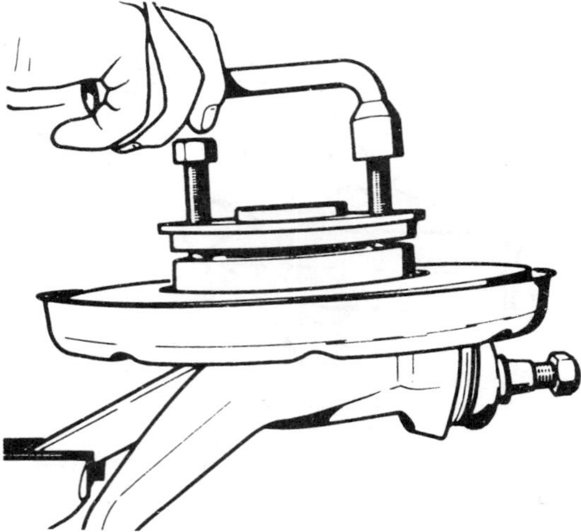

Fig. 7.12. Extracting the front wheel hub assembly

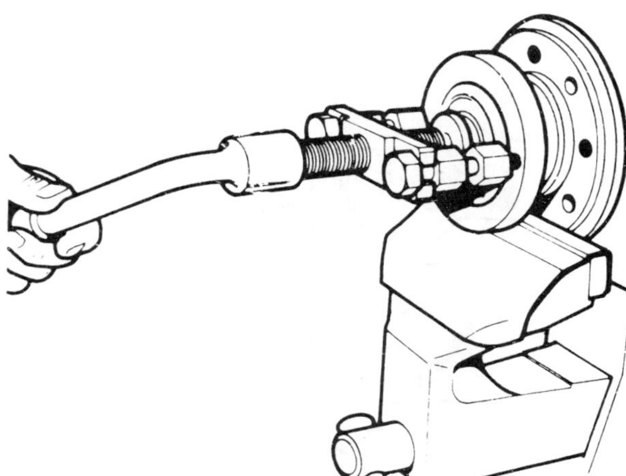

Fig. 7.13. Extracting the hub bearing ring

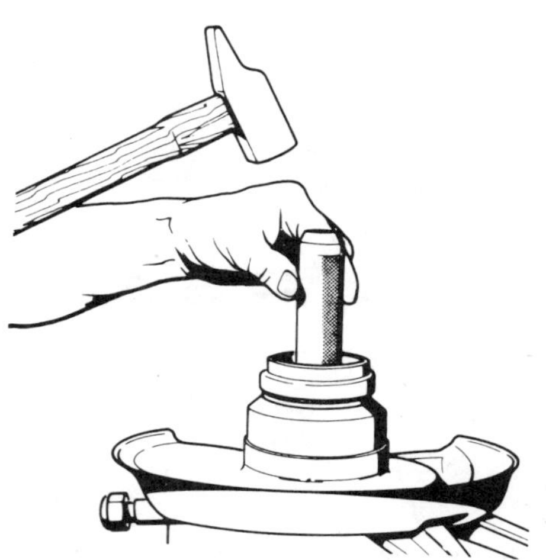

Fig. 7.14. Fitting a new oil seal to the stub axle carrier

16 Fit the smaller mouth of the bellows into the groove in the intermediate shaft and secure with the rubber retaining ring.
17 Check the movement of the assembled joint for tight spots.
18 Where the rubber bellows only are to be renewed then they and the rubber retaining ring should be cut off. Without any further dismantling, the new bellows and ring may be slid into position.

(b) Outer universal type joint

19 The outer, (roadwheel end) universal type joint is not repairable and if defective must be renewed.
20 If the driveshaft has covered a considerable mileage without any repair work being carried out on the inner constant velocity joint, the most sensible policy is to obtain a complete new driveshaft assembly from your Chrysler Dealer.
21 Alternatively, if the constant velocity joint is known to be in good condition, only the outer universal joint complete with intermediate shaft need be renewed.

4 Hubs and outer bearings - dismantling and reassembly

1 Disconnect the outer splined end of the driveshaft from the hub assembly using the method described for driveshaft removal in Chapter 1, Section 4. It is not necessary to withdraw the inner end of the shaft from the differential unit.
2 Do not forget to temporarily refit the lower balljoint securing nut.
3 Remove the disc brake caliper as described in Chapter 9.
4 Unscrew and remove the four bolts which secure the brake disc to the wheel hub.
5 Rotate the brake disc just enough so that two suitable bolts may be screwed into the hub (roadwheel fixing) bolts holes and by bearing against the brake disc with equal pressure extract the hub (Fig. 7.12).
6 Withdraw the hub together with the oil seal and bearing.
7 Remove the brake disc and then pick out the bearing balls from their plastic cage and then remove the cage.
8 Pull the bearing ring from the hub with the aid of a suitable extractor (Fig. 7.13).
9 Fit the new bearing balls and their plastic cage to the bearing ring. Pack with wheel bearing grease and then fit the assembly to the stub axle carrier.
10 Fit a new oil seal (lip upwards) into the stub axle carrier, using a suitable tubular drift (Fig. 7.14).
11 Refit the brake disc to the hub, tightening the securing bolts to the specified torque wrench setting.
12 Using a suitable bolt and plates, draw the disc/hub assembly into position on the stub axle carrier (Fig. 7.15).
13 Fit the driveshaft into the stub axle carrier and reconnect the lower balljoint and the track rod to steering arm balljoint.
14 Refit the brake caliper ensuring that a new locking plate is used.
15 Fit a new hub nut, tightening it onto the thrust washer to a torque of 145 lb f ft (20.05 kg f m). Peen the nut.
16 Refit the telescopic damper, refit the roadwheel and lower the car.

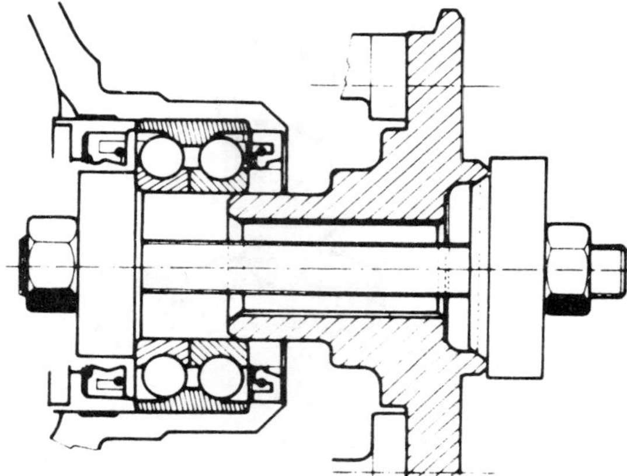

Fig. 7.15. Drawing the hub and disc assembly into the stub axle carrier

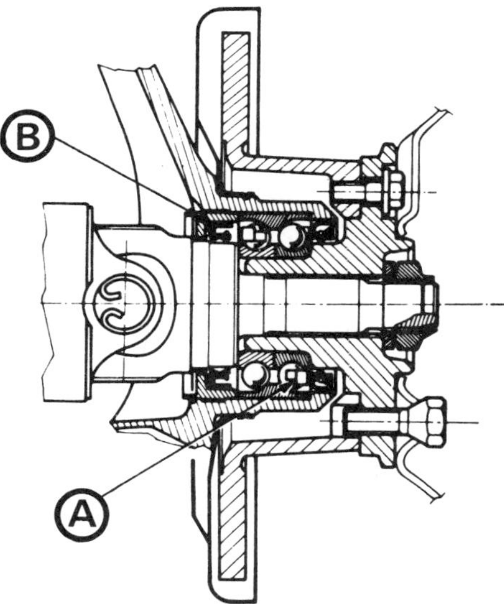

Fig. 7.16. Cross-section through front hub assembly

A Outer bearing *B Inner bearing*

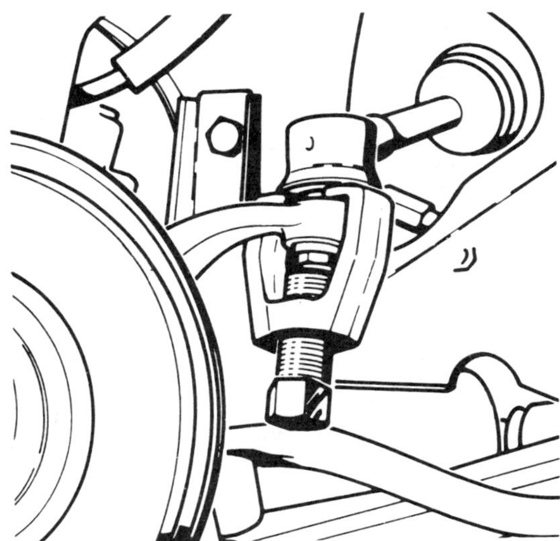

Fig. 7.17. Disconnecting the steering arm ball joint using an extractor

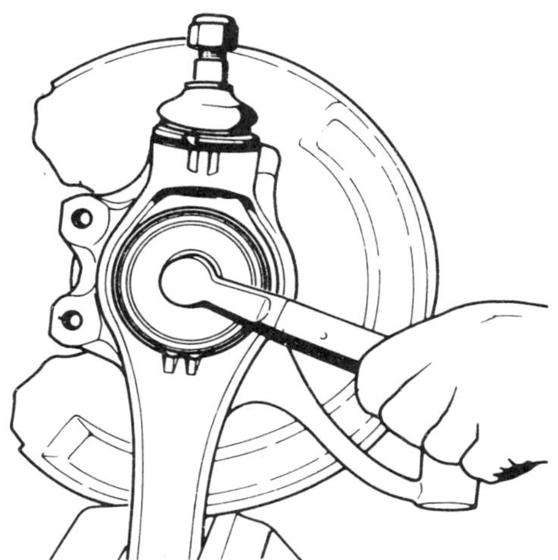

Fig. 7.18. Using special tool to remove the sleeve nut from rear of axle carrier

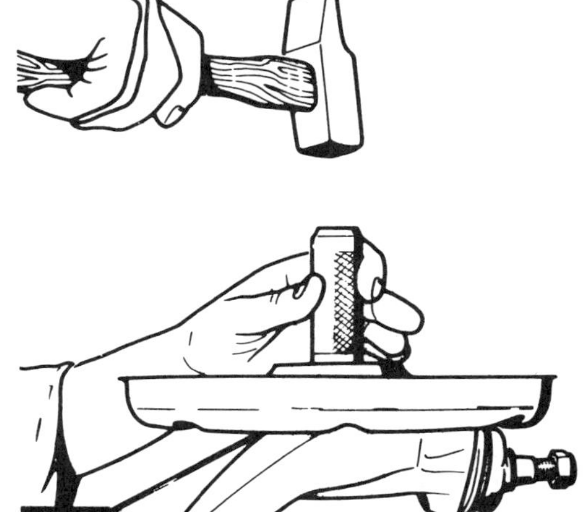

Fig. 7.19. Driving out the inner hub bearing

5 Inner hub bearings - dismantling and reassembly

1 These bearings are retained within the stub axle carrier by a screwed sleeve and it is recommended that the stub axle carrier be removed from the vehicle if the inner bearings are to be renewed. Location of the inner and outer hub bearings are shown in Fig. 7.16.
2 First slacken the roadwheel bolts and the large centre hub nut.
3 Jack up the car, support it on axle stands and remove the front wheel.
4 Remove the brake caliper as described in Chapter 9.
5 Remove the telescopic damper and fit the suspension restraining tool (Part No. CF 0004) in its place. Refer to Chapter 1, Section 4 if necessary.
Note: This tool **must** be used for this operation. Do not attempt to support the lower suspension arm using only a jack as described for driveshaft removal.
6 Disconnect the upper and lower stub axle carrier balljoints using a suitable extractor as described in Sections 6 and 7 of this Chapter.
7 Disconnect the steering arm balljoint using an extractor (Fig. 7.17)..
8 Remove the large centre driveshaft nut, hold the driveshaft in place to avoid stretching the rubber gaiters and withdraw the complete hub, disc and axle carrier assembly from the shaft.
9 Relieve the peening on the threaded sleeve at the rear of the stub axle carrier and unscrew the sleeve nut which will require the use of removal tool 20908H or one made up to perform the operation (Fig. 7.18).
10 Using a tubular drift, drive the inner bearing from the stub axle carrier (Fig. 7.19).
11 Before fitting the new bearing, pack it with the recommended grease. Pull it into position in the stub axle carrier by using a bolt and two plates as shown in Fig. 7.20.
12 Use a new threaded sleeve as the required peening on refitting makes the original unsuitable for re-use.
13 Fit a new oil seal into the threaded sleeve and screw it and its nut into position using the fitting tool. The lip of the oil seal will face inwards when correctly positioned, and the sleeve nut must be tightened to a torque of 214 lb f ft (29.59 kg f m).
14 Lock the sleeve in position by peening it at two points.
15 Carry out the operations described in paragraphs 9 to 14 in Section 4.
15 Reconnect the stub axle carrier upper balljoint.

6 Stub axle carrier lower balljoint - renewal

1 Remove the stub axle carrier as described in Section 5.
2 Place the carrier in a vice with the balljoint uppermost.

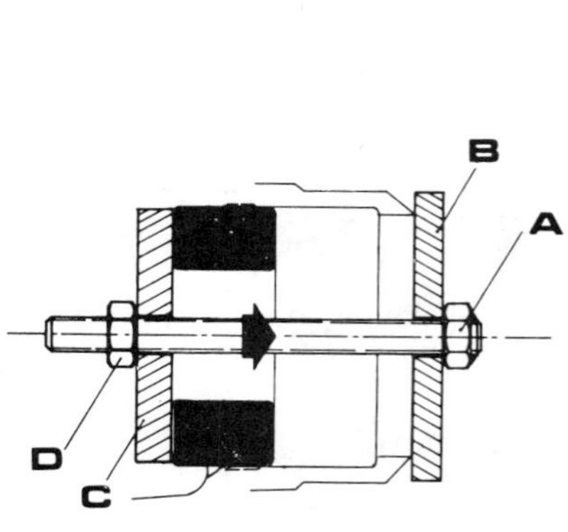

Fig. 7.20. Pulling the inner hub bearing into position using plates (B and C), a threaded bolt and nuts (A and D)

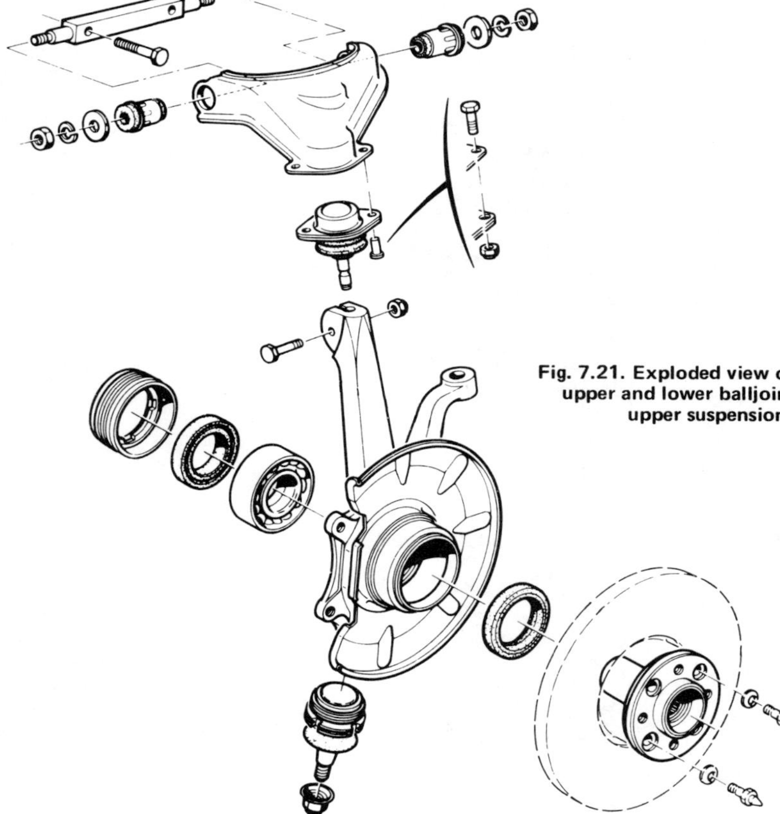

Fig. 7.21. Exploded view of axle carrier, upper and lower balljoints and upper suspension arm

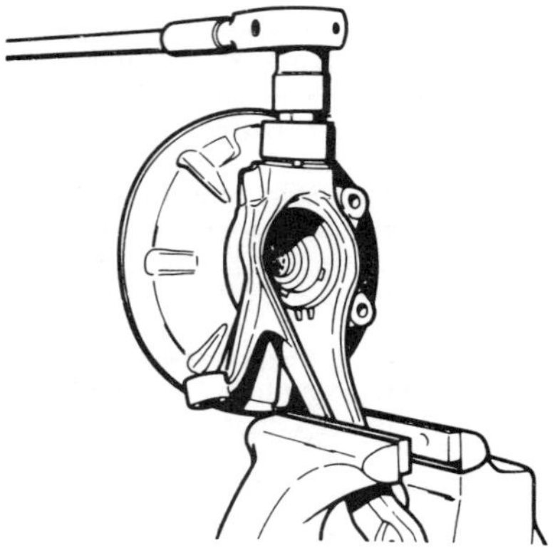

Fig. 7.22. Removing the lower balljoint using special tool

3 Relieve the peening at the balljoint slotted ring.
4 A removal tool must now be employed (21811C and CJ0019) and it should be positioned and secured by the balljoint nut (Fig. 7.22).
5 Unscrew and remove the balljoint.
6 Fitting the new balljoint is a reversal of removal but take care not to pinch or tear the rubber cover. Tighten the slotted securing ring to a torque of 250 lb f ft (34.58 kg f m).
7 Remove the tool and peen the ring at one notch to lock it.
8 Refit the stub axle carrier as described in Section 5.

7 Upper suspension arm balljoint - renewal

1 Refer to Fig. 7.21. Remove the damper and fit the restraining rod to offset the torque of the torsion bar.
2 Jack up the front of the vehicle and support it on stands. Remove the roadwheel.
3 Remove the nut from the cotter pin that secures the balljoint pin to the axle carrier (Fig. 7.23) and drive out the cotter pin.
4 Drill out the rivets securing the top of the balljoint assembly to the upper suspension arm and withdraw the balljoint pin from the axle carrier.
5 The new balljoint will be supplied as a kit complete with nuts and bolts as replacement for the rivets.
6 Fitting is a reversal of dismantling but ensure that the balljoint securing nuts are located on the top surface of the upper suspension arm and are correctly locked.

8 Shock absorber - removal, testing and refitting

1 The telescopic type hydraulic shock absorber cannot be repaired and in the event of evidence occurring of bad cornering, steering wander or an unusually soft ride then the units should be removed from the car and tested.
2 Disconnect the upper and lower mountings, details of which are shown in Fig. 7.24.
3 Secure the lower damper mounting in a vice and in the vertical position. Operate the damper for the full length of its travel ten times. There should be good resistance in both directions of travel. If the action is jerky, or there is no resistance at all, renew the unit.
4 Refitting is a reversal of removal but note carefully the fitting sequence of the mountings.

9 Anti-roll bar - removal, inspection and refitting

1 The anti-roll bar is connected at each end to the lower suspension arm by a long bolt, nut and bush assembly as shown in Fig. 7.25.

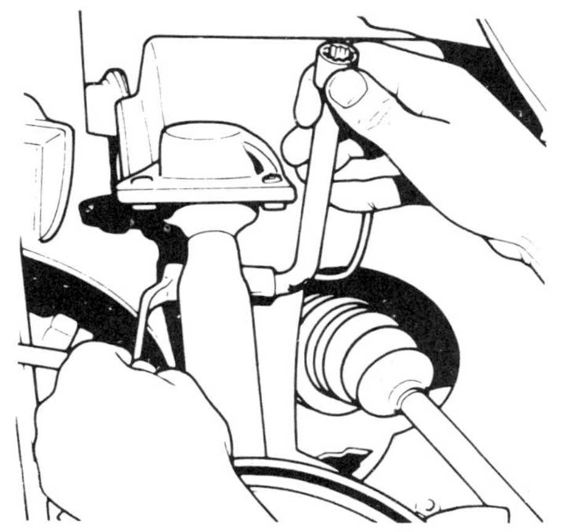
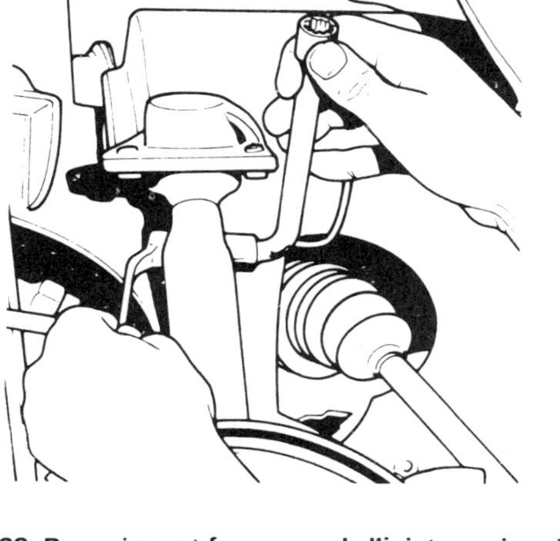

Fig. 7.23. Removing nut from upper balljoint securing pin

2 It is also supported by rubber bushes retained by semi-circular clamps.

3 Removal of the bar is carried out by withdrawing the anchor bolts and the rubber bush clamps.

4 Inspect the bar closely for cracks and also the rubber mounting components for deterioration. Renew as appropriate.

5 Refitting is a reversal of removal but do not exceed a tightening torque for the anchor or clamp nuts of 18 lb f ft (2.48 kg f m).

10 Torsion bars - removal and refitting

1 Jack up the front of the car and place axle stands beneath the crossmember or jacking points so that the suspension arms are suspended.

2 The tension of the torsion bar must now be relieved by using the special Chrysler tool No. 20916Q. As the bars are under considerable tension it is unwise to try and make do with a lever or similar.

3 Fit the special tool on the rear end of the torsion bar sleeve located in the centre crossmember and push the tool handle upwards to release the load on the adjusting rod and remove the adjusting rod pivot bolt.

4 Slowly allow the torsion bar lever to swing down and remove the tool (Fig. 7.27).

5 Remove the bolt securing the sleeve retaining fork to the rear of the crossmember and remove the fork (Fig. 7.28).

6 Pull the sleeve and lever assembly rearwards off the end of the torsion bar (Fig. 7.29).

7 Withdraw the torsion bar from the lower suspension arm by moving it to the rear and then remove it from the car by pulling it forward out of the crossmember. Recover the rubber thrust washer.

8 Refitting is a reversal of removal but grease the splines of the torsion bar and the hubs into which it locates with wheel bearing type grease. The left-hand torsion bar is identified by a white paint spot, the right-hand by an orange paint spot. The torsion bars are not interchangeable. Never mark a torsion bar by scratching or filing as this may cause premature failure.

9 With the torsion bars correctly fitted in the vehicle, the tensioning arms should be fitted to the torsion bars at the angle shown in the diagram (Fig. 7.29).

10 A template should be used to obtain this initial setting and then the procedure described in the following Section must be carried out.

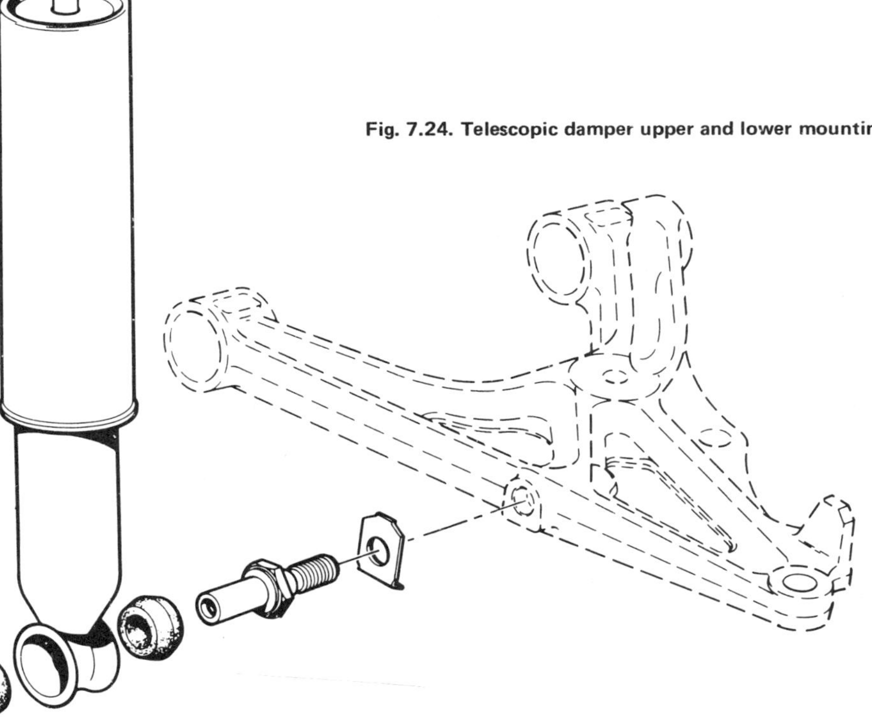

Fig. 7.24. Telescopic damper upper and lower mountings

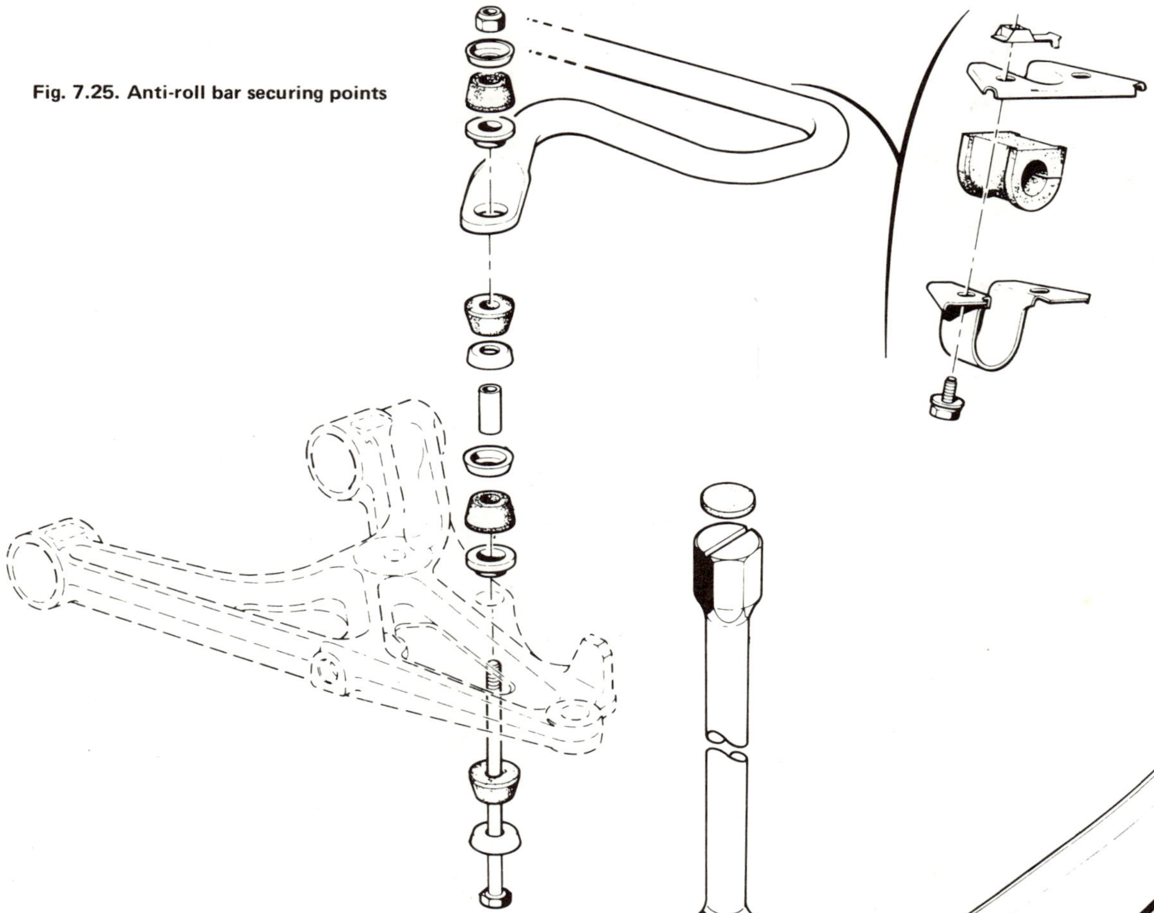

Fig. 7.25. Anti-roll bar securing points

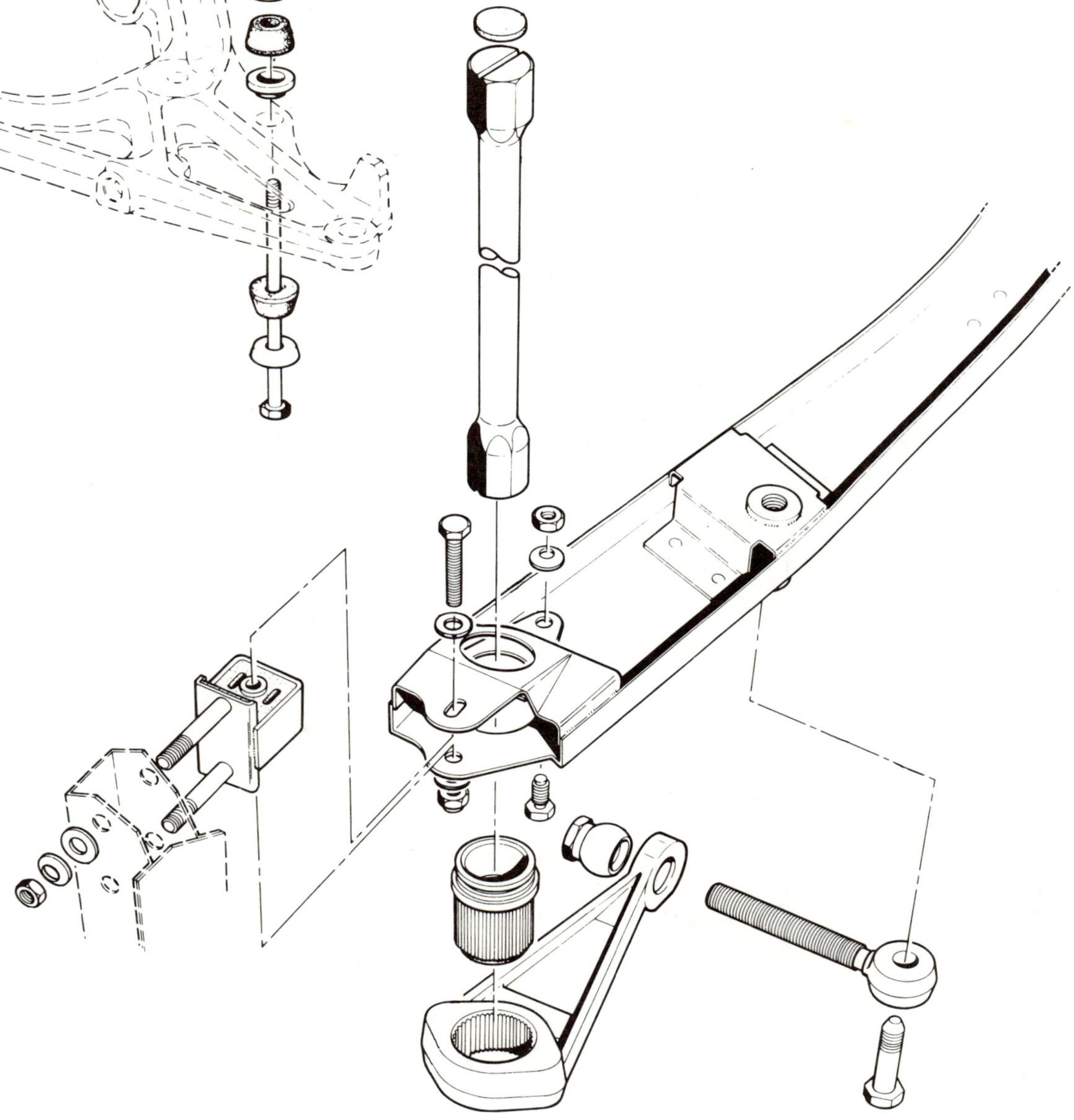

Fig. 7.26. Exploded view of torsion bar rear attachment point

11 Torsion bars - setting

1 If the torsion bars have been removed or renewed they must be set to the correct tension to ensure the safe handling characteristics of the car are maintained.
2 A special gauge bar (Part No. 21818K) and height stands, (Part No. CJ0017 for right-hand drive cars and Part No. 20917 for left-hand drive cars) are required when setting the height of the vehicle, and it is essential that after the torsion bar(s) have been fitted as described in the previous Section, the car is taken to the local Chrysler Dealer to have the car's height correctly set.
3 For the owner who is able to hire or borrow these tools, the method of setting the car's height is as follows.
4 Place the car on a level surface with the tyres correctly inflated and the fuel tank full.
5 Set the steering in the straight ahead position and disconnect both lower telescopic damper mountings.
6 Slacken the pivot bolts on the torsion bar adjusting rods.
7 Turn the adjusting nuts anticlockwise until they are flush with the ends of the adjusting rods.
8 Fit the gauge bar beneath the front of the car as far forward as possible with the brackets hooked over the torsion bars (Fig. 7.31).
9 Place the height stands on either side of the gauge bar and bounce the front of the car several times to settle it.
10 Turn each adjusting nut alternately two turns at a time to raise the car until both sides of the gauge bar are exactly level with the groove in each height stand (Fig. 7.30).
Caution: If, before commencing adjustment, the gauge bar is higher than the height stands even after bouncing the vehicle; do **not** slacken the adjusting nuts any further in an attempt to lower the car. Leave the adjustment as it is and re-check the height after the car has covered a small mileage and the suspension has settled.
11 When the height is satisfactory, tighten the adjusting rod pivot nuts, remove the height gauge bar from beneath the vehicle and reconnect the telescopic dampers.

12 Lower suspension arm - servicing

1 If wear is evident in the lower suspension arm bushes or spindle, then the assembly must be removed and the worn components renewed. Jack up the car and remove the wheel.
2 Refer to Fig. 7.30 and note the sequence of fitting of the various components **before** dismantling.
3 Remove the torsion bar as described in Section 10.
4 Disconnect the anti-roll bar at its attachment to the lower suspension arm. Also disconnect the mounting on the side being worked on.

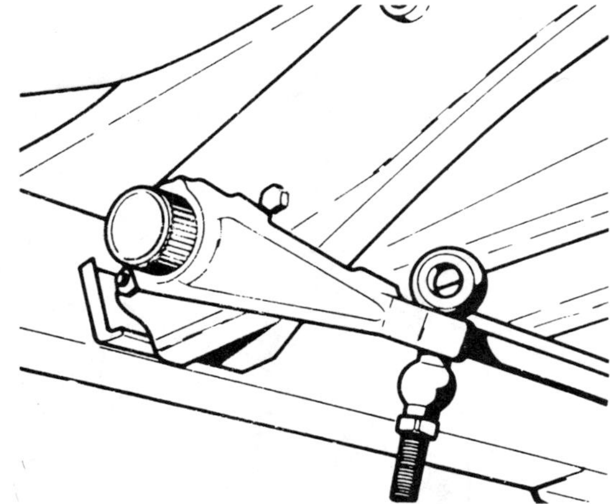

Fig. 7.27. Torsion bar adjusting rod removed

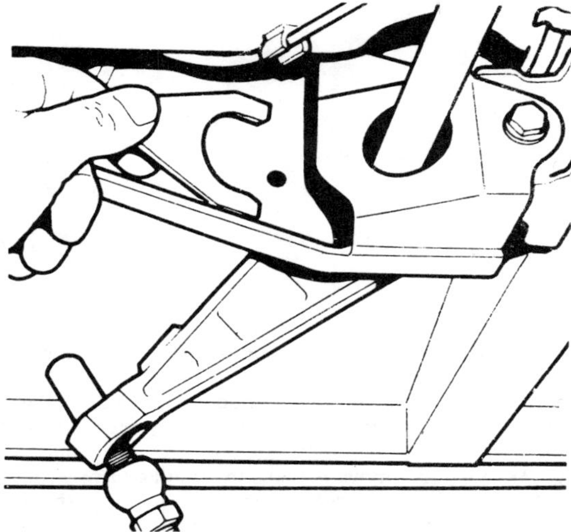

Fig. 7.28. Removing the torsion bar sleeve retaining plate from crossmember

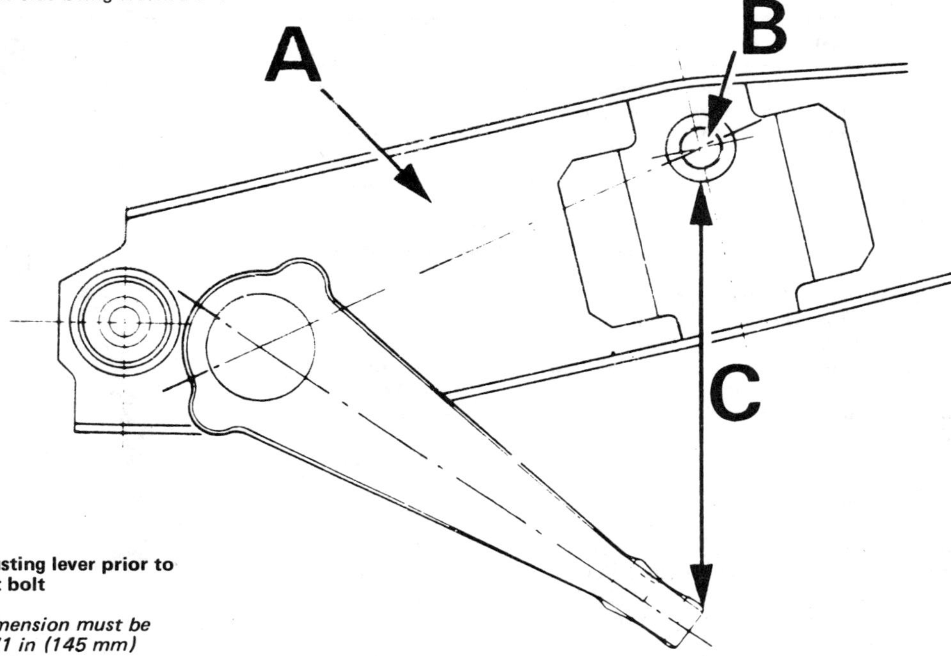

Fig. 7.29. Correct position of torsion bar adjusting lever prior to refitting the adjusting rod pivot bolt

A Crossmember
B Pivot bolt hole
C Dimension must be 5.71 in (145 mm)

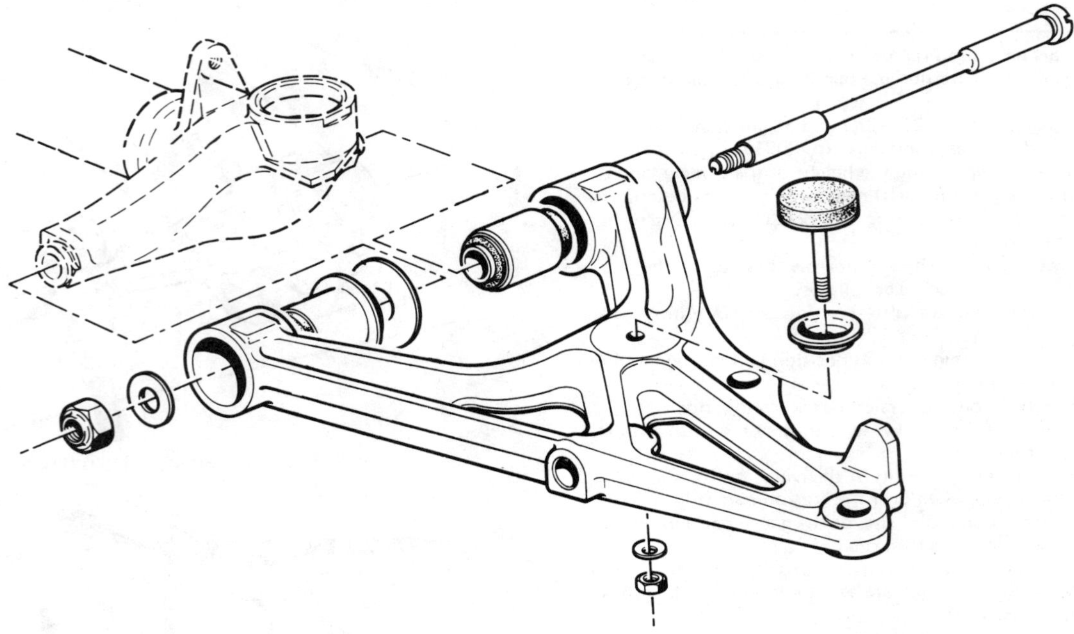

Fig. 7.30. Lower suspension arm and bushes

5 Disconnect the stub axle carrier lower balljoint as described in Section 6.
6 Unscrew and remove the nut from the spindle, withdraw the spindle, remove the bush components and lift away the suspension arm.
Note: If, while trying to remove the nut, the spindle turns with it, remove the torsion bar thrust washer from the rear end of the suspension arm and hold the spindle with a screwdriver (Fig. 7.33).
7 Refitting is a reversal of removal but tighten the spindle nut to the specified torque wrench setting.
8 Fit and set the torsion bar as described earlier in this Chapter.

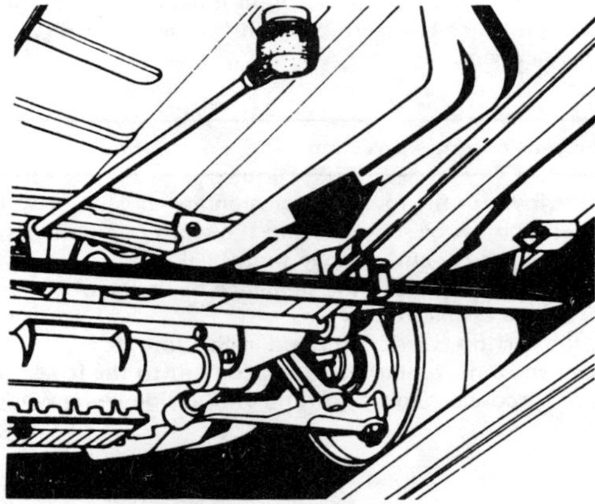

Fig. 7.31. Correct location of height gauge

13 Upper suspension arm - servicing

1 It will be necessary to remove the upper suspension arm when there is evidence of wear in the swivel bushes or pin. Its removal will also make easier the renewal of the balljoint assembly described in Section 7.
2 Remove the shock absorber and fit the restraining rod previously described.
3 Jack up the front of the vehicle and support it on stands.
4 Place a restraining strap round the stub axle carrier and secure it to the damper substitute rod.
5 Disconnect the upper suspension arm to stub axle carrier balljoint.
6 Remove the two bolts which secure the upper suspension arm to the upper crossmember (Fig. 7.34).
7 As the upper suspension arm is withdrawn, note carefully the location of the spacers and retain them for future refitting. These spacers are used to adjust the front wheel camber as described in Chapter 8.
8 Unscrew and remove the swivel pin nut and remove the various components.
9 Renew bushes and other parts as necessary.
10 Refitting is a reversal of removal. Tighten the swivel pin nut to a torque of 41 lb f ft (5.67 kg f m) and ensure that the spacers are returned to their original positions before tightening the upper suspension arm securing bolts to 37 lb f ft (5.12 kg f m).
11 Whenever the suspension has been removed and refitted, the steering geometry should be checked as described in Chapter 8.

Fig. 7.32. Setting the vehicle height against the special height stand

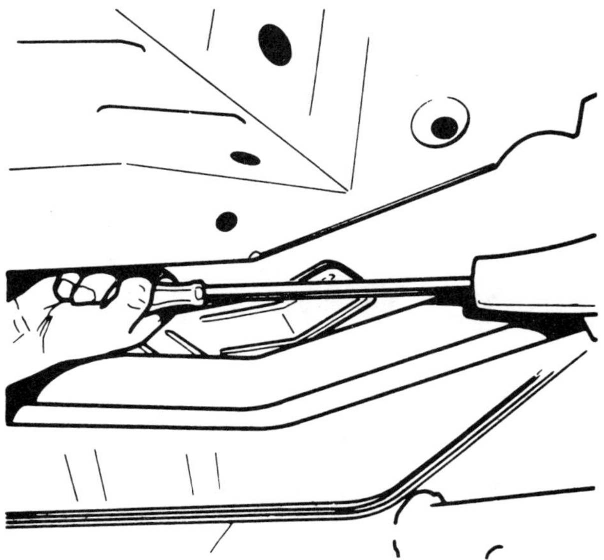

Fig. 7.33. Holding the slotted end of the lower suspension arm spindle

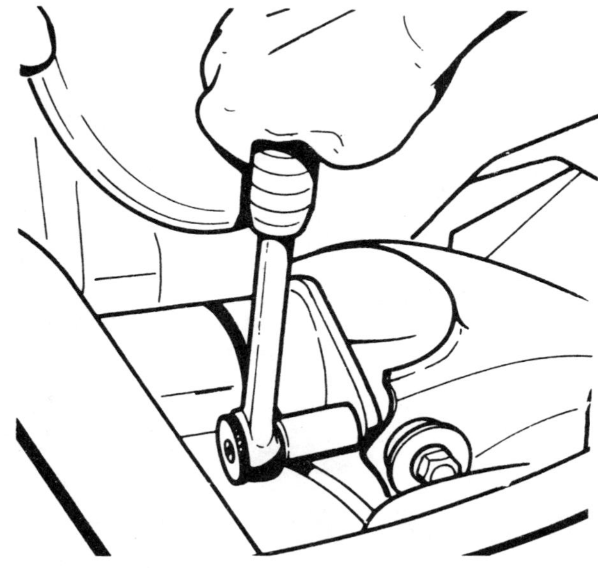

Fig. 7.34. Removing the bolts securing the upper suspension arm to the crossmember (LH side shown)

14 Upper suspension crossmember - removal and refitting

1 Substitute restraining rods for both front dampers (see Section 2).
2 Jack up the front of the vehicle and support on stands, and remove both front wheels.
3 Remove the nut from the cotter pin securing the upper balljoint to the axle carrier and drive out the cotter pin. Withdraw the balljoint from the axle carrier. Repeat the operation on the other side of the car.
4 Remove the air cleaner and hot air intake (See Chapter 3).
5 Remove the two bolts securing the right-hand side upper suspension arm to the crossmember, retain the camber adjustment shims.

Note: On left-hand drive cars, remove the left-hand side upper suspension arm retaining bolts.
6 Lift away the upper suspension arm and spindle assembly.
7 Detach any wires or pipes clipped to the crossmember.
8 Remove the bolts securing the crossmember to the two reinforcement brackets and to the longitudinal members.
9 Manoeuvre the crossmember, complete with left-hand side upper suspension arm, out under the left-hand side front wing. On left-hand drive cars the crossmember is withdrawn out under the right-hand side front wing.
10 Refit the crossmember using the reverse procedure to that of removal. If the suspension arm spindle nuts have been slackened do not retighten them until the weight of the car is on the front wheels.
11 Do not forget to refit the camber adjustment spacers.

15 Fault diagnosis - front suspension; driveshafts and hubs

Symptom	Reason/s	Remedy
General wear	Tyre pressures uneven	Check and adjust if necessary.
	Shock absorbers inoperative	Test and renew.
	Suspension balljoints worn	Disconnect and renew.
	Wheels out of balance	Balance wheels correctly.
	Hub bearings worn	Renew.
Poor road holding	Torsion bars, incorrectly set	Reset correctly.
	Incorrect steering geometry	See Chapter 8.
	Brakes binding on one side of vehicle	Dismantle and service (Chapter 10).

Chapter 8 Steering

For modifications, and information applicable to later models, see Supplement at end of manual

Contents

Specifications

Type ... Rack and pinion

Turning circle ... 36 ft (11 m)

Wheel alignment

Front wheel toe-out (vehicle unladen)	3/64 to 1/8 in (1 to 3 mm)
Camber	$0^o \pm 30'$
Caster	$2^o 30' \pm 30'$
Steering axis inclination	$15^o 20' \pm 10'$

Pinion shaft endfloat ... 0.004 in (0.1 mm)

Wheels and tyres

Roadwheels	Pressed steel (or alloy optional on some models), 5J13FH	
Tyres:		
Size	Radial 155SR13 or 165SR13 according to model	
Pressures (cold):	**Front**	**Rear**
Up to 4 occupants without luggage	26 lbf/in^2 (1.8 bar)	26 lbf/in^2 (1.8 bar)
4 occupants with luggage, or sustained high speed driving	28 lbf/in^2 (1.9 bar)	29 lbf/in^2 (2.0 bar)

Torque wrench settings

	lb f ft	**kg f m**
Steering ball joint nut	22	3.04
Lock nut, track rod	33	4.56
Universal joint to upper inner column	11	1.52
Steering outer column to support	18	2.49
Lock nut, rack damper	44	6.09
Pinion self-locking nut	18	2.49
Nut, coupling pinch bolt	7	0.97
Steering rack to frame	15	2.07
Securing nut, steering hand wheel	40	5.53
Shock absorber lower mounting, front	18	2.49
Shock absorber lower mounting, rear	18	2.49
Upper suspension arm pivot to frame	37	5.12

1 General description

The Alpine is fitted with the rack and pinion type steering gear.

Turning movement from the steering wheel is transmitted through a universally jointed shaft to the steering gear pinion.

The angle of the steering shaft and the two universal joints ensure that in the event of a frontal collision the shaft will collapse sideways rather than moving rearwards, thus reducing the possibility of chest injury to the driver.

A balljoint and tie-rod (or track rod) is attached to each end of the steering rod and these terminate in a second adjustable balljoint which enables the front wheel track to be correctly set.

The track rods are connected to the steering arms which are integral with the stub axle carriers.

Shims are used between the rack and pinion steering gear mounting points and brackets and these have a direct bearing upon the track setting during movements of the front suspension.

2 Maintenance

1 No routine maintenance is required but check that the rubber bellows on the steering gear assembly and the balljoints have not deteriorated or split. Should this be evident, then renew these components.

2 Should the steering gear become stiff after a high mileage then 25 cc of recommended grease should be introduced to the steering unit through the rack damper aperture, see Section 5.

3 Keep all connecting bolts tightened to the torque wrench settings given in the Specifications.

3 Steering gear - adjustment

1 If there is play in the steering or a knocking is felt through the steering wheel when driving over bumpy surfaces, then the steering rack probably requires adjustment.

2 Jack up the front of the car, support it on axle stands and remove both front wheels.

3 Remove the nuts from both track rod balljoints and, using a suitable extractor, detach the balljoints from the steering arms.

4 Release the damper plug lock nut.

5 Remove the nut and the used locking tab washer.

6 Now tighten the damper plug until it bears against the damper but do not force.

7 Mark the location of the damper plug in relation to the steering gear housing. Loosen the damper plug fractionally by 0.118 in (3.0 mm).

8 Turn the steering wheel slowly from lock to lock and check for tight spots. Should one be found then carry out the damper adjustment procedure with the steering remaining in the 'tight spot' position. Again check from lock to lock.

9 Fit a new locking tab ring, fit the locknut and tighten it to the specified torque. Bend over the locking tab.

10 Reconnect the track rods and lower the jacks.

4 Steering gear - removal and refitting

1 Jack up the front of the car and support securely on stands or blocks.

2 Remove the securing nuts and using a suitable extractor, disconnect both track rod balljoints from the steering arms.

3 Disconnect the gearshift linkage balljoints from the relay levers on the steering gear and final drive casing and remove the relay lever assembly from the steering gear (refer to Chapter 6 if necessary).

4 Remove the nut from the pinchbolt on the steering gear pinion coupling and extract the pinchbolt (see Fig. 8.2).

5 Remove the four bolts securing the steering gear to the mounting brackets (Figs. 8.3 and 8.4). Make a note of the number of shims between the steering gear housing and each bracket.

6 Remove the three securing bolts from the mounting bracket on the left-hand end of the steering gear and remove the bracket.

7 Slide the steering gear assembly through the left-hand side wing

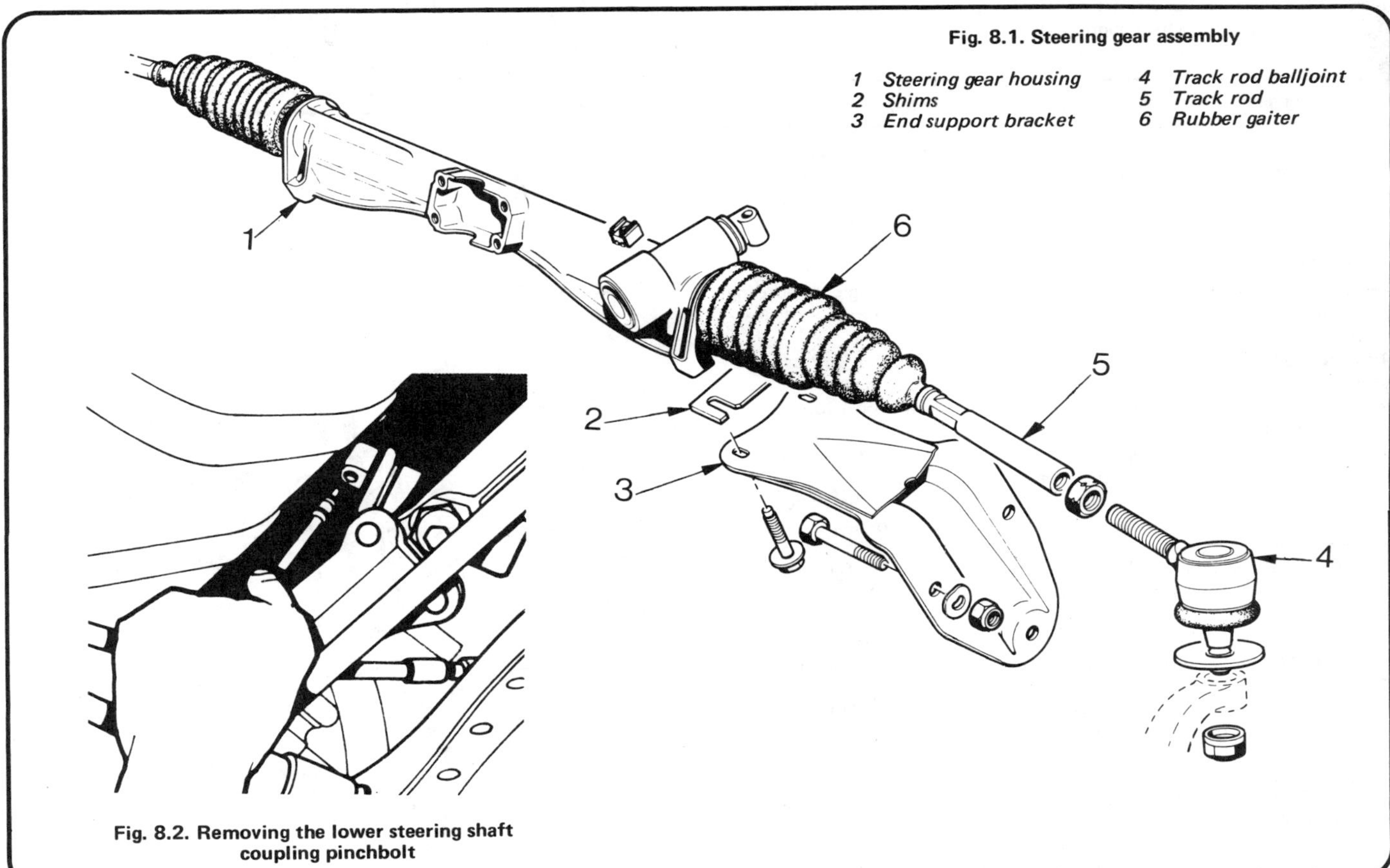

Fig. 8.1. Steering gear assembly

1 Steering gear housing
2 Shims
3 End support bracket
4 Track rod balljoint
5 Track rod
6 Rubber gaiter

Fig. 8.2. Removing the lower steering shaft coupling pinchbolt

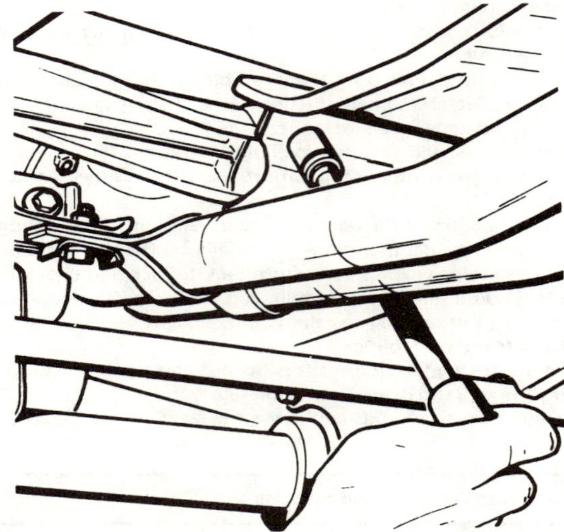

Fig. 8.3. Removing the RH side steering gear securing bolts

valance aperture passing the right-hand end of the steering gear over the anti-roll bar to withdraw it.

Note: On left-hand drive cars, remove the bracket from the right-hand end of the steering gear housing. Slide the steering gear to the right and rotate it so that the pinion coupling is facing downwards. Pass the left-hand end of the steering gear assembly over the anti-roll bar and withdraw it (Fig. 8.5).

8 Refitting is a reversal of removal where the original steering gear is being refitted in the vehicle. Refit the original shims in their correct location. After major steering dismantling, always check the tracking of the front wheels as described in Section 10 of this Chapter.

9 Where a new or factory reconditioned steering gear unit is fitted, the calculation of fitting shim thickness must be left to your Chrysler dealer due to the specialised nature of the equipment required and its effect upon the front wheel alignment.

5 Steering gear - dismantling, servicing and reassembly

1 Having removed the steering gear as described in the preceding Section consider whether in view of the mileage covered it would be more economical to obtain a factory exchange unit rather than recondition the existing unit. Where only minor components require renewal or the internal parts require cleaning and packing with fresh grease, then proceed as follows.

2 Remove the gearshift linkage bracket from the steering gear housing.

3 Secure the unit in a vice fitted with jaw protectors and do not overtighten.

4 Remove the spring clips and carefully pull both rubber gaiters back over the track rods.

5 With the steering rack in approximately the central position, measure the distance between the flange on the track rod coupling locknut and the end of the steering gear housing nearest the pinion shaft. Make a note of this measurement.

6 Without moving the rack, scribe a line on the pinion shaft and rack. This will enable the pinion to be refitted in the correct position relative to the rack.

7 Drift out the sealing cap from the pinion housing (Fig. 8.6), hold the pinion shaft and unscrew the self-locking nut from the end of the shaft (Fig. 8.7).

8 Tap back the lockwasher tab, unscrew the locknut and remove the damper adjusting plug, spring and slipper. Retrieve any shims that are fitted.

9 Drive the pinion from its location by using a brass or copper drift.

10 Hold the flats on the track rods with a spanner and slacken the locknuts. Unscrew both track rods from the rack assembly.

11 Withdraw the rack from the pinion end of the housing.

12 The pinion bearing may be renewed by removing the circlips and shims (Fig. 8.9).

13 The bearing is renewable after extracting its securing circlip and then driving out the anti-friction bush using a stepped drift up to the

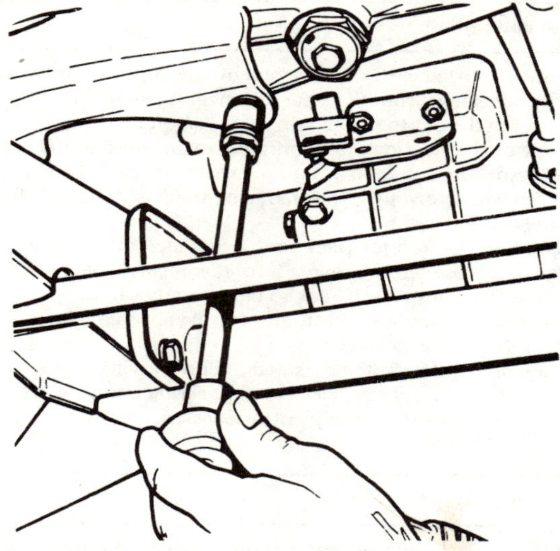

Fig. 8.4. Removing the LH side steering gear securing bolts

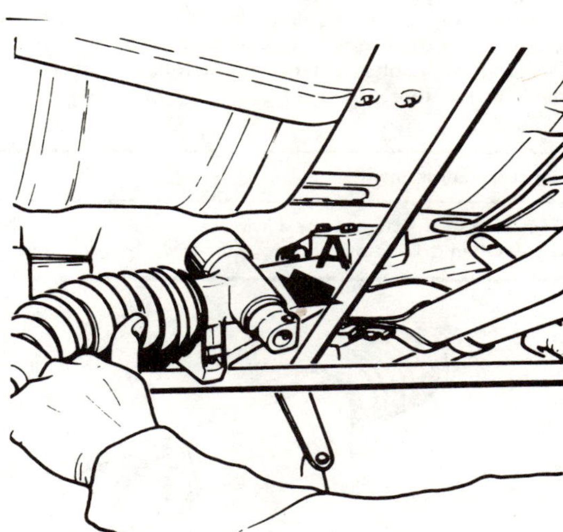

Fig. 8.5. Withdrawing the steering gear assembly (LH drive models)

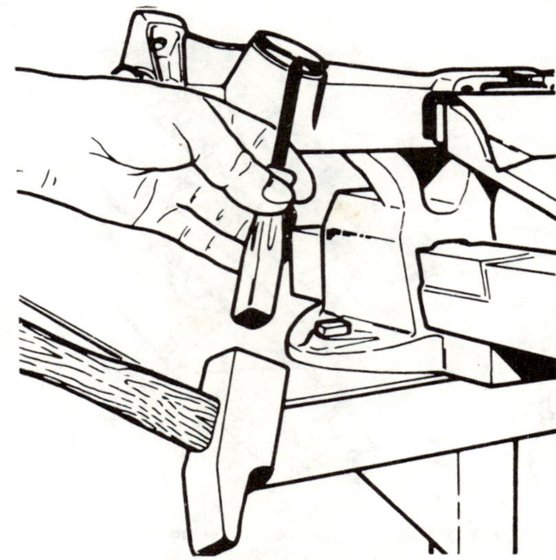

Fig. 8.6. Removing the pinion shaft sealing cap

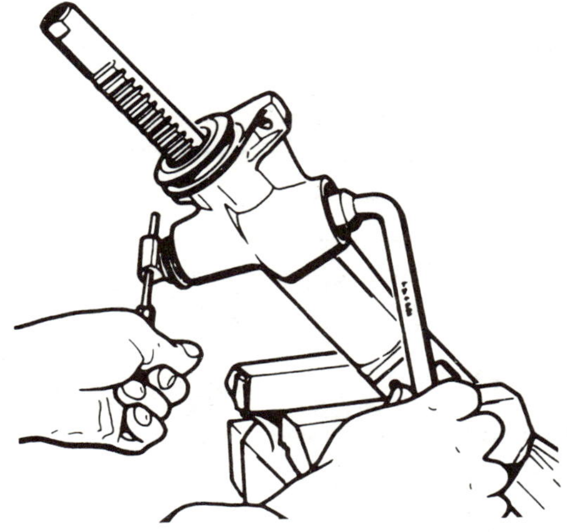

Fig. 8.7. Unscrewing the pinion shaft securing nut

Fig. 8.8. Dimensions of drift for removing the pinion bearing (measurements in mm)

dimensions given in the diagram, Fig. 8.8.

14 Extract the circlip, washer and the rubber bush. Remove the outer track.

15 Thoroughly wash all components in paraffin and dry them with a piece of non-fluffy cloth.

16 Examine all components for wear and renew as appropriate.

17 Commence reassembly by inserting the outer track so that its slot is opposite the lubrication groove.

18 Fit the rubber bush and grease it with the recommenced grade of grease and then locate the washer and circlip.

19 Using the tool previously described, fit the anti-friction ring so that it is flush with the surface of the steering gear housing. Grease the anti-friction bush.

20 Fit the pinion bearing assembly by including an excessive number of shims so that the circlip will not fit into its groove. Remove one shim at a time until the circlip will just locate correctly. This will provide an endfloat of 0.1 mm (0.004 in).

21 Grease the rack assembly and slide it into the housing.

22 Measure out 80 cc of the recommended grade of grease and insert it into the steering gear by using the damper as an injector. During this operation, move the rack from lock to lock over the full length of its travel. Do not exceed the stipulated quantity of grease or the bellows may burst in service.

23 Fit the rack damper assembly and adjust it as described in Section 3 of this Chapter.

24 Screw both track rods, complete with rubber gaiters, into each end of the rack assembly.

25 Tighten the locknuts and check that the distance between the flange of the locknut and the inner face of the balljoint is 1/8 in (3 mm). Adjust as necessary until this dimension is achieved (Fig. 8.10).

26 Set the rack so that the end nearest the pinion is protruding from the housing by the dimension noted during dismantling.

27 Fit a new 'O' ring to the pinion shaft and with the rack held in the correct position, insert the pinion into the housing so that the scribe marks made previously are aligned.

28 Fit the washer and nut onto the end of the pinion shaft and tighten the nut to the specified torque wrench setting.

29 Fill the pinion housing with grease and tap the sealing cap back into position.

30 Refit the rubber gaiters to the steering gear housing and secure with the spring clips.

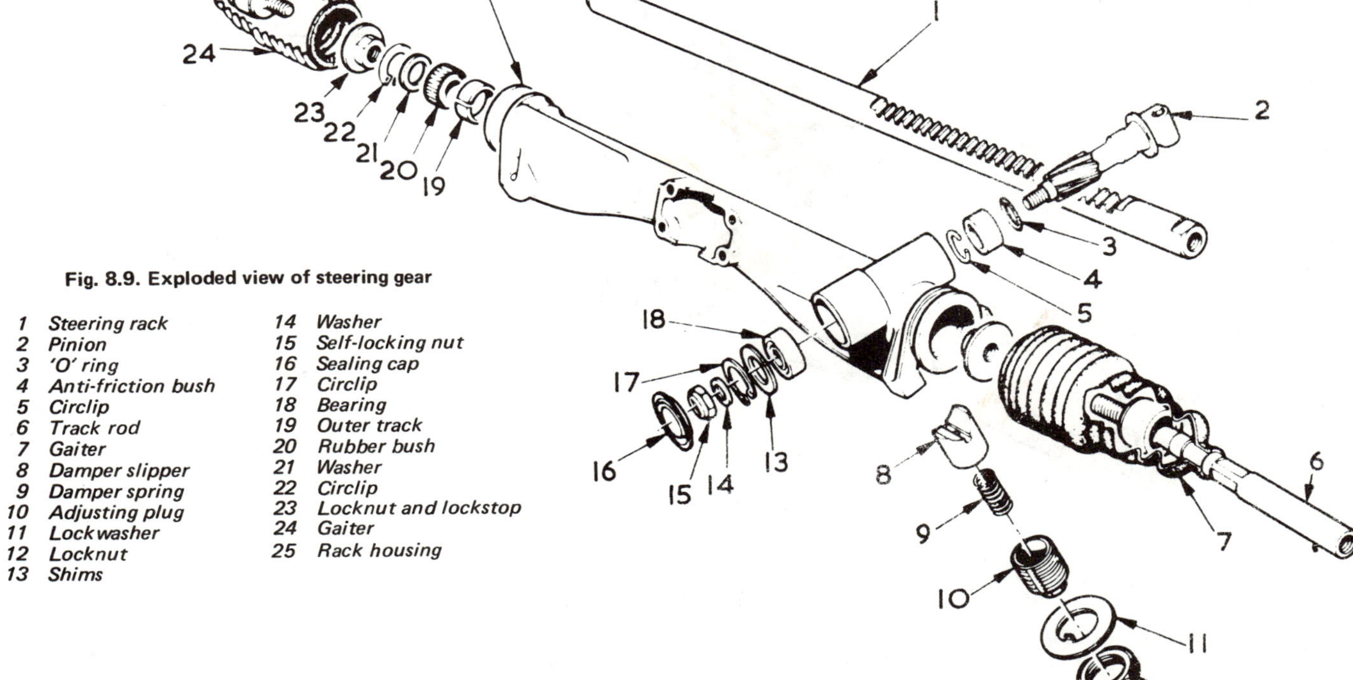

Fig. 8.9. Exploded view of steering gear

1 Steering rack
2 Pinion
3 'O' ring
4 Anti-friction bush
5 Circlip
6 Track rod
7 Gaiter
8 Damper slipper
9 Damper spring
10 Adjusting plug
11 Lockwasher
12 Locknut
13 Shims
14 Washer
15 Self-locking nut
16 Sealing cap
17 Circlip
18 Bearing
19 Outer track
20 Rubber bush
21 Washer
22 Circlip
23 Locknut and lockstop
24 Gaiter
25 Rack housing

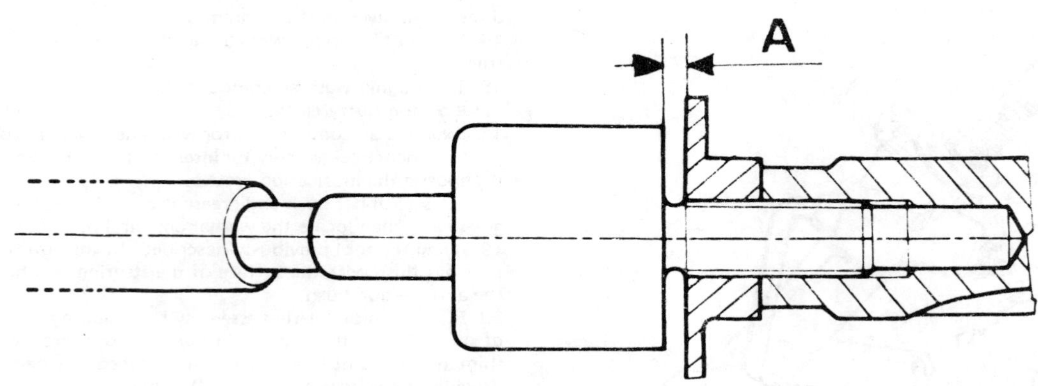

Fig. 8.10. Correct dimension between track rod inner balljoint and locknut (A = 1/8 in - 3 mm)

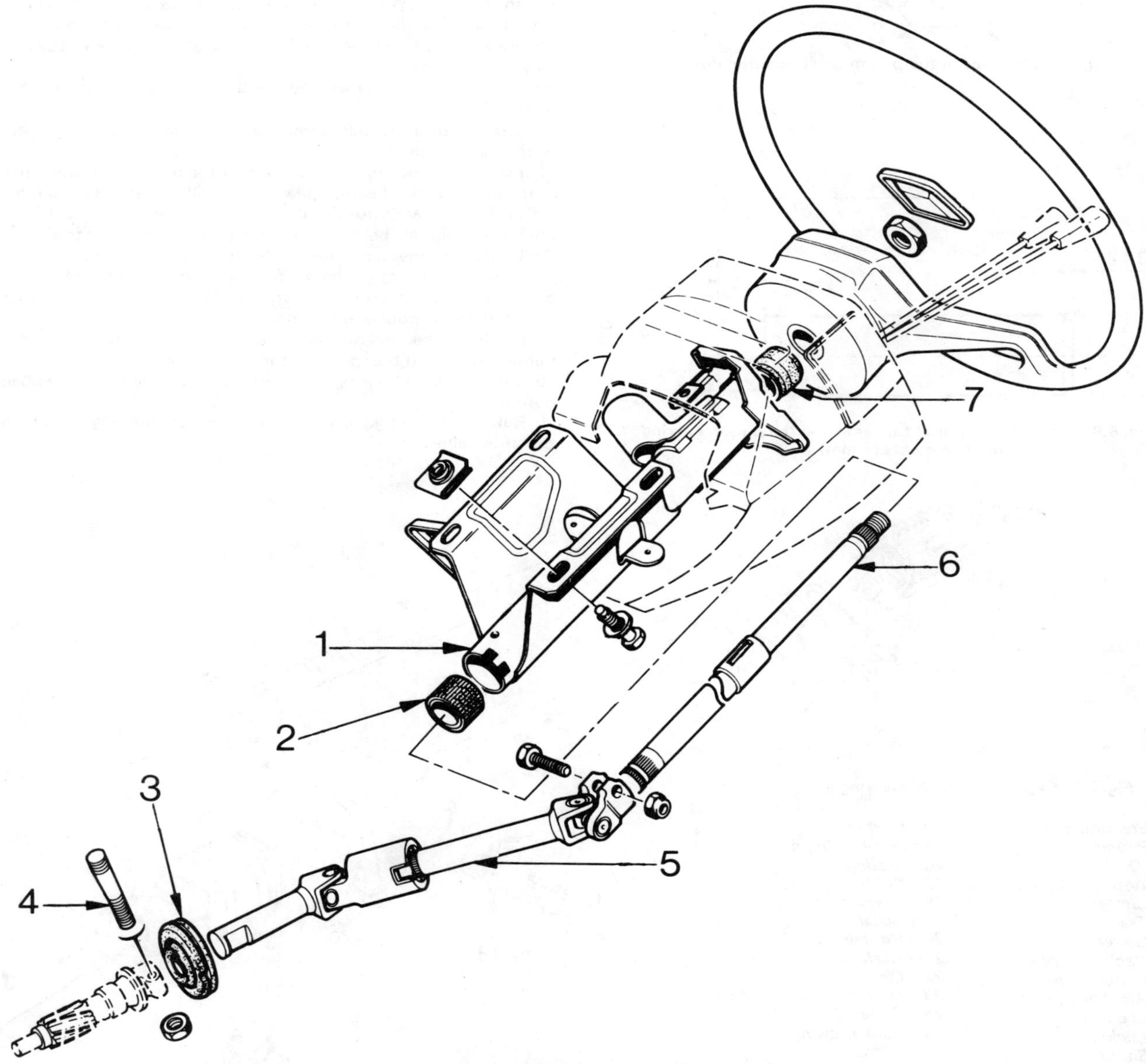

Fig. 8.11. Exploded view of steering column

1 Steering column housing
2 Lower bearing
3 Bulkhead grommet
4 Lower shaft pinchbolt
5 Lower shaft and couplings
6 Upper shaft
7 Upper bearing

6 Steering column - removal and refitting

1 With the front of the car over a pit or supported on axle stands, remove the pinchbolt securing the lower end of the steering shaft to the steering gear pinion.
2 Working inside the car, remove the lower half of the steering column cowl (where fitted).
3 Disconnect the battery earth terminal.
4 Disconnect the electrical multi-plugs that connect the steering column wiring to the main harness.
5 Support the weight of the steering column assembly and remove the four bolts securing the column to the facia bracket (see Fig. 8.12).
6 Carefully pull the column rearwards until the lower end of the steering shaft is detached from the steering gear pinion, and then withdraw the complete column and shaft assembly from the car.
7 To refit the column, set the front wheels to the straight ahead position.
8 Insert the steering shaft through the rubber grommet in the lower bulkhead and guide the shaft onto the steering gear pinion.
9 Temporarily secure the column to the facia bracket with the four bolts and tighten the lower shaft pinchbolt.
10 Adjust the position of the column in the slotted holes to provide adequate clearance between the steering wheel and the cowl, and also to obtain the maximum clearance between the steering shaft and the clutch pedal. Tighten the column retaining bolts. Reconnect the multi-plugs.
11 Refit the lower cowl panel.

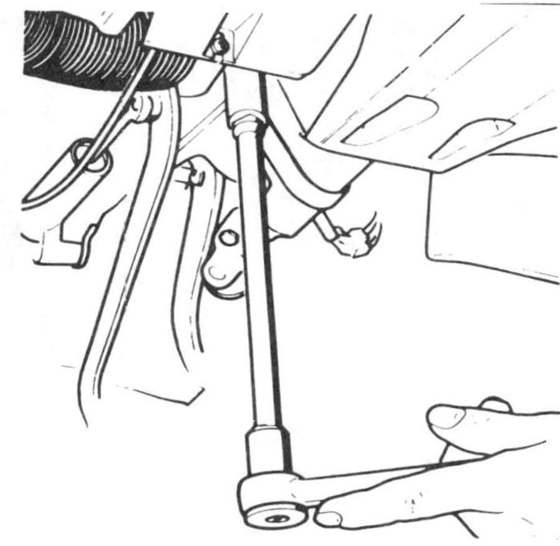

Fig. 8.12. Removing the steering column securing bolts

7 Steering column flexible couplings - renewal

1 The two flexible couplings on the steering shaft are not repairable and if they become worn they must be renewed.
2 Withdraw the lower steering shaft assembly from the steering gear pinion as described in the previous Section.
3 Remove the pinchbolt securing the upper universal joint to the steering shaft.
4 Remove the lower shaft complete with the universal joints and obtain a new shaft assembly.
5 Ensure the front wheels are in the straight ahead position and the steering wheel spokes are facing downwards, and slide the upper universal joint onto the splines of the steering shaft.
6 Refit the pinchbolt and tighten it to the specified torque.
7 Refit the steering shaft assembly to the steering gear pinion as described in the previous Section.

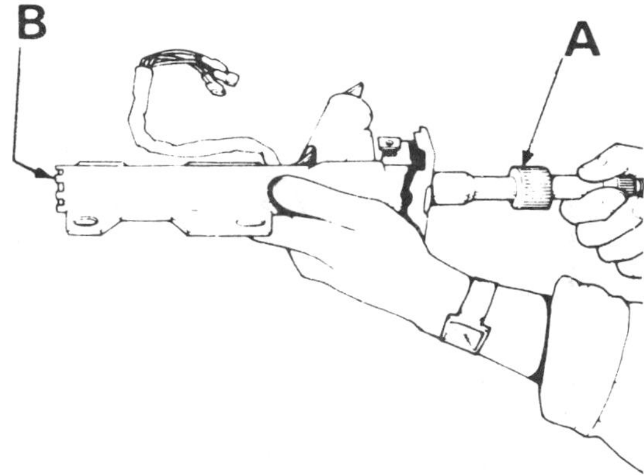

Fig. 8.13. Removing the steering shaft and upper bearing

A Upper bearing *B Lower bearing retaining tabs*

8 Steering column - dismantling and reassembly

1 The main reason for dismantling the steering column is to renew the upper and lower bearings, and to do this it is necessary to first remove the complete steering column assembly as described in Section 6.
2 Prise out the motif from the centre of the steering wheel (photo) and remove the securing nut using a socket.
3 Pull the wheel off the steering shaft splines.
4 Remove the screws retaining the steering column cowl together and lift away the two halves of the cowl.
5 Mark the position of the upper universal joint in relation to the steering shaft, remove the pinchbolt and withdraw the lower shaft complete with universal joints.
6 Tap the bottom of the shaft with a soft-faced hammer and withdraw the shaft and top bearing from the top of the steering column (Fig. 8.13).
7 Bend back the tabs securing the bottom bearing in the column and use the steering shaft to push the bearing out of the column.
8 If necessary remove the top bearing inner race using a puller.
9 Examine the bearings and shaft for wear and renew where necessary.
10 Lubricate the bearings with a molybdenum disulphide based grease before refitting them.
11 Retain the bottom bearing in position by bending in the tags on the bottom of the column. Push the top bearing into place using a thin bladed screwdriver between the shaft and column.

8.2 Steering wheel centre motif removed

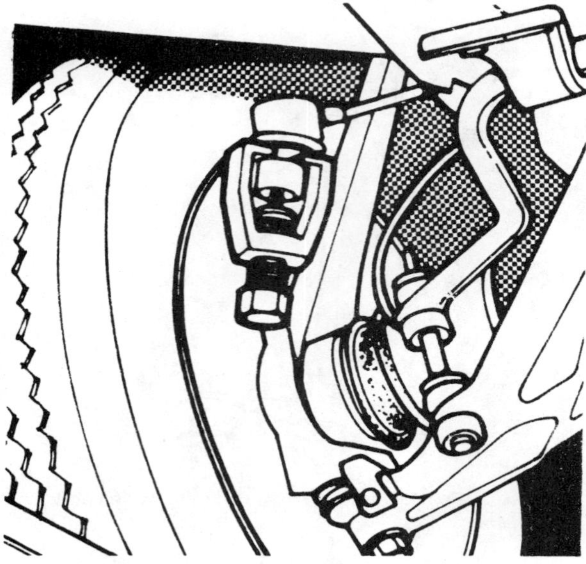

Fig. 8.14. Removing a track rod balljoint using an extractor

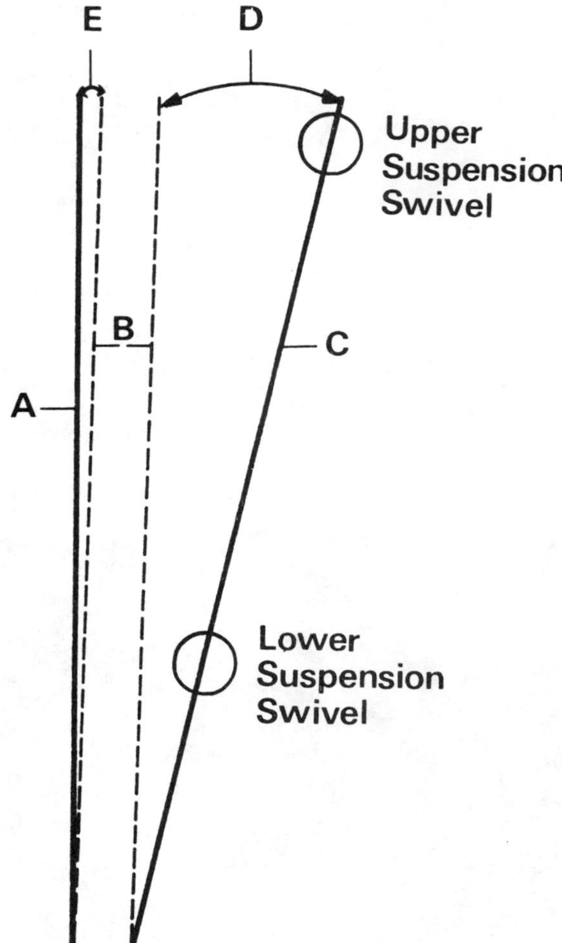

Fig. 8.16. Diagram of steering angles

A Wheel camber
B Vertical lines
C Steering axis inclination
D Angle of inclination
E Positive camber angle

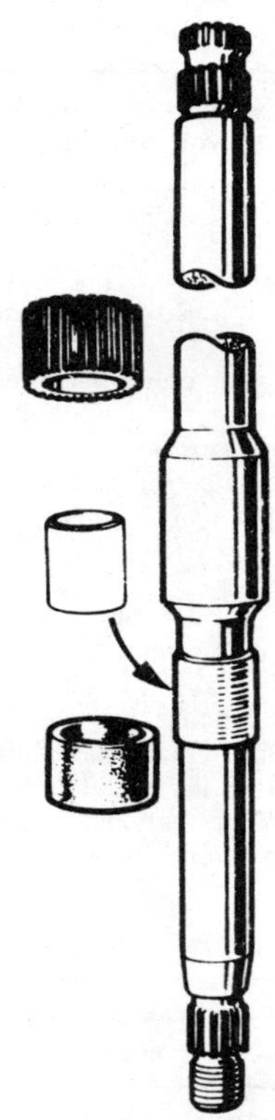

Fig. 8.15. Steering shaft and bearings

12 Refit the lower shaft and universal joints ensuring the marks made previously are aligned.
13 Refit the steering column into the car using the reverse procedure to removal.

9 Track rod ends - renewal

1 To gain easier access, first jack up the front of the car, support it on axle stands and remove the front wheels.
2 Remove the nut securing the balljoint to the steering arm and using an extractor, disconnect the balljoint tapered pin from the arm (Fig. 8.14).
3 Mark with paint the amount of thread exposed between the end of the balljoint and locknut, slacken the locknut and unscrew the balljoint from the trackrod end.
4 Screw the nut onto the new balljoint until it is in approximately the same position as the original one.
5 Lubricate the threads and screw the balljoint into the track rod until the locknut abuts the end of the rod. Hold the balljoint in the correct position and tighten the locknut.
6 Refit the balljoint to the steering arm and tighten the securing nut.
7 Repeat the procedure on the other track rod balljoint if necessary.
8 The method described will provide an approximate setting of the original front wheel tracking but before the track rod lock nuts are fully tightened, the track must be checked as described in Section 11.

10 Steering geometry

1 Accurate front wheel alignment is essential for good steering and minimum tyre wear.
2 Wheel alignment embraces four factors:

Camber which is the angle at which the front wheels are set from the vertical, when viewed from the front of the vehicle. Positive camber is the amount (in degrees) that the wheels are inclined outwards from the vertical at their tops.

Castor is the angle between the steering axis and a vertical line when viewed from each side of the vehicle. Positive castor is when the steering axis is included rearwards at the top.

Steering axis inclination is the angle, when viewed from the front of the vehicle, between the vertical and an imaginary line drawn between the upper and lower stub axle carrier swivel balljoints.

Toe-in is the amount by which the distance between the front inside edges of the roadwheels (measured at hub height) is less than the diametrically opposite distance measured between the rear inside edges of the roadwheels. Toe-out is specified for early Alpine models and this is a greater measurement at the front inside edges of the wheel.

Checking and adjustment of toe-out (tracking) is described in the next Section.

The steering angles just described are either set during manufacture or in the case of the Camber angle, this can be altered by varying the shims described in the preceding Chapter. Any variation of the Camber angle should be left to your Chrysler dealer who will have the necessary equipment for measuring these critical angles.

Two steering angles are shown in diagrammatic form in Fig. 8.16 and reference should be made to the Specifications Section of this Chapter for the precise angles (in degrees).

11 Front wheel track - checking and adjustment

1 Although it is preferable to leave these operations to your Chrysler dealer, an approximate setting may be made which will be useful after renewal of any of the steering or suspension components and will at least permit the car to be driven to the dealers for a more accurate check to be made, if necessary.
2 Place the vehicle on level ground with the tyres correctly inflated. Position the wheels in the 'straight ahead' position.
3 Obtain or make an alignment gauge. One may be easily made up from a piece of tubing or bar, suitably cranked to clear the engine/transmission unit with a bolt at one end to permit adjustment of its overall length.
4 With the gauge, measure the distance between the two inner wheel rims (at hub height) at the fronts of the roadwheels.
5 Mark the tyre and then roll the vehicle forward so that the mark on the tyre rotates 180°.
6 Measure the distance between the inner edges of the wheel rims at hub height at the rear of the roadwheels. This second measurement should vary from that taken previously by being 3/64 to 1/8 in (1 to 3 mm) less to give the correct toe-out angle.
7 If the dimensions are incorrect, slacken the track rod end locknuts.
8 Turn the tie rods by equal amounts ¼ turn at a time and recheck the toe-out measurement.
9 When the toe-out measurement is correct tighten the locknuts.

12 Wheels and tyres

1 The roadwheels fitted to the Alpine models are of pressed steel type, four bolt fixing.
2 To minimise tyre wear and to prevent steering wobble or vibration it is imperative to have the wheels balanced when the tyres are first fitted.
3 From time to time, check the security of the securing bolts and check that the bolt head recess in the wheel has not become enlarged or elongated. Where this is the case, the roadwheel must be renewed.
4 Any deviation from even tread wear will indicate the need for rebalancing of the wheels or checking and adjusting of the steering angles.

13 Fault diagnosis - steering

Before diagnosing faults from the following chart, check that any irregularities are not caused by:

1 Binding brakes
2 Incorrect 'mix' of radial or crossply tyres
3 Incorrect tyre pressures
4 Misalignment of bodyframe.

Symptom	Reason/s	Remedy
Steering wheel can be moved before any sign of roadwheel movement is apparent	Wear in linkage, gear or column couplings or track rod ends	Check for movement in all joints and gear and renew as required.
Car wanders and difficult to hold to a straight line	As above or wheel alignment incorrect or hub bearings loose or worn. Upper or lower suspension balljoints worn	Adjust or renew as necessary.
Steering stiff and heavy	Incorrect wheel alignment Seizure of suspension or steering balljoints	Adjust or renew.
Wheel wobble and vibration	Roadwheels out of balance Roadwheels buckled Incorrect wheel alignment Wear in steering joints Wheel nuts loose	Balance wheels. Renew wheels. Check and adjust. Renew. Tighten.

Chapter 9 Rear suspension and hubs

Contents

Specifications

Type

Independent trailing arm and coil spring with anti-roll bar and telescopic dampers

Coil spring

Free length	13.625 in (346 mm)

Torque wrench settings

	lb f ft	kg fm
Crossmember brackets to body	33	4.56
Rear crossmember pivots to brackets	44	6.08
Suspension arm pivots to crossmember	48	6.63
Damper lower mountings	15	2.07
Damper upper mountings	11	1.52
Anti-roll bar mountings	15	2.07

1 General description

The rear suspension is independent and comprises trailing arms supported on coil springs with telescopic shockabsorbers and an anti-roll bar.

The forward ends of each trailing arm pivot on rubber bushes attached to a tubular crossmember. The crossmember in turn is attached to the body underframe through rubber mountings.

The anti-roll bar is attached to a brake equaliser device which has the effect of reducing the possibility of the rear wheels locking under heavy braking conditions.

Caution: When jacking up the rear of the car using a trolley jack, ensure that the pad of the jack is positioned under the jacking points only. Attempts to raise the car by means of the rear crossmember or trailing arms may cause damage.

2 Rear hubs - removal, inspection and refitting

1 Jack up the rear of the car, support it on blocks or axle stands and remove the roadwheel.
2 Remove the retaining screws and pull off the brake drum. If difficulty is experienced, slacken the handbrake cable adjuster and automatic brake adjuster (refer to Chapter 10).
3 Prise the grease cap from the end of the hub using a screwdriver.
4 Raise the peening on the large hub nut using a small chisel and remove the nut and thrust washer.
5 Carefully withdraw the complete hub and bearings ensuring that the outer bearing does not drop out onto the floor.
6 Prise out the oil seal from the inner end of the hub and withdraw the roller bearing (Fig. 9.1).
7 If the bearings are being renewed due to wear, the inner tracks in the hub must also be renewed. Drive out the inner tracks using suitably sized pieces of tubing. Refit the new tracks ensuring the thicker end of the track faces in towards the centre of the hub (Fig. 9.3).
8 Wash out the hub in clean petrol to remove all traces of old grease.
9 Spread some of the recommended grade of grease evenly around the inside of the hub between the two bearings.
10 Grease the inner (larger) roller bearing and insert it in the hub.
11 Tap in a new oil seal ensuring that the lip faces towards the bearing.
12 Refit the hub to the axle, grease the outer (smaller) roller bearing and insert it in the hub.
13 Fit the thrust washer, engaging its tongue correctly, screw on the hub nut and tighten it to a torque of 11 lb f ft (1.52 kg fm) at the same time rotating the hub. Back off the nut until the hub turns freely without any end-float. Peen the nut collar into the stub axle groove to retain it in the predetermined position.
14 Refit the brake drum and securing screws, fill the grease cap half full with grease and tap it into position in the hub.
15 Re-adjust the handbrake cable if it was slackened off. Fit the roadwheel and lower the jack.

3 Rear springs - removal and refitting

1 Raise the rear of the car and place axle stands under the side frame members so that the suspension is suspended.
2 Place a small jack below the rear end of the suspension arm and raise it sufficiently to take the weight of the arm (Fig. 9.4).

Fig. 9.1. Removing the inner wheel bearing

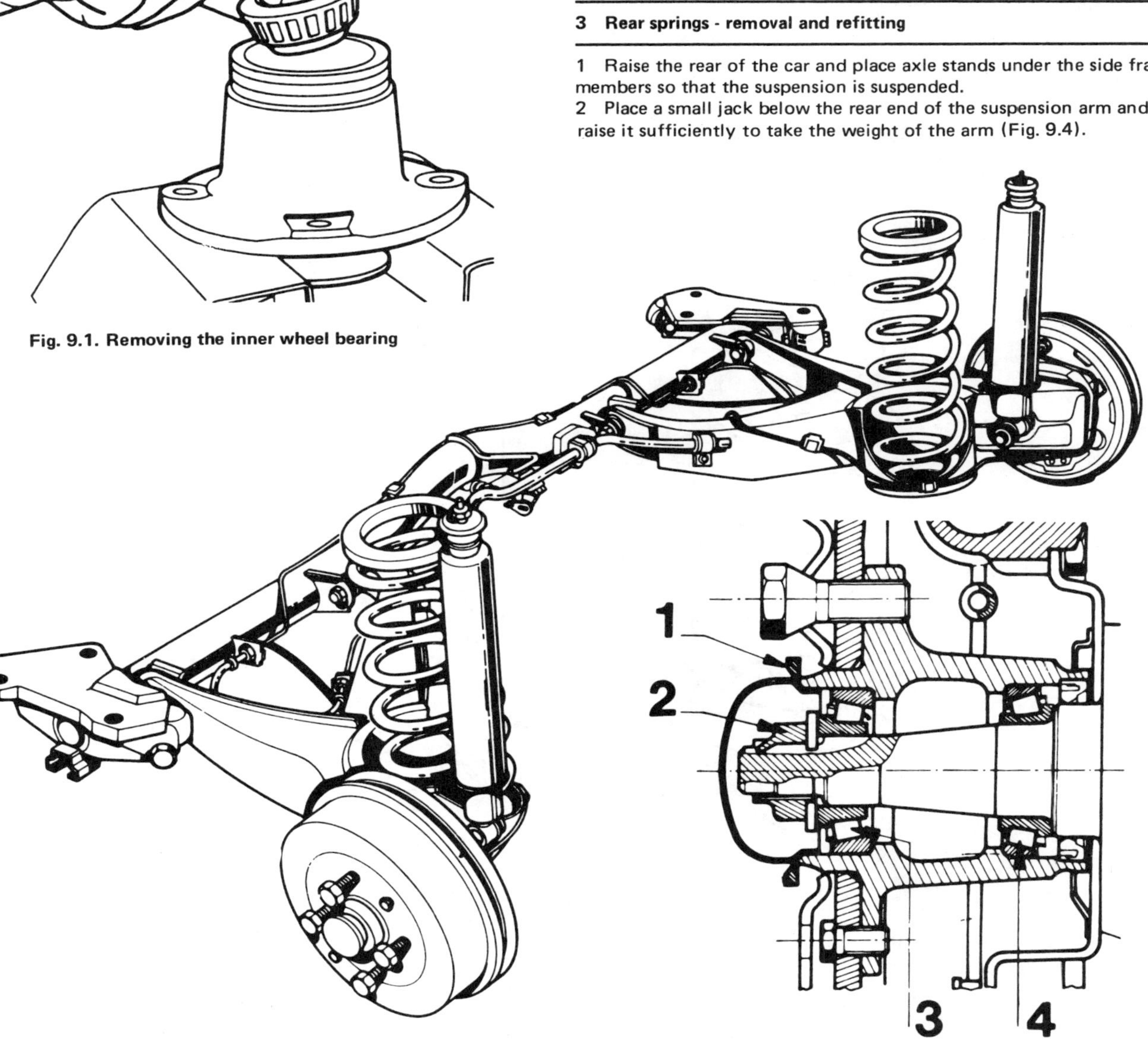

Fig. 9.2. Layout of the rear suspension system

Fig. 9.3. Cross-sectional view of rear hub assembly

1 Grease cap
2 Hub nut
3 Outer bearing
4 Inner bearing

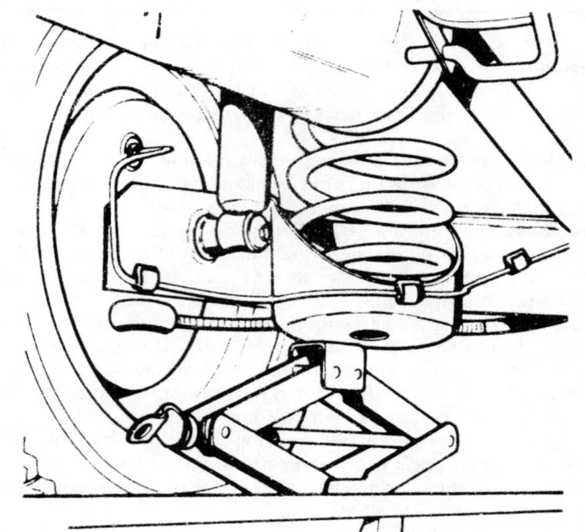

Fig. 9.4. Supporting the suspension arm prior to removing the spring

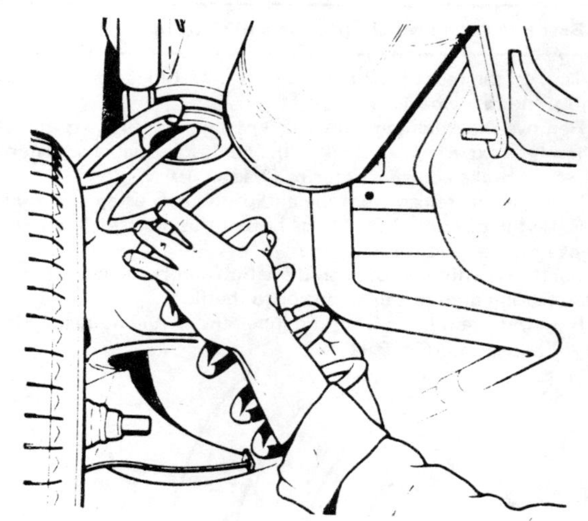

Fig. 9.5. Withdrawing the coil spring

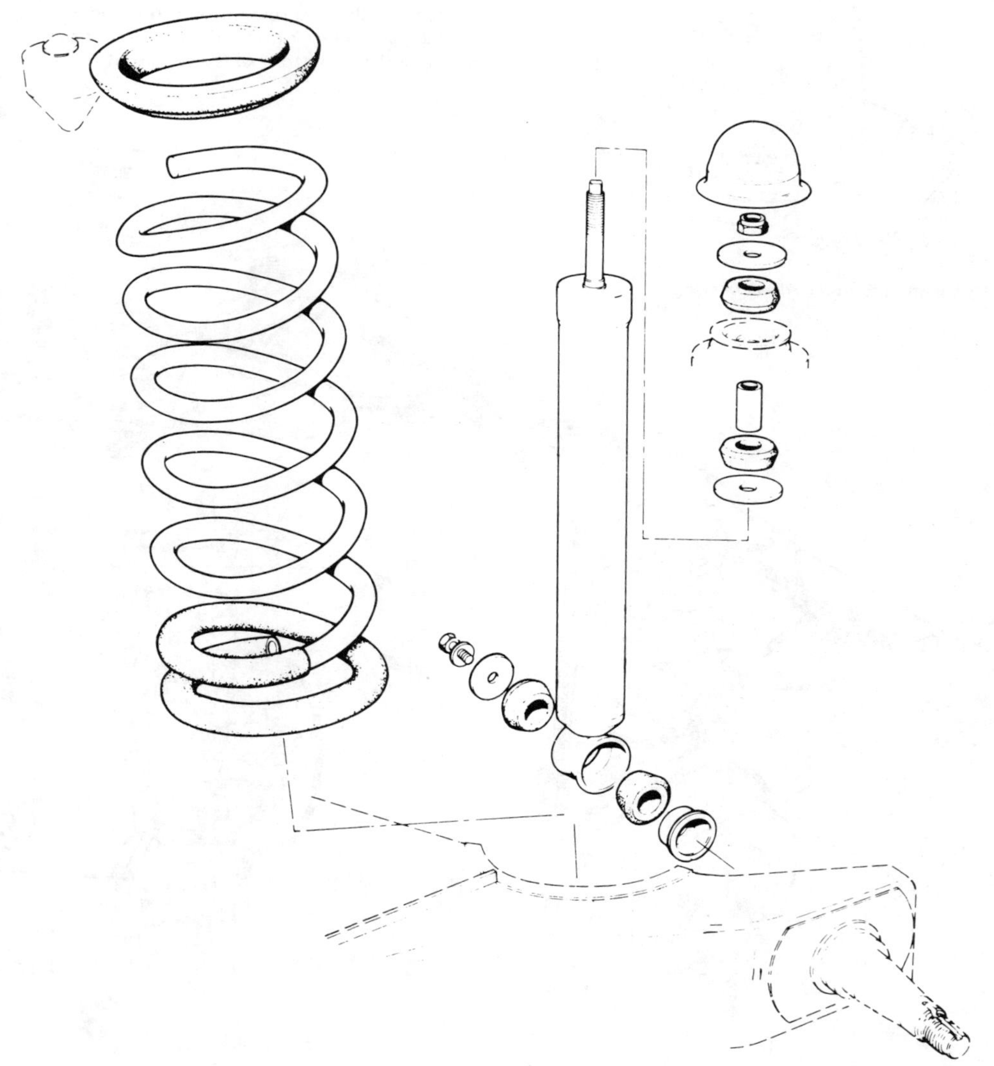

Fig. 9.6. Exploded view of damper and bushes

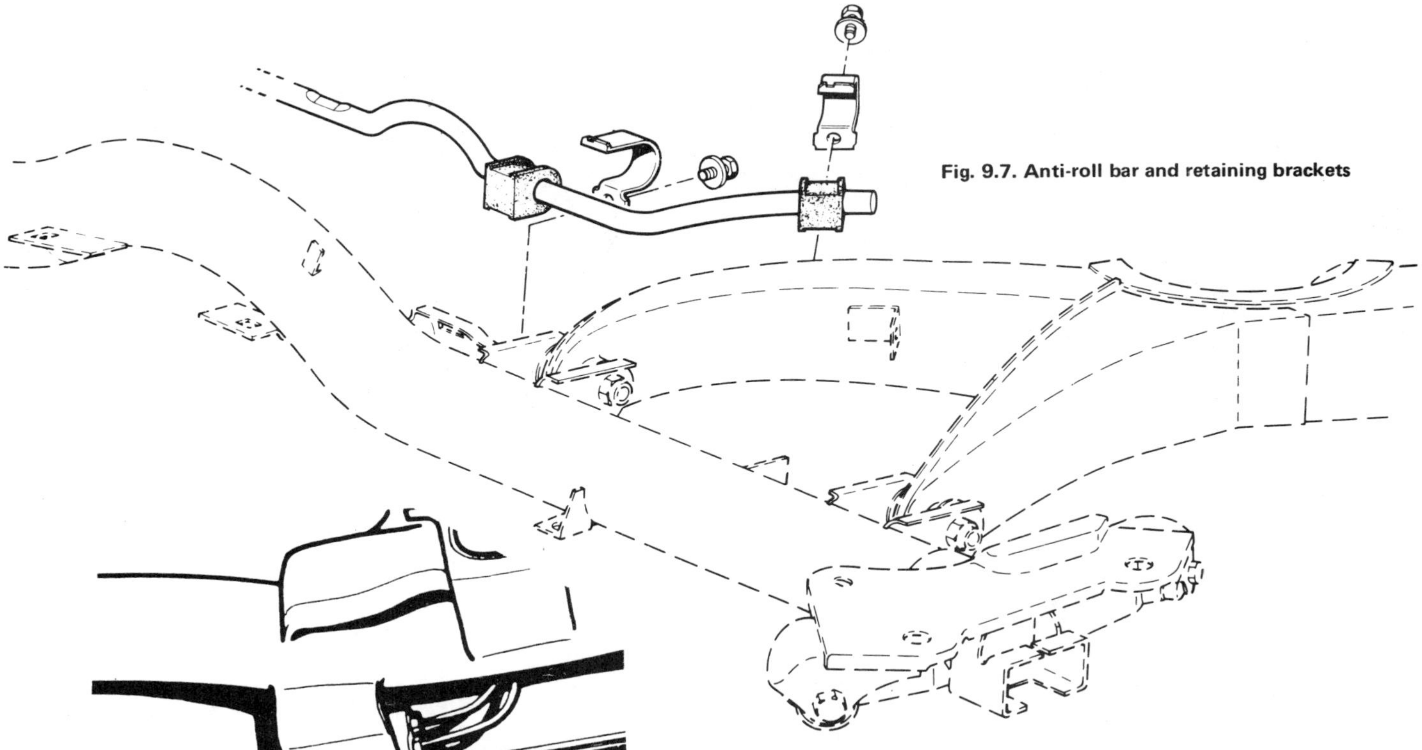

Fig. 9.7. Anti-roll bar and retaining brackets

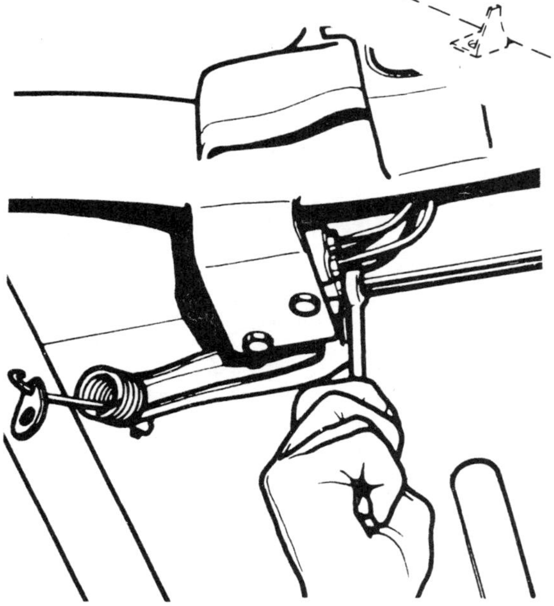

Fig. 9.8. Disconnecting the inlet pipe from the brake pressure valve

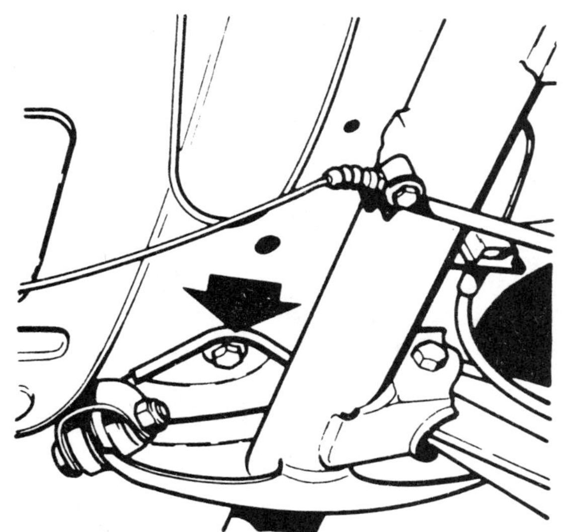

Fig. 9.9. Location of crossmember inner securing bolt (Right-hand shown)

3 Remove the damper lower mounting bolt.
4 Slowly lower the jack and remove it from beneath the suspension arm. Pull the arm downwards as far as possible and withdraw the spring (Fig. 9.5).
5 Refit the spring using the reverse procedure to that of removal.

4 Shockabsorbers - removal, testing and refitting

1 From inside the car remove the protective cap and undo the top shockabsorbers retaining nut. Retrieve the rubber bushes and spacers (Fig. 9.6).
2 Remove the lower shockabsorber retaining bolt, withdraw the shockabsorber unit and retrieve the rubber bushes and spacers.
3 Test the shockabsorbers as described in Chapter 7 for the front shockabsorbers.
4 Refit the shockabsorbers using the reverse procedure to that of removal, ensuring that the rubber bushes and spacers are refitted in the correct sequence.

5 Anti-roll bar - removal and refitting

1 Place the car over a pit or jack up the rear of the car and support it on axle stands.
2 Secure the brake pressure valve operating arm to the valve body with wire or string, then unhook the spring from the anti-roll bar.
3 Remove the four brackets securing the roll bar to the rear suspension arms and crossmember and withdraw the bar (Fig. 9.6).
4 Renew the rubber bushes where necessary and refit the anti-roll bar using the reversal of the removal procedure.
5 Check the setting of the brake pressure valve as described in Chapter 10.

6 Rear suspension assembly - removal and refitting

1 If the rear suspension arm bushes are worn, the best policy is to remove the complete rear suspension assembly for servicing as it is quite likely that the crossmember mounting bushes will also require renewal.
2 Jack up the rear of the car and support it on firmly based axle

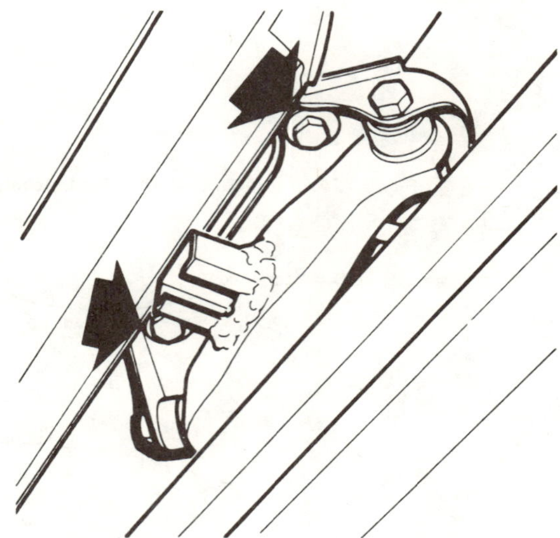

Fig. 9.10 Location of the two outer crossmember securing bolts (Right-hand shown)

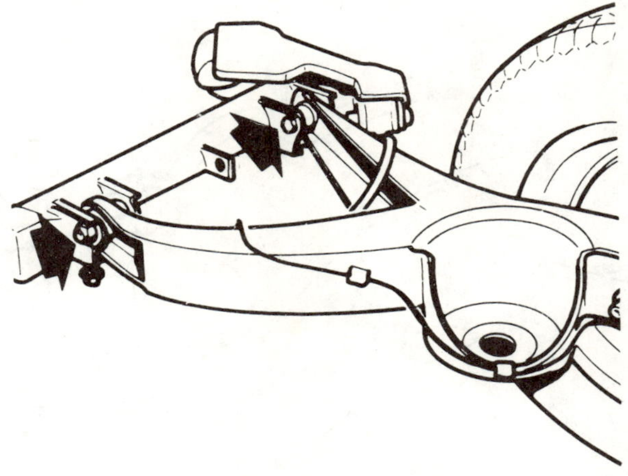

Fig. 9.11. Suspension arm pivot bolts

stands or blocks placed beneath the side frame members. Do not remove the roadwheels.
3 Disconnect the handbrake cable at the adjuster quadrant.
4 Remove the anti-roll bar as described in Section 5.
5 Remove both shockabsorber lower mounting bolts.
6 Remove both rear coil springs as described in Section 3.
7 Disconnect the brake fluid inlet pipe from the brake pressure valve (Fig. 9.8) and plug the pipe and valve aperture to prevent dirt ingress and loss of fluid.
8 Disconnect the exhaust pipe supports from the brackets on the two crossmembers and sidemember and lower the pipe and silencer carefully to the ground.
9 Support the weight of the rear suspension crossmember by placing a jack beneath the centre of it.
10 Remove the three bolts securing the suspension crossmember support brackets to each side of the car (see Figs. 9.9 and 9.10).
11 Carefully lower the jack from beneath the suspension crossmember and roll the complete suspension assembly out from under the rear of the car.
12 Refit the rear suspension assembly using the reverse procedure to that of removal.
13 Bleed the braking system and adjust the brake pressure valve setting as described in Chapter 10.

7 Rear suspension bushes - renewal

1 Remove the complete rear suspension assembly as described in the previous Section.
2 Disconnect the flexible brake hoses and plug the ends to prevent dirt ingress.
3 Remove the roadwheels from the suspension assembly.
4 Remove the pivot bolts from the suspension arm bushes and withdraw the suspension arms from the crossmember (Fig. 9.11).
5 Undo the bolts securing the support brackets to each end of the crossmember and remove the brackets (Fig. 9.13).
6 Make up an extractor tool as shown in Fig. 9.12 and withdraw the bushes from each end of the crossmember and the suspension arms. **Note:** If difficulty is experienced in removing the bushes, carefully cut through the inner steel sleeve and rubber bush with a hacksaw blade and then collapse the outer sleeve and drive it out using a suitable drift. Take great care not to cut or damage the actual crossmember or suspension arm assembly.
7 Draw the new bushes into their housing using the extractor tool made up for removal, or drive the bushes in using a piece of flat hardwood and a hammer.

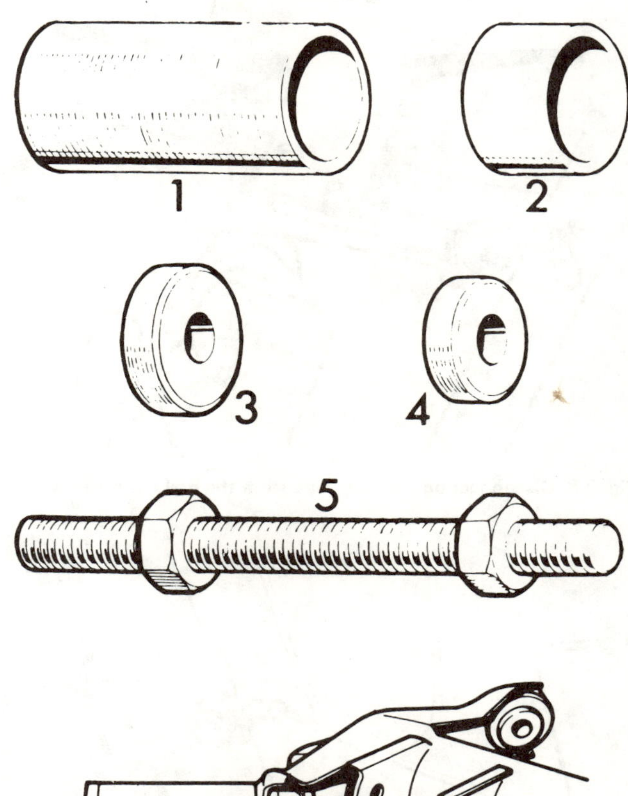

Fig. 9.12. Tool that can be made up for extracting the rear suspension bushes

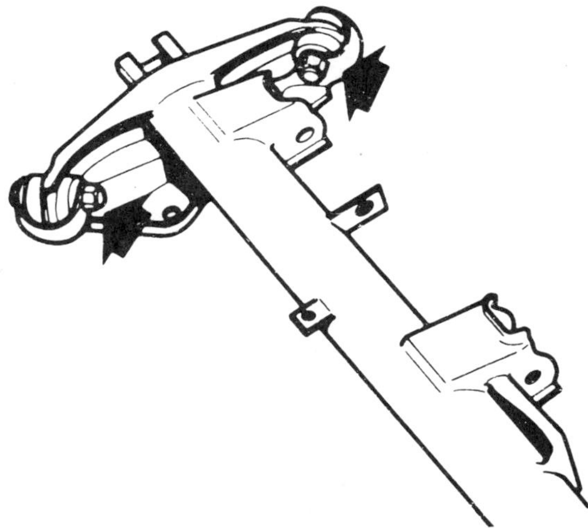

Fig. 9.13. Crossmember bracket securing bolts

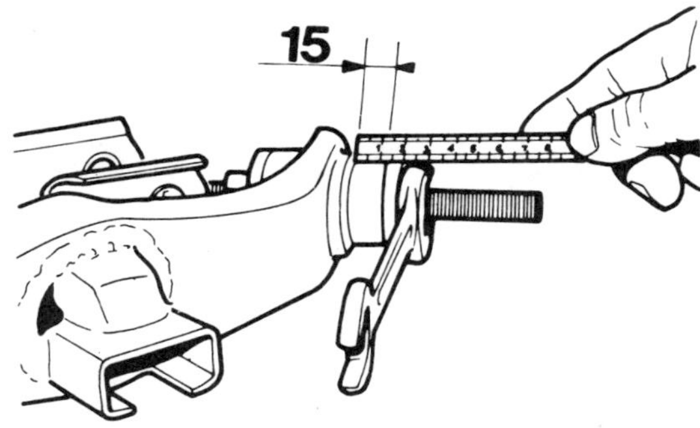

Fig. 9.14. Correct position of rear crossmember bush dimension = 0.60 in (15 mm)

8 The crossmember bushes must be fitted so that there is 0.5 in (15 mm) protruding from the boss on the bush housing (Fig. 9.14).
9 The bushes can be lubricated with petroleum jelly (Vaseline) to assist in refitting, but never use oil or grease.
10 Refit the suspension arms and crossmember support brackets using the reverse of the removal procedure, but do **not** fully tighten the suspension arm pivot bolts at this stage.
11 Refit the complete rear suspension assembly to the car as described in Section 6.
12 Before tightening the rear suspension pivot bolts, bounce the rear of the car up and down several times to settle the coil springs. Run the car over an inspection pit; get a couple of medium weight friends to sit in the rear compartment to load the car and then tighten the suspension arm pivot bolts to the specified torque wrench setting.

13 It is recommended that the car is then taken round to your local Chrysler dealer who will have the special equipment necessary to check the height of the rear suspension.

8 Rear wheel alignment

1 The suspension angles at the rear are all set during manufacture and no adjustment is possible.
2 Should the track or other angles be suspect and be reflected in abnormal wear to the rear tyres (and there is no distortion of the suspension components due to collision damage), the hub bearings and the trailing arm bonded rubber bushes should be checked for wear and renewed if necessary.

9 Fault diagnosis - rear suspension and hubs

Symptom	Reason/s	Remedy
Excessive tyre wear	Wear in hub bearings	Renew bearings.
	Dampers inoperative	Renew.
	Wear in trailing arm bushes	Arrange renewal with Chrysler dealer.
	Tyre pressures too low	Inflate to specified pressures (cold).
Wheel wobble and vibration	Roadwheels out of balance	Re-balance on or off car.
	Roadwheels buckled	Renew.

Chapter 10 Braking system

For modifications, and information applicable to later models, see Supplement at end of manual

Contents

Specifications

Type ... ... Front disc and rear drum operated by dual hydraulic system with servo assistance. Mechanical handbrake, rear wheels only

Front brakes

Manufacturer	Teves (Ate)
Type	Disc, with single piston floating caliper
Piston diameter	1.890 in (48 mm)
Minimum thickness of disc after resurfacing	0.34 in (10 mm)
Minimum thickness of pad and backing plate	0.28 in (7 mm)

Rear brakes

Manufacturer	Girling or DBA or Teves
Type	Drum, leading and trailing shoe, self-adjusting
Wheel cylinder inner diameter	0.81 in (20.6 mm)
Drum diameter (nominal)	9 in (228.6 mm)
Maximum oversize	0.040 in (1 mm)

Master cylinder

Manufacturer	Teves or Automotive Products (Lockheed)

Type	Split circuit, incorporating differential pressure warning actuator
Cylinder internal diameter	0.75 in (19 mm)
Stroke (both cylinders)	0.59 in (15 mm)

Brake servo

Manufacturer	Lockheed 'Master-Vac' or Girling FD (flexing diaphragm)

Stop light switch

Type	Mechanical

Torque wrench settings

	lb f ft	kg f m
Front brakes		
Caliper fixing bolts	54	7.47
Bleed screw	5	0.69
Brake disc to hub	37	5.12
Flexible hose to caliper	7	0.97
Brake pipe union nut	7	0.97
Driveshaft nut to hub	144	19.91
Lower suspension balljoint nut	55	7.61
Rear brakes		
Wheel cylinder to back plate	7	0.97
Back plate to suspension arm	22	3.04
Drum to hub	11	1.52
Bleed screw	4	0.55
Brake pipe to wheel cylinder	7	0.97
Brake controls		
Brake servo fixing nuts	7	0.97
Master cylinder to brake servo	7	0.97
Brake pressure reducing valve fixings	16	2.21
Brake pipe union nuts to master cylinder and brake pressure reducing valve	7	0.97
Locknut, handbrake adjuster	7	0.97
Handbrake fixings to body	13	1.8
Locknut, brake pressure reducing valve adjusting screw	7	0.97
Bracket, handbrake outer cable to suspension arm	17	2.35

1 General description

The braking system is hydraulically operated with servo assisted disc brakes on the front wheels and drum brakes on the rear.

The handbrake operates mechanically on the rear wheels only. Both front and rear brakes are self-adjusting.

The master cylinder is the dual circuit tandem type. The primary cylinder supplies the rear brakes and the secondary cylinder supplies the front brakes. Both systems are independent so that in the event of a hydraulic leak occurring in one system, the other system will remain fully operative although the brake pedal travel will increase.

A pressure differential switch is fitted to the master cylinder and if a pressure loss occurs in either the front or rear braking system, the switch will illuminate a warning light on the instrument panel. The warning light circuit has a press-to-test facility which should be checked at regular intervals.

A brake equaliser valve is mounted on the rear suspension cross-member and is actuated by the position of the anti-roll bar under varying loads. The valve automatically varies the braking effort to the rear wheels according to the vehicle load. Adjustment of the valve is critical to ensure the correct braking forces are applied under all loading and road surface conditions.

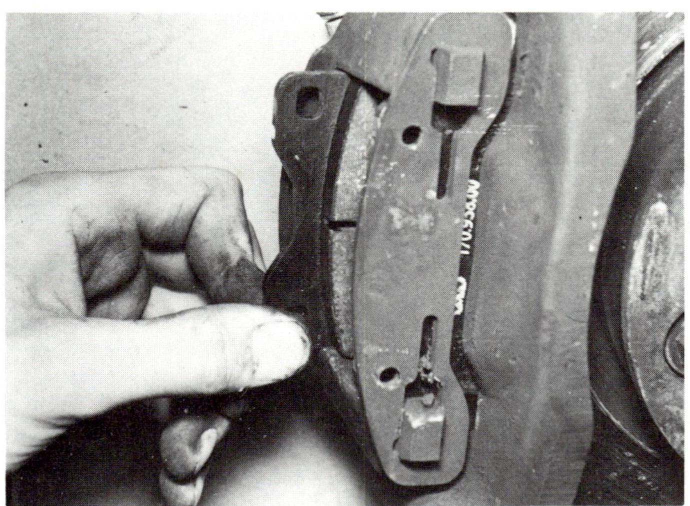

2.5 Removing the inner brake pad

2.7 Removing the outer brake pad

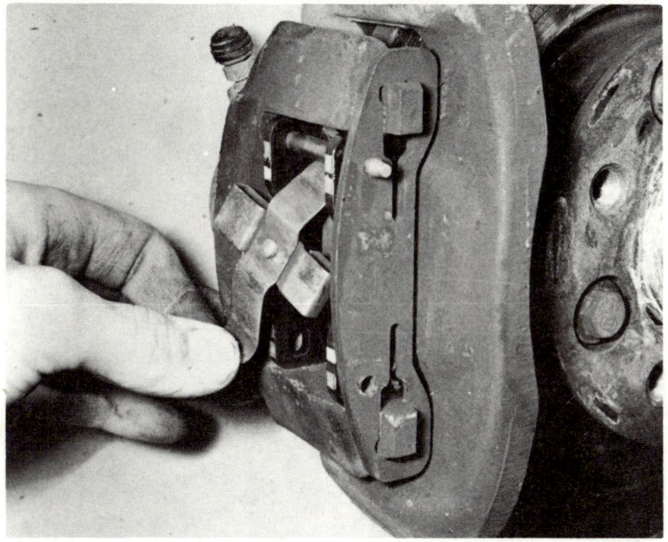

2.14 Refitting the spring clip

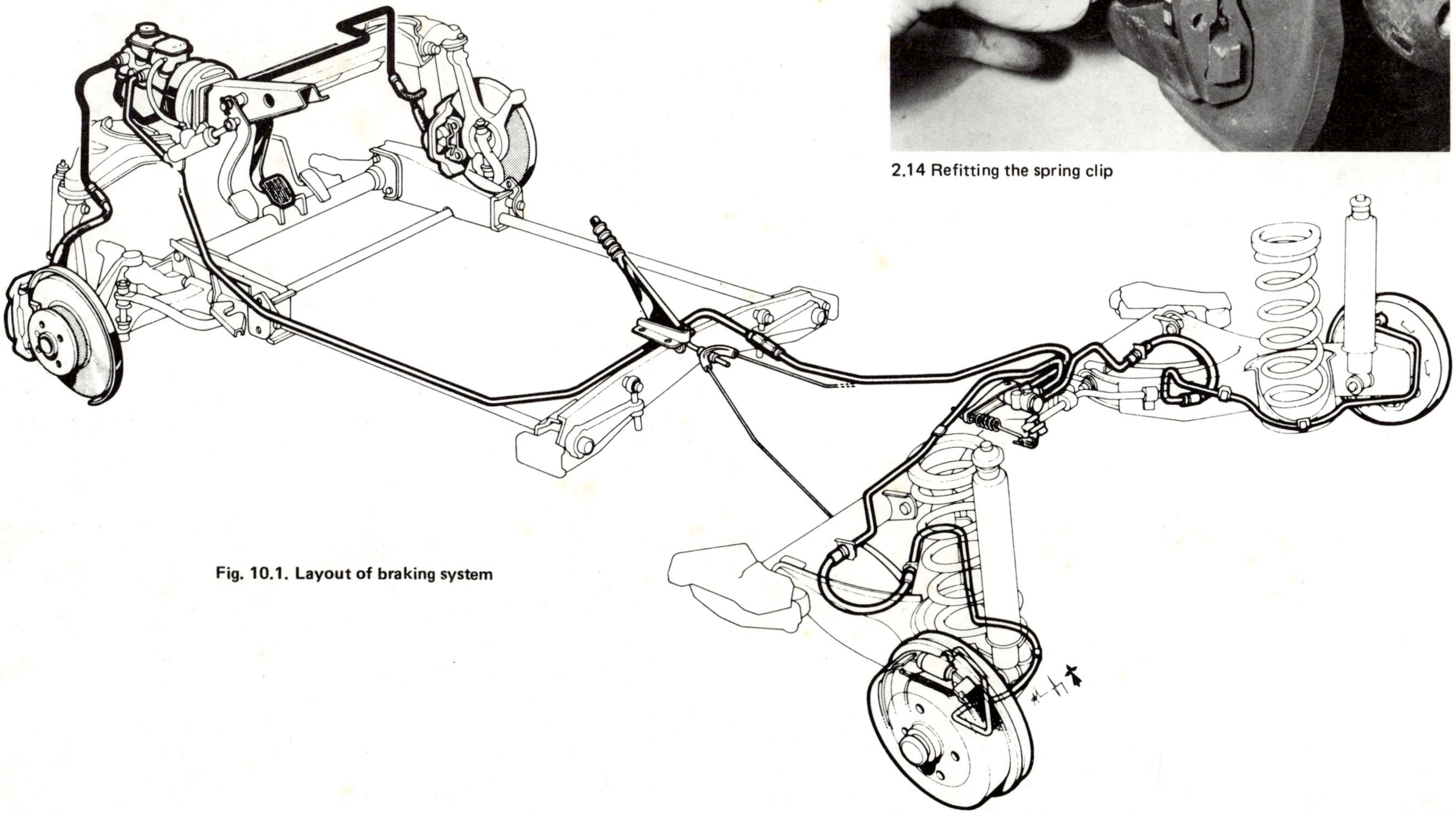

Fig. 10.1. Layout of braking system

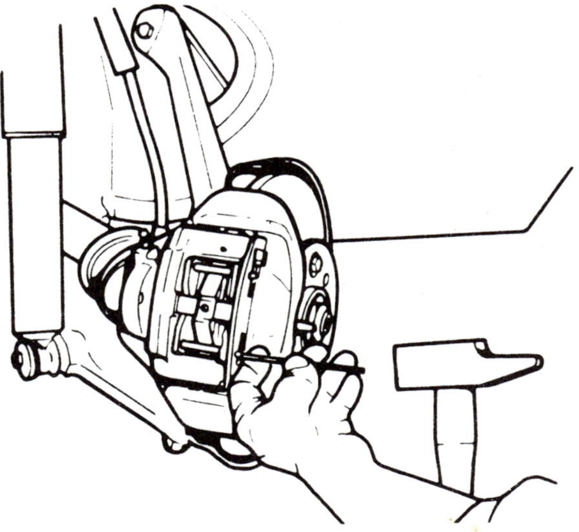

Fig. 10.2. Driving out the brake pad retaining pins

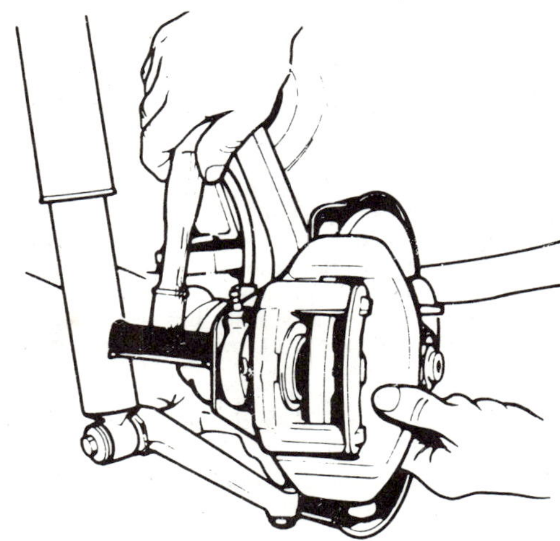

Fig. 10.3. Tapping caliper assembly across prior to removing the inner pad

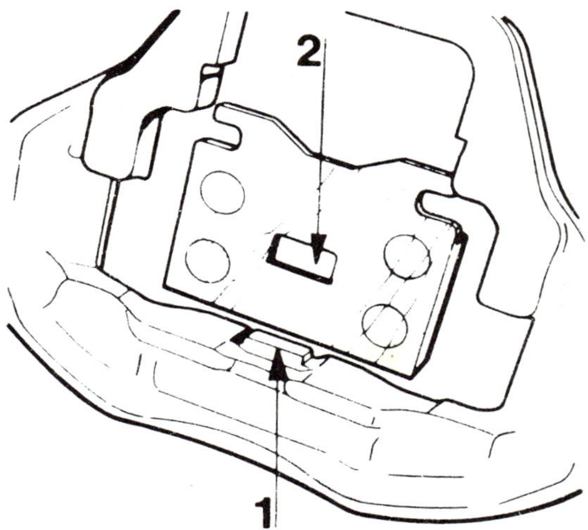

Fig. 10.4. Brake pad locating points

1 Tongue on caliper *2 Groove in pad*

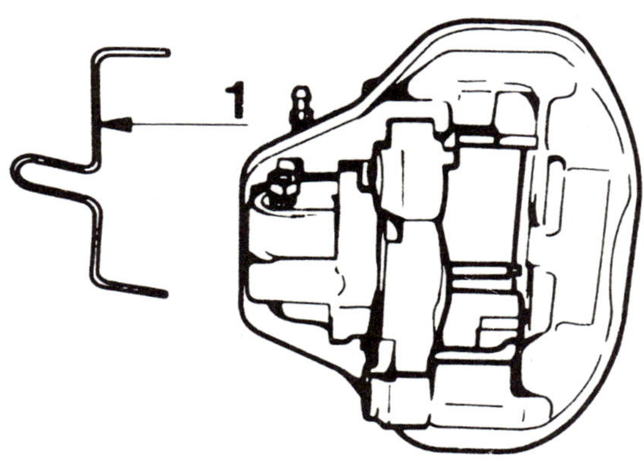

Fig. 10.5. Caliper retaining spring (1)

2 Disc brake pads - inspection, removal and refitting

1 Although the disc brakes are self-adjusting, pad wear should be checked every 3,000 miles.
2 Jack up the vehicle and remove the roadwheels.
3 Inspect the caliper unit for signs of oil or grease. Where these are evident, then the source must be found and the leak cured. Almost certain it will be due to a defective hub oil seal or leaking caliper unit piston seals.
4 Tap out the two disc pad retaining pins from the caliper (Fig. 10.2) and remove the spring.
5 Remove the inner pad first (photo).
6 Gently tap the inner side of the caliper so that it moves across approximately 3/16 in (5 mm) (Fig. 11.3).
7 The outer pad which has a locating slot in it may now be withdrawn (photo).
8 Clean the inside of the caliper assembly and check that the piston dust cover is in good condition. If there is any sign of fluid leakage around the piston, the caliper must be overhauled as described in Section 3.
9 Check the thickness of the friction pad (not including its metal backing plate). If the thickness is 3/32 in or less, renew the pads. Pads must be renewed as complete front wheel sets of four to maintain even braking.
10 If new pads are to be fitted, the caliper unit pistons must be pushed back into the cylinder to accommodate the thicker pads. Use a flat lever to do this but ensure that the piston is pressed squarely in and only sufficiently to enable the inner pad to enter the caliper unit opening. During this operation, the fluid reservoir level will rise and it may be necessary to syphon some off.
11 If difficulty is experienced in retracting the pistons, it may be necessary to open the bleed screw to release the hydraulic pressure. Take care not to let air enter the system otherwise if will be necessary to bleed the system as described in Section 10.
12 Insert the outer pad first, ensuring that the slot in the metal backing of the pad is correctly engaged with the tongue in the caliper (Fig. 10.4).
13 Insert the inner pad into the caliper assembly.
14 Slide the top retaining pin through the holes in the caliper and pads and hook the end of the spring under it (photo).
15 Push the other end of the spring down and slide in the second retaining pin.
16 Press the footbrake several times to centralise the pads and top up the reservoir if necessary.

3 Front disc brake caliper - removal, overhaul and refitting

1 Jack up the front of the car and remove the brake pads as described in the previous Section.

2 Unscrew and remove the bolts which secure the caliper to the stub axle.
3 Lift the caliper from its location and, if no further operations are to be carried out to the unit, tie it up to the suspension to prevent strain on the flexible hydraulic hoses.
4 If the caliper is to be overhauled or renewed, remove the cap from the master cylinder reservoir and place a piece of polythene sheet over the opening and then press the cap on again. This will create a vacuum and prevent loss of fluid when the hydraulic line is disconnected.
5 The flexible hose to the caliper can either be removed as described in Section 11, or a simpler method is to hold the union at the caliper end of the hose in a spanner and unscrew the caliper from the hose. Make sure when using this method that the hose is not twisted. Retain the sealing washer and plug the end of the hose to prevent loss of fluid.
6 Remove the spring clip shown in Fig. 10.5.
7 Slide the bracket out of the yoke assembly.
8 Tap the yoke assembly firmly on a wooden bench to dislodge the cylinder from the yoke. The caliper assembly is now dismantled into its three main components as shown in Fig. 10.7.
9 To dismantle the cylinder assembly, grip it in a vice with padded jaws and remove the dust cover securing clip (Fig. 10.6).
10 Remove the dust cover and tap the cylinder on a wooden bench to remove the piston and seal.
11 Inspect the piston and cylinder for signs of score marks or excessive scuffing and, if evident, obtain a new cylinder and piston assembly complete with seals.
12 If the piston and cylinder are in good condition, purchase a seal overhaul kit from your Chrysler Dealer.
13 Lubricate the new piston seal with brake fluid and fit it into the groove in the cylinder using finger pressure only.
14 Lubricate the piston and insert it into the cylinder.
15 Fit the dust cover over the end of the piston and cylinder and secure it in place with the spring clip.
16 Clean the yoke and bracket assembly and refit the cylinder into the yoke.

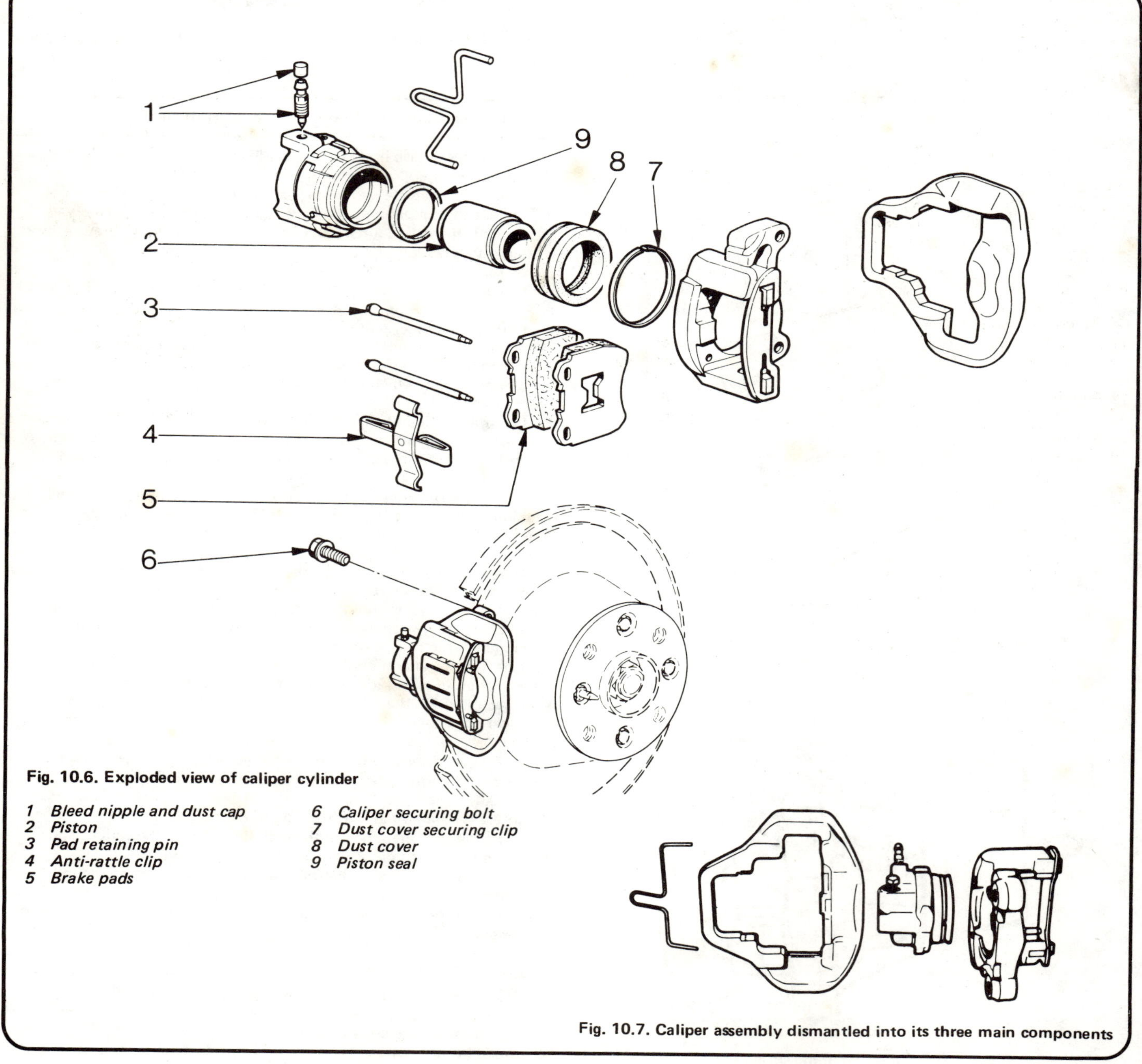

Fig. 10.6. Exploded view of caliper cylinder

1 Bleed nipple and dust cap
2 Piston
3 Pad retaining pin
4 Anti-rattle clip
5 Brake pads
6 Caliper securing bolt
7 Dust cover securing clip
8 Dust cover
9 Piston seal

Fig. 10.7. Caliper assembly dismantled into its three main components

17 Insert the bracket into the yoke, fit it over the end of the cylinder and secure it in place with the spring.
18 Reconnect the flexible brake hose to the caliper and refit the caliper to the stub axle carrier. Tighten the bolts to the specified torque wrench setting.
19 Refit the disc pads as described in Section 2.
20 Remove the polythene sheet from the reservoir and bleed the front braking circuit as described in Section 10. Refit the roadwheel and lower the car. Tighten the wheel nuts fully when the wheels are on the ground.

4 Rear brake shoes - inspection, removal and refitting

1 Every 10,000 miles the rear brake linings should be examined for wear.
2 Jack up the rear of the car, support it on axle stands and remove the roadwheels.
3 Remove the two retaining screws and pull off the drum. If the drum sticks, insert a thin screwdriver through the aperture in the rear of the back plate and release the brake adjusting pawl (Fig. 10.8)
4 Examine the linings and if they have worn down to a thickness of 0.1 in (2.5 mm) or less, or if they are contaminated with oil or grease a new set of shoes and linings should be obtained.
5 Two different types of rear brakes are fitted to the Alpine; either the Girling type (Fig. 10.10) or the DBA type (Fig. 10.11).

Shoe removal - Girling type

6 Unhook the adjuster lever spring from the forward brake shoe and remove it (photo).
7 Withdraw the adjuster lever from its pivot pin.
8 Rotate the serrated nut in the appropriate direction to reduce the adjuster pushrod to its minimum length (photo).
9 Grip the end of each spring retainer with a pair of pliers and rotate them through 90°. Remove the retainer springs and pins.
10 Push the handbrake lever forward and disengage the end of the handbrake cable (photo).
11 Unhook the ends of the lower return spring from each shoe.
12 Lift the lower end of each shoe away from the bottom retaining plate.
13 Pull the top of the shoes away from the wheel cylinder taking care not to tear the rubber boots, and remove both shoes complete with adjuster and top return spring.
14 Place a strong rubber band around the wheel cylinder to prevent the pistons being pushed out and subsequent loss of brake fluid.
15 Unhook the top spring from each shoe and remove the adjuster.
16 Remove the spring clip from the pivot pin on the trailing shoe and withdraw the handbrake lever.

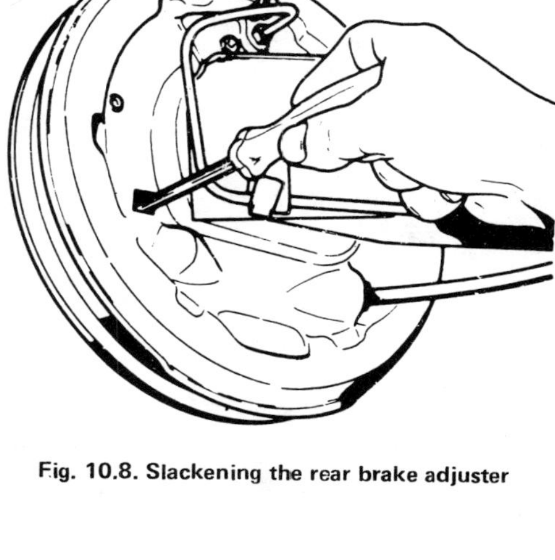

Fig. 10.8. Slackening the rear brake adjuster

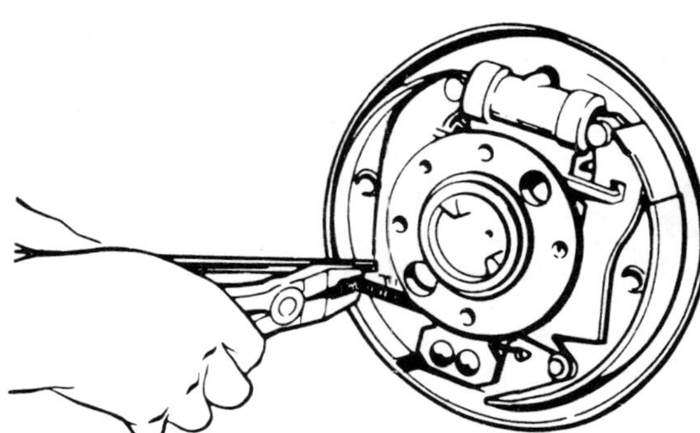

Fig. 10.9. Disconnecting the handbrake cable

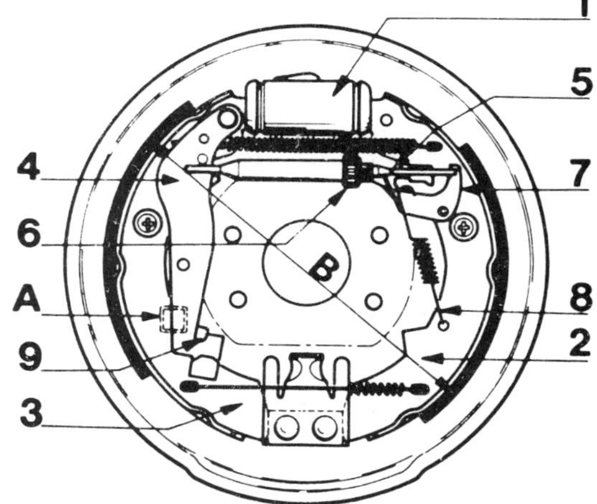

Fig. 10.10 Girling rear brake assembly

1 Wheel cylinder
2 Leading (primary) shoe
3 Trailing (secondary) shoe
4 Handbrake lever
5 Adjuster push rod
6 Serrated nut
7 Adjuster lever
8 Spring, adjuster lever
9 Stop, handbrake lever
A Aperture, adjuster release
B Overall diameter of shoes prior to fitting drum - 227 to 227.9 mm (8 15/16 to 8 31/32 in)

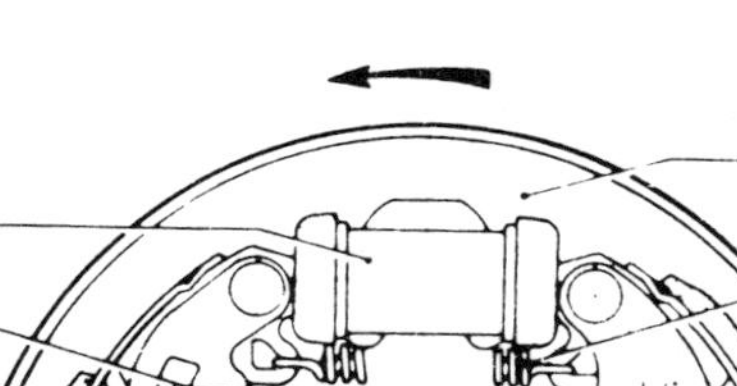

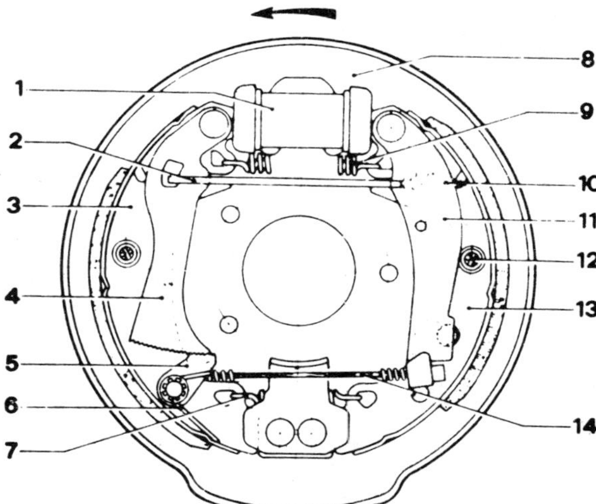

Fig. 10.11. DBA rear brake assembly

1 Wheel cylinder
2 Link rod
3 Primary (leading) shoe
4 Adjuster lever
5 Adjuster pawl
6 Lower shoe return spring
7 Lower shoe return spring
8 Backplate
9 Upper shoe return spring
10 Spring, link rod and handbrake lever
11 Handbrake lever
12 Shoe steady spring
13 Secondary (trailing) shoe
14 Handbrake cable

4.6 View of Girling type rear brake

4.8 Brake adjuster push rod and nut (Girling type brakes)

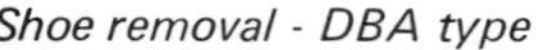

Shoe removal - DBA type

17 Remove the shoe steady springs by pressing them in with a screwdriver and turning until they disengage from the back plate.

18 Unhook the handbrake cable from the brake operating lever (Fig. 10.9).

19 Using a strong screwdriver, lever the lower ends of the shoes from behind the bottom retaining plate and pull the upper ends of both shoes from the wheel cylinder.

20 Push the shoes in towards the hub, unhook the top and bottom return springs and withdraw the shoes complete with adjuster lever and link rod (Fig. 10.12).

21 Fit an elastic band around the wheel cylinder to prevent the pistons coming out.

22 Remove the spring clip and withdraw the adjuster lever from the leading shoe.

23 Unhook the small spring and remove the link rod from the trailing shoe.

24 Remove the spring clip and withdraw the handbrake lever from the trailing shoe.

25 Thoroughly clean all traces of dust from the shoes, backplate, and drum, using a stiff brush. **Do not blow the dust clear as it is asbestos based and should not be inhaled.** Brake dust can cause judder and squeal and therefore it is important to clean away all traces.

26 Check that each piston is free in its cylinder and that the rubber dust covers are undamaged and in position. Check also that there are no hydraulic fluid leaks.

27 Prior to refitting the brake shoes, smear a trace of Castrol PM Brake Grease to the ends of the brake shoes, the rear shoe/handbrake lever pivot, the steady platform anchor posts and the adjuster surface. Do not allow any of the grease to come into contact with the brake linings or rubber boots.

28 The refitting procedure is the reversal of removal. The two pull off springs should preferably be renewed every time new shoes are fitted and and must be refitted into their original web holes.

29 On the Girling type brakes, turn the serrated nut on the pushrod to expand the shoes to the point where the brake drum will just slide over them.

30 In the case of the DBA brakes, push the adjuster lever towards the leading shoe and engage the first few teeth with the pawl.

31 Refit the brake drums and tighten the retaining screw. Push the brake pedal fully down at least ten times to enable the automatic adjusting mechanism to reset the shoes to the correct positions.

32 Check the operation of the handbrake and if necessary adjust the cable as described in Section 14.

33 Refit the roadwheels, lower the car to the ground and tighten the wheel bolts.

4.10 Handbrake cable attachment point (Girling type brakes)

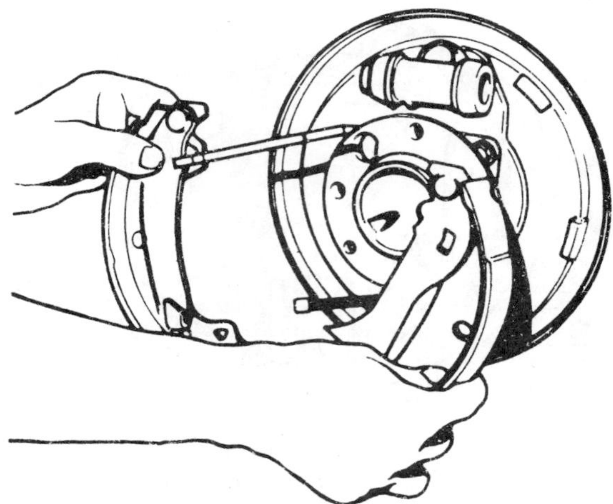

Fig. 10.12. Removing the brake shoes (DBA type)

5 Rear brake wheel cylinder - removal and overhaul

1 Two types of wheel cylinders are fitted to the Alpine (See Figs.

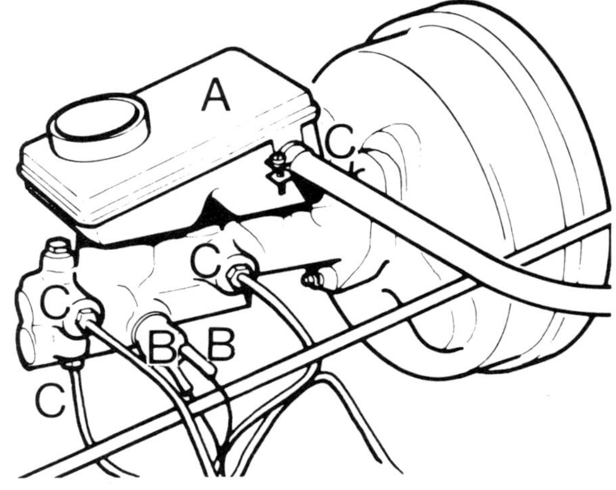

Fig. 10.13. Master cylinder assembly

A Reservoir
B Pressure switch leads
C Fluid outlet pipes

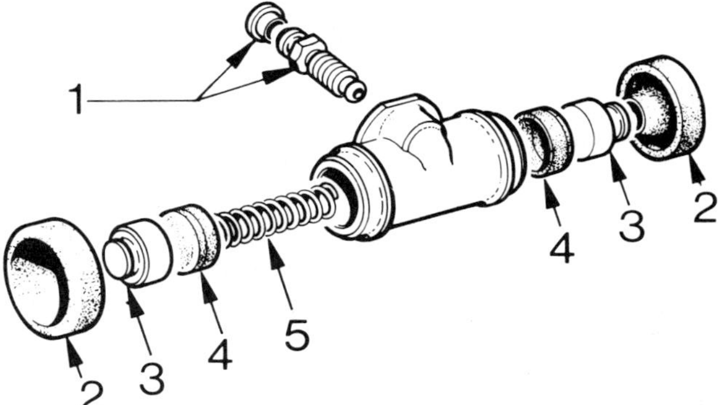

Fig. 10.14. Exploded view of Bendix type rear wheel brake cylinder

1 Bleed nipple and cap
2 Dust cover
3 Piston
4 Seal
5 Spring

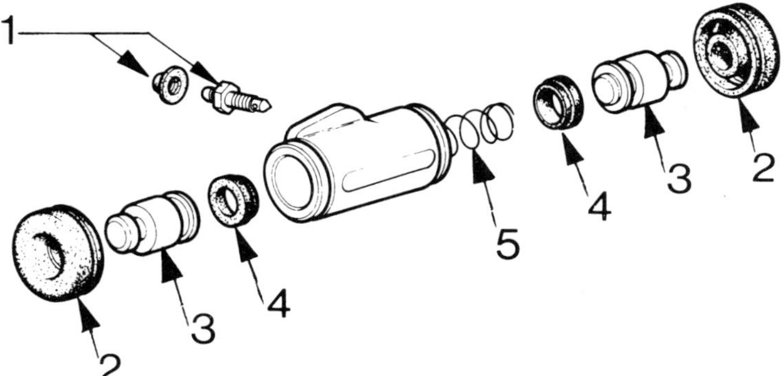

Fig. 10.15. Exploded view of Girling type rear wheel brake cylinder

1 Bleed nipple and cap
2 Dust cover
3 Piston
4 Seal
5 Spring

10.14 and 10.15) and although the method of removal and overhaul is virtually identical they are not interchangeable. When purchasing a new cylinder assembly take the old one along to your Chrysler Dealer to ensure that the correct type is obtained.
2 Jack up the rear of the car, remove the roadwheel.
3 Remove the drum and brake shoes as described in Section 4.
4 Disconnect the hydraulic fluid hose from the wheel cylinder and plug the hose.
5 Remove the two securing bolts which retain the wheel cylinder to the backplate.
6 Remove the wheel cylinder.
7 Peel off the rubber dust excluders from each end of the unit and then eject the internal components - pistons, spring and seals. This may be done by tapping the cylinder carefully on a piece of wood until the pistons emerge or by removing the bleed nipple and applying a tyre pump to the nipple orifice.
8 Wash all components in hydraulic fluid and examine the piston and cylinder internal surfaces for scratches, scoring or bright areas. Where these are evident, renew the complete wheel cylinder assembly on an exchange basis.
9 Obtain a repair kit which will contain all the necessary seals. Observe absolute cleanliness during the following operations.
10 Lubricate the bore of the cylinder with clean hydraulic fluid and insert one seal, one piston and one dust excluderat one end of the unit. Insert the spring from the opposite end, followed by the seal, piston and dust cover. Note that the lips of both seals face inwards.
11 Refit the cylinder to the backplate, refit the shoes and drum. Reconnect the fluid hose.
12 Bleed the hydraulic system as described in Section 10.
13 Refit the roadwheel, lower the car to the ground and tighten the wheel bolts.

6 Master cylinder - removal and refitting

1 Two types of master cylinder are fitted, either the Teves or Lockheed. However, the method of removal and refitting is the same for both types.
Caution: Brake fluid has the same effect as paint stripper and care must be taken to avoid dripping any on the paintwork.
2 Refer to Fig. 10.13 and disconnect the fluid outlet pipes from the reservoir and master cylinder. Plug the ends of the pipes to reduce fluid loss and prevent dirt ingress.
3 Disconnect the two wires from the pressure differential switch on the side of the master cylinder.
4 Remove the nuts and washers securing the master cylinder to the servo unit and withdraw the cylinder assembly.
5 Refit the master cylinder using the reverse procedure to that of removal and bleed the system as described in Section 10.

7 Master cylinder - servicing

1 Although the Teves and Lockheed type master cylinders differ slightly in design the method of dismantling and inspection is virtually

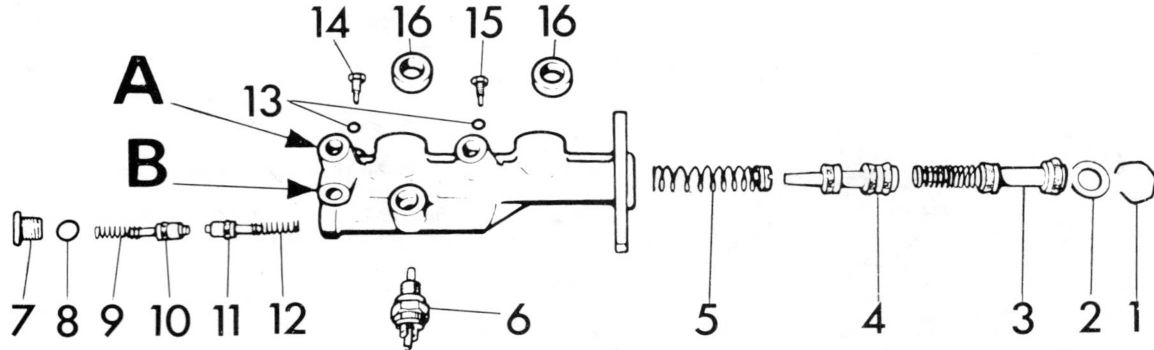

Fig. 10.16. Exploded view of Teves type master cylinder

1 Circlip
2 Stop washer
3 Primary piston
4 Secondary piston
5 Piston spring
6 Switch, warning
7 Plug
8 Seal
9 Spring
10 Piston
11 Piston
12 Spring
13 Seal, stop screw
14 Front stop screw
15 Rear stop screw
16 Sealing rings
AB Outlet ports, front brakes

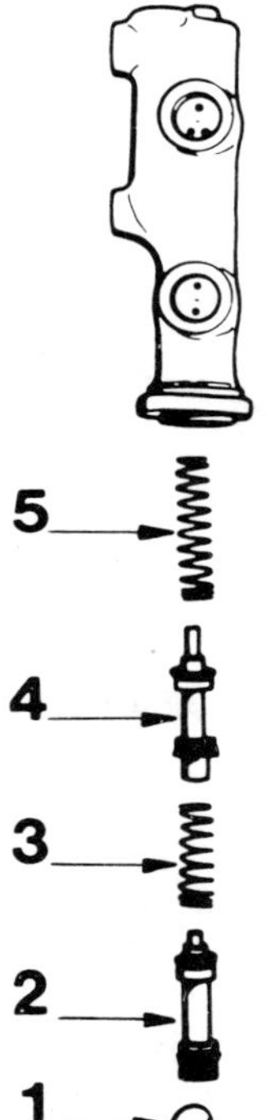

Fig. 10.17. Master cylinder pistons (Lockheed)

1 Circlip
2 Primary piston
3 Primary spring
4 Secondary piston
5 Secondary spring

the same, and providing reference is made to the appropriate illustrations no problems should be encountered.

2 Remove the reservoir filler cap and drain the fluid into a suitable container.

3 Pull the reservoir off the master cylinder and withdraw the sealing rings from the inlet ports.

4 Push the primary piston down the cylinder bore just enough to remove the circlip and withdraw the washer (if fitted), primary piston and spring.

5 Push the secondary piston down the cylinder bore slightly and remove the stop screw(s). Withdraw the secondary piston and spring. If necessary tap the end of the cylinder on a wooden bench to dislodge the piston.

6 Unscrew the pressure differential warning switch from the cylinder assembly.

7 Unscrew the plug from the end of the cylinder and withdraw the warning switch actuating piston(s) assembly, (see Figs. 10.16 and 10.20).

8 Make a careful note of which way round all the rubber seals are fitted before removing and discarding them.

9 Note that on the Lockheed type master cylinder a washer is fitted under the piston seals.

10 Thoroughly wash all parts in either methylated spirit or clean approved hydraulic fluid and place in order ready for inspection.

11 Examine the bore of the master cylinder carefully for any signs of scoring, ridges or corrosion, and if it is found to be smooth all over, new seals can be fitted. If there is any doubt as to the condition of the bore, then a new cylinder must be fitted.

12 If examination of the seals shows them to be apparently oversize or very loose on their seats, suspect oil contamination in the system. Oil will swell these rubber seals, and if one is found to be swollen it is reasonable to assume that all seals in the braking system will require attention.

13 Before reassembly again wash all parts in methylated spirit or clean approved hydraulic fluid. Do not use any other type of oil or cleaning fluid or the seals will be damaged.

14 Commence reassembling by lubricating the bores with clean hydraulic fluid.

15 Smear the new secondary piston seals with hydraulic fluid and fit these to the secondary piston. Make sure that they are fitted the correct way round.

16 Smear the new primary piston seals with hydraulic fluid and fit these to the primary pistons. Make sure that they are fitted the correct way round.

17 Position the master cylinder between soft faces and clamp in a vice in such a manner that the main bore is inclined with the open end downwards.

18 Insert the spring and then the secondary piston assembly, second spring and the primary pistons assembly. To avoid any damage to the cup seals a flattened needle should be passed around the lip of each seal to assist entry into the cylinder bore.

19 Reposition the master cylinder so that it is now vertical with the open end upwards and place the top washer in position. Depress the primary piston slightly and fit the circlip.

20 Next fully depress the primary piston and fit the stop screw(s).

21 Fit new seals to the warning switch actuator pistons, and insert them into the cylinder housing. Refit the end plug and sealing ring.

22 Fit a new seal to the warning switch and screw it into the cylinder housing.

23 Fit new seals into the fluid inlet ports and press the reservoir into the correct position on the master cylinder.

24 The master cylinder is now ready for refitting to the car as described in Section 6.

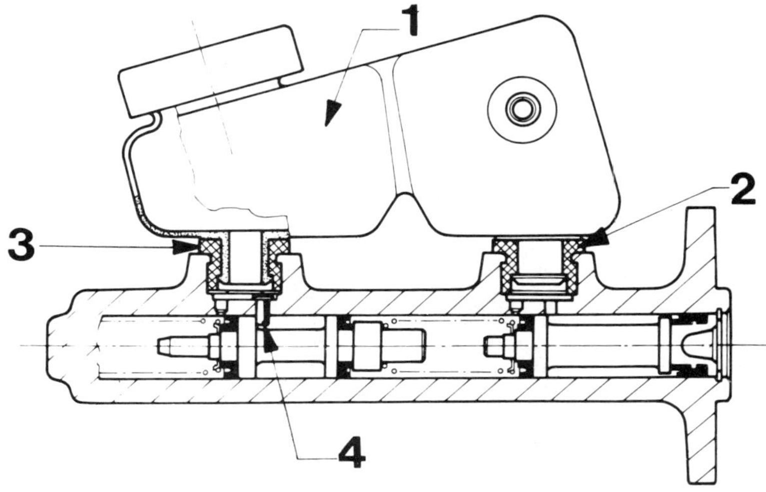

Fig. 10.18. Cross-sectional view of Lockheed type master cylinder

1 Reservoir
2 Inlet port seal (primary)
3 Inlet port seal (secondary)
4 Secondary piston stop pin

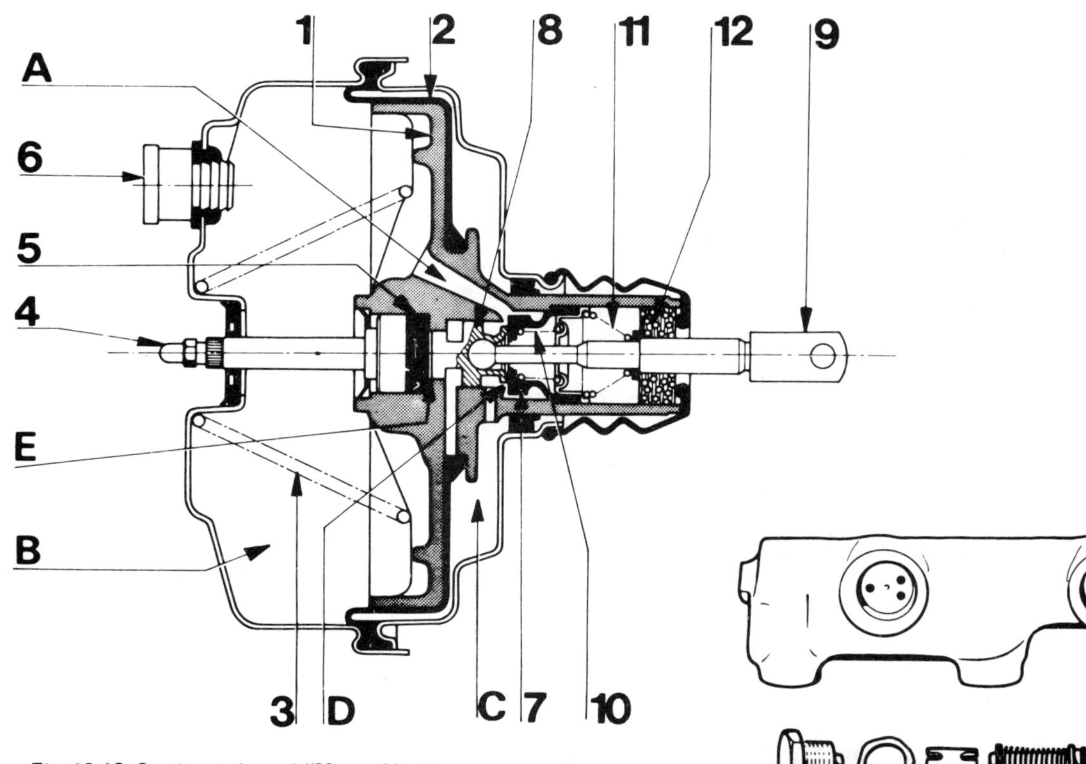

Fig. 10.19. Sectional view of "Master Vac" type servo unit

1 Valve body
2 Diaphragm
3 Return spring
4 Push rod (master cylinder)
5 Reaction disc
6 Non-return valve
7 Air valve
8 Plunger
9 Push rod (brake pedal)
10 Air valve spring
11 Valve rod spring
12 Air filter
A Vacuum passage
B Vacuum chamber
C Atmospheric chamber
D Atmospheric passage
E Valve body shoulder

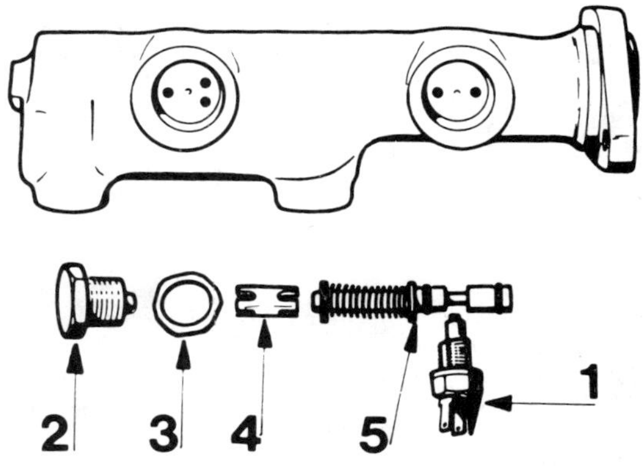

Fig. 10.20. Master cylinder pressure switch actuator components (Lockheed)

1 Switch
2 Plug
3 Sealing ring
4 Distance piece
5 Piston

8 Brake servo unit - description and maintenance

Description

1 A vacuum servo unit is fitted into the brake hydraulic circuit in series with the master cylinder, to provide power assistance to the driver when the brake pedal is depressed.

The unit operates by vacuum obtained from the induction manifold and comprises, basically, a booster diaphragm and a non-return valve.

The servo unit and hydraulic master cylinder are connected together so that the servo unit piston rod acts as the master cylinder pushrod. The driver's braking effort is transmitted through another pushrod to the servo unit piston and its built-in control system. The servo unit piston does not fit tightly into the cylinder, but has a strong diaphragm to keep its edges in constant contact with the cylinder wall so assuring an air tight seal between the two parts. The forward chamber is held under vacuum conditions created in the inlet manifold of the engine and, during periods when the brake pedal is not in use, the controls open a passage to the rear chamber so placing it under vacuum. When the brake pedal is depressed, the vacuum passage to the rear chamber is cut off and the chamber opened to atmospheric pressure. The consequent rush of air pushes the servo piston forward in the vacuum chamber and operates the main pushrod to the master cylinder. The controls are designed so that assistance is given under all conditions and, when the brakes are not required, vacuum in the rear chamber is established when the brake pedal is released. Air from the atmosphere entering the rear chamber is passed through a small air filter.

Two types of servo units are fitted to the Alpine, (see Figs. 10.19 and 10.21). However the method of operation, removal and servicing procedures are identical.

Maintenance and adjustments

2 The brake servo unit operation can be checked easily without any special tools. Proceed as follows:

3 Stop the engine and clear the servo of any vacuum by depressing the brake pedal several times.

4 Once the servo is cleared, keep the brake pedal depressed and start the engine. If the servo unit is in proper working order, the brake pedal should move further downwards, under even foot pressure, due to the effect of the inlet manifold vacuum on the servo diaphragms.

5 If the brake pedal does not move further downwards the servo system is not operating properly, and the vacuum hoses from the inlet manifold to the servo should be inspected. The vacuum control valve should also be checked. This valve is in the vacuum hose to prevent air flowing into the vacuum side of the servo from the inlet manifold when the engine stops. It is in effect a one way valve.

6 If the brake servo operates properly in the test, but still gives less effective service on the road, the air filter through which air flows into

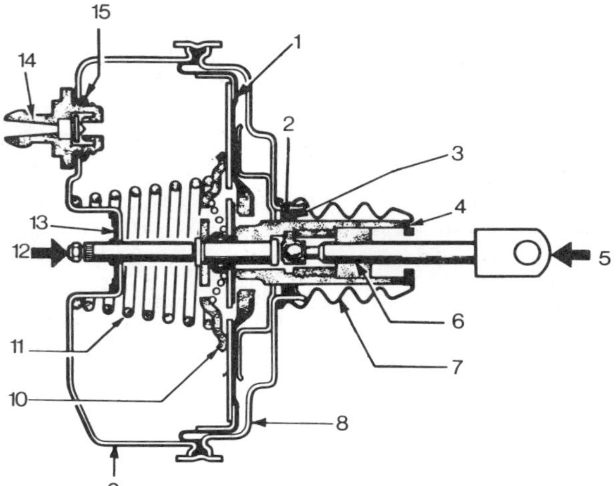

Fig. 10.21. Sectional view of Girling "FD" type servo unit

1 Diaphragm
2 Seal
3 Retainer
4 Filter retainer
5 Push rod (pedal)
6 Air filter
7 Dust cover
8 Rear shell
9 Front shell
10 Fulcrum plate
11 Return spring
12 Push rod (master cylinder)
13 Seal and support plate
14 Non-return valve
15 Grommet

10.2A Brake bleeding kit automatic dispenser fitted to the brake fluid reservoir

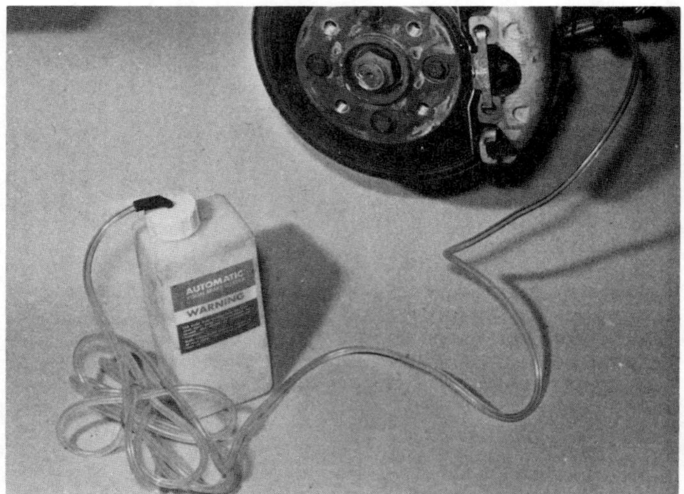

10.2B Brake bleeding kit container and visual line connected to a front brake caliper

the servo should be inspected. A dirty filter will limit the formation of a difference in pressure across the servo diaphragm.
7 The servo unit itself cannot be repaired and therefore a complete renewal is necessary if the measures described are not effective.
8 *Air filter replacement:* From inside the car, detach the brake pedal from the servo pushrod, then remove the rubber boot over the pushrod housing and air filter.
9 Hook out the filter and push in a new one. Refit the retainer and rubber boot.
10 *Non-return valve:* The non-return valve is fitted on the front face of the servo unit and is connected to the vacuum inlet hose.
11 It is not repairable and if faulty should be renewed.

9 Brake servo unit - removal and refitting

1 Refer to Section 6 and remove the brake master cylinder.
2 Slacken the hose clip and detach the vacuum hose from the connector on the non-return valve.
3 Remove the clip retaining the servo to the brake pedal pushrod clevis pin. Extract the clip, lift away the plain washer and withdraw the clevis pin.
4 Undo and remove the four nuts and spring washers that secure the servo unit to the mounting bracket. Lift away the servo unit.
5 Refitting the servo unit is the reverse sequence to removal. It is important that the brake hydraulic system be completely bled as described in the following Section.

10 Bleeding the hydraulic system

1 If any of the hydraulic components in the braking system have been removed or disconnected, or if the fluid level in the master cylinder has been allowed to fall appreciably, it is inevitable that air will have been introduced into the system. The removal of all this air from the hydraulic system is essential if the brakes are to function correctly, and the process of removing it is known as bleeding.
2 There are a number of one-man, do-it-yourself, brake bleeding kits currently available from motor accessory shops. It is recommended that one of these kits should be used wherever possible as they greatly simplify the bleeding operation and also reduce the risk of expelled air and fluid being drawn back into the system (photos).
3 If one of these kits is not available then it will be necessary to gather together a clean jar and a suitable length of clear plastic tubing which is a tight fit over the bleed screw, and also to engage the help of an assistant.
4 Before commencing the bleeding operation, check that all rigid pipes and flexible hoses are in good condition and that all hydraulic unions are tight. Take great care not to allow hydraulic fluid to come into contact with the vehicle paintwork, otherwise the finish will be seriously damaged. Wash off any spilled fluid immediately with cold water.
5 If hydraulic fluid has been lost from the master cylinder, due to a leak in the system, ensure that the cause is traced and rectified before proceeding further or a serious malfunction of the braking system may occur.
6 To bleed the system, clean the area around the bleed screw at the wheel cylinder to be bled. If the hydraulic system has only been partially disconnected and suitable precautions were taken to prevent further loss of fluid, it should only be necessary to bleed that part of the system. However, if the entire system is to be bled, start at the wheel furthest away from the master cylinder.
7 Remove the master cylinder filler cap and top up the reservoir. Periodically check the fluid level during the bleeding operation and top up as necessary.
8 If a one-man brake bleeding kit is being used, connect the outlet tube to the bleed screw and then open the screw half a turn. If possible position the unit so that it can be viewed from the car, then depress the brake pedal to the floor and slowly release it. The one-way valve in the kit will prevent dispelled air from returning to the system at the end of each stroke. Repeat this operation until clean hydraulic fluid, free from air bubbles, can be seen coming through the tube. Now tighten the bleed screw and remove the outlet tube.
9 If a one-man brake bleeding kit is not available, connect one end of the plastic tubing to the bleed screw and immerse the other end in the jam jar containing sufficient clean hydraulic fluid to keep the end of the tube submerged. Open the bleed screw half a turn and have your assistant depress the brake pedal to the floor and then slowly release it. Tighten the bleed screw at the end of each downstroke to prevent expelled air and fluid from being drawn back into the system. Repeat this operation until clean hydraulic fluid, free from air bubbles, can be seen coming through the tube. Now tighten the bleed screw and remove the plastic tube.
10 If the entire system is being bled the procedures described above should now be repeated at each wheel, finishing at the wheel nearest to the master cylinder. Do not forget to recheck the fluid level in the master cylinder at regular intervals and top up as necessary.
11 If the failure indicator lamp (where fitted) lights up during these operations when the ignition switch is on, close the nipple which is open, and open one in the other brake circuit. Apply a steady pressure at the foot pedal until the lamp goes out, then release the pedal and close the nipple.
12 When completed, recheck the fluid level in the master cylinder, top up if necessary and refit the cap. Check the 'feel' of the brake pedal which should be firm and free from any 'sponginess' which would indicate air still present in the system.
13 Discard any expelled hydraulic fluid as it is likely to be contaminated with moisture, air and dirt which makes it unsuitable for further use.

11 Hydraulic pipes and hoses - inspection, removal and refitting

Important: All the hydraulic pipe and flexible hose connections on the Alpine have a M10 x 1 mm metric thread as opposed to the 3/8 in UNF thread used on other models. The metric pipe unions, hoses etc, are identified by a black or olive paint mark and care must be taken to ensure that new items have the correct size thread.
1 Periodically, and certainly well in advance of the DoE Test, all brake pipes, connections and unions should be completely and carefully examined.
2 Examine first all the unions for signs of leaks. Then look at the flexible hoses for signs of fraying and chafing (as well as for leaks). This is only a preliminary inspection of the flexible hoses, as exterior condition does not necessarily indicate interior condition which will be considered later.
3 The steel pipes must be examined equally carefully. They must be thoroughly cleaned and examined for signs of dents or other percussive damage, rust and corrosion. Rust and corrosion should be scraped off, and, if the depth of pitting in the pipes is significant, they will require renewal. This is most likely in those areas underneath the chassis and along the rear suspension arms where the pipes are exposed to the full force of road and weather conditions.
4 If any section of pipe is to be removed, first take off the fluid reservoir cap, line it with a piece of polythene film to make it airtight and screw it back on. This will minimise the amount of fluid dripping out of the system when the pipes are removed.

12.1 Rear brake pressure equalising valve

5 Rigid pipe removal is usually quite straightforward. The unions at each end are undone and the pipe drawn out of the connection. The clips which may hold it to the car body are bent back and it is then removed. Underneath the car the exposed union can be particularly stubborn, defying the efforts of an open ended spanner. As few people will have the special split ring spanner required, a self-grip wrench is the only answer. If the pipe is being renewed, new unions will be provided. If not, then one will have to put up with the possibility of burring over the flats on the unions and of using a self-grip wrench for refitting.

6 Flexible hoses are always fitted to a rigid support bracket where they join a rigid pipe, the bracket being fixed to the chassis or rear suspension arm. The rigid pipe unions must first be removed from the flexible union. Then the locknut securing the flexible pipe to the bracket must be unscrewed, releasing the end of the pipe from the bracket. As these connections are usually exposed they are, more often than not, rusted up and a penetrating fluid is virtually essential to aid removal. When undoing them, both halves must be supported as the bracket is not strong enough to support the torque required to undo the nut and can be snapped off easily.

7 Once the flexible hose is removed, examine the internal bore. If clear of fluid it should be possible to see through it. Any specks of rubber that come out, or there are signs of restriction in the bore, mean that the inner lining is breaking up and the hose must be renewed.

8 Rigid pipes which need renewing can usually be purchased at your local garage where they have the pipe, unions and special tools to make them up. All that they need to know is the pipe length required and the type of flare used at the ends of the pipe. These may be different at each end of the same pipe. If possible, it is a good idea to take the old pipe along as a pattern.

9 Refitting of the pipes is a straightforward reversal of the removal procedure. It is best to get all the sets (bends) made prior to fitting. Also, any acute bends should be put in by the garage on a bending machine otherwise there is the possibility of kinking them, and restricting the bore area and thus, fluid flow.

10 With the pipes refitted, remove the polythene from the reservoir cap and bleed the system as described in Section 10.

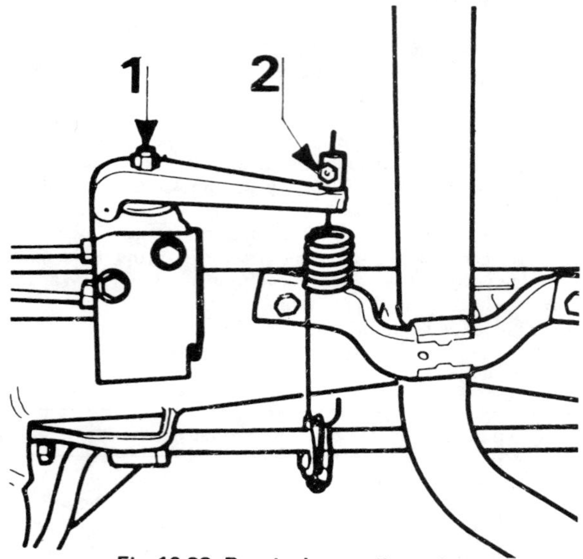

Fig. 10.22. Rear brake equaliser valve

1 Adjusting screw *2 Spring stop screw*

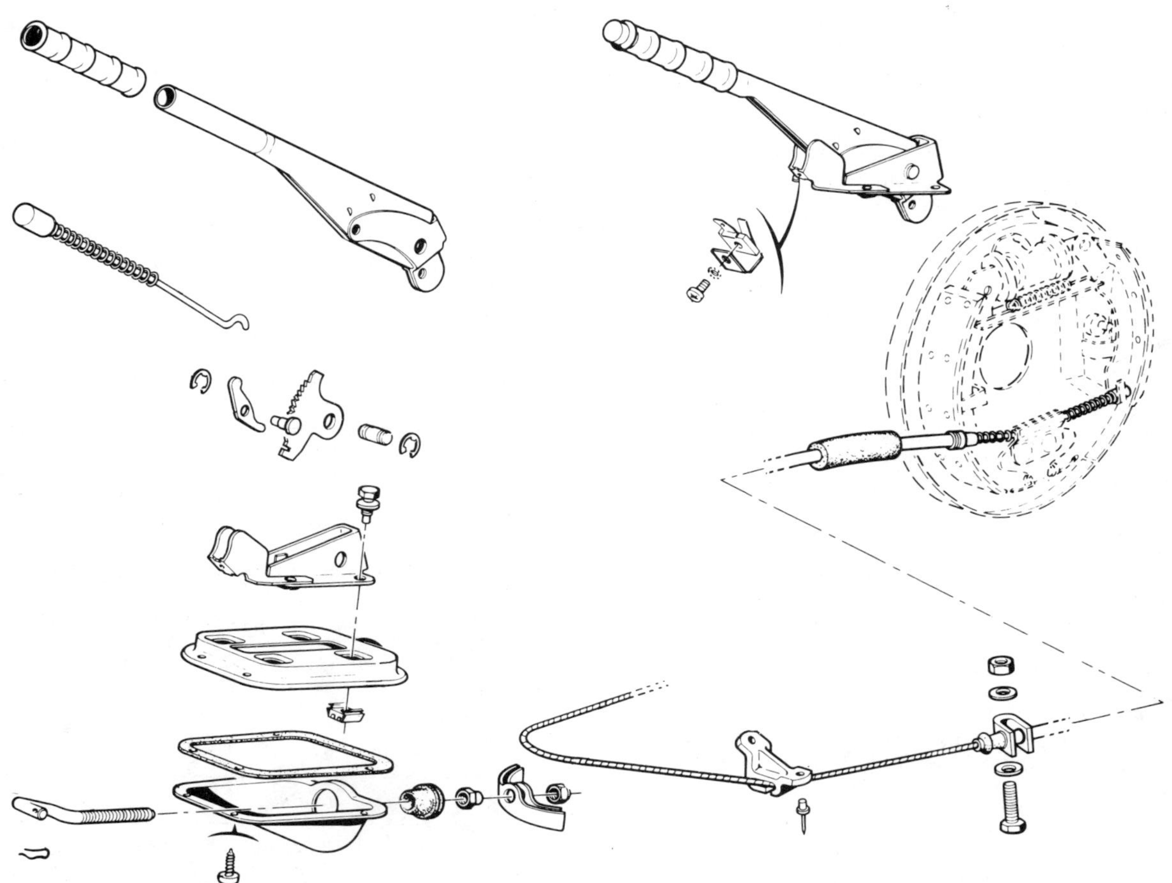

Fig. 10.23. Handbrake lever components and cable

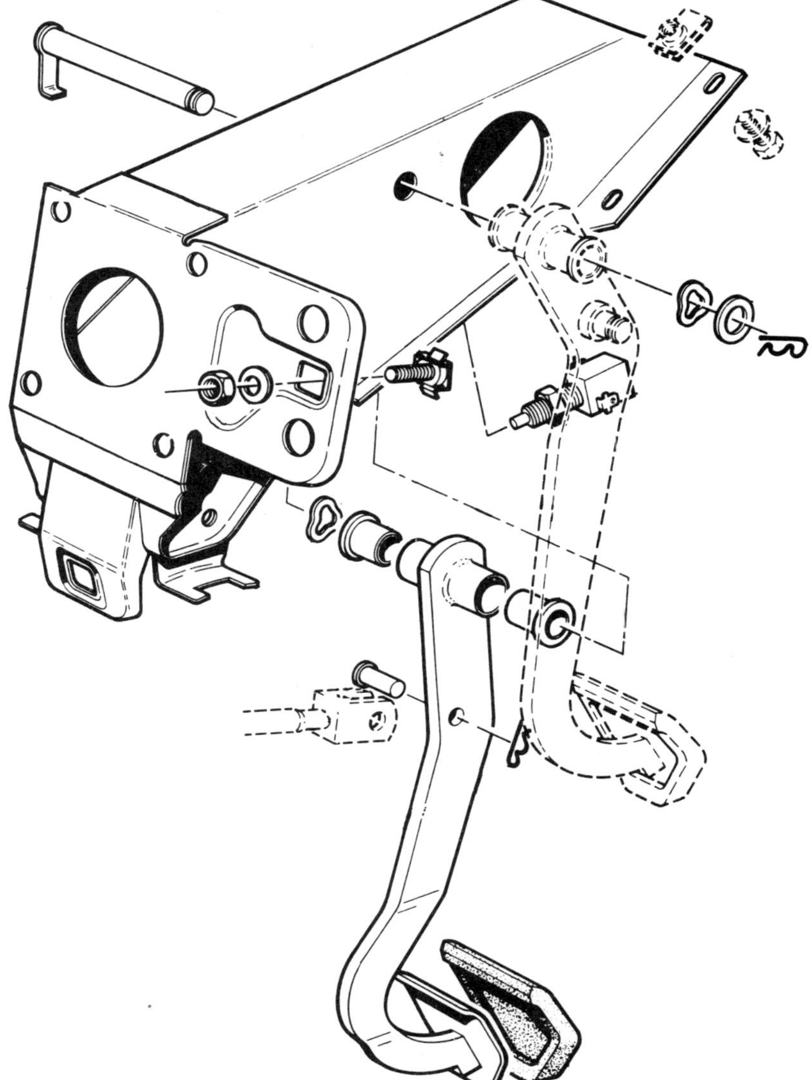

Fig. 10.24. Footbrake pedal components

12 Brake equaliser valve - checking and adjusting

1 The brake equaliser valve assembly exerts a compensating and regulating effect upon the front and rear braking effort according to the vehicle load. The equaliser valve is mounted on the rear suspension crossmember and is actuated by the pull of a spring fitted between the valve operating lever and an arm on the rear anti-roll bar (photo). It is essential for safe and effective braking that the following operations are correctly carried out.
2 First disconnect both rear shock absorber lower mountings.
3 With the car unladen check that the spring is fairly taut but that the coils are closed up and not in a stretched condition.
4 To adjust the spring tension, slacken the locknut and turn the adjusting screw in or out to reposition the spring. Do **not** make any attempt to alter the position of the spring stop screw (See Fig. 10.22).
5 The length of the spring is very important and must not be altered in any way.
6 If it is felt that the rear wheels are overbraking, slacken the locknut and turn the adjustment screw out a small amount. Tighten the locknut and retest the brakes.
7 If the rear wheel braking effort appears to be insufficient, turn the screw inwards a small amount and then recheck the brakes.
8 The valve is not repairable and if faulty must be renewed. This will necessitate bleeding the rear braking system (see Section 10).

13 Handbrake - adjustment

1 The handbrake is normally automatically adjusted when the rear drum brakes are adjusted. However, after extended service, the cable may stretch and the following procedure should be carried out. This method should also be used when a new cable or handbrake assembly has been fitted.
2 Fully release the handbrake and then pull the handbrake up five notches (clicks).
3 Release the locknut (A) on the cable equalising yoke and adjust nut (B) (Fig. 10.25).
4 Adjustment is correct when the cables are taut and the rear wheels are fully locked when the lever is pulled up six notches (clicks).
5 When adjustment is complete, hold nut (B) quite still and tighten the locknut (A).

14 Handbrake assembly - renewal

1 The handbrake assembly comprises three sub assemblies, the lever, the operating rod and yoke and the cable.
2 A broken or overstretched cable can only be renewed as a unit as fittings and ends are not detachable.
3 To remove the cable, first detach the two ends from the rear brake

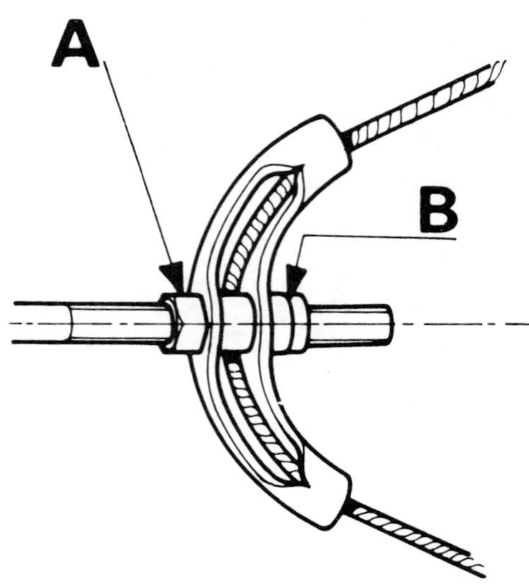

Fig. 10.25. Brake cable adjusting nuts ('A' front nut, 'B' rear nut)

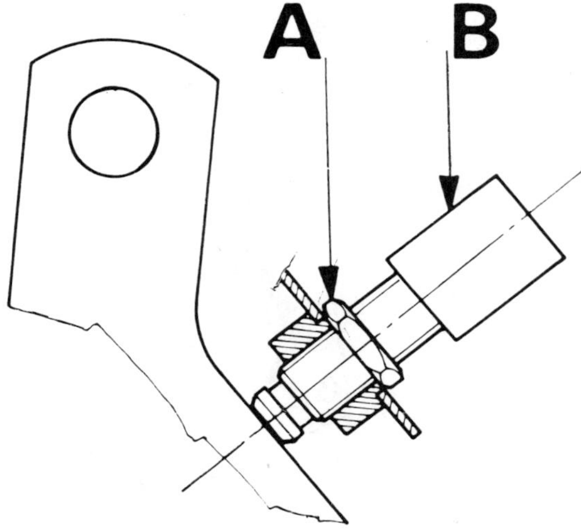

Fig. 10.26. Stop light switch

A Locknut *B Switch*

shoes as described in Section 4.
4 Pull the ends of the cable through the rear brake backplate and remove the two clips securing the cable to the underside of the car.
5 Detach the equalising yoke from the handbrake operating rod, unhook the cable from the front clips and remove it from beneath the car.
6 Refit the new cable using the reverse procedure to that of removal, and adjust it as described in the previous Section.
7 Wear or damage to the rod and yoke can be repaired by renewal of components affected.
8 It is possible to renew the lever ratchet as a separate component after dismantling by removal of the circlips and crosspins. Occasional application of grease to the ratchet and quadrant notches will help to reduce wear. Application of the handbrake with the ratchet button depressed will also increase the life of the lever components.
9 Details of the handbrake assembly are shown in Fig. 10.23.

15 Brake operating pedal - removal and refitting

1 The brake pedal operates on a common cross shaft with the clutch pedal (Fig. 10.24).
2 The method of removal and refitting is as described for the clutch pedal in Chapter 5, Section 3.

16 Brake stop light switch - adjustment

1 The brake stop light switch is located on a bracket above the brake pedal lever.
2 To adjust the function of the switch, first depress the brake pedal several times to exhaust the brake servo unit and then switch on the ignition.
3 Get a friend to watch the rear lights and then depress the pedal and check that the stop lights come on after the pedal has moved down between 1/8 - 3/4 in (4 - 20 mm).
4 Release the pedal fully and check that the stop lights extinguish.
5 Adjust the switch if necessary by slackening the locknut and screwing the switch in or out of its bracket as required. Tighten the locknut when the setting is correct (Fig. 10.26).

17 Discs and drums - reconditioning

1 After extended mileage it is possible that the brake discs and drums will become scored. Any skimming must be carried out professionally and within the tolerances specified in the Specifications.
2 Ovality in brake drums may be corrected by skimming, but excessive run-out in discs is best obviated by renewing the disc.
3 For instructions on removing the discs, refer to Chapter 7.

18 Fault diagnosis - braking system

Symptom	Reason/s
Pedal travels almost to floor before brakes operate	Brake fluid level too low. Caliper leaking. Master cylinder leaking (bubbles in master cylinder fluid) Brake flexible hose leaking. Brake line fractured. Brake system unions loose. Pad or shoe linings over 75% worn. Rear brake badly out of adjustment (automatic adjusters seized)
Brake pedal feels springy	New linings not yet bedded-in. Brake discs or drums badly worn or cracked. Master cylinder securing nuts loose.
Brake pedal feels 'spongy' and 'soggy'	Caliper or wheel cylinder leaking. Master cylinder leaking (bubbles in master cylinder fluid). Brake pipe line or flexible hose leaking. Unions in brake system loose.
Excessive effort required to brake car	Faulty vacuum servo unit. Pad or shoe linings badly worn. New pads or shoes recently fitted - not yet bedded-in. Harder linings fitted than standard causing increase in pedal pressure. Linings and brake drums contaminated with oil, grease or hydraulic fluid.
Brake uneven and pulling to one side	Linings and discs or drums contaminated with oil, grease or hydraulic fluid. Tyre pressures unequal. Brake caliper loose. Brake pads or shoes fitted incorrectly. Different type of linings fitted at each wheel. Anchorage for front suspension or rear suspension loose. Brake discs or drums badly worn, cracked or distorted.
Brakes tend to bind, drag or lock-on	Rear brakes over adjusted. Air in system. Handbrake cables over-tightened. Wheel cylinder or caliper pistons seized.
Brake warning light comes on and stays on	Leak in front or rear hydraulic circuit. Check master cylinder reservoirs.

Chapter 11 Electrical system

For modifications, and information applicable to later models, see Supplement at end of manual

Contents

Specifications

System type	12 volt negative earth
Battery	
Type	Lead acid
Capacity	40 amp/hr at 20 hr discharge rate
Alternator	
Make	Paris-Rhone, Ducellier or Motorola
Output	
Paris-Rhone	35 or 40 amps

Ducellier	35 amps
Motorola	40 amps
Field resistance	7 ohms (all types)

Alternator control unit

Make	Ducellier or Paris-Rhone
Regulating voltage	14.6 to 15.1 volts (35 amp unit), 13.8 to 14.4 volts (40 amp unit)
Condenser capacity (35 amp unit only)	470 mfd

Starter motor

Make	Ducellier, Paris-Rhone or Bosch
Type	Pre-engaged
No. of teeth (drive pinion)	9
No. of teeth (ring gear)	112
Pinion end-float	0.020 to 0.090 in (0.51 to 2.29 mm)

Windscreen wiper motor

Make	Bosch, Marchal or Siem
Speeds	Two

Bulb ratings	**Wattage**
Headlamps	40 - 45 55 - 60 halogen
Side lamps	4
Front indicators	21
Rear indicators	21
Reverse lamps	21
Rear fog lamp	21
Stop/tail lamps	21 - 5
Rear number plate lamp	5 + 5
Interior courtesy lamps	5
Luggage compartment lamp	4
Heater illumination lamp	3

Torque wrench settings	**lb f ft**	**kg fm**
Starter motor securing bolts	15	2.07
Alternator to mounting bracket	30	4.15
Mounting bracket to crankcase	15	2.07
Strap to alternator	15	2.07

1 General description

The electrical system is of the 12 volt negative earth type. The battery is charged by a belt-driven alternator which is controlled by a voltage regulator.

The starter motor fitted as standard is of the pre-engaged type.

Although repair procedures and methods are fully described in this Chapter, in view of the long life of the major electrical components, it is recommended that when a major fault does develop, consideration should be given to exchanging the unit for a factory reconditioned assembly rather than renewing individual components of a well worn unit.

When fitting electrical accessories to cars with a negative earth system, it is important, if they contain silicon diodes, or transistors, that they are connected correctly, otherwise serious damage may result to the components concerned. Items such as radios, tape recorders, electronic ignition systems, electronic tachometer, automatic dipping etc, should all be checked for correct polarity.

It is important that the battery is always disconnected if it is to be charged when an alternator is fitted. Also if body repairs are to be carried out using electronic arc welding equipment the alternator must be disconnected otherwise serious damage can be caused to the more delicate instruments. When the battery has to be disconnected it must always be reconnected with the negative terminal earthed.

2 Battery - removal and refitting

1 The battery should be removed once every three months for cleaning and testing. Disconnect the battery by slackening the clamp bolts on the terminals and lifting away the clamps. It is important to disconnect the negative (earth) lead first, and reconnect it last.

2 Undo and remove the nut and washer securing the battery clamp and lift away the clamp. Carefully lift the battery from its carrier and hold it upright to ensure that none of the electrolyte is spilled.

3 Refitting is a direct reversal of the removal procedure. Smear the terminals and clamps with petroleum jelly (vaseline) to prevent corrosion. **Never** use ordinary grease.

3 Battery - maintenance and inspection

1 Normal weekly battery maintenance consists of checking the electrolyte level of each cell to ensure that the separators are covered by 0.25 in (6.5 mm) of electrolyte. If the level has fallen, top-up the battery and check the securing bolts, the clamp plate, tray and leads for corrosion. If overfilled or any electrolyte is spilt, immediately wipe away the excess as electrolyte attacks and corrodes any metal it comes into contact with very rapidly.

2 As well as keeping the terminals clean and covered with petroleum jelly (vaseline), the top of the battery, and especially the top of the cells, should be kept clean and dry. This helps prevent corrosion and ensures that the battery does not become partially discharged by leakage through dampness and dirt.

3 Once every three months remove the battery and inspect the battery securing bolts, the clamp plate, tray and leads for corrosion. If any corrosion is found, clean off the deposit with ammonia and paint over the clean metal with an anti-rust and acid paint.

4 At the same time inspect the battery case for cracks. If a crack is found, clean and plug it with one of the proprietary compounds made for this purpose. If leakage through the cracks has been excessive then it will be necessary to refill the appropriate cell with fresh electrolyte as detailed later. Cracks are frequently caused to the top of the battery cases by pouring in distilled water in the middle of winter *after* instead of *before* a run. This gives the water no chance to mix with the electrolyte and so the former freezes and splits the battery case.

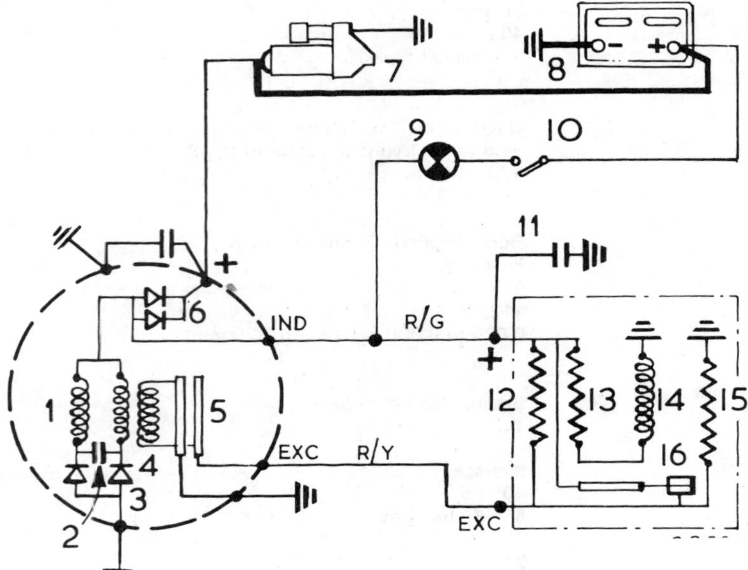

Fig. 11.1. Charging circuit diagram (single phase alternators)

1 Stator windings
2 Capacitor
3 Rectifier diodes
4 Rotor winding
5 Slip rings
6 Isolator diodes
7 Starter motor
8 Battery
9 Warning lamp
10 Ignition switch
11 Capacitor
12 Field resistor
13 Thermal resistor
14 Relay coil
15 Absorption resistor
16 VR contacts

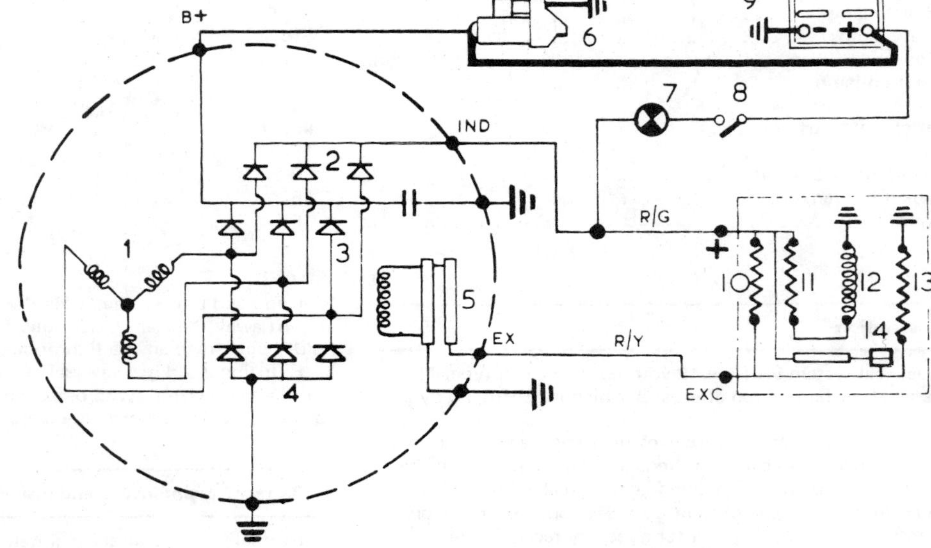

Fig. 11.2. Charging circuit diagram (three phase alternators)

1 Stator windings
2 Field diodes
3 Positive diodes
4 Negative diodes
5 Slip rings
6 Starter motor
7 Warning lamp
8 Ignition switch
9 Battery
10 Field resistor
11 Thermal resistor
12 Relay coil
13 Absorption resistor
14 VR contacts

5 If topping up the battery becomes excessive and the case has been inspected for cracks that could cause leakage, but none are found, the battery is being overcharged and the voltage regulator will have to be checked and if necessary renewed.

6 With the battery on the bench at the three monthly interval check, measure the specific gravity with a hydrometer to determine the state of charge and condition of the electrolyte. There should be very little variation between the different cells and if a variation in excess of 0.025 is present it will be due to either:

a) Loss of electrolyte from the battery at some time caused by spillage or a leak, resulting in a drop in the specific gravity of the electrolyte, when the deficiency was replaced with distilled water instead of fresh electrolyte.

b) An internal short circuit caused by buckling of the plates or a similar malady pointing to the likelihood of total battery failure in the near future.

7 The specific gravity of the electrolyte for fully charged conditions at the electrolyte temperature indicated, is listed in Table A. The specific gravity of a fully discharged battery at different temperatures of the electrolyte is given in Table B.

Table A

Specific gravity - battery fully charged

1.268 at	*100°F or 38°C*	*electrolyte*	*temperature*
1.272 at	*90°F or 32°C*	*"*	*"*
1.276 at	*80°F or 27°C*	*"*	*"*
1.280 at	*70°F or 21°C*	*"*	*"*
1.284 at	*60°F or 16°C*	*"*	*"*
1.288 at	*50°F or 10°C*	*"*	*"*
1.292 at	*40°F or 4°C*	*"*	*"*
1.296 at	*30°F or -1.5°C*	*"*	*"*

Table B

Specific gravity - battery fully discharged

1.098 at	*100°F or 38°C*	*electrolyte*	*temperature*
1.102 at	*90°F or 32°C*	*"*	*"*
1.106 at	*80°F or 27°C*	*"*	*"*
1.110 at	*70°F or 21°C*	*"*	*"*
1.114 at	*60°F or 16°C*	*"*	*"*
1.118 at	*50°F or 10°C*	*"*	*"*
1.122 at	*40°F or 4°C*	*"*	*"*
1.126 at	*30°F or -1.5°C*	*"*	*"*

4 Battery - electrolyte replenishment

1 If the battery is in a fully charged state and one of the cells maintains a specific gravity reading which is 0.025 or more lower than the others, and a check of each cell has been made with a voltage meter to check for short circuits (a four to seven second test should give a steady reading of between 1.2 and 1.8 volts), then it is likely that electrolyte has been lost from the cell at some time with the low reading.
2 Top up the cell with a solution of 1 part sulphuric acid to 2.5 parts of water. If the cell is already fully topped-up draw some electrolyte out of it with a hydrometer.
3 When mixing the sulphuric acid and water **never add water to sulphuric acid** - always pour the acid slowly onto the water in a glass container. **If water is added to sulphuric acid it will explode.**
4 Continue to top up the cell with the freshly made electrolyte and then recharge the battery and check the hydrometer readings.

5 Battery - charging

1 In winter time when heavy demand is placed upon the battery such as when starting from cold, and a lot of electrical equipment is continually in use, it is a good idea to occasionally have the battery fully charged from an external source at the rate of 3.5 to 4 amps.
2 Continue to charge the battery at this rate until no further rise in specific gravity is noted over a four hour period.
3 Alternatively, a trickle charge, charging at the rate of 1.5 amps can be safely used overnight.
4 Specially rapid 'boost' charges which are claimed to restore the power of the battery in 1 to 2 hours are not recommended as they can cause serious damage to the battery plates through overheating.
5 **While charging the battery note that the temperature of the electrolyte should never exceed 100°F (37.8°C).**

6 Alternator driving belt - checking and adjustment

1 The correct belt tension must be maintained at all times, to ensure the correct charging rate and to avoid strain on the alternator bearings.
2 Remove the drive belt protective shield and slacken the alternator mounting bolts and the bolts which secure the adjustment strap (Fig. 11.3).
3 Prise the alternator away from the engine block and tighten the adjustment strap bolts when there is a **total** free movement of ½ in (13 mm) at the centre of the longest run of the belt.
4 Make periodic inspections of the belt and renew it when any sign of fraying is evident.

7 Alternator - fault finding and repair

Due to the specialist knowledge and equipment required to test or service an alternator it is recommended that if the performance is suspect, the car be taken to an auto electrician who will have the facilities for such work. Because of this recommendation, information is limited to the inspection and renewal of the brushes. Should the alternator not charge or the system be suspect the following points may be checked before seeking further assistance.
1 Check the fan belt tension.
2 Check the battery.
3 Check all electrical cable connections for cleanliness and security.

8 Alternator - removal and refitting

1 Disconnect the leads from the battery terminals.
2 Remove the drivebelt protective shield.
3 Slacken the alternator mounting bolts and adjustment strap bolts.
4 Push the alternator in towards the engine and slip the driving belt from the crankshaft, water pump and alternator pulleys.
5 Disconnect the four leads from the rear end of the alternator.
6 On single phase type alternators (Paris-Rhone and Ducellier) remove the pivot bolt, lower the alternator sufficiently to turn it, so the pulley is uppermost, and then lift it up between the distributor and chassis frame.
7 In the case of the three phase (Motorola) alternator, first remove the pivot bolt and then the mounting bracket from the engine. Retrieve the special washer from the bracket retaining bolt nearest the timing cover end of the engine. Lower the bracket down past the engine, followed by the alternator.
8 Refit the alternator using the reversal of the removal procedure.
9 Adjust the drivebelt tension as described in Section 6 and ensure that the leads are reconnected to the alternator as follows:

Black lead to earth (–)
Red lead to positive (+)
Red/yellow lead to 'EXC'
Red/green lead to 'IND'

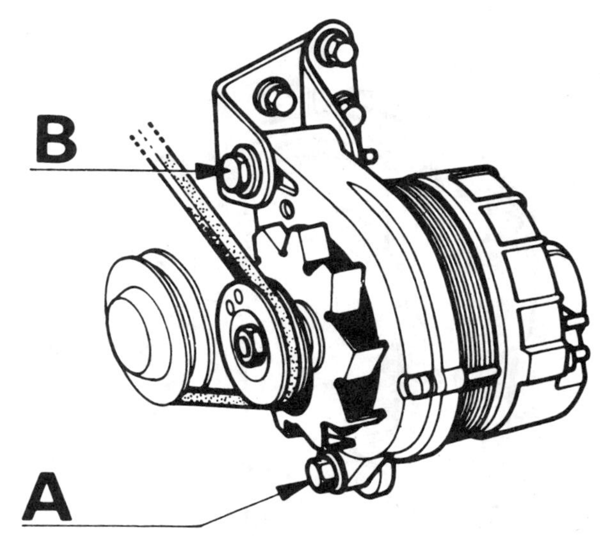

Fig. 11.3. Alternator securing bolts

A Adjusting bolt *B Pivot bolt*

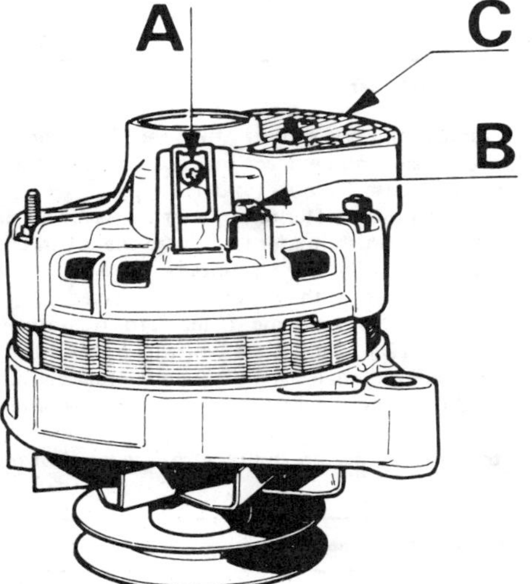

Fig. 11.4. Rear casing of Ducellier alternator

A Brush holder securing screws
B Brush holder securing screws
C Rear cover

Fig. 11.5. Exploded view of Ducellier alternator

1 Pulley nut and washer
2 Spacer
3 Pulley
4 Spacers
5 Fan
6 Through bolt
7 Drive end bracket
8 Bearing drive end
9 Stator laminations
10 Rotor assembly
11 Bearing, slip ring end
12 'O' ring
13 Diode cover
14 Slip ring end bracket
15 Earth connection, stator
16 Diode connections, stator
17 Bearing retainer
18 Spacers
19 Terminal nut
20 Plain washer
21 Locknut
22 Plain washer
23 Insulating washer
24 Diode heat sink
25 Crimped washer
26 Insulating washer
27 Insulator
28 Insulating block
29 Terminal screw
30 Brush terminal
31 Slip ring brush
32 Brush holder

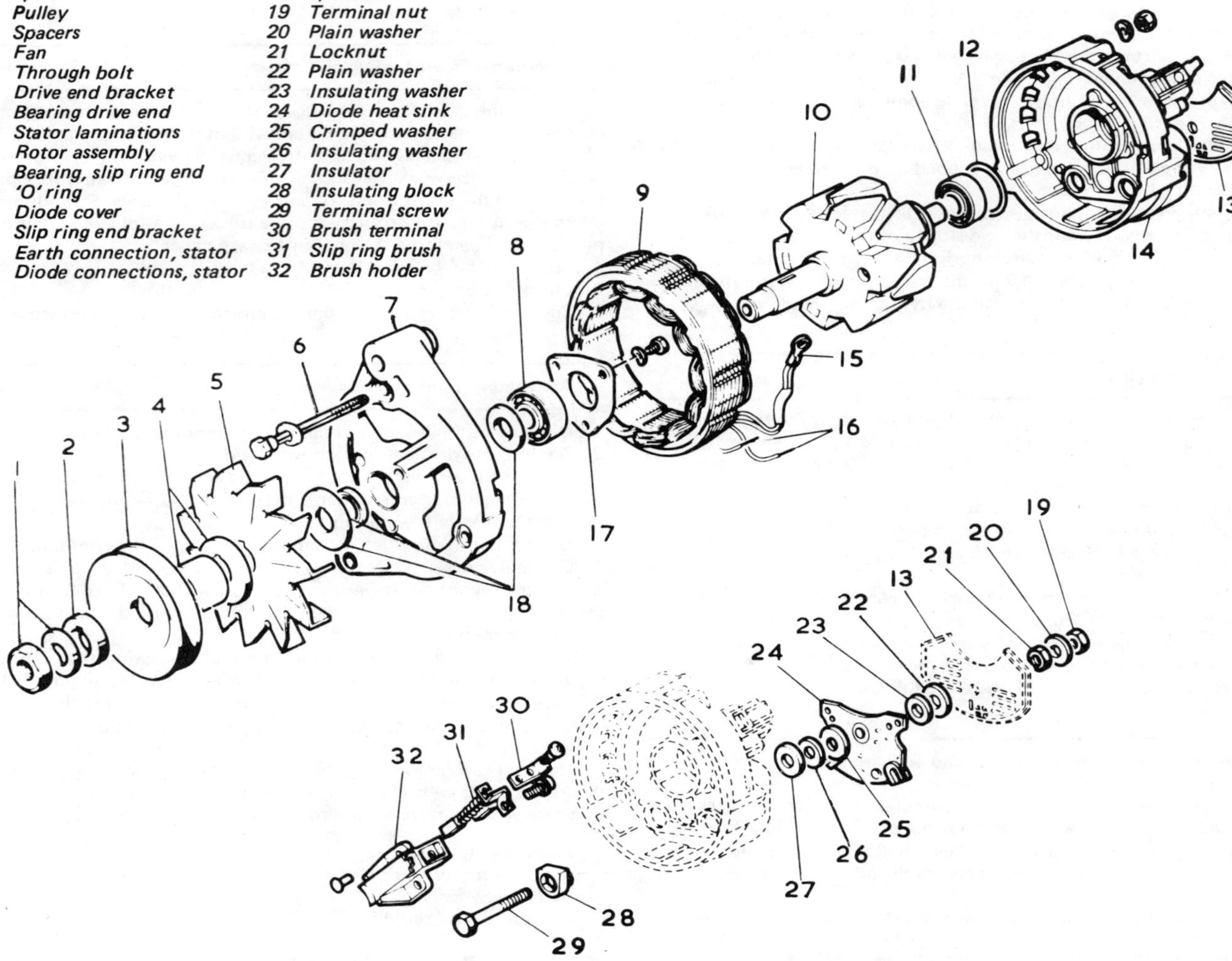

9 Alternator (Ducellier) - servicing

1 Unscrew and remove the two screws which secure the brush holder to the alternator rear casing and extract the brush holder (Fig. 11.4).
2 Withdraw the earth brush from its holder.
3 Withdraw the second brush after removal of the securing screw.
4 Clean the brushes and brush holder with a rag soaked in clean fuel and check that they slide easily in their holders.
5 If the brushes are well worn, they should be renewed.
6 This should be the limit of servicing. If the rotor or stator are damaged or the front or rear bearings are worn then it is quite uneconomical to consider a repair even if the individual components were obtainable and it is recommended that a factory reconditioned unit is obtained.
7 Refitting of the brushes and brush holder is a reversal of removal.

10 Alternator (Paris-Rhone) - servicing

1 The procedure for brush removal is similar to that described in the preceding Section. An exploded view of the alternator is shown in Fig. 11.8.
2 The screws which secure the brush holder are shown in Fig. 11.6.
3 Pull the brushes from their spring holders.
4 Refer to paragraphs 4, 5, 6 and 7 of the preceding Section.

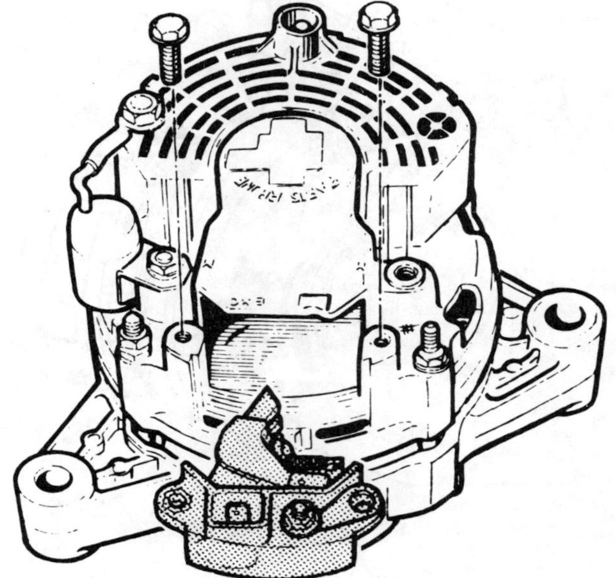

Fig. 11.6. Brush holder and retaining screws on Paris - Rhone alternator

11 Alternator (Motorola) - servicing

1 The heavy duty, three phase Motorola alternator is fitted for operation in cold climates.
2 To renew the brushes undo the two screws securing the cover plate to the rear end of the alternator and lift away the plate taking care not to strain the connecting wire (Fig. 11.7).
3 Remove the securing screws and withdraw the brushholder, taking care not to damage the brushes.
4 Clean or renew the brushes as described in Section 9. An exploded view of the Motorola alternator is shown in Fig. 11.9.

12 Alternator voltage regulator - description

1 The voltage regulator comprises a vibrating contact that controls the alternator output by limiting the current in the field windings.
2 The regulator, which is located inside the engine compartment on the left-hand wing valance is a sealed unit and if it is suspected of being faulty must be renewed.

13 Starter motor - general description

1 The starter motor is of the pre-engaged type which is designed to ensure that the pinion is engaged with the ring gear on the flywheel (torque converter-automatic) before the starter motor is energised.
2 Conversely, the drive does not disengage, until the starter motor is de-energised.
3 The starter motor may be of Ducellier make, Fig. 11.10 or Paris-Rhone Fig. 11.11.

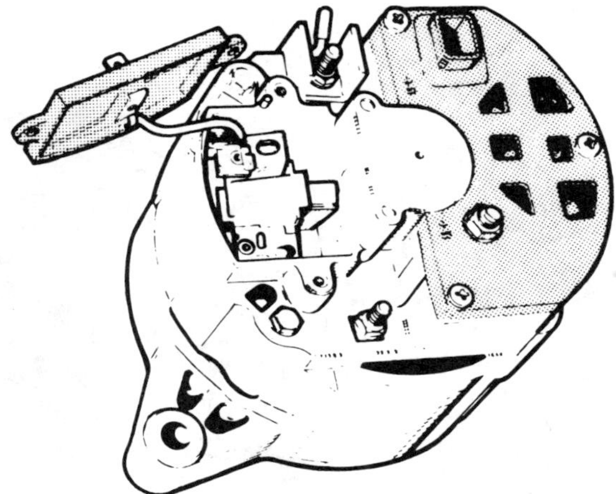

Fig. 11.7. Brush holder cover plate on the Motorola alternator

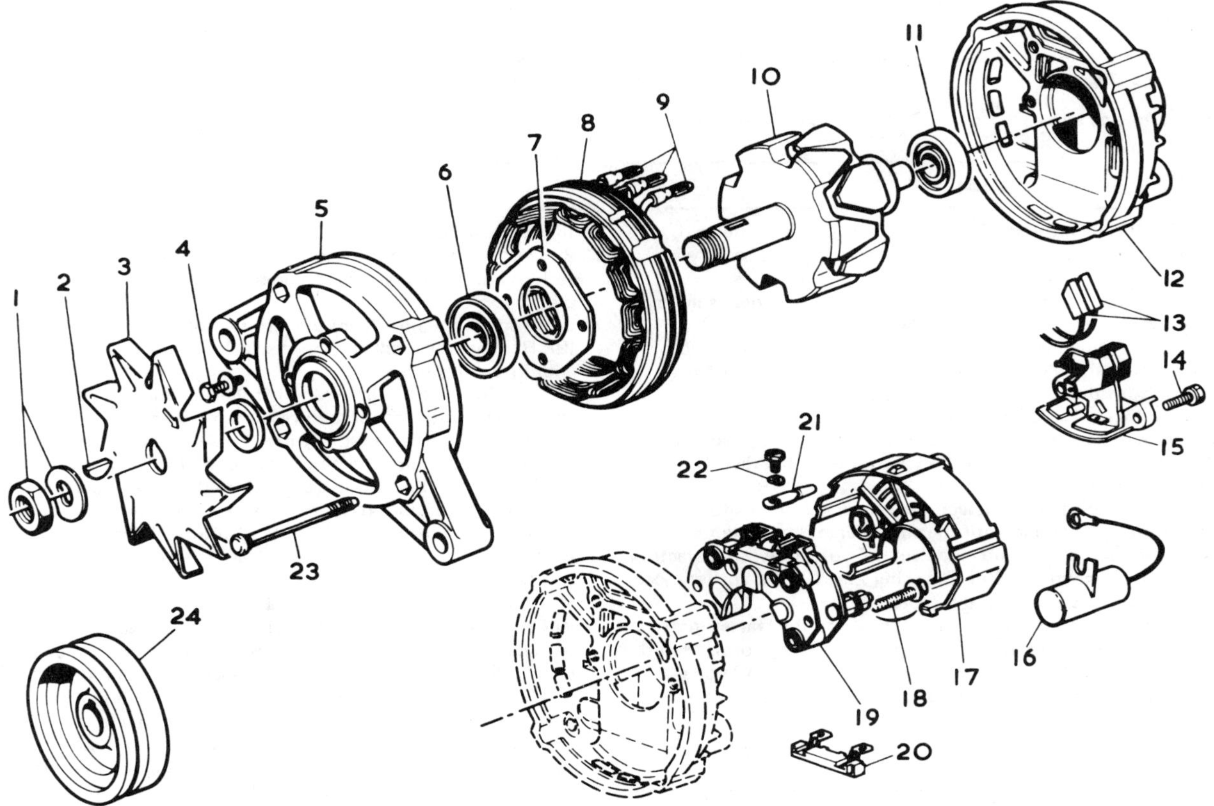

Fig. 11.8. Exploded view of Paris - Rhone alternator

1 Nut and washer
2 Driving key
3 Fan
4 Screw, bearing retainer
5 Drive end bracket
6 Bearing, drive end
7 Bearing retainer
8 Stator laminations
9 Stator leads
10 Rotor assembly
11 Bearing, slip ring end
12 Slip ring end bracket
13 Brushes
14 Screw, brush holder
15 Brush holder
16 Capacitor (when fitted)
17 Diode cover
18 Screw, diode heat sink
19 Diode heat sink assembly
20 Capacitor
21 Terminal, warning lamp
22 Terminal screw
23 Through bolt
24 Pulley

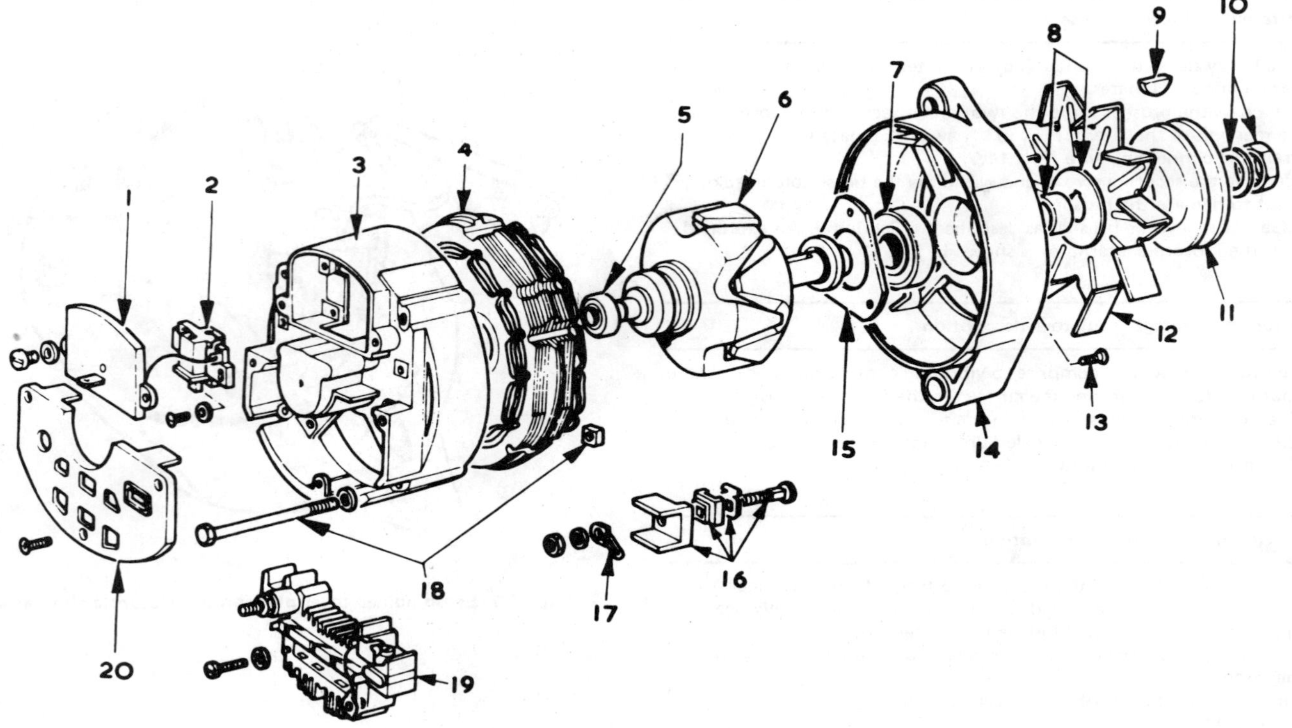

Fig. 11.9. Exploded view of Motorola alternator

1 Brush holder plate
2 Brush holder
3 Bracket, slip ring end
4 Stator
5 Bearing, slip ring end
6 Rotor assembly
7 Bearing, drive end
8 Spacer and washer
9 Driving key
10 Pulley nut and washer
11 Pulley
12 Fan
13 Retainer screw
14 Bracket, drive end
15 Bearing retainer
16 Terminal screw
17 Terminal, warning light
18 Through bolt and nut
19 Diode and heat sink assembly
20 Rear cover

14 Starter motor - testing in vehicle

1 If unsatisfactory operation of the starter motor is experienced, check that the battery connections are tight and that the battery is charged.
2 Check the security of cable terminals at the starter motor solenoid and the leads at the combined ignition starter switch.
3 Do not confuse a jammed starter drive with an inoperative motor. If a distinct click is heard when the starter switch is operated but the motor will not turn then it is certain to be due to a jammed starter drive and the vehicle should be rocked in gear to release it.
4 Where the foregoing possibilities have been eliminated, proceed to test the starter but first disconnect the LT lead from the distributor to prevent the engine from firing.
5 Connect a 0-20 voltmeter between the starter terminal and earth, operate the starter switch and with the engine cranking, note the reading. A minimum voltage (indicated) of 4.5 volts proves satisfactory cable and switch connections. Slow cranking speed of the starter motor at this voltage indicates a fault in the motor.
6 Connect the voltmeter between the battery and the starter motor terminal and with the starter motor actuated, the voltage drop should not exceed more than half the indicated battery voltage (12 volts). Where this is exceeded it indicates excessive resistance in the starter circuit.

15 Starter motor - removal and refitting

1 Disconnect the battery leads.
2 Disconnect the leads from the starter motor terminals.
3 Unscrew the securing bolts which hold the starter motor to the clutch bellhousing and remove the sump bracket bolt.
4 Withdraw the starter motor from beneath the vehicle.
5 Refitting is a reversal of removal.

16 Starter (Ducellier) - dismantling, servicing and reassembly

1 Remove the end cover (2) (Fig. 11.10).
2 Unscrew and remove the armature end bolt (2) and withdraw the various armature shaft end components and the brush holders and plate.
3 Unscrew the nuts from the two tie bolts (7) and pull off the body (9). Remove the circlips from the fork bearing pin (location arrowed) and drift out the pin.
4 Withdraw the armature and lift the solenoid connecting fork out of engagement with the pinion drive assembly (5).
5 The pinion drive assembly may be dismantled by compressing the end collar (13) and detaching the circlip (14).
6 Dismantling beyond this stage should not be undertaken as if bearings or field coils require attention it will be more economical to exchange the unit for a factory reconditioned one, apart from the fact that special tools are needed to release the pole screws and to remove and fit new bearings.
7 Check for wear in the starter brushes and renew them if their overall length is less than 9/16 in (14 mm). The brushes must be unsoldered and the new ones soldered into position. Take care to localise the heat during the operation and do not damage the insulation of the field coils.
8 Clean the commutator on the armature with a fuel soaked cloth. Any pitting or burning may be removed with fine glass paper, not emery. If the commutator is so badly pitted that it would require skimming on a lathe to remove it, renew the complete starter.
9 Checking of suspect field coils for continuity should be carried out by connecting a battery and bulb between the starter terminal and each brush in turn.
10 Reassembly is largely a reversal of dismantling but use a new collar (13) and circlip (14) and stake the collar rim in several places to retain the circlip.
11 Ensure that all starter drive components are clean and lightly oil them with thin oil when assembled.
12 Check the starter drive pinion end-float as described in Section 18.

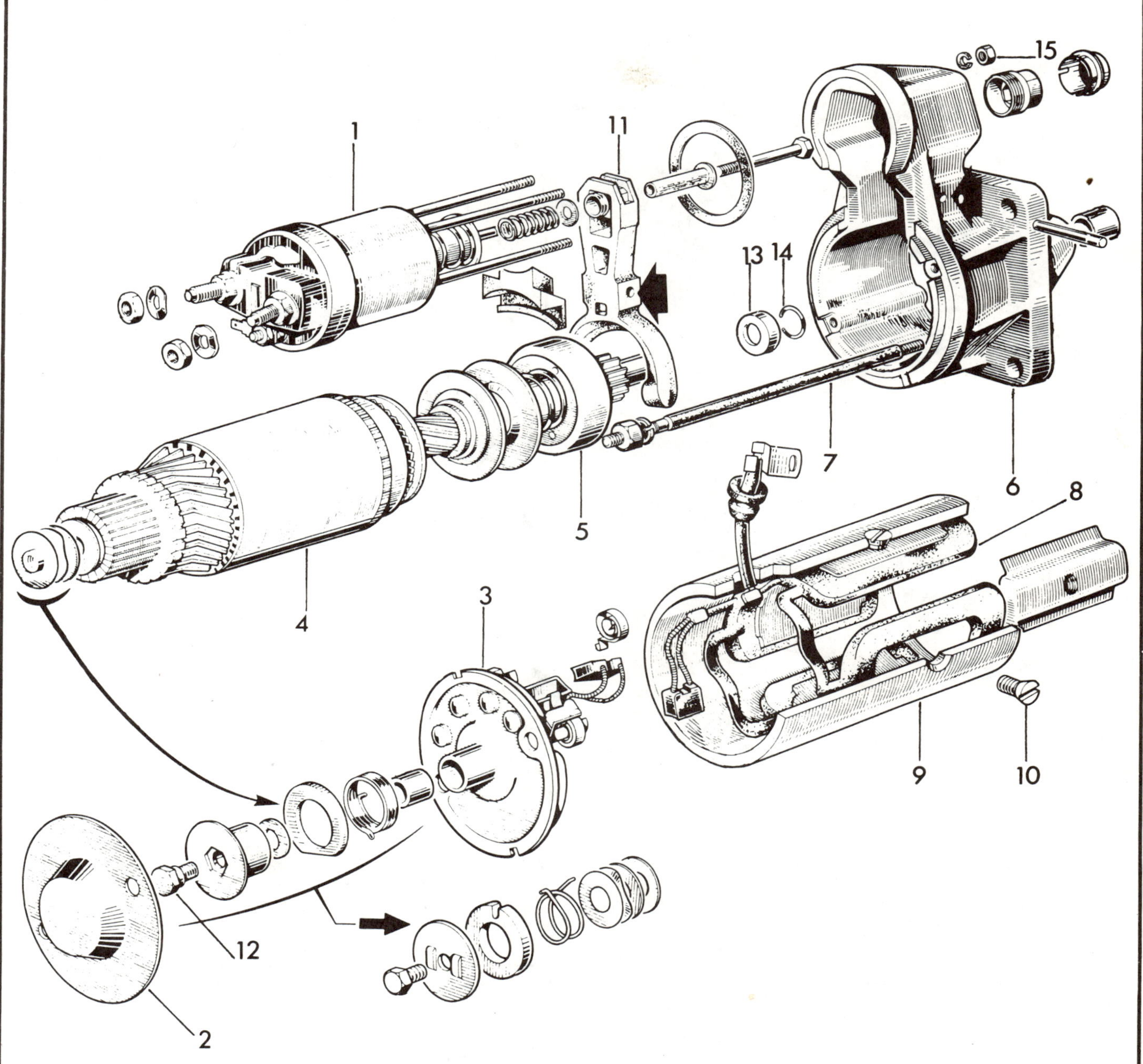

Fig. 11.10. Exploded view of Ducellier starter motor

1 Solenoid body	*5 Pinion drive*	*9 Body*	*13 Collar*
2 End cover	*6 Bearing cover*	*10 Pole screw*	*14 Circlip*
3 Brush holders	*7 Tie bolt*	*11 Engagement fork*	*15 Adjuster nut*
4 Armature	*8 Field coil*	*12 Bolt*	

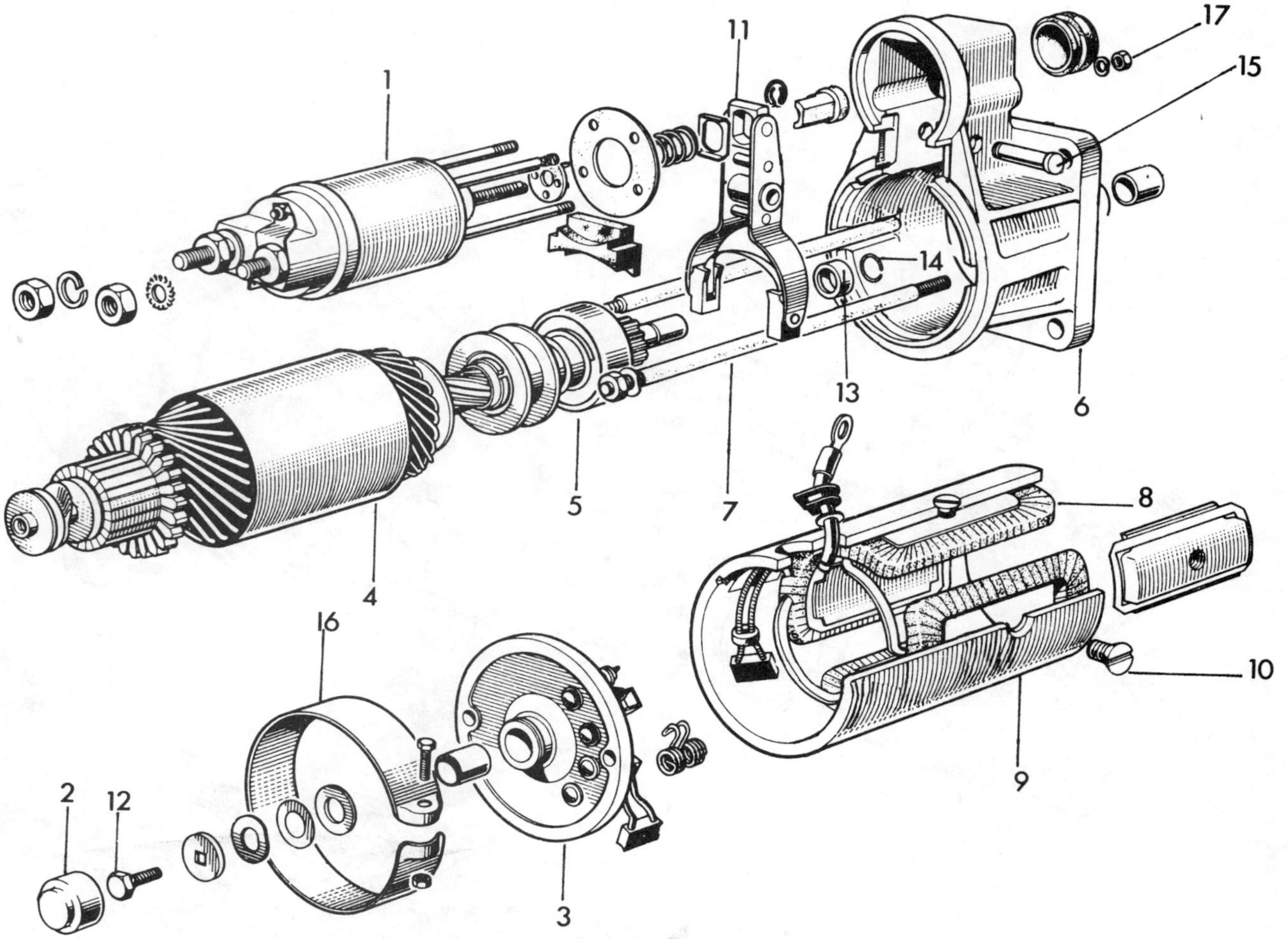

Fig. 11.11. Exploded view of Paris - Rhone starter motor

1 *Solenoid body*
2 *End cover*
3 *Brush holder*
4 *Armature*
5 *Pinion drive*
6 *Bearing cover*
7 *Tie bolt*
8 *Field coil*
9 *Body*
10 *Pole screw*
11 *Engagement fork*
12 *Bolt*
13 *Collar*
14 *Circlip*
15 *Selector fork pin*
16 *Inspection slip band*
17 *Adjuster nut*

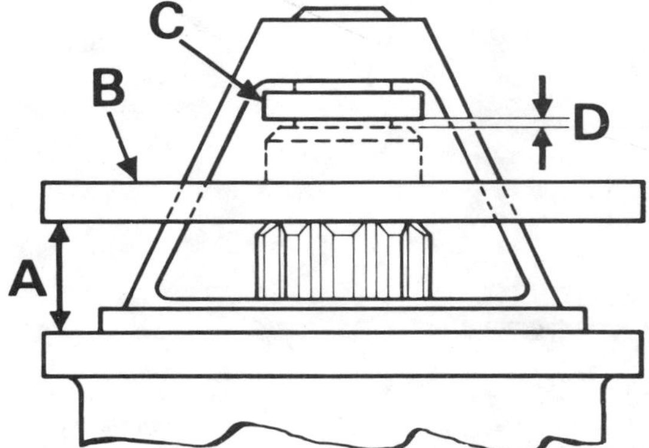

Fig. 11.12. Checking the starter motor pinion clearances

A *Dimension with solenoid at rest 14 mm (9/16 in)*
B *Straight edge*
C *Pinion stop*
D *Clearance with solenoid energised 0.5 to 1.5 mm (0.019 to 0.058 in)*

17 Starter (Paris-Rhone) - dismantling, servicing and reassembly

1 Refer to the previous Section and carry out the operations described but in conjunction with Fig. 11.11.

18 Starter motor drive pinion end-float - checking and adjusting

1 Disconnect the field coil wire from the solenoid terminal.
2 Using a 6 volt dry battery, energise the solenoid which will cause the pinion drive assembly to move forward into its engaged position. Press the pinion towards the motor to take up any end-float and using feeler gauges, check the clearance between the end of the pinion and the face of the thrust collar. The clearance should be 0.019 to 0.058 in (0.5 to 1.5 mm).
3 With the solenoid de-energised, check that the dimension between the outer edge of the pinion and the motor flange is 0.56 in (14 mm). See Fig. 11.12.
4 If adjustment is necessary, screw the nut in or out as required.

19 Flasher circuit - fault tracing and rectification

1 A flasher unit is located beneath the instrument panel next to the steering column. It plugs into a multiconnector and is retained by a

single screw.

2 If the flasher unit works twice as fast as usual when indicating either right or left turns, this is an indication that there is a broken filament in the front or rear indicator bulb on the side operating quickly.

3 If the external flashers are working but the internal flasher warning light has ceased to function, check the filament of the warning bulb and renew as necessary.

4 With the aid of the wiring diagram check all the flasher circuit connections if a flasher bulb is sound but does not work.

5 With the ignition switched on check that the current is reaching the flasher unit by connecting a voltmeter between the 'plus' terminal and earth. If it is found that current is reaching the unit, connect the two flasher unit terminals together and operate the direction indicator switch. If one of the flasher warning lights comes on this proves that the flasher unit itself is at fault and must be renewed as it is not possible to dismantle and repair it.

20 Windscreen wiper arms and blades - removal and refitting

1 Before removing a wiper arm, turn the windscreen wiper switch on and off to ensure the arms are in their normal parked position parallel with the bottom of the windscreen.

2 To remove the arm, pivot the cap back to expose the retaining nut (photo).

3 Unscrew the nut using a 13 mm spanner and pull the wiper arm off the splined drive spindle.

4 When refitting an arm, position it so it is in the correct relative parked position and then press the arm head onto the splined drive until it is fully home on the splines.

5 Renew the windscreen wiper blades at intervals of 12000 miles or annually or whenever they cease to wipe the screen effectively.

21 Windscreen wiper mechanism - fault diagnosis and rectification

1 Should the windscreen wipers fail, or work very slowly, then check the terminals on the motor for loose connections and make sure the insulation of all wiring has not been damaged thus causing a short circuit. If this is in order then check the current the motor is taking by connecting an ammeter in the circuit and turning on the wiper switch. Consumption should be between 2.3 and 3.1 amps.

2 If no current is passing through the motor, check that the switch is operating correctly.

3 If the wiper motor takes a very high current check the wiper blades for freedom of movement. If this is satisfactory check the gearbox cover and gear assembly for damage.

4 If the motor takes a very low current ensure that the battery is fully charged. Check the brush gear and ensure the brushes are bearing on the commutator. If not, check the brushes for freedom of movement and, if necessary, renew the tension springs. If the brushes are very worn they should be replaced with new ones. Check the armature by substitution if this part is suspect.

20.2 Wiper arm retaining nut

22 Windscreen wiper motor - removal and refitting

1 The wiper motor is attached to the rear bulkhead in the engine compartment..

2 Disconnect the battery leads.

3 Disconnect the multi-pin electrical plug from the motor.

4 Remove the three bolts securing the motor to the rear bulkhead panel.

5 Withdraw the motor slightly and hook a piece of wire round the wiper linkage arm to prevent it falling down behind the panel.

6 Prise off the linkage balljoint from the arm on the motor and withdraw the motor.

7 To remove the linkage first remove the wiper arms as described in Section 20. Undo the securing screws and lift away the scuttle panel below the windscreen.

8 Remove the nuts, washers and rubber seals securing the spindles to

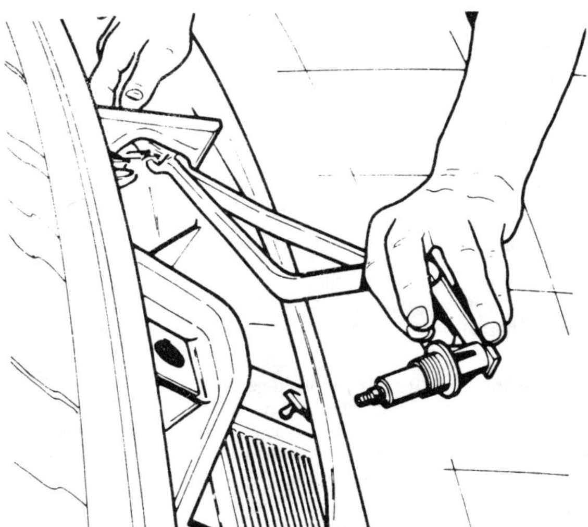

Fig. 11.13. Removing the wiper linkage

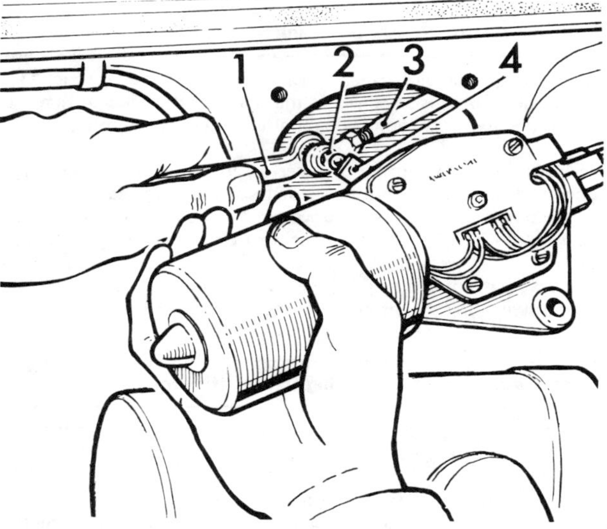

Fig. 11.14. Refitting the wiper linkage balljoint to the motor

1 Ring spanner
2 Balljoint
3 Linkage arm
4 Motor arm

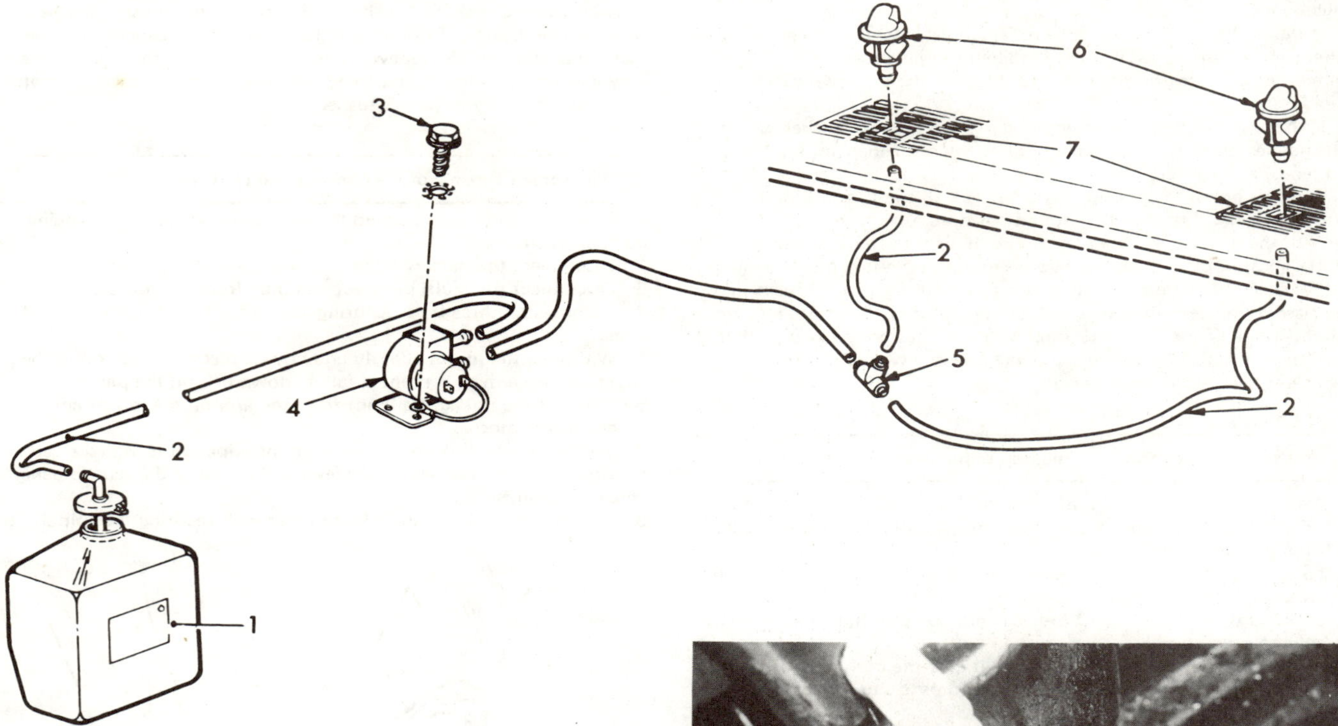

Fig. 11.15. Layout of windscreen washer system

1 *Reservoir*
2 *Flexible tubes*
3 *Pump fixing screw*
4 *Pump*
5 *"T" piece*
6 *Jet nozzles*
7 *Grilles*

the body brackets and lift out the linkage assembly (Fig. 11.13).
9 Any servicing of the motor should be limited to the operations described in Section 21 otherwise the motor should be exchanged for a factory reconditioned unit. Any wear in the linkage should be rectified by renewal of the components concerned.
10 Refit the motor and linkage using the reverse procedure to that of removal. Use a ring spanner to press the linkage balljoint back onto the wiper motor arm (see Fig. 11.14).

23 Windscreen washer - description and servicing

1 The layout of the windscreen washer system is shown in Fig. 11.15.
2 The washer motor is attached to the right-hand valance inside the engine compartment. It cannot be repaired and if faulty must be renewed.
3 Servicing should be limited to occasionally checking the security of the tubes and connectors and the security of the electrical leads to the switch and pump motor. Keep the washer fluid container topped up as described in the Routine Maintenance Section at the front of this manual.
4 The spray pattern can be adjusted by inserting a pin in the jet orifice and swivelling the jet(s) in the required direction.

24 Horns - description, fault tracing and rectification

1 Twin horns are located in the engine compartment adjacent to the radiator. The horns are operated by the multi-purpose switch on the steering column.
2 If the horns fail to operate, disconnect the terminal from each horn and connect a test light between the horn feed wire and earth. Switch on the ignition and get an assistant to operate the horn stalk and check that the test lamp illuminates.
3 If the lamp lights but the horns do not operate, the horns are faulty. They are not repairable and must be renewed.

25.1 Headlight bulb electrical plug

25.2 Removing the headlight bulb

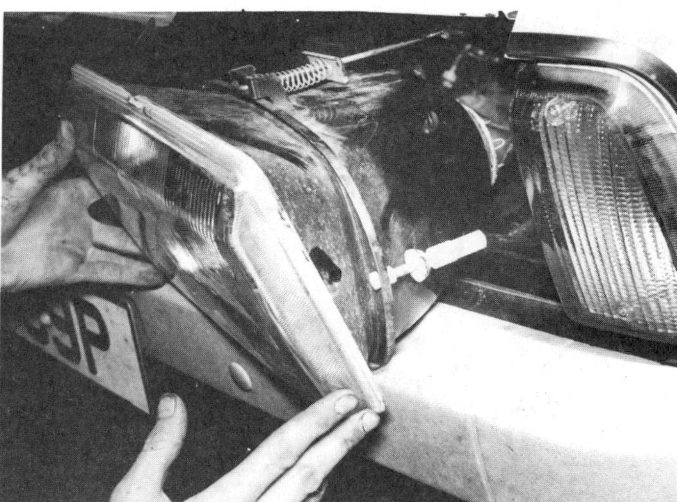
26.5 Withdrawing the headlight assembly

28.1 Removing the front side lamp bulb holder

4 If the test lamp fails to light, check that the wiring is not loose or disconnected. Clean the horn terminals and re-test.

25 Headlights - bulb renewal

1 Open the bonnet and pull the electrical plug off the rear of the headlight unit (photo).
2 Release the spring clips and withdraw the bulb (photo).
Note: On some Halogen type bulbs it is necessary to turn the bulb outer ring so that the tags are aligned with the notches.
3 Refit the bulb using the reverse procedure to that of removal.

26 Headlight unit - removal and refitting

1 Open the bonnet and disconnect the battery leads.
2 Disconnect the headlight plug and sidelamp from the rear of the headlight unit.
3 Press the upper front of the headlight unit inwards and disengage the nylon adjusting knob from its slot.
4 Push the two spring clips upwards to release them from the side adjusting knobs.
5 Withdraw the headlight unit from the front of the car (photo).
Note: If the car is fitted with headlight washer/wipers the front grille must be removed prior to removing the headlight unit (refer to Chapter 12).
6 Refit the headlight unit using the reversal of the removal procedure and adjust the alignment as described in the following Section.

27 Headlights - alignment

1 To compensate for the different height of the vehicle when it is either loaded or unloaded, there is a two position nylon control knob at the rear of the headlight unit (see Fig. 11.16). When resetting the control knob press the headlight lens inwards slightly to relieve the tension on the spring.
2 Completely resetting the headlight alignment is achieved by rotating the three nylon knobs on the rear of the light unit as shown in Fig. 11.17.
3 It is always advisable to have the headlamps aligned on proper optical beam setting equipment but if this is not available the following procedures may be used.
4 Position the car on level ground 30 feet in front of a dark wall or board. The wall or board must be at right angles to the centre line of the car.

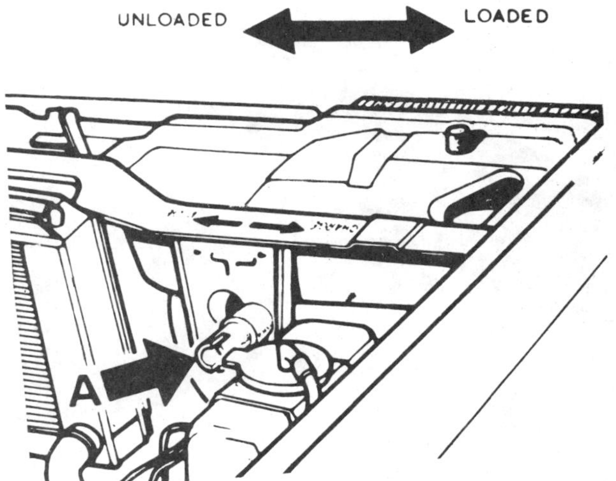

Fig. 11.16. Headlight adjusting knob "A"

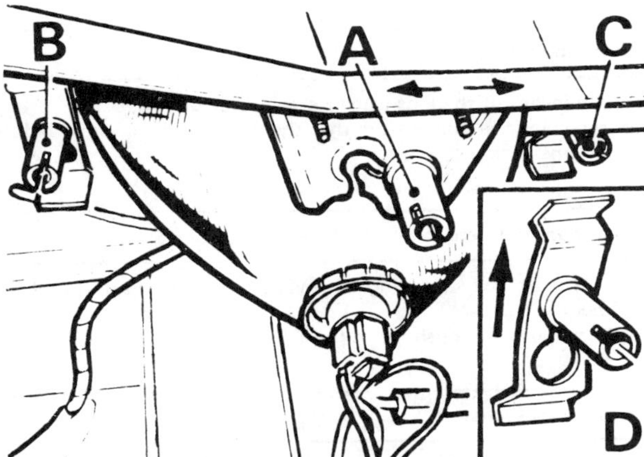

Fig. 11.17. Headlight beam resetting knobs

A Vertical adjuster
B Lateral adjusters
C Lateral adjuster
D Headlight securing clip

5 Draw a vertical line on the board in line with the centre line of the car.
6 Bounce the car on its suspension to ensure correct settlement and then measure the height between the ground and the centre of the headlamps.
7 Draw a horizontal line across the board at this measured height. On this horizontal line mark a cross at a point equal to half the distance between the headlamp centres either side of the vertical centre line.
8 Adjust the headlight position by rotating the lower nylon knobs in the required direction for lateral alignment, and the top knob for vertical alignment. The knobs have slotted ends and can be turned using a coin.
9 If a beamsetter is used the top knob must be in the unloaded position prior to adjustment.

28 Front sidelamp - bulb renewal

1 Access to the bulb is gained by pulling the bulb holder from the lower rear end of the headlight unit (photo).
2 Renew the bulb with one of the same wattage and push the holder back into place.

29 Front indicator - bulb renewal

1 Access is obtained simply by removing the two lens securing screws (photo).
2 Renew the bulbs with ones of the same type and wattage. Do not overtighten the lens securing screws.

30 Rear lamp cluster - bulb renewal

1 Remove the three retaining screws and withdraw the rear lens (photo).
2 Renew any defective bulbs with ones of the same wattage, do not overtighten the lens retaining screws.
3 The complete bulb holder assembly can be removed if required by pressing down four of the six retaining clips and withdrawing the unit from the car body aperture (photo).

31 Number plate lamp - bulb renewal

1 Remove the two screws and withdraw the light unit just enough to disconnect the two leads.
2 Remove the rubber seal and renew the defective bulb with one of the same wattage (photo).

32 Interior lamp - bulb renewal

1 Carefully lever out the lens using a small screwdriver.
2 After renewing the bulb ensure that the wire is tucked inside before pushing the lens back into place.

33 Luggage compartment lamp - bulb renewal

1 Remove the lens by squeezing the ends together between thumb and forefinger.
2 Renew the bulb and push the lens back into place.

34 Push switches - renewal

1 First pull off the knobs from the air control levers and the temperature control knob.
2 Remove the retaining screws and carefully withdraw the centre panel just enough to provide access to the rear of the push switches (photo).
3 Disconnect the leads from the rear of the switch. Press in the two retainers on the switch body and withdraw it from the front of the

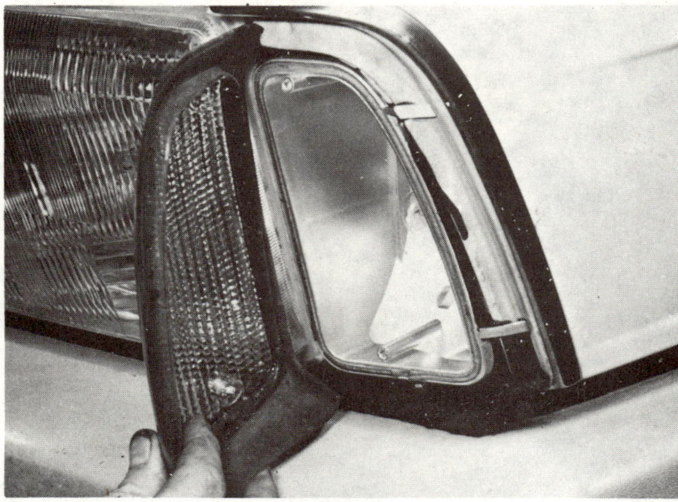

29.1 Removing a front indicator lens

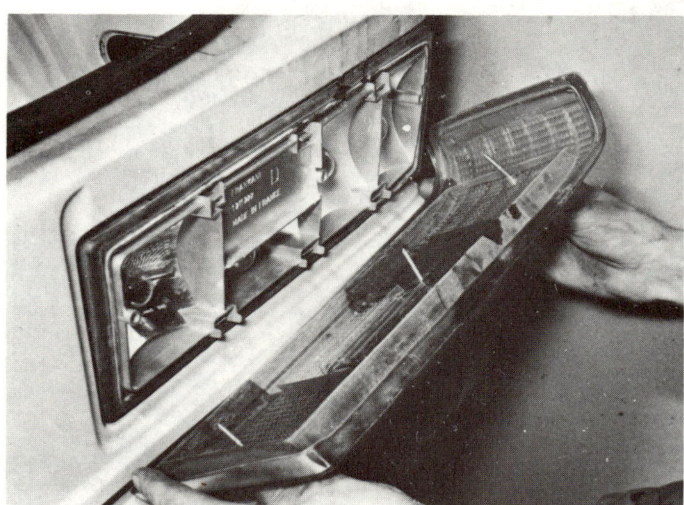

30.1 Removing a rear light cluster lens

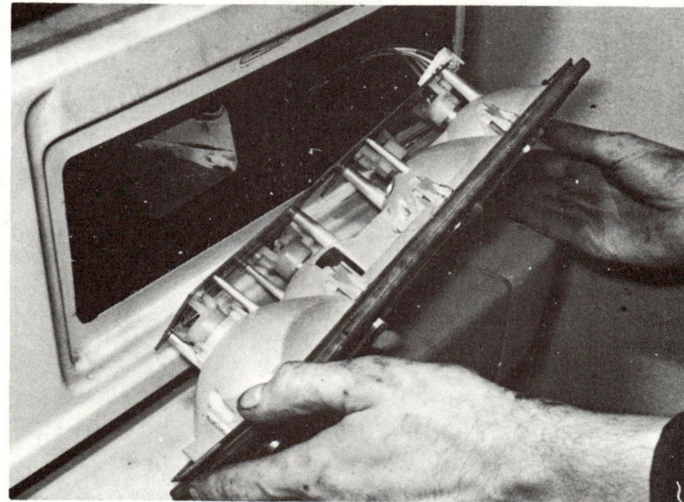

30.3 Withdrawing the complete rear light cluster assembly

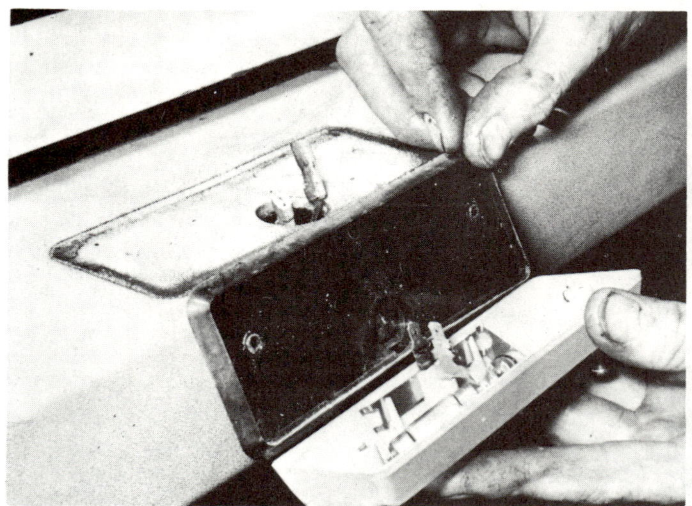
31.2 Replacing the number plate bulbs

34.2 Withdrawing the centre switch panel

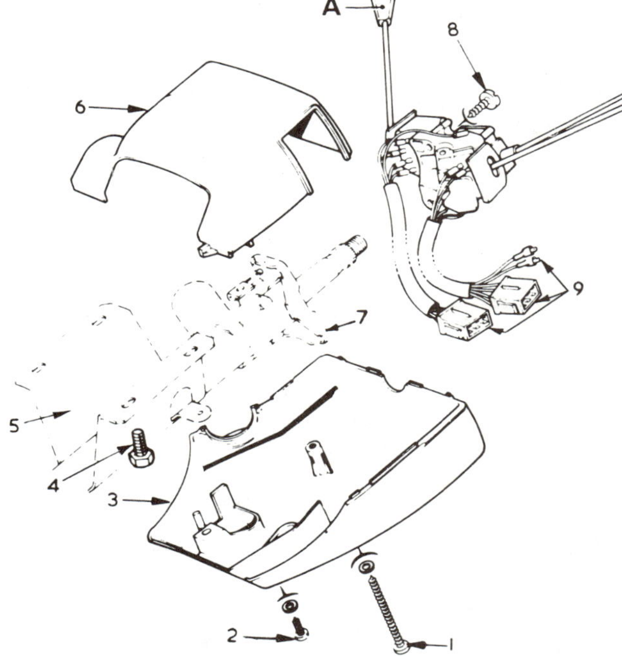

Fig. 11.18. Steering column combination switch

1 Securing screw, (long)
2 Securing screw, (short)
3 Lower shroud
4 Set screw
5 Column bracket
6 Upper shroud
7 Switch mounting bracket
8 Switch securing screw
9 Switch connectors

A = Wipers, washer B = Indicators C = Lights and horns

panel.

4 The switches are not repairable and if faulty must be renewed. Refit the switches and centre panel using the reversal of the removal procedure.

35 Steering column combination switch - removal and refitting

1 Disconnect the battery leads.
2 Prise off the centre motif from the steering wheel and remove the securing nut.
3 Mark the position of the steering wheel in relation to the splined shaft and then withdraw the wheel from the shaft.
4 Undo the retaining screws and remove the lower half of the steering column shroud (photo).
5 Remove the four bolts from the column securing bracket and lower the column sufficiently to withdraw the top half of the shroud (refer to Chapter 8 if necessary).
6 Remove the three retaining screws, disconnect the two multi-pin plugs and the two line connectors and withdraw the switch (Fig. 11.18).
7 The combination switch is not repairable and if any part of it is faulty the complete switch assembly must be renewed.
8 Refit the switch using the reverse procedure to that of removal.

35.4 Lower half of the steering column shroud removed

37.1 Location of the fuse holder unit

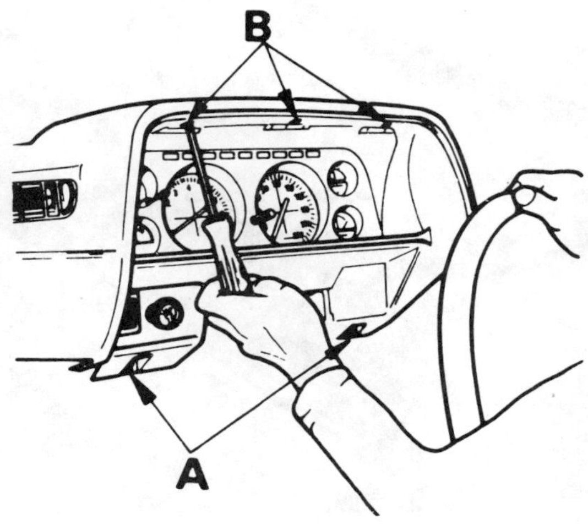

Fig. 11.19. Instrument panel securing screws

A Two lower screws — *B Three upper screws*

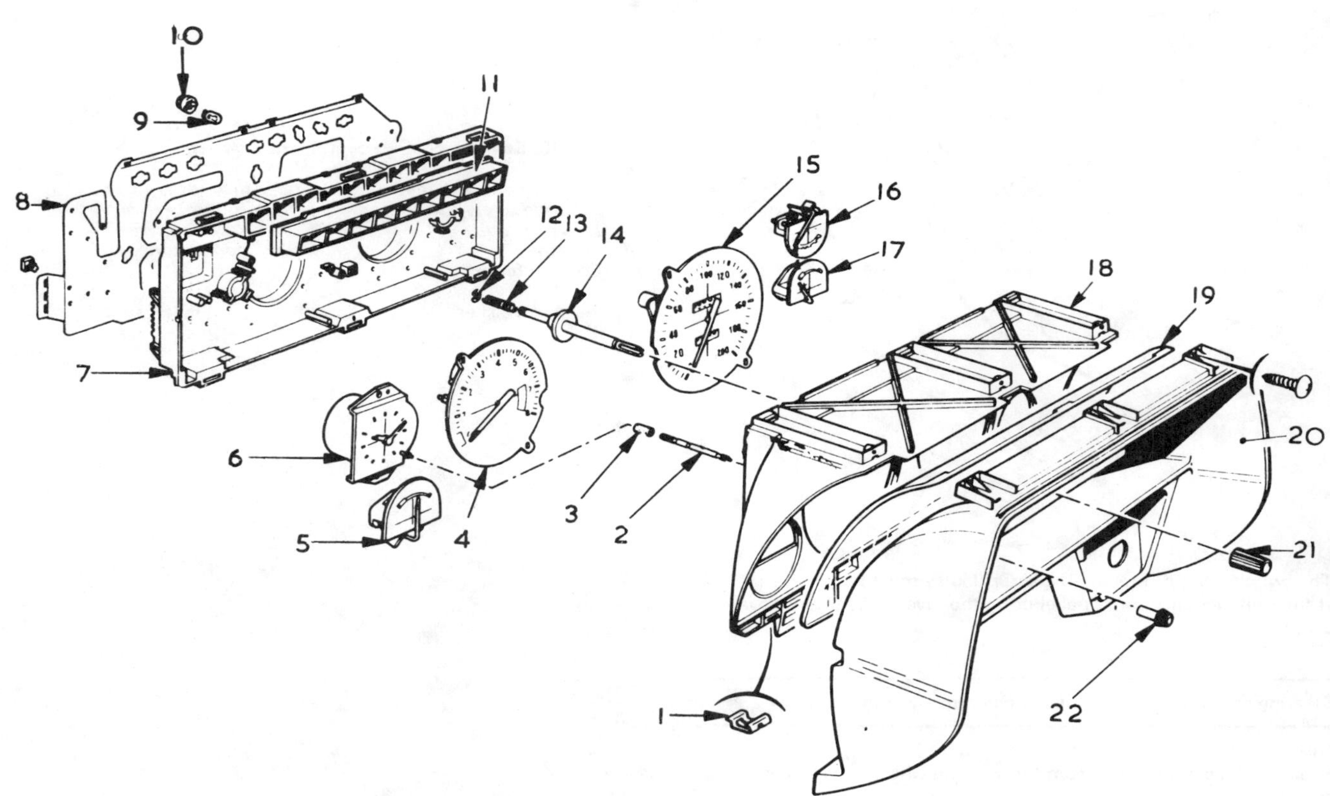

Fig. 11.20. Exploded view of six-dial instrument panel

1	*Spring clip, (cowl to case)*	*7*	*Instrument case*	*13*	*Spring*	*18*	*Inner cowl*
2	*Clock reset spindle*	*8*	*Printed circuit*	*14*	*Trip reset spindle*	*19*	*Curved window*
3	*Bush*	*9*	*Bulb*	*15*	*Speedometer*	*20*	*Outer cowl*
4	*Tachometer*	*10*	*Bulb holder*	*16*	*Fuel gauge*	*21*	*Trip reset knob*
5	*Water temperature gauge*	*11*	*Warning light lenses*	*17*	*Oil pressure*	*22*	*Clock reset knob*
6	*Electric clock*	*12*	*Circlip*				

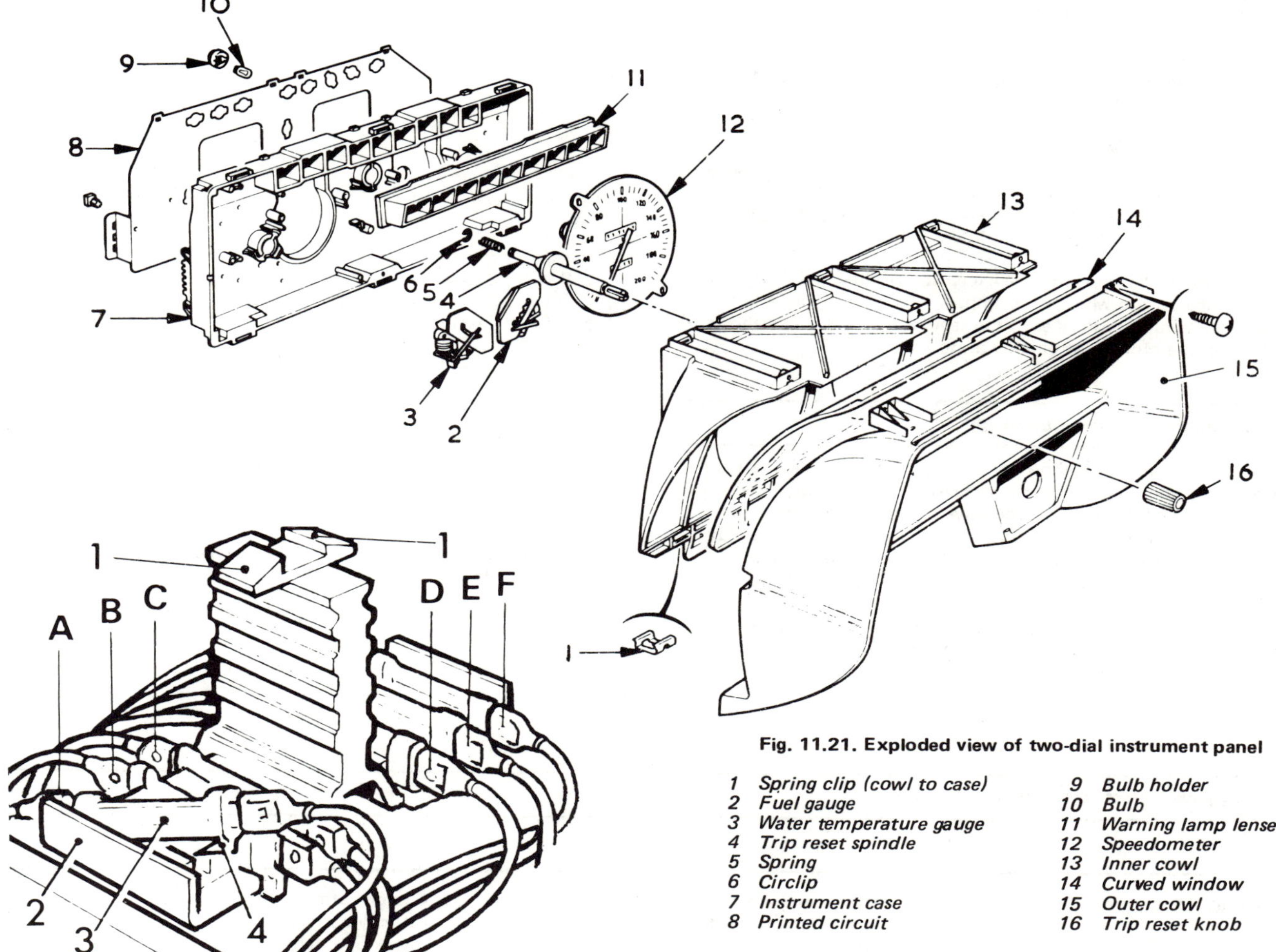

Fig. 11.21. Exploded view of two-dial instrument panel

1	*Spring clip (cowl to case)*	9	*Bulb holder*
2	*Fuel gauge*	10	*Bulb*
3	*Water temperature gauge*	11	*Warning lamp lenses*
4	*Trip reset spindle*	12	*Speedometer*
5	*Spring*	13	*Inner cowl*
6	*Circlip*	14	*Curved window*
7	*Instrument case*	15	*Outer cowl*
8	*Printed circuit*	16	*Trip reset knob*

Fig. 11.22. Fuse holder assembly

1	*Retaining clips*	3	*Fuse*
2	*Fuse holder*	4	*Fuse locating claw*

(A, B, C, D, E and F = Fuse identification)

36 Instrument panel - removal and refitting

1 The Alpine is fitted with either a two dial or six dial instrument panel depending on the model. However, the method of removal and dismantling is similar for both types of panel.
2 First disconnect both battery terminals.
3 Disconnect the choke cable from the carburettor, remove the cable grommet from the rear bulkhead and pull the cable through from the inside of the car.
4 Remove the trim panel below the instrument panel (if fitted).
5 Remove the four bolts securing the steering column bracket and lower the complete column assembly.
6 Refer to Fig. 11.19 and remove the upper and lower retaining screws from the front of the instrument panel.
7 Detach the speedometer drive cable from the rear of the panel.
8 Partially withdraw the panel and disconnect the two multi-pin plugs from the printed circuit board.
9 Disconnect the two terminals from the panel light dimmer switch (if fitted).
10 Carefully lift the instrument panel assembly away from the car.
11 To dismantle the instrument panel assembly refer to Figs. 11.20 or 11.21 as appropriate and proceed as follows.
12 If a clock is fitted, pull off the hand reset knob.
13 Remove the outer cowl retaining screws and remove the cowl and curved lens from the main instrument panel assembly.
14 Remove the six spring clips and withdraw the inner cowl.
15 Remove the warning lights lens carrier.
16 The instruments and gauges can be removed by undoing the nuts or bolts from the rear of the panel and withdrawing them from the front. The clock (if fitted) has a single retaining nut at the rear and two screws on the front of the panel.
17 To remove the printed circuit board, unplug all the warning light and panel lighting bulb holders. Unclip the flexible board from the locating pegs in the casing, detach the clock connecting tab and carefully withdraw the board.
18 To clean the board use methylated spirits and a piece of clean cloth. Examine the metallic connecting strips for breaks or fractures.
19 Reassemble and refit the instrument panel using the reverse procedure to that of removal. **Note:** It is possible to refit the warning and lighting bulbs without removing the instrument panel, by reaching up from beneath the dash panel.

37 Fuses

1 The fuse block is located in the engine compartment on the right-hand side wing valance (see photo).
2 The fuses can be removed by unclipping the holder from the wing and pressing the two projections inwards to open the holder (Fig. 11.22).

3 Each fuse is lettered and protects the following circuits:

Fuse A – Side lamps, rear lamps, number plate lamp, instrument panel lamps, heater controls, ashtray and centre panel illumination.
Fuse B – Rear fog lamp
Fuse C – Heater blower motor.
Fuse D – Wiper and washer motors, stop lights, and reverse lights.
Fuse E – Rear window heater and cigar lighter (if fitted).
Fuse F – Electric clock, interior lights, flasher unit, glove box light and luggage compartment light.

4 All the fuses are 10 amp with the exception of the cigar lighter and rear window heater fuse which is 16 amp.
5 Always renew fuses with ones of the same rating and establish the reason for the fuse blowing as quickly as possible.

38 Electric window lift - description

An electric motor and window lift linkage is fitted to each left and right-hand front door. Two rocker switches on the facia panel control the window mechanism, which is operative only when the ignition switch is in the 'M' (ignition on) position. A thermal overload cutout is fitted to the electric motors. The wiring harness is protected by a flexible rubber sheath at the point where it is routed into the door casings.

When a window reaches either extremity of travel, the rocker switch must be released. Failure to do this will cause the lift motor overload cutout to operate, putting the system temporarily out of action.

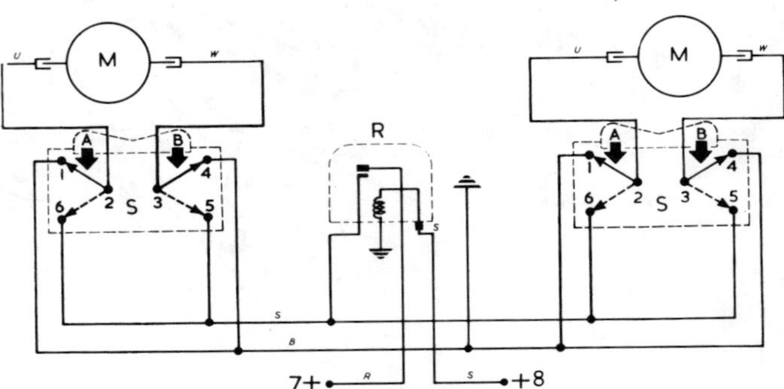

Fig. 11.23. Circuit diagram for electric windows

M Lift motors
S Rocker switches
7+ Power supply to motors
+8 To ignition switch

Colour code U = Blue, W = White, S = Slate, B = Black, R = Red

39 Fault diagnosis - electrical system

Symptom	Reason/s	Remedy
No current at starter motor	Battery discharged	Charge battery.
	Battery defective internally	Fit new battery.
	Battery terminal leads loose or earth lead not securely attached to body	Check and tighten leads.
	Loose or broken connections in starter motor circuit	Check all connections and tighten any that are loose.
	Starter motor switch or solenoid faulty	Test and replace faulty components with new.
Current at starter motor: faulty motor	Starter motor pinion jammed in mesh with flywheel gear ring	Disengage pinion by turning squared end of armature shaft.
	Starter brushes badly worn, sticking, or brush wires loose	Examine brushes, renew as necessary, tighten down brush wires.
	Commutator dirty, worn or burnt	Clean commutator, recut if badly burnt.
	Starter motor armature faulty	Overhaul starter motor, fit new armature.
	Field coils earthed	Overhaul starter motor.
Electrical defects	Battery in discharged condition	Charge battery.
	Starter brushes badly worn, sticking, or brush wires loose	Examine brushes, renew as necessary, tighten down brush wires.
	Loose wires in starter motor circuit	Check wiring and tighten as necessary.
Dirt or oil on drive gear	Starter motor pinion sticking on the screwed sleeve	Remove starter motor, clean starter motor drive.
Mechanical damage	Pinion or flywheel gear teeth broken or worn	Fit new gear ring to flywheel, and new pinion to starter motor drive.
Lack of attention or mechanical damage	Pinion or flywheel gear teeth broken or worn	Fit new gear teeth to flywheel, or new pinion to starter motor drive.
	Starter drive main spring broken	Dismantle and fit new main spring.
	Starter motor retaining bolts loose	Tighten starter motor securing bolts. Fit new spring washer if necessary.
Wear or damage	Battery defective internally	Remove and fit new battery.
	Electrolyte level too low or electrolyte too weak due to leakage	Top up electrolyte level to just above plates.

Symptom	Reason/s	Remedy
	Plate separators no longer fully effective	Remove and fit new battery.
	Battery plates severely sulphated	Remove and fit new battery.
Insufficient current flow to keep battery charged	Alternator belt slipping	Check belt for wear, replace if necessary, and tighten.
	Battery terminal connections loose or corroded	Check terminals for tightness, and remove all corrosion.
	Alternator not charging properly	Remove and overhaul alternator.
	Short in lighting circuit causing continual battery drain	Trace and rectify.
	Alternator control unit not working correctly	Renew control unit.
Alternator not charging	Drivebelt loose and slipping, or broken.	Check, renew and tighten as necessary.
	Brushes worn, sticking, broken or dirty	Examine, clean or renew brushes as necessary.
	Brush springs weak or broken	Examine and test. Renew as necessary.
	Commutator dirty, greasy, worn, or burnt	Clean commutator and undercut segment separators.
	Armature badly worn or armature shaft bent	Fit new or reconditioned armature.
	Commutator bars shorting	Undercut segment separations.
	Alternator bearings badly worn	Fit exchange unit.
	Alternator field coils burnt, open, or shorted	Remove and fit rebuilt unit.
	Commutator no longer circular	Recut commutator and undercut segment separators.
	Open circuit in wiring of cut-out and regulator unit	Remove, examine and renew as necessary. Take car to specialist Auto-Electrician.
Fuel gauge		
Fuel gauge gives no reading	Fuel tank empty!	Fill fuel tank.
	Electric cable between tank sender unit and gauge earthed or loose	Check cable for earthing and joints for tightness.
	Fuel gauge case not earthed	Ensure case is well earthed.
	Fuel gauge supply cable interrupted	Check and renew cable if necessary.
	Fuel gauge unit broken	Renew fuel gauge.
Fuel gauge registers full all the time	Electric cable between tank unit and gauge broken or disconnected	Check over cable and repair as necessary.
Horn		
Horn operates all the time	Horn push either earthed or stuck down	Disconnect battery earth. Check and rectify source of trouble.
	Horn cable to horn push earthed	Disconnect battery earth. Check and rectify source of trouble.
Horn fails to operate	Blown fuse	Check and renew if broken. Ascertain cause.
	Cable or cable connections loose, broken or disconnected	Check all connections for tightness and cables for breaks.
	Horn has an internal fault	Remove and overhaul horn.
Horn emits intermittent or unsatisfactory noise	Faulty or loose wiring	Check wiring.
Lights		
Lights do not come on	If engine not running, battery discharged.	Push-start car, charge battery.
	Light bulb filament burnt out or bulbs broken	Test bulbs in live bulb holder.
	Wire connections loose, disconnected or broken	Check all connections for tightness and wire cable for breaks.
	Light switch shorting or otherwise faulty	By-pass light switch to ascertain if fault is in switch and fit new switch as appropriate.
Lights come on but fade out	If engine not running battery discharged	Push-start car, and charge battery.
Lights give very poor illumination	Lamp glasses dirty	Clean glasses.
	Reflector tarnished or dirty	Fit new reflectors.
	Lamps badly out of adjustment	Adjust lamps correctly.
	Incorrect bulb with too low wattage fitted	Remove bulb and replace with correct grade.
	Existing bulbs old and badly discoloured	Renew bulb units.
	Electrical wiring too thin not allowing full current to pass	Re-wire lighting system.
Lights work erratically - flashing on and off, especially over bumps	Battery terminals or earth connection loose	Tighten battery terminals and earth connection.
	Lights not earthing properly	Examine and rectify.

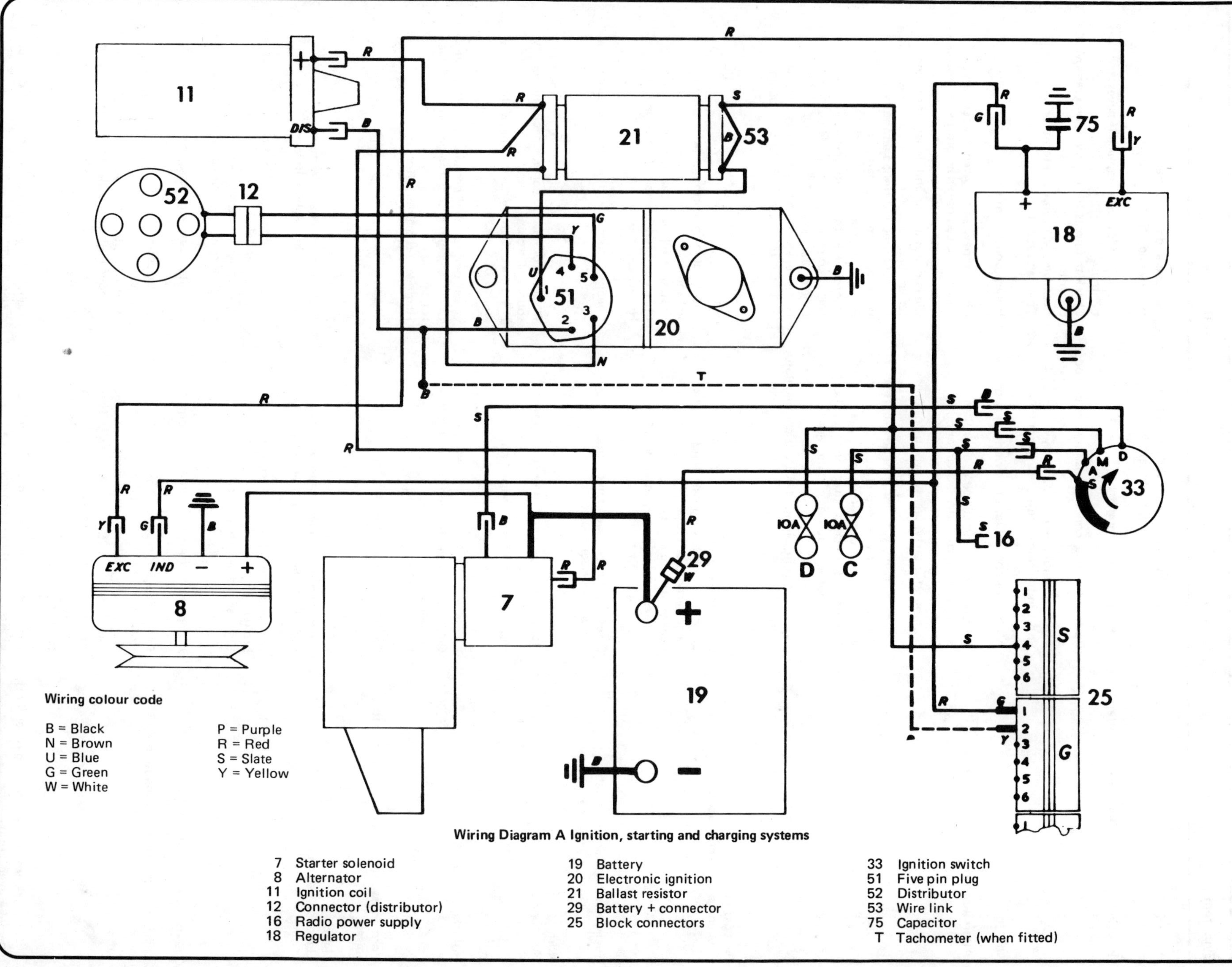

Wiring Diagram A Ignition, starting and charging systems

7 Starter solenoid
8 Alternator
11 Ignition coil
12 Connector (distributor)
16 Radio power supply
18 Regulator
19 Battery
20 Electronic ignition
21 Ballast resistor
29 Battery + connector
25 Block connectors
33 Ignition switch
51 Five pin plug
52 Distributor
53 Wire link
75 Capacitor
T Tachometer (when fitted)

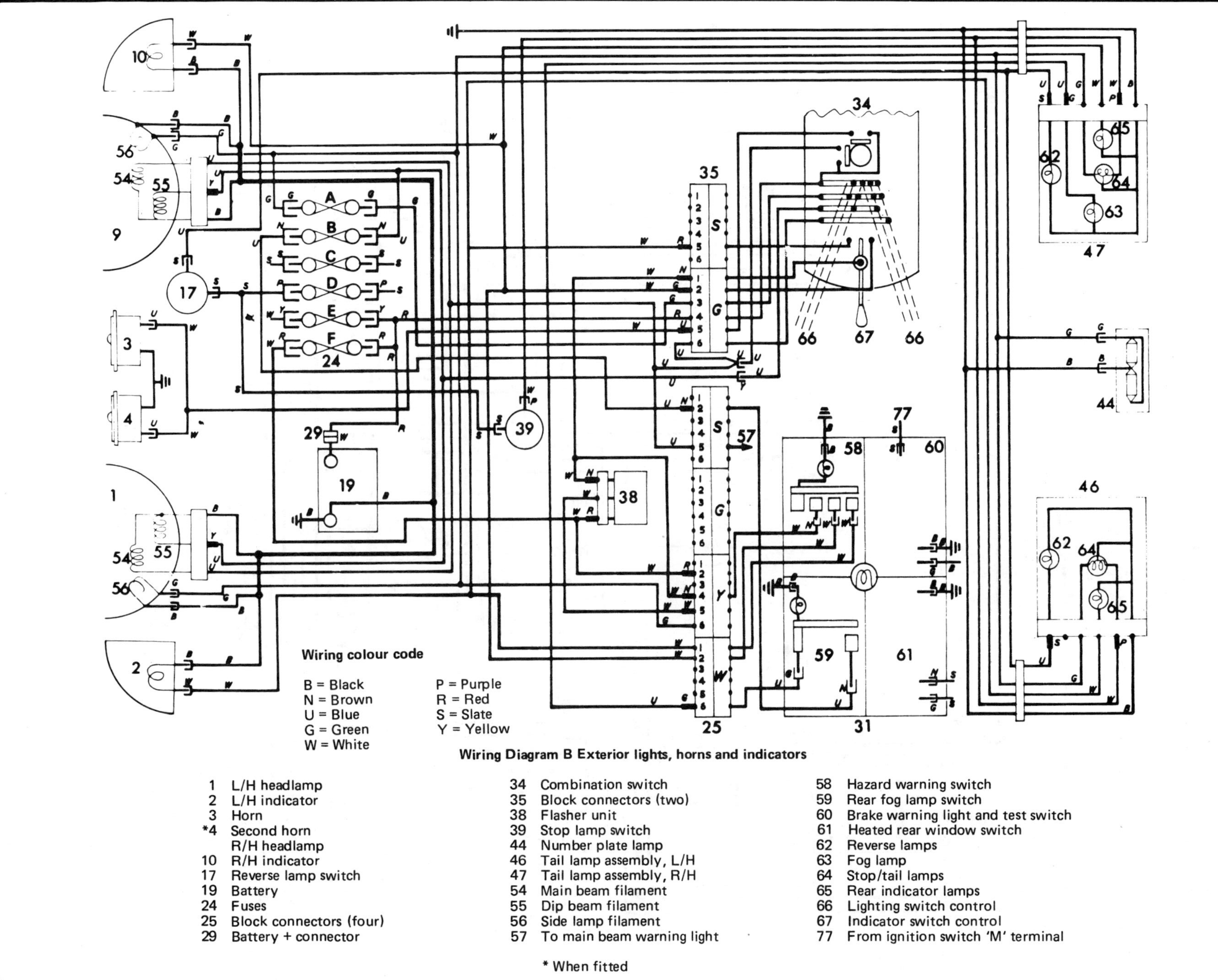

Wiring Diagram B Exterior lights, horns and indicators

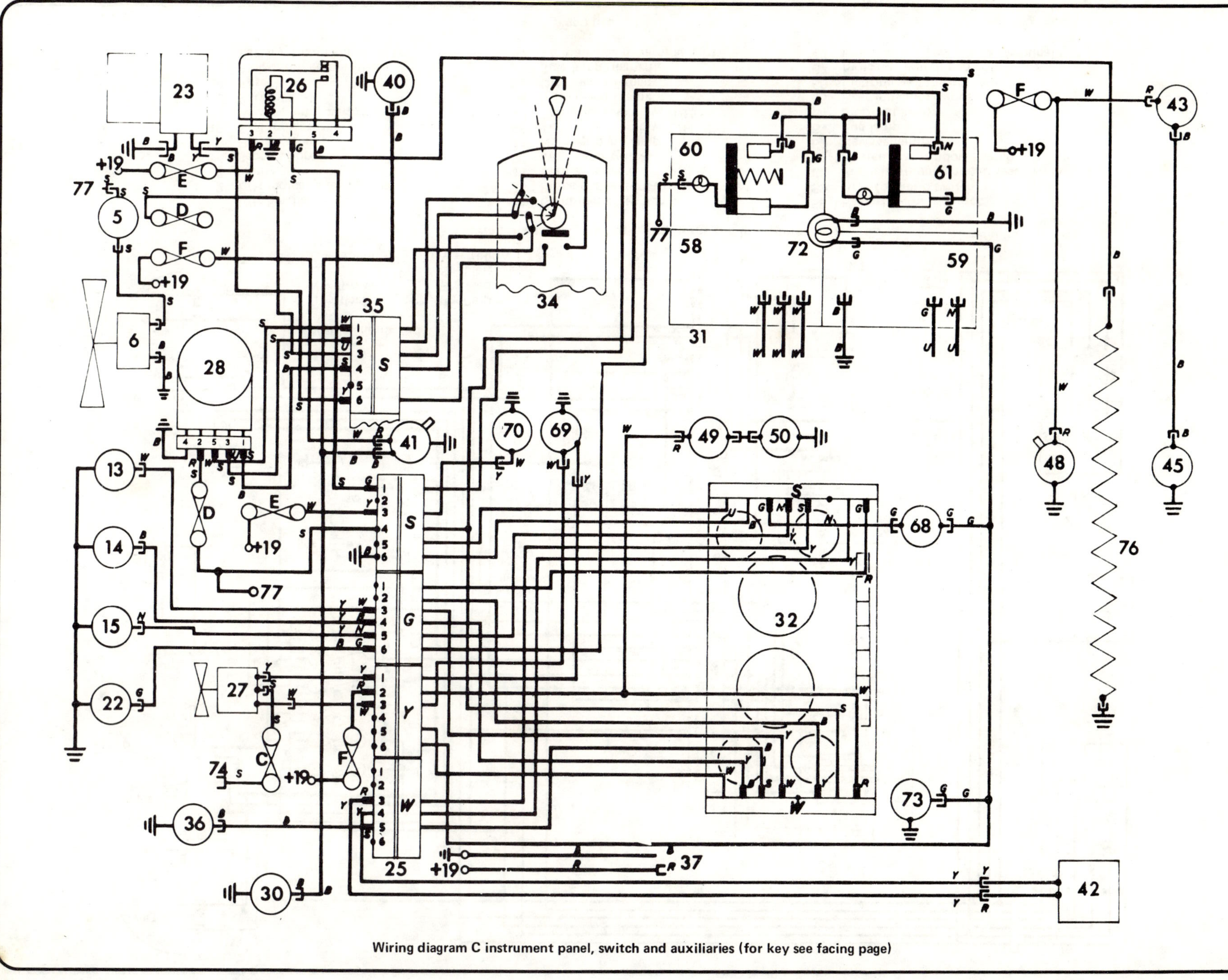

Wiring diagram C instrument panel, switch and auxiliaries (for key see facing page)

Wiring Diagram C Instrument panel, switches and auxiliaries

5 Radiator
6 Radiator fan
13 Water temperature sender
14 Oil warning switch
*15 Oil pressure sender
19 Battery positive supply
22 Braking pressure differential warning actuator
23 Screenwash pump
25 Four block connectors
26 Heated rear window relay
27 Heater blower fan
28 Screen wiper motor
30 Courtesy switch
32 Instrument panel printed circuit
34 Combination switch
*36 Handbrake warning switch
*37 Accessory connectors
40 Courtesy switch
41 Front interior lamp
42 Fuel gauge tank unit
*43 Luggage compartment lamp
*45 Luggage compartment lamp switch
*48 Rear interior lamp
*49 Glove box lamp
*50 Glove box lamp switch
58 Hazard warning switch
59 Rear fog lamp switch
60 Brake warning check switch
61 Heated rear window switch
*68 Instrument panel rheostat
69 Heater blower switch
70 Cigar lighter
71 Wash/wipe control
72 Centre panel lamp
73 Switch illumination lamp
74 From ignition switch 'A' terminal
76 Heated rear window
77 From ignition switch 'M' terminal

* When fitted

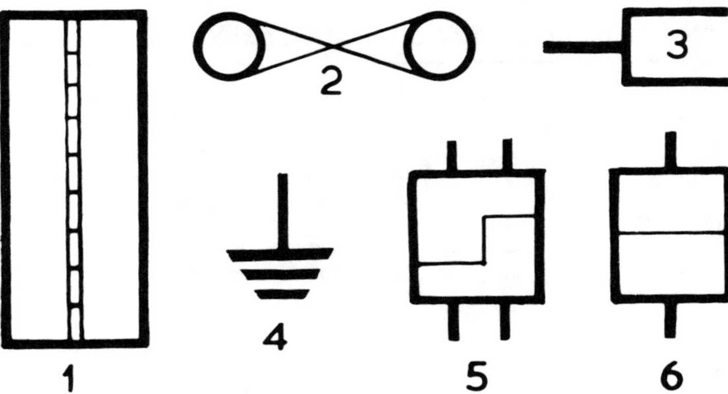

Symbol identification

1 Block connector
2 Fuse
3 Connector
4 Earth
5 Two pin connector
6 Single pin connector

Chapter 12 Bodywork and fittings

For modifications, and information applicable to later models, see Supplement at end of manual

Contents

1 General description

The vehicle body structure is a welded fabrication of many individual shaped panels to form a 'monocoque' bodyshell. Certain areas are strengthened locally to provide for suspension system, steering system, engine support anchorages and transmission. The resultant structure is very strong and rigid (Fig. 12.1).

It is as well to remember that monocoque structures have no load paths and all metal is stressed to an extent. It is essential therefore to maintain the whole bodyshell both top and underside, inside and outside, clean and corrosion free. Every effort should be made to keep the underside of the car as clear of mud and dirt accumulations as possible. If you were fortunate enough to acquire a new car then it is advisable to have it rust proofed and undersealed at one of the specialist workshops who guarantee their work.

2 Maintenance - exterior

The general condition of a car's bodywork is the one thing that significantly affects its value. Maintenance is easy but needs to be regular and particular. Neglect - particularly after minor damage - can quickly lead to further deterioration and costly repair bills. It is important to keep watch on those parts of the bodywork not immediately visible, for example the underside, inside all the wheel arches and the lower part of the engine compartment.

The basic maintenance routine for the bodywork is washing, preferably with a lot of water from a hose. This will remove all the loose solids which may have stuck to the car. It is important to flush these off in such a way as to prevent grit from scratching the finish. The wheel arches and underbody need washing in the same way to remove any accumulated mud which will retain moisture and tend to encourage rust. Paradoxically enough, the best time to clean the underbody and wheel arches is in the wet weather when the mud is thoroughly wet and soft. In very wet weather the underbody is usually cleaned of large accumulations automatically and this is a good time for inspection.

Periodically it is a good idea to have the whole of the underside of the car steam cleaned, engine compartment included, for removal of accumulation of oily grime which sometimes collects thickly in areas near the engine and gearbox, so that a thorough inspection can be carried out to see what minor repairs and renovations are necessary.

If steam facilities are not available there are one or two excellent grease solvents available which can be brush applied. The dirt can then be simply hosed off. Any signs of rust on the underside panels and chassis members must be attended to immediately. Thorough wire brushing followed by treatment with an anti-rust compound, primer and underbody sealer will prevent continued deterioration. If not dealt with the car could eventually become structurally unsound, and therefore, unsafe.

After washing the paintwork wipe it off with a chamois leather to give a clear unspotted finish. A coat of clear wax polish will give added protection against chemical pollutants in the air and will survive several subsequent washings. If the paintwork sheen has dulled or oxidised use a cleaner/polisher combination to restore the brilliance of the shine. This requires a little more effort but it usually is because regular washing has been neglected. Always check that door and drain holes and pipes are completely clear so that water can drain out. Brightwork should be treated the same way as paintwork. Windscreens and windows can be kept clear of smeary film which often appears, if a little ammonia is added to the water. If glass work is scratched, a good rub with a proprietary metal polish will often clean it. Never use any form of wax or other paint/chromium polish on glass.

3 Maintenance - interior

The flooring cover, usually carpet, should be brushed or vacuum cleaned regularly to keep it free from grit. If badly stained, remove it from the car for scrubbing and sponging and make quite sure that it is dry before refitting. Seats and interior trim panels can be kept clean with a wipe over with a damp cloth. If they do become stained (which can be more apparent on light coloured upholstery) use a little liquid detergent and soft nailbrush to scour the grime out of the grain of the material. Do not forget to keep the headlining clean in the same way as the upholstery. When using liquid cleaners inside the car do not over-wet the surfaces being cleaned. Excessive damp could get into the upholstery seams and padded interior, causing stains, offensive odours or even rot. If the inside of the car gets wet accidently it is worthwhile taking some trouble to dry it out properly. **Do not** use oil or electric heaters inside the car for this purpose. If, when removing mats for cleaning, there are signs of damp underneath, all the interior of the car floor should be uncovered and the point of water entry found. It may be only a missing grommet, but it could be a rusted through floor panel and this demands immediate attention as described in the previous Section. More often than not both sides on the panel will require treatment.

4 Minor body damage - repair

See photo sequences on pages 158 and 159

Repair of minor scratches in the car's bodywork

If the scratch is very superficial, and does not penetrate to the metal of the bodywork - repair is very simple. Lightly rub the area of the scratch with a paintwork renovator, or a very fine cutting paste, to remove loose paint from the scratch and to clear the surrounding bodywork of wax polish. Rinse the area with clean water.

Apply touch-up paint to the scratch using a thin paint brush; continue to apply thin layers of paint until the surface of the paint in the scratch is level with the surrounding paintwork. Allow the new paint at least two weeks to harden, then, blend it into the surrounding paintwork by rubbing the paintwork in the scratch area with a paintwork renovator, or a very fine cutting paste. Finally apply wax polish.

Where a scratch has penetrated right through to the metal of the bodywork, causing the metal to rust, a different repair technique is required. Remove any loose rust from the bottom of the scratch with a penknife; then apply rust inhibiting paint to prevent the formation of rust in the future. Using a rubber or nylon applicator fill the scratch with body-stopper paste. If required, this paste can be mixed with cellulose thinners to provide a very thin paste which is ideal for filling narrow scratches. Before the stopper paste in the scratch hardens, wrap a piece of smooth cotton rag around the tip of a finger. Dip the finger in cellulose thinners and then quickly sweep it across the surface of the stopper-paste, this will ensure that it is slightly hollowed. The scratch can now be painted over as described earlier in this Section.

Repair of dents in the car's bodywork

When deep denting of the car's bodywork has taken place, the

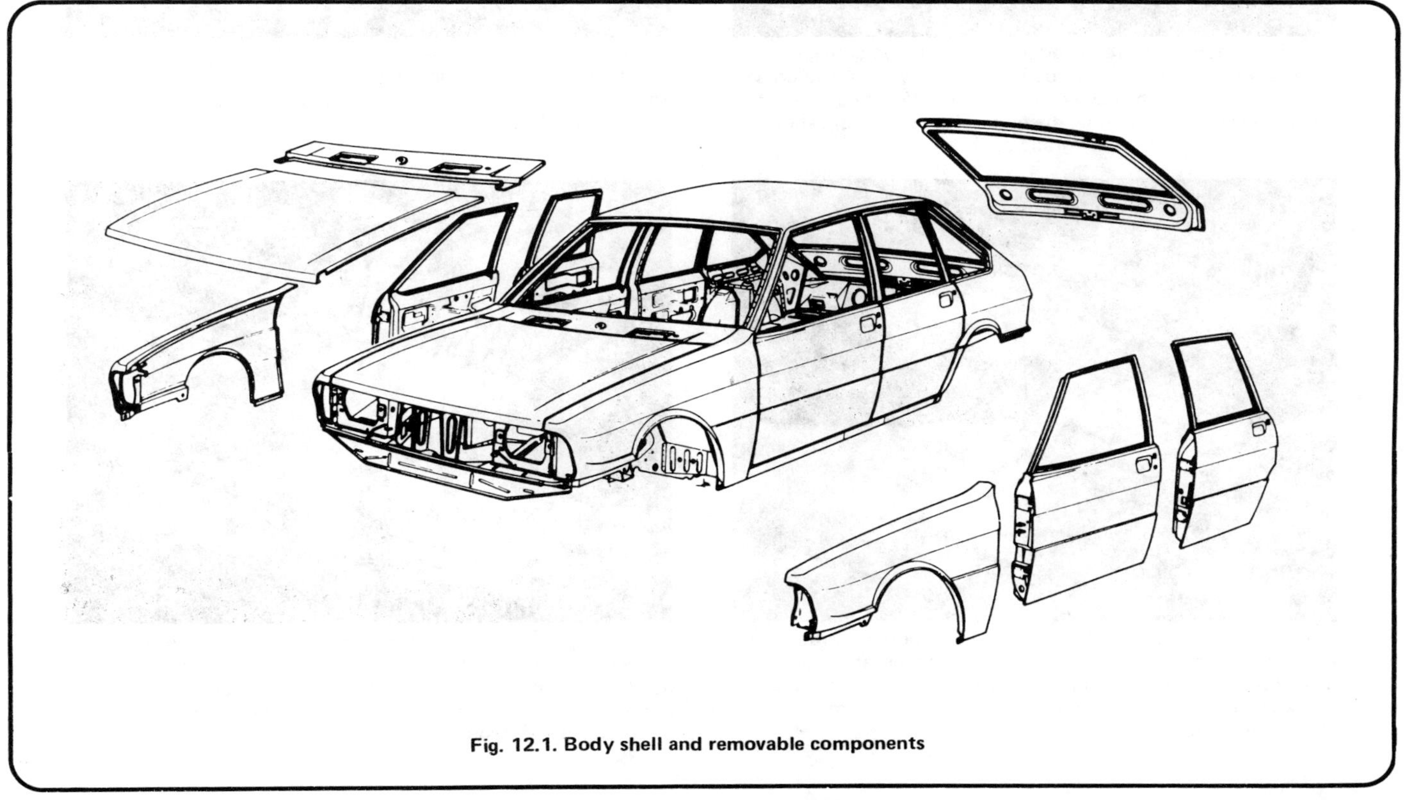

Fig. 12.1. Body shell and removable components

This sequence of photographs deals with the repair of the dent and paintwork damage shown in this photo. The procedure will be similar for the repair of a hole. It should be noted that the procedures given here are simplified – more explicit instructions will be found in the text

In the case of a dent the first job – after removing surrounding trim – is to hammer out the dent where access is possible. This will minimise filling. Here, the large dent having been hammered out, the damaged area is being made slightly concave

Now all paint must be removed from the damaged area, by rubbing with coarse abrasive paper. Alternatively, a wire brush or abrasive pad can be used in a power drill. Where the repair area meets good paintwork, the edge of the paintwork should be 'feathered', using a finer grade of abrasive paper

In the case of a hole caused by rusting, all damaged sheet-metal should be cut away before proceeding to this stage. Here, the damaged area is being treated with rust remover and inhibitor before being filled

Mix the body filler according to its manufacturer's instructions. In the case of corrosion damage, it will be necessary to block off any large holes before filling – this can be done with aluminium or plastic mesh, or aluminium tape. Make sure the area is absolutely clean before ...

... applying the filler. Filler should be applied with a flexible applicator, as shown, for best results; the wooden spatula being used for confined areas. Apply thin layers of filler at 20-minute intervals, until the surface of the filler is slightly proud of the surrounding bodywork

Initial shaping can be done with a Surform plane or Dreadnought file. Then, using progressively finer grades of wet-and-dry paper, wrapped around a sanding block, and copious amounts of clean water, rub down the filler until really smooth and flat. Again, feather the edges of adjoining paintwork

The whole repair area can now be sprayed or brush-painted with primer. If spraying, ensure adjoining areas are protected from over-spray. Note that at least one inch of the surrounding sound paintwork should be coated with primer. Primer has a 'thick' consistency, so will find small imperfections

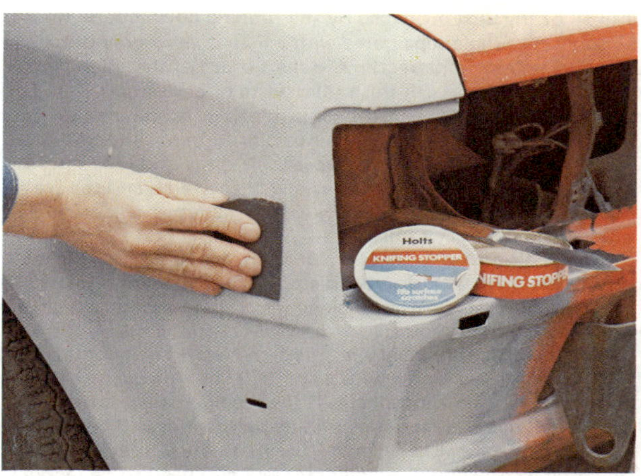

Again, using plenty of water, rub down the primer with a fine grade wet-and-dry paper (400 grade is probably best) until it is really smooth and well blended into the surrounding paintwork. Any remaining imperfections can now be filled by carefully applied knifing stopper paste

When the stopper has hardened, rub down the repair area again before applying the final coat of primer. Before rubbing down this last coat of primer, ensure the repair area is blemish-free – use more stopper if necessary. To ensure that the surface of the primer is really smooth use some finishing compound

The top coat can now be applied. When working out of doors, pick a dry, warm and wind-free day. Ensure surrounding areas are protected from over-spray. Agitate the aerosol thoroughly, then spray the centre of the repair area, working outwards with a circular motion. Apply the paint as several thin coats

After a period of about two weeks, which the paint needs to harden fully, the surface of the repaired area can be 'cut' with a mild cutting compound prior to wax polishing. When carrying out bodywork repairs, remember that the quality of the finished job is proportional to the time and effort expended

first task is to pull the dent out, until the affected bodywork almost attains its original shape. There is little point in trying to restore the original shape completely, as the metal in the damaged area will have stretched on impact and cannot be reshaped fully to its original contour. It is better to bring the level of the dent up to a point which is about 1/8 inch (3 mm) below the level of the surrounding bodywork. In cases where the dent is very shallow anyway, it is not worth trying to pull it out at all.

If the underside of the dent is accessible, it can be hammered out gently from behind, using a mallet with a wooden or plastic head. Whilst doing this, hold a suitable block of wood firmly against the outside of the dent. This block will absorb the impact from the hammer blows and thus prevent a large area of bodywork from being 'belled-out'.

Should the dent be in a section of the bodywork which has double skin or some other factor making it inaccessible from behind, a different technique is called for. Drill several small holes through the metal inside the dent area - particularly in the deeper sections. Then screw long self-tapping screws into the holes just sufficiently for them to gain a good purchase in the metal. Now the dent can be pulled out by pulling on the protruding heads of the screws with a pair of pliers.

The next stage of the repair is the removal of the paint from the damaged area, and from an inch or so of the surrounding 'sound' bodywork. This is accomplished most easily by using a wire brush or abrasive pad on a power drill, although it can be done just as effectively by hand using sheets of abrasive paper. To complete the preparations for filling, score the surface of the bare metal with a screwdriver or the tang of a file, or alternatively, drill small holes in the affected area. This will provide a really good 'key' for the filler paste.

To complete the repair see the Section on filling and respraying.

Repair of rust holes or gashes in the car's bodywork

Remove all paint from the affected area and from an inch or so of the surrounding 'sound' bodywork, using an abrasive pad or a wire brush on a power drill. If these are not available a few sheets of abrasive paper will do the job just as effectively. With the paint removed you will be able to gauge the severity of the corrosion and therefore decide whether to renew the whole panel (if possible) or to repair the affected area. It is often quicker and more satisfactory to fit a new panel than to attempt to repair large areas of corrosion.

Remove all fittings from the affected area except those which will act as a guide to the original shape of the damaged bodywork (eg, headlamp shells etc). Then, using tin snips or a hacksaw blade, remove all loose metal and any other metal badly affected by corrosion. Hammer the edges of the hole inwards in order to create a slight depression for the filler paste.

Wire brush the affected area to remove the powdery rust from the surface of the remaining metal. Paint the affected area with rust inhibiting paint; if the back of the rusted area is accessible treat this also.

Before filling can take place it will be necessary to block the hole in some way. This can be achieved by the use of aluminium or plastic mesh, or aluminium tape.

Aluminium or plastic mesh is probably the best material to use for a large hole. Cut a piece to the approximate size and shape of the hole to be filled, then position it in the hole so that its edges are below the level of the surrounding bodywork. It can be retained in position by several blobs of filler paste around its periphery.

Aluminium tape should be used for small or very narrow holes. Pull a piece off the roll and trim it to the approximate size and shape required, then pull off the backing paper (if used) and stick the tape over the hole; it can be overlapped if the thickness of one piece is insufficient. Burnish down the edges of the tape with the handle of a screwdriver or similar, to ensure that the tape is securely attached to the metal underneath.

Having blocked off the hole the affected area must now be filled and sprayed - see Section on bodywork filling and re-spraying.

Bodywork repairs - filling and re-spraying

Before using this Section, see Sections on dent, deep scratch, rust hole, and gash repairs.

Many types of bodyfiller are available, but generally speaking those proprietary kits which contain a tin of filler paste and a tube of resin hardener are best for this type of repair. A wide, flexible plastic or nylon applicator will be found invaluable for imparting a smooth and well contoured finish to the surface of the filler.

Mix up a little filler on a clean piece of card or board - use the hardener sparingly (follow the maker's instructions on the pack), otherwise the filler will set very rapidly.

Using the applicator, apply the filler paste to the prepared area; draw the applicator across the surface of the filler to achieve the correct contour and to level the filler surface. As soon as a contour that approximates the correct one is achieved, stop working the paste - if you carry on too long the paste will become sticky and begin to 'pick-up' on the applicator.

Continue to add thin layers of filler paste at twenty-minute intervals until the level of the filler is just 'proud' of the surrounding bodywork.

Once the filler has hardened, the excess can be removed using a plane or file. From then on, progressively finer grades of abrasive paper should be used, starting with a 40 grade 'wet-and-dry' paper. Always wrap the abrasive paper around a flat rubber, cork, or wooden block - otherwise the surface of the filler will not be completely flat. During the smoothing of the filler surface the 'wet-and-dry' paper should be periodically rinsed in water - this will ensure that a very smooth finish is imparted to the filler at the final stage.

At this stage the 'dent' should be surrounded by a ring of bare metal, which in turn should be encircled by the finely 'feathered' edge of the good paintwork. Rinse the repair area with clean water, until all of the dust produced by the rubbing down operation is gone.

Spray the whole repair area with a light coat of grey primer - this will show up any imperfections in the surface of the filler. Repair these imperfections with fresh filler paste or bodystopper, and once more smooth the surface with abrasive paper. If bodystopper is used, it can be mixed with cellulose thinners to form a really thin paste which is ideal for filling small holes. Repeat this spray and repair procedure until you are satisfied that the surface of the filler, and the feathered edge of the paintwork are perfect. Clean the repair area with clean water and allow to dry fully.

The repair area is now ready for spraying. Paint spraying must be carried out in a warm, dry, windless and dust free atmosphere. This condition can be created artificially if you have access to a large indoor working area, but if you are forced to work in the open, you will have to pick your day very carefully. If you are working indoors, dousing the floor in the work area with water will 'lay' the dust which would otherwise be in the atmosphere. If the repair area is confined to one body panel, mask off the surrounding panels; this will help to minimise the effects of a slight mis-match in paint colours. Bodywork fittings (eg, chrome strips, door handles etc) will also need to be masked off. Use genuine masking tape and several thicknesses of newspaper for the masking operation.

Before commencing to spray, agitate the aersol can thoroughly, then spray a test area (an old tin, or similar) until the technique is mastered. Cover the repair area with a thick coat of primer; the thickness should be built up using several thin layers of paint rather than one thick one. Using 400 grade 'wet-and-dry' paper, rub down the surface of the primer until it is really smooth. While doing this, the work area should be thoroughly doused with water, and the wet-and-dry paper periodically rinsed in water. Allow to dry before spraying on more paint.

Spray on the top coat, again building up the thickness by using several thin layers of paint. Start spraying in the centre of the repair area and then using a circular motion, work outwards until the whole area and about 2 inches of the surrounding original paintwork is covered. Remove all masking material 10 to 15 minutes after spraying on the final coat of paint. Allow the new paint at least 2 weeks to harden fully, then, using a paintwork renovator or a very fine cutting paste, blend the edges of the new paint into the existing paintwork. Finally, apply wax polish.

5 Major body repairs

Where serious damage has occurred or large areas need renewal due to neglect, it means certainly that completely new sections or panels will need welding in and this is best left to professionals. If the damage is due to impact it will also be necessary to completely check the alignment of the body shell structure. Due to the principle of construction the strength and shape of the whole can be affected by damage to a part. In such instances the services of a Chrysler agent with specialist checking jigs are essential. If a body is left misaligned it is first of all

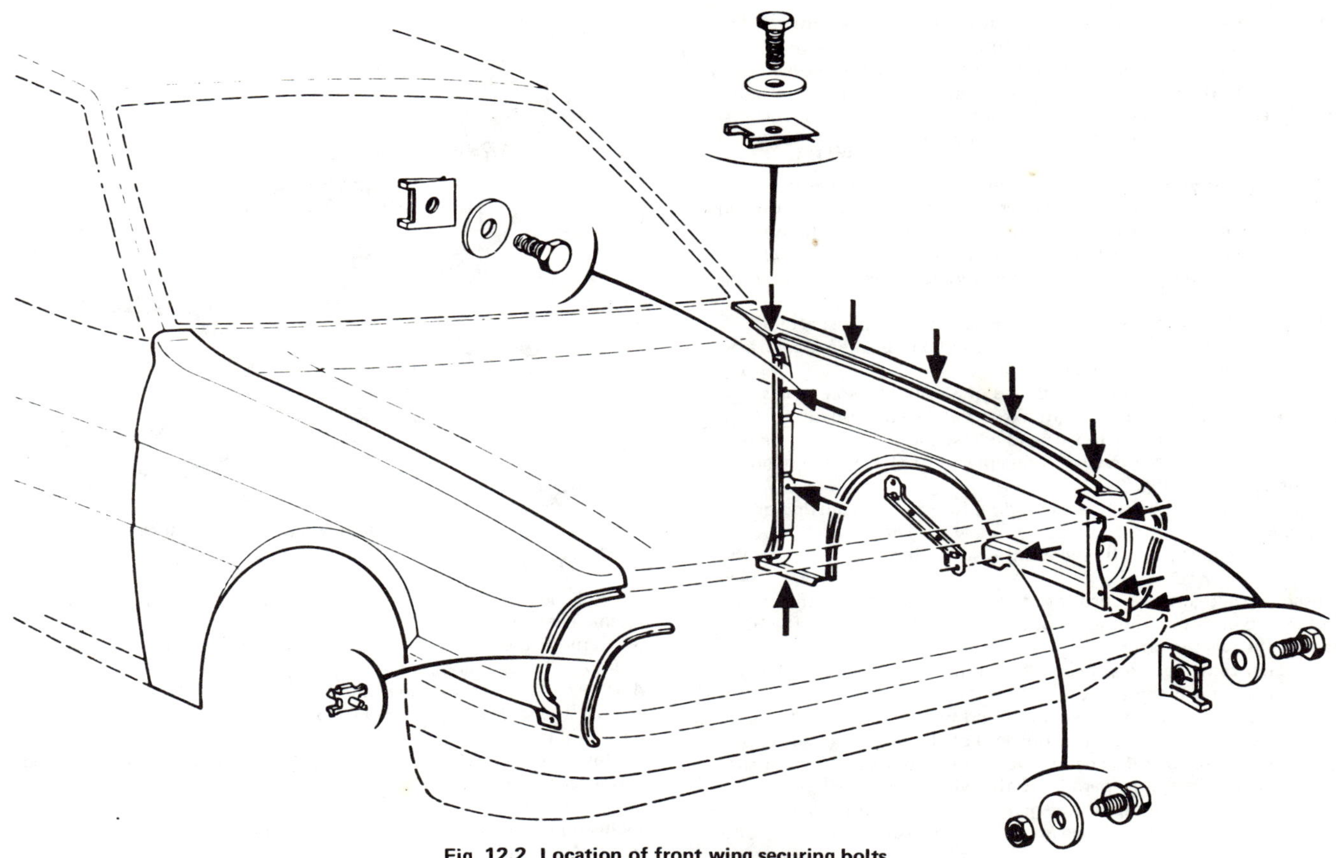

Fig. 12.2. Location of front wing securing bolts

dangerous as the car will not handle properly and secondly uneven stresses will be imposed on the steering, engine and transmission, causing abnormal wear or complete failure. Tyre wear may also be excessive.

6 Maintenance - hinges and locks

1 Oil the hinges of the bonnet, tailgate and doors with a drop or two of light oil periodically. A good time is after the car has been washed.
2 Oil the bonnet release catch pivot pin and the safety catch pivot pin periodically.
3 Do not over lubricate door latches and strikers. Normally a little oil on the rotary cam spindle alone is sufficient.

7 Doors - tracing rattles and their rectification

1 Check first that the door is not loose at the hinges and that the latch is holding the door firmly in position. Check also that the door lines up with the aperture in the body.
2 If the hinges are loose or the door is out of alignment it will be necessary to reset the hinge positions.
3 If the latch is holding the door properly it should hold the door tightly when fully latched and the door should line up with the body. If it is out of alignment it needs adjustment. If loose, some part of the lock mechanism must be worn out and requiring renewal.
4 Other rattles from the door would be caused by wear or looseness in the window winder, the glass channels and sill strips or the door buttons and interior latch release mechanism.

8 Wings - removal and refitting

1 The front wings are bolted in position and are fairly easy to remove and refit.
2 First disconnect the battery.
3 Remove the wiper arms and the heater panel at the rear of the bonnet (refer to Section 22).
4 Remove the headlight and the front direction indicator assembly (see Chapter 11).
5 Remove the front bumper as described in Section 17.
6 Jack up the front of the car, support it on axle stands and remove the front wheel.
7 Refer to Fig. 12.2 and remove the bolts and support stay securing the front wing to the bodyshell.
8 It may be necessary to cut through the wing-to-body sealing compound before the wing can be removed.
9 Clean all trace of sealing mastic from the body mating flanges.
10 Place a bead of sealing compound on the whole length of the body to wing mating flange and then locate the front wing in position. Screw in the securing bolts, with their threads well greased, finger tight.
11 Move the wing slightly as required to obtain an exact and flush fit with adjacent body panels and then tighten the securing bolts.
12 As new wings are supplied in primer, the external surface will now have to be sprayed in cellulose to match the body. If the vehicle is reasonably new this can be carried out using a colour matched aerosol spray but if the body paint is badly faded it is advisable to leave the refinishing to a professional bodyshop.
13 The procedure for a rear wing is similar but before the damaged wing can be removed, the rear lamp cluster, the rear bumper and the fuel filler cap cover must be removed.

9 Windscreen glass - removal and refitting

1 Where a windscreen is to be renewed due to shattering, the facia air vents should be covered before attempting removal. Adhesive sheeting is useful to stick to the outside of the glass to enable large areas of crystallised glass to be removed.
2 Remove both wiper arms as described in Chapter 11.
3 Remove the interior mirror by pushing it upwards out of the bracket.

4 Where the screen is to be removed intact then an assistant will be required. First release the rubber surround from the bodywork by running a blunt, small screwdriver around and under the rubber weatherstrip both inside and outside the car. This operation will break the adhesion of the sealer originally used. Take care not to damage the paintwork or cut the rubber surround with the screw-driver.

5 Have your assistant push the inner lip of the rubber surround off the flange of the windscreen body aperture. Once the rubber surround starts to peel off the flange, the screen may be forced gently outwards by careful hand pressure. The second person should support and remove the screen complete with rubber surround and metal beading as it comes out.

6 If you are having to renew your windscreen due to a shattered screen, remove all traces of sealing compound and broken glass from the weatherstrip and body flange.

7 Now is the time to remove all pieces of glass if the screen has shattered. Use a vacuum cleaner to extract as much as possible. Switch on the heater boost motor and adjust the screen controls to screen defrost but watch out for flying pieces of glass which might blow out of the ducting.

8 Carefully inspect the rubber moulding for signs of splitting or deterioration.

9 To refit the glass, first fit the weatherstrip onto the glass with the joint at the lower edge.

10 Insert a piece of thick cord into the channel of the weatherstrip with the two ends protruding by at least 12 in (305 mm) at the top centre of the weatherstrip.

11 Mix a concentrated soap and water solution and apply to the flange of the windscreen aperture.

12 Offer the screen up to the aperture and with an assistant to press the rubber surround hard against one end of the cord, move round the windscreen, so drawing the lip over the windscreen flange of the body. Keep the draw cord parallel to the windscreen. Using the palms of the hands, thump on the glass from the outside to assist the lip in passing over the flange and to seat the screen correctly onto the aperture.

13 To ensure a good watertight joint apply some Seelastik SR51 between the weatherstrip and the body and press the weatherstrip against the body to give a good seal.

14 Any excess Seelastik may be removed with a petrol moistened cloth.

10 Door locks - removal, refitting and adjustment

1 First wind the window up to the closed position.

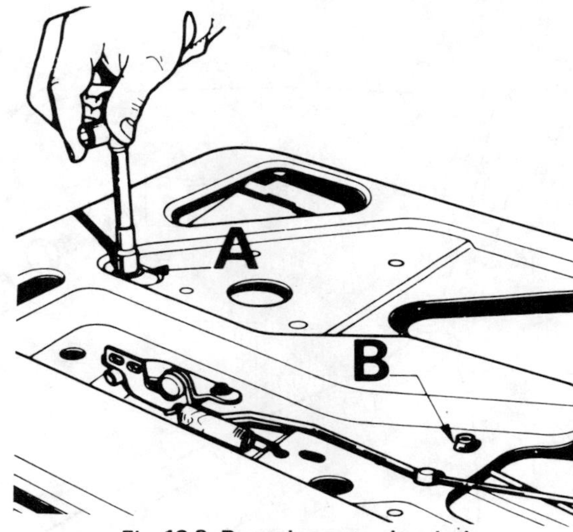

Fig. 12.3. Removing a rear door lock

A Inner retaining bolt *B Anti-rattle bush*

2 Pull the inside door latch open and remove the screw retaining it to the lever (photo).

3 Remove the arm rests which are secured to the interior panels of the doors by self-tapping screws (photos).

4 Press the window winder handle escutcheon plate inwards and remove the spring clip which retains the handle to the winder mech-anism shaft (photo).

5 Insert a screwdriver between the door interior panel and the door and lever the panel clips from their locations. Once the first securing clip has been displaced, use the fingers instead of the screwdriver and by giving the panel a sharp jerk, the remaining clips can be removed in succession.

6 Carefully remove the polythene sheet water barrier now exposed.

7 Unscrew the locking pushbutton from the top of the door.

8 Disconnect the three control rods from the lock mechanism, remove the retaining screws and withdraw the lock through the door aperture.

9 The inside (remote control) door latch can be removed by releasing the return spring and retaining screws and withdrawing the handle assembly complete with control rod (photo).

10 The procedure for removing the rear door locks is the same, with the addition that the anti-rattle bush and retaining bolt must also be

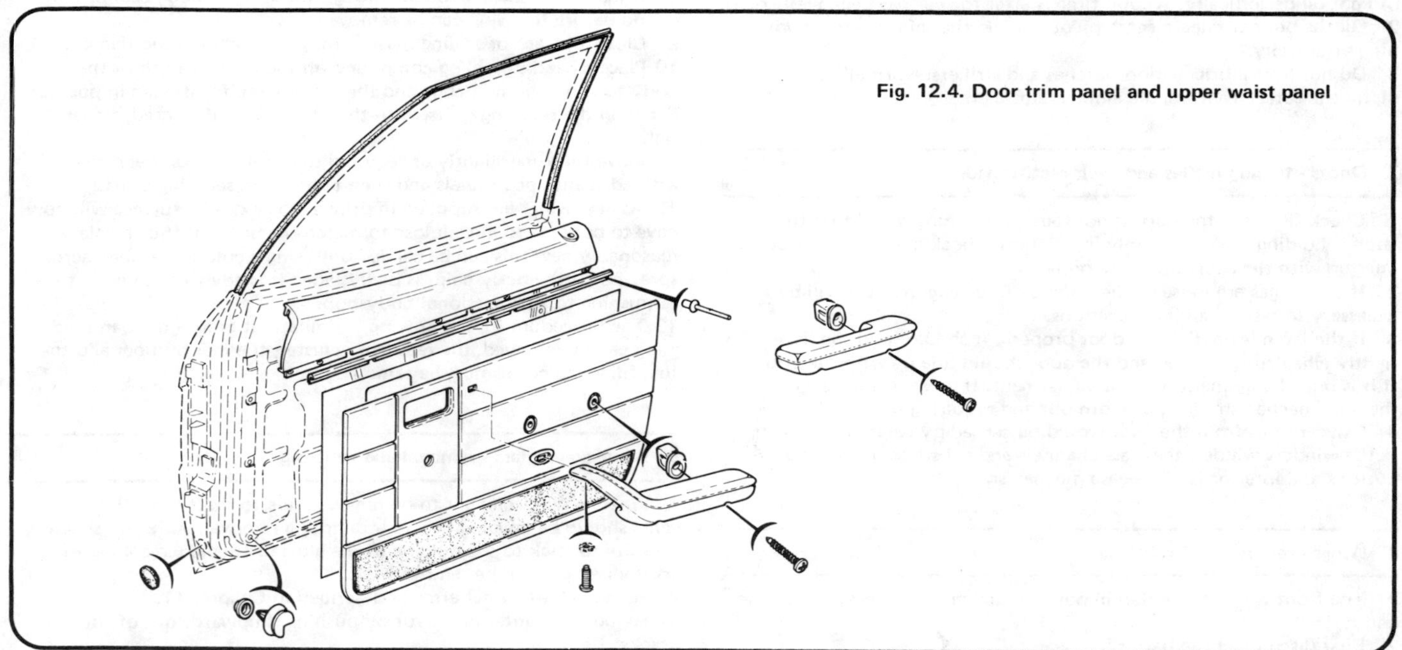

Fig. 12.4. Door trim panel and upper waist panel

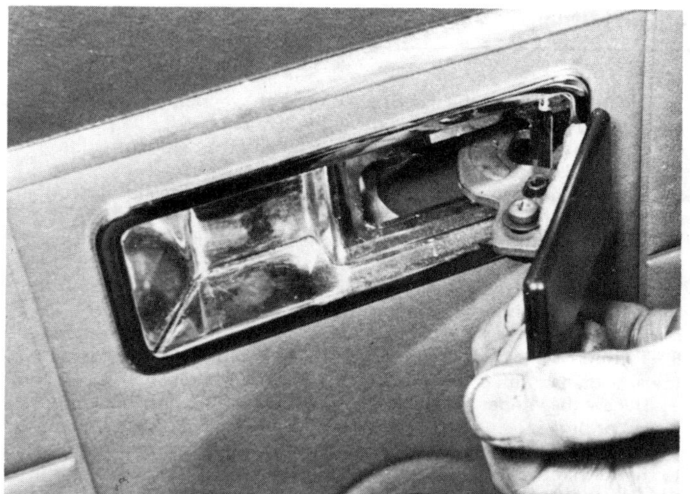

10.2 Inside door handle retaining screw

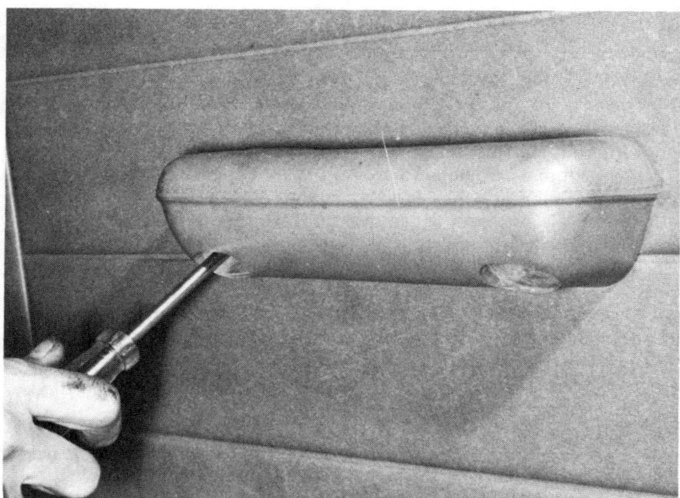

10.3A Removing the arm rest

10.3B Removing the extended arm rest upper screw

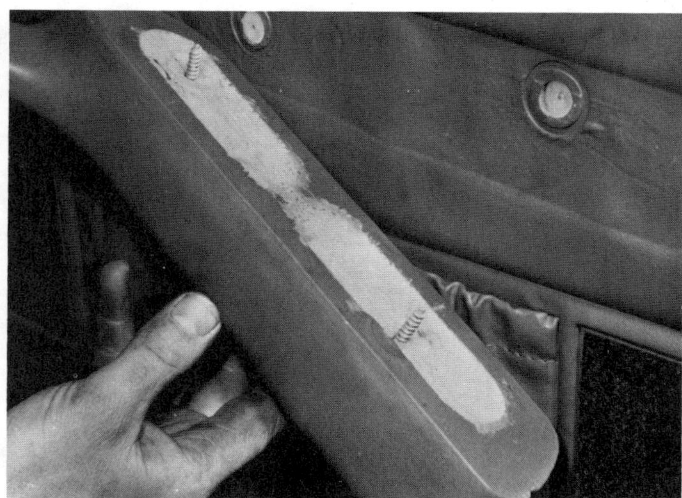

10.3C Removing the extended arm rest

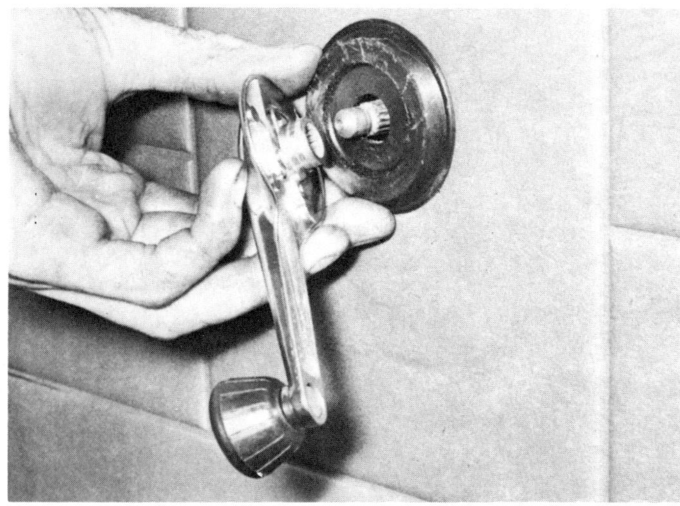

10.4 Removing the window winder handle

10.9 Inside door latch retaining screws and return spring

removed (see Fig. 12.3).
11 The exterior door handle, push button and cylinder lock are retained by a nut, screw and clip, all accessible through one of the door interior apertures.
12 It is seldom worthwhile to attempt to repair a door lock or remote control mechanism, renew as an assembly.
13 Refitting is a reversal of removal but take care to adjust the stroke of the external push button in relation to the lock operating plate. Ensure that the polythene sheet beneath the interior trim panel is refitted.
14 Check the door closure. Adjust the position of the striker plate on the door pillar if necessary, so that the door closure is firm and rattle free and the edge of the door is flush with the surrounding body panel surfaces when fully closed.

11 Window winder mechanism and glass - removal and refitting

1 Carry out operations as described in the preceding Section, paragraphs 1 to 6 and 9.
2 Remove the bolts securing the lower glass channel to the carrier (Fig. 12.5).
3 Secure the glass in the closed position with wooden wedges to prevent it dropping down when the guides are removed.
4 Remove the upper and lower bolts retaining the regulator guide (See Figs. 12.8 and 12.9).
5 Remove the bolts securing the winder (regulator) mechanism to the door.
6 Carefully withdraw the winder mechanism through the door aperture.
7 To remove the glass it is necessary to drill out the 'pop' rivets from the waist trim finisher and remove the waist trim panel. Withdraw the inner seal.
8 To remove the glass rotate it through 90° so that the front is towards the bottom of the door and carefully withdraw it (Fig. 12.10).
9 Renew the winder mechanism if the return spring or if the cable is frayed or broken.
10 Refit the glass and winder mechanism using the reverse procedure to that of removal.
11 Adjust the position of the upper and lower guide rail retaining bolts so that the glass can be raised and lowered smoothly and squarely.
12 Adjust the cable tension by means of the adjustable stop on the outer cable to obtain the correct movement of the glass (Fig. 12.11).

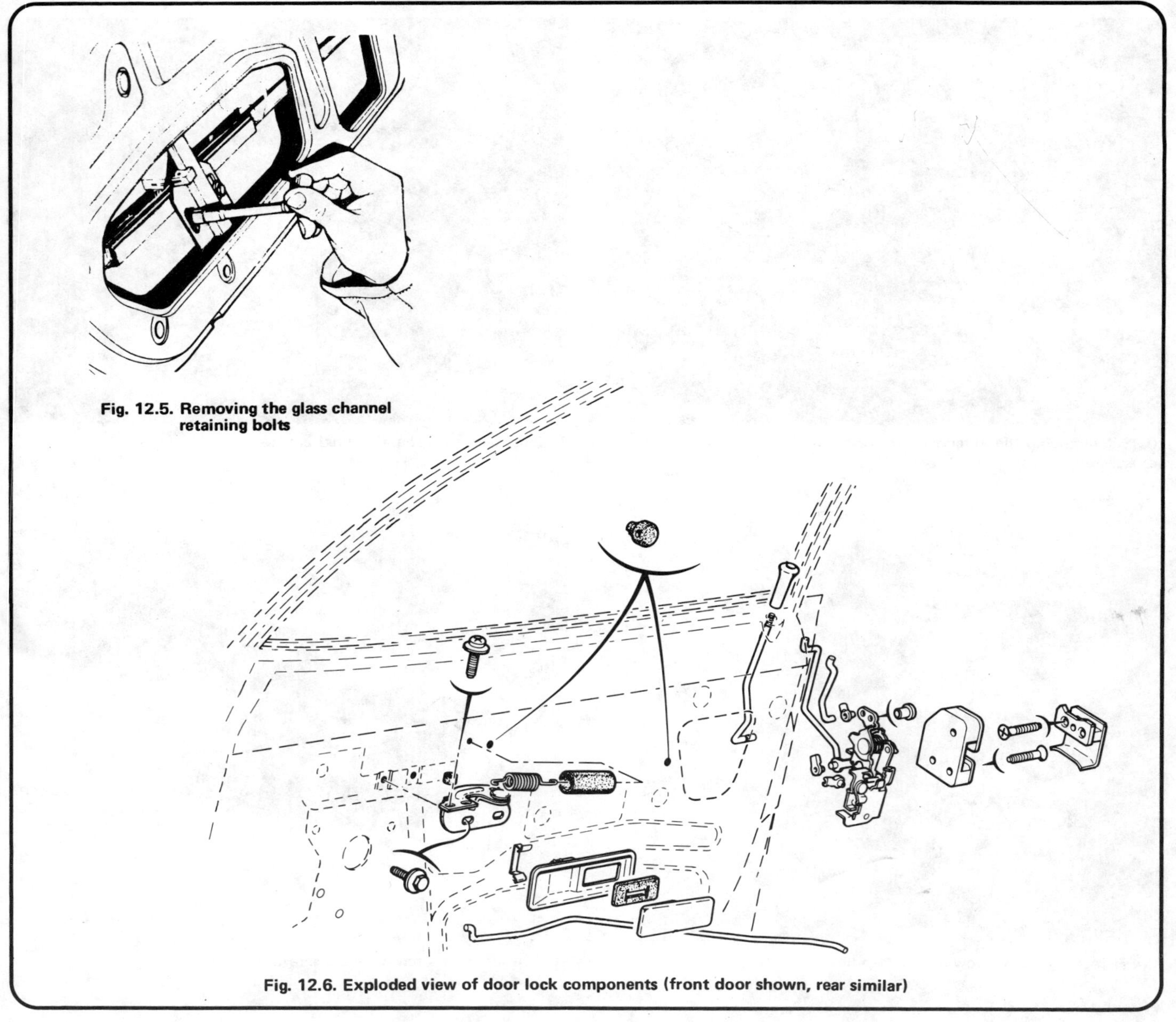

Fig. 12.5. Removing the glass channel retaining bolts

Fig. 12.6. Exploded view of door lock components (front door shown, rear similar)

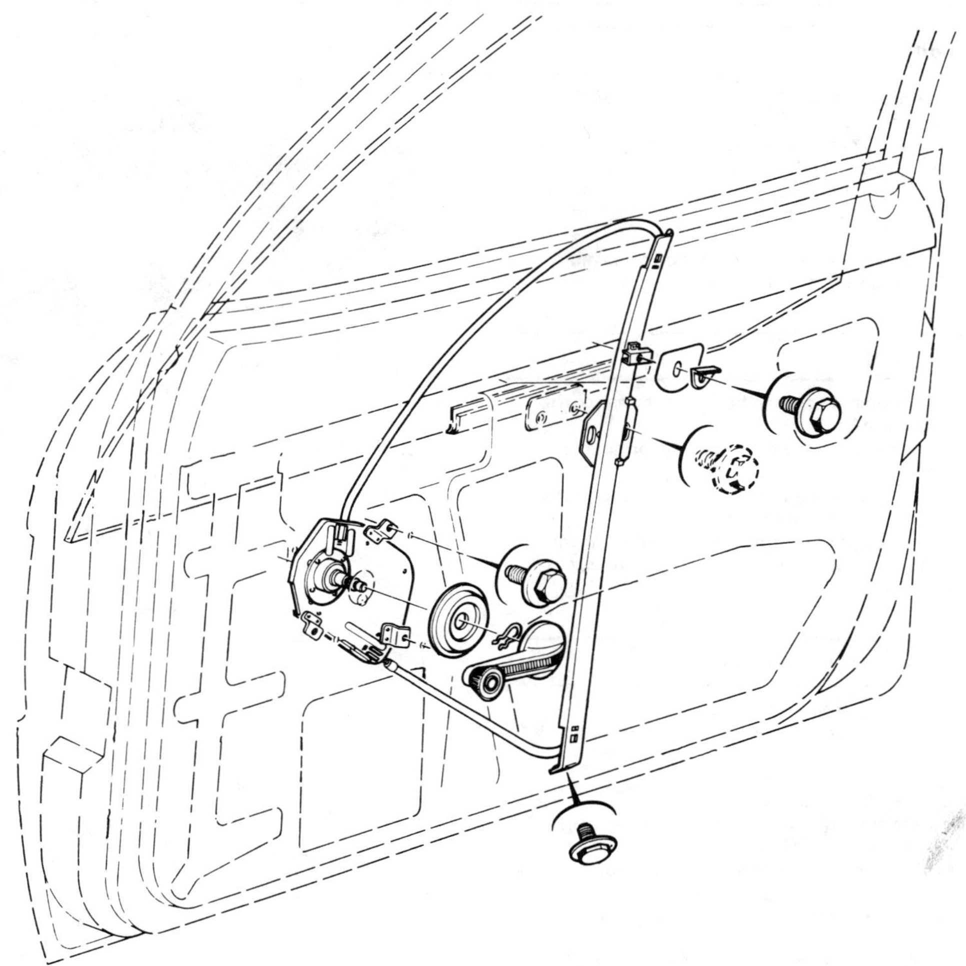

Fig. 12.7. Exploded view of window winder components

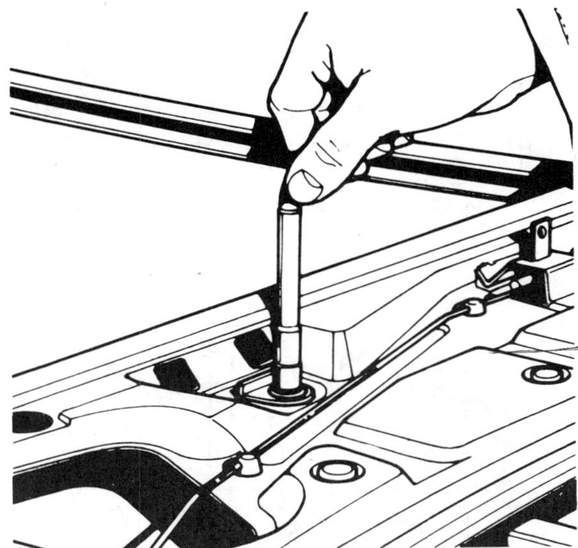

Fig. 12.8. Removing the regulator guide rail upper retaining bolt

Fig. 12.9. Removing the regulator guide rail lower retaining bolt

12 Doors - removal and refitting

1 First it is necessary to carefully grind off the riveted end of the check strap pin and remove the pin.
2 Get a friend to support the weight of the door and using a pin punch, drive out the roll pins from each hinge.
3 The position of the door can only be adjusted by bending the hinges in the required direction using the special Chrysler tool no. CO.0016.
4 Refit the door using the reverse procedure to that of removal. Fit the check strap using a new pin and rivet it in position.

13 Bonnet - removal and refitting

1 Open the bonnet and place pieces of cloth below each rear corner of the bonnet to protect the paintwork.
2 Lightly pencil a line around each hinge to ensure the bonnet is refitted in the correct position.
3 With the help of a friend, support the weight of the bonnet, remove the hinge bolts and carefully lift away the bonnet.
4 Refit the bonnet using the reverse procedure to that of removal. If necessary adjust the position of the hinges so that the bonnet fits correctly when closed.

14 Bonnet lock and cable - removal and refitting

Note: If the bonnet release cable has broken the bonnet can be opened by inserting a long screwdriver through the radiator grille and operating the release catch.
1 Remove the radiator grille as described in Section 15.
2 Disconnect the release cable from the lock.
3 Remove the screw securing the interior bonnet release handle to the bracket and withdraw the handle complete with cable.
4 The lock assembly can be removed by unscrewing the four retaining bolts and lifting out the lock assembly and stiffener plates.
5 Refit the cable and lock assembly using the reverse procedure to that of removal.
IMPORTANT: When fully closed there should be 3/16 in (5 mm) clearance between the underside of the bonnet and the top of the headlights. This clearance can be obtained by slackening the locknut and screwing the striker rod on the bonnet either in or out as required.

15 Radiator grille - removal and refitting

1 The grille is constructed of moulded plastic and is retained in

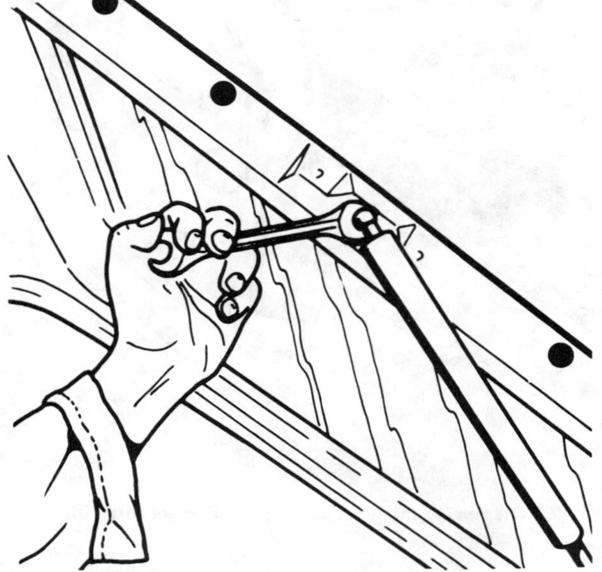

Fig. 12.12. Removing the tailgate support struts

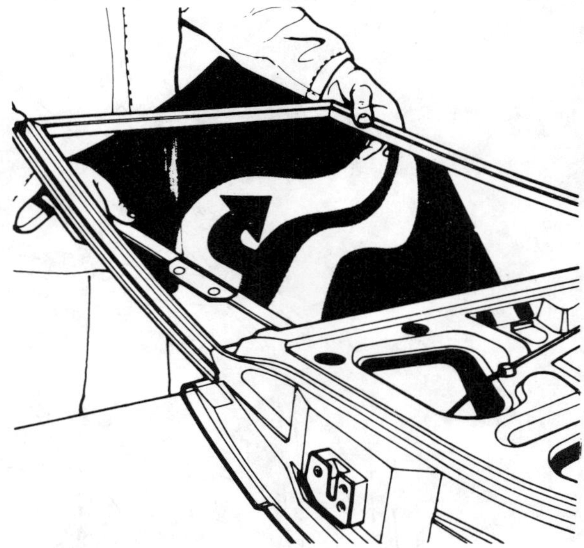

Fig. 12.10. Withdrawing the door glass

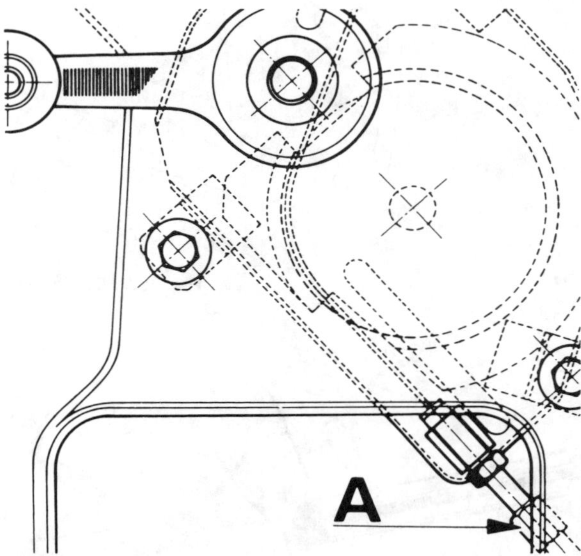

Fig. 12.11. Window winder cable adjustable stop 'A'

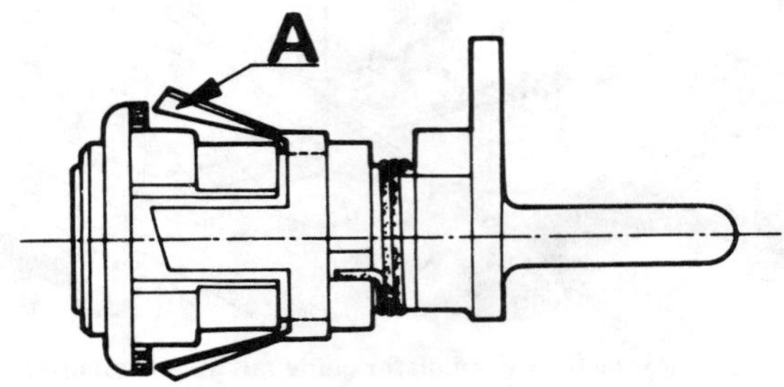

Fig. 12.13. Tailgate lock barrel assembly showing spring retaining catches 'A'

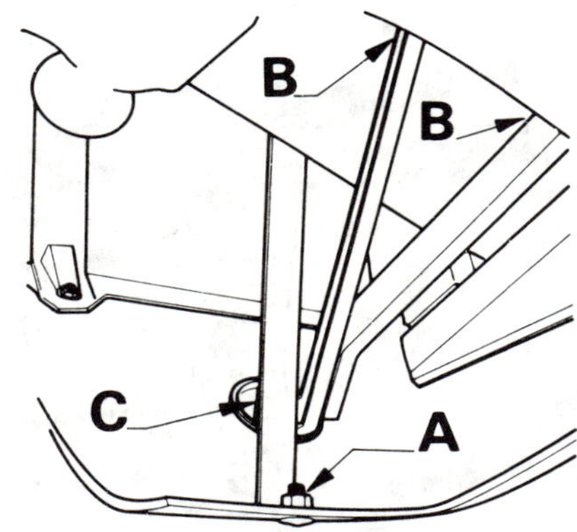

Fig. 12.14. Front bumper side support struts

A Bottom retaining bolt
B Support struts
C Side retaining bolt

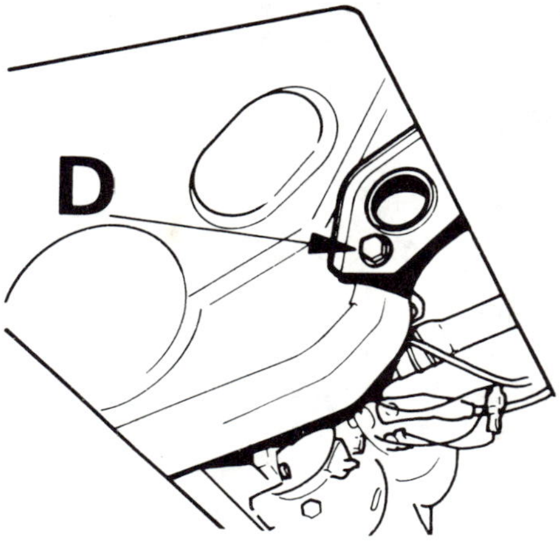

Fig. 12.15. Front bumper centre support bracket and retaining bolt 'D'

position by three pegs at the lower end which are located in rubber bushes and a single retaining screw at the top.
2 Open the bonnet and remove the retaining screw from the bonnet lock mounting panel.
3 Push the retaining clips down and lift out the grille.
4 Refit the grille using the reversal of the removal procedure.

16 Tailgate - removal and refitting

1 Disconnect the battery and detach the two wires to the rear window heating elements.
2 Support the tailgate in the open position and detach the upper mountings of the two gas-filled support struts (Fig. 12.12).
3 Carefully peel back the aperture seal adjacent to the tailgate hinges.
4 Get a friend to support the weight of the tailgate and drive out the roll pin from each hinge using a pin punch.
5 If required the lock barrel can be removed from the tailgate by pressing the spring catches and withdrawing the lock (Fig. 12.13).
6 Refit the tailgate using the reverse procedure to that of removal. Adjust the position of the tailgate in the body aperture by slackening the bolts securing the hinges to the body. Tighten the bolts when the tailgate is correctly positioned.

17 Front bumper - removal and refitting

1 Remove the two bolts on each side of the bumper, securing the support struts (Fig. 12.14).
2 Remove the two bolts on each side securing the side struts to the body frame. Remove the side struts complete with stiffener plates.
3 Remove the bolt on the right-hand side securing the drivebelt cover to the centre bracket (Fig. 12.15).
4 Withdraw the front bumper from the car.
5 Refit the bumper using the reverse procedure to that of removal.

18 Rear bumper - removal and refitting

1 Disconnect the battery and detach the two wires to the number plate (refer to Chapter 11 if necessary).

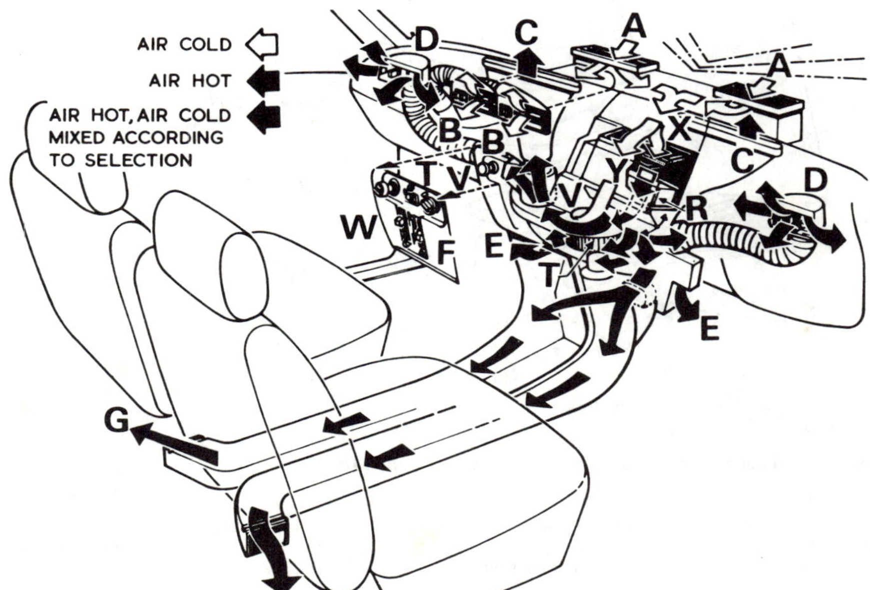

Fig. 12.16. Diagram showing heated air flow

A Fresh air inlets
B Fresh air outlets, not connected to blower
C Heated air outlets, windscreen demisting
D Heated air outlets, front door glass (adjusted by diffusers)
E Heated air outlets to front compartment
F Air flap and control of heated air to front and rear compartments
G Heated air outlets to rear compartment (on models with centre console)
R Heater matrix
T Two speed blower motor and control switch
V Air mixing flaps and control
W Demisting flaps and ducts control lever
X Air inlet to matrix
Y Fresh air vent (unheated

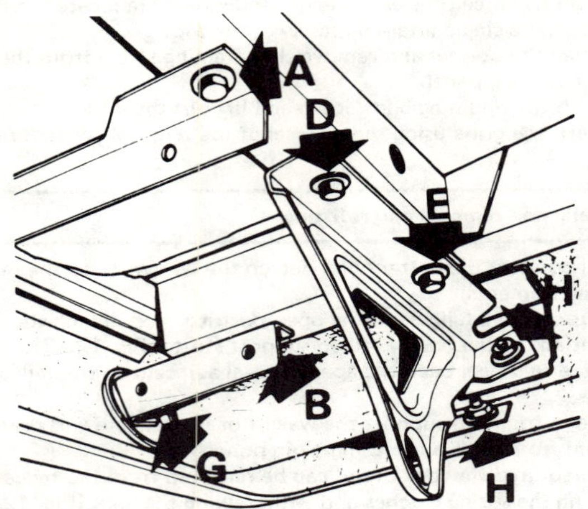

Fig. 12.17. Rear bumper brackets and retaining bolts (see text for key)

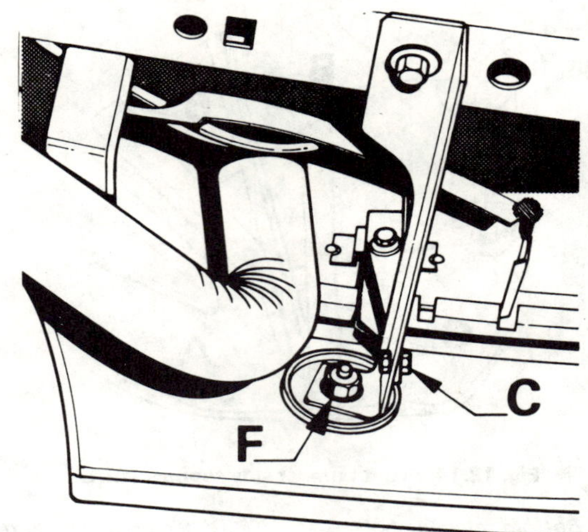

Fig. 12.18. Forward bumper to bracket securing bolts (see text for key)

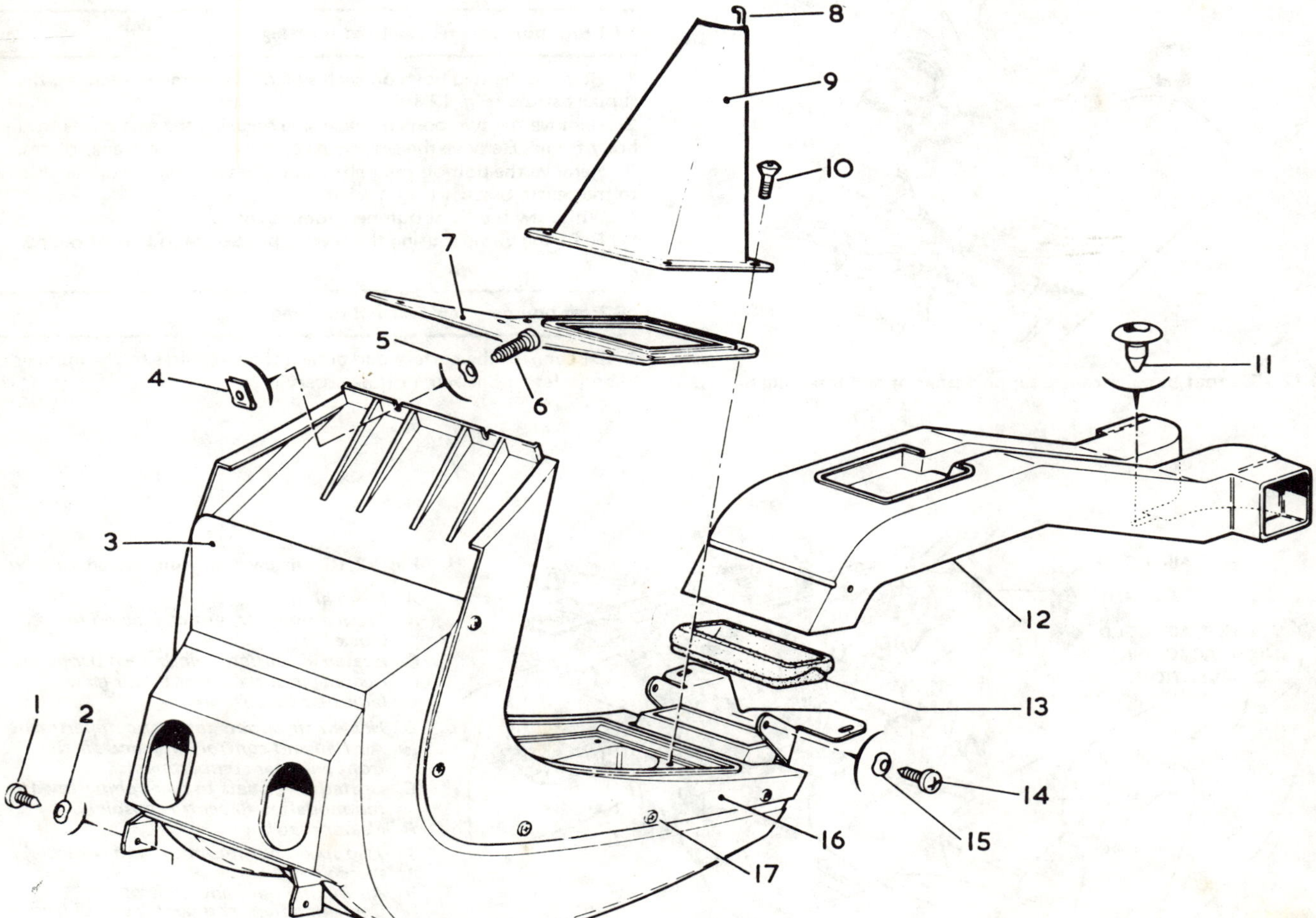

Fig. 12.19. Centre console and rear compartment heater ducts (when fitted)

1 *Screw*	6 *Screw*	10 *Screw*	14 *Screw*
2 *Washer*	7 *Console floor*	11 *Nylon pin*	15 *Washer*
3 *Console - lower portion*	8 *Gaiter grip pin*	12 *Rear duct*	16 *Console - upper portion*
4 *Spire nut*	9 *Gaiter*	13 *Rubber seal*	17 *Screw*
5 *Washer*			

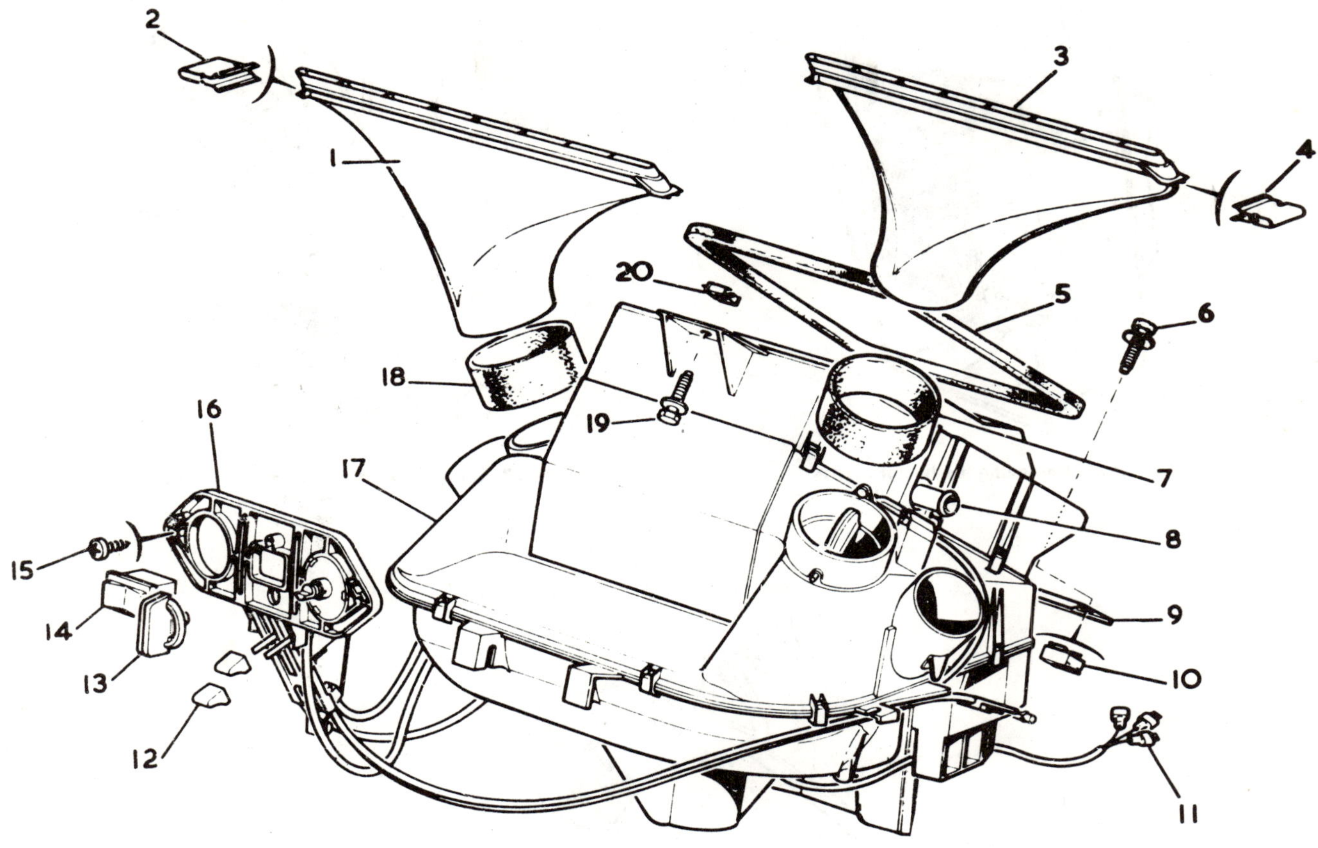

Fig. 12.20. Heater unit and controls

1 *Demister duct*
2 *Spring clip*
3 *Demister duct*
4 *Spring clip*
5 *Seal*
6 *Bolt*
7 *Sleeve connection*
8 *Heater hose connection*
9 *Clamp plate*
10 *Cage nut*
11 *Blower motor connections*
12 *Control knobs*
13 *Temperature control*
14 *Blower switch*
15 *Screw*
16 *Heater panel*
17 *Heater unit*
18 *Sleeve connection*
19 *Screw*
20 *Cage nut*

2 Place wooden blocks beneath each side of the bumper to support the weight and, referring to Figs. 12.17 and 12.18, remove bolts A, B, C, D and E securing the support bracket to the frame members.
3 Withdraw the bumper complete with bracket from the car.
4 To remove the support brackets from the bumper, detach bolts F, G and H.
5 Refit the bumper and support brackets using the reverse procedure to that of removal. Check the bumper alignment before fully tightening the securing bolts.

19 Heater unit - general description

The heater unit is mounted below the instrument panel. Air is drawn in through the intake grilles on the panel below the windscreen and, depending on the position of the heater control levers, either passes through the water heated matrix to provide hot air to the car interior, or bypasses the matrix to provide cold air.

The temperature of the air entering the car can be set to any intermediate heat range by means of the temperature control knob on the centre switch panel.

A slide lever also on the centre panel enables the air flow into the car to be completely shut off if required and a second slide lever provides selection of windscreen or car interior heating.

A three speed blower motor provides a boosted air flow if required.

A diagram of the heater unit air flow is shown in Fig. 12.16.

20 Heater unit - removal and refitting

1 Disconnect the battery negative terminal, and drain the cooling system as described in Chapter 2.
2 Drain the heater matrix by disconnecting the hose at the cylinder block.
3 Remove the radio, where fitted.
4 Remove the glovebox lid, striker, and glovebox pocket. Disconnect the switch and lamp leads.
5 Remove the passenger side parcel shelf.
6 Remove the heater cover plates.
7 Prise out the driver's side face vent finisher. Remove the screw and prise out the passenger side face vent finisher.
8 Unbolt the facia crash roll.
9 Remove the front seats and console.
10 Remove the knobs from the heater control, cigarette lighter, and temperature control. Unbolt the switch panel, disconnect the wiring, pass the wires through the control panel, and allow the switch panel to hang from the wires.
11 Remove the heater control panel and push it under the facia.
12 Remove the driver's side sound insulation (if fitted).
13 Disconnect the passenger side air hoses and water hose.
14 Loosen the engine-side heater mounting bolts.
15 Remove the facia crash roll lower rail-to-heater screws.
16 Unscrew the remaining securing bolt and disconnect the air hoses on the driver's side.

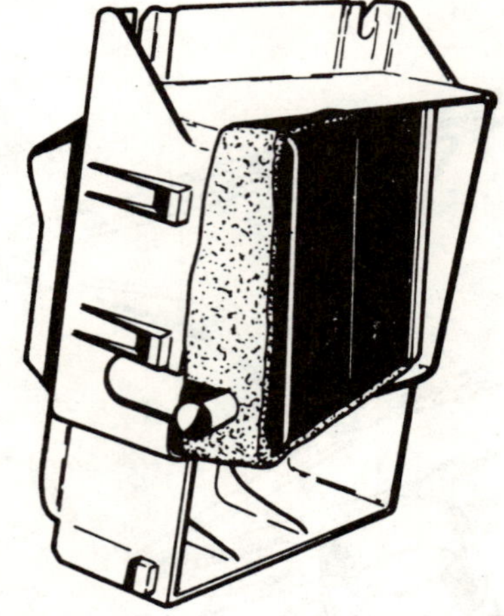

Fig. 12.21. Removing the heater matrix

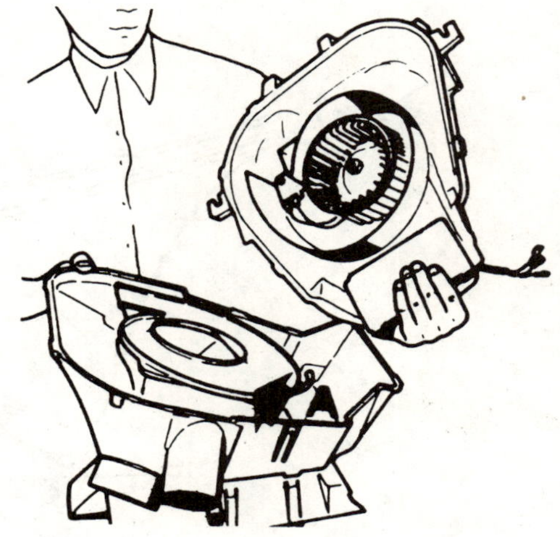

Fig. 12.22. Separating the two halves of the heater unit casing

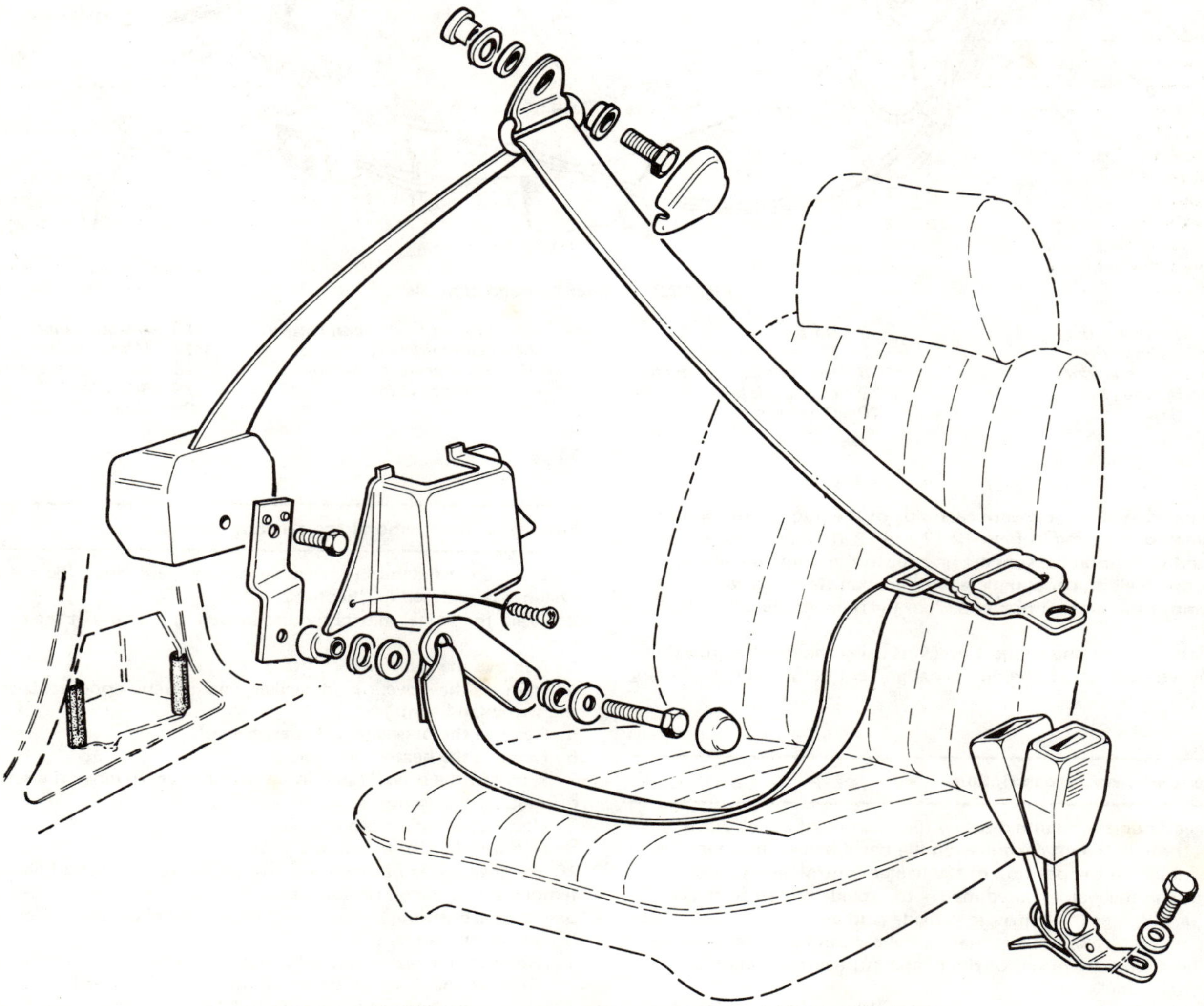

Fig. 12.23. Layout of the front seat safety harness (inertia reel type shown)

Fig. 12.24. Removing the rear air intake panel retaining screws

17 Withdraw the heater so that the supply wires can be disconnected. Detach and plug the driver's side heater hose.
18 Withdraw the heater through the passenger door aperture while an assistant lifts the facia crash roll.
19 Refitting is a reversal of removal. Fill the cooling system as described in Chapter 2.

21 Heater unit - dismantling and reassembly

1 Release the six retaining clips securing the upper casing, remove the casing and lift out the heater matrix (Fig. 12.21).
2 Release the eight clips securing the casing and separate the two halves of the casing (Fig. 12.22).
3 Lift out the fan deflector plate.
4 Mark the position of the fan on the motor spindle, remove the securing nut and withdraw the fan from the motor.
5 Remove the two motor securing nuts, detach the two supply wires and lift out the motor.
6 Release the two spring clips and remove the support bracket from the motor.
7 The motor is not repairable and if faulty should be renewed.
8 Check the matrix. Where there is evidence of water leakage from this component, renew it, attempts to solder it are seldom satisfactory. It is possible to unblock the matrix by reverse flushing with a cold water hose but the use of descaling or cleansing compounds is not recommended as the water tubes are so narrow that any sediment resulting from such treatment will probably only clog the matrix in another position.
9 Refitting is a reversal of dismantling but check that the controls are correctly connected and adjusted.

22 Air intake panel - removal and refitting

1 Remove the bonnet as described in Section 13.
2 Remove both windscreen arms and blades as described in Chapter 11.
3 Carefully ease the front of each intake grille up out of the panel. Prise the adhesive pads at the rear of each grille away from the panel, disconnect the plastic washer pipes and remove the grilles.
4 Remove the four retaining screws from the grille apertures (Fig. 12.24).
5 Remove the six retaining screws from the forward edge of the panel and carefully lift out the panel.
6 Refit the intake panel and grilles using the reverse procedure to that of removal. Do not forget to refit the washer pipes before pushing the grilles back into the panel.

23 Seat belts

1 Seat belts of the three point fixing type are fitted.
2 Components of the belts and anchorages are shown in Fig. 12.23.
3 On no account must the sequence of fitting of the anchorage plates, distance pieces or washers be changed nor the location of the fixing points be moved.
4 Periodically, inspect the straps for fraying or general deterioration and renew if necessary with ones of identical type.

Routine maintenance – additional information

As from July 1985 it is acceptable to extend the service intervals for vehicles operating in most territories.

Service A was 5000 miles (8000 km), now 6000 miles (10 000 km)
Service B was 10 000 miles (16 000 km), now 12 000 miles (19 000 km)
Service C was 15 000 miles (24 000 km) now 18 000 miles (28 000 km)
Service D was 30 000 miles (48 000 km) now 36 000 miles (58 000 km)

These extended service intervals are recommended where economy is of paramount importance, otherwise the shorter intervals given in the introductory Section of this Manual should be adhered to.

For all models, add the following to the maintenance tasks.

Every 12 000 miles (19 000 km) or annually

Check power steering fluid level and top up
Check and adjust ignition timing

Every 24 000 miles (38 000 km) or two yearly

Renew brake hydraulic fluid by bleeding
Renew air cleaner element (rectangular type) and clean oil separator gauge

Chapter 13 Supplement: Revisions and information on later models and Solara

Contents

1 Introduction

Since its introduction in 1976 the Alpine has had a number of modifications and improvements in order to keep pace with current technical and servicing innovations. This Supplement deals mainly with those incorporated on Series 7 models onward, but minor modifications which do not affect the procedures described in Chapters 1 to 12 are not included.

Particularly where an owner has a model other than Series 6, it is suggested that this Supplement is referred to before the main Chapters of the manual; this will ensure that any relevant information can be collected and accommodated into the procedures given in Chapters 1 to 12.

The Supplement also contains information on the Talbot Solara, Minx and Rapier, in particular where this differs from, or is in addition to, information given in Chapters 1 to 12.

Chrysler/Talbot UK Ltd is now known as Peugeot Talbot Motor Company Ltd, and all references to Chrysler should be regarded as references to Peugeot Talbot.

2 Specifications

The specifications listed here are revised and supplementary to the main specifications given at the beginning of each Chapter

Engine

Engine data (1592 cc type)

Bore	3.17 in (80.6 mm)
Stroke	3.07 in (78.0 mm)
Compression ratio	9.35:1
Maximum rpm	6100 in 4th gear
Piston diameters:	
Class A	3.1727 to 3.1730 in (80.5875 to 80.5950 mm)
Class B	3.1730 to 3.1733 in (80.5950 to 80.6025 mm)
Class C	3.1733 to 3.1736 in (80.6025 to 80.6100 mm)
Class D	3.1736 to 3.1739 in (80.6100 to 80.6175 mm)
Piston clearance in bore*	0.0011 to 0.0019 in (0.027 to 0.047 mm)
Piston ring gaps:*	
Top ring	0.012 to 0.018 in (0.30 to 0.45 mm)
Second ring	0.012 to 0.018 in (0.30 to 0.45 mm)
Oil control ring	0.010 to 0.016 in (0.25 to 0.40 mm)
Gudgeon pin length	2.9094 in (73.9 mm)

Crankshaft and connecting rods:	
Connecting rod side play*	0.006 to 0.009 in (0.145 to 0.225 mm)
Pushrod length	8.523 in (216.5 mm)
Valve clearances (cold) – modified camshaft and followers, refer to Section 3 of this Supplement:	
Inlet and exhaust	0.25 mm (0.010 in)

Note: *Items marked * are also applicable to all 8-Series engines onwards*

Cooling system

Capacity (automatic transmission)	11.1 pints (6.3 litres)

Carburation

Carburettor type and data (commencing 6-Series)

Engine type	**1294 cc single carb, low compression**
Engine code	6 G 1A
From engine No.	67 830 021
Compression ratio	8.2:1
Carburettor make	Solex
Type	32 BISA 5
Ref No	86
Choke tube	27
Main jet	135 ± 2.5
Air correction jet	175 ± 5
Emulsion tube	E2
Econostat fuel jet	45
Idling fuel jet	39 to 45 with damper
Idling air jet	180
Progression holes	4 x 0.6
Pump injector	45
Pump stroke	3 mm
Float needle	1.5
Float weight	5.7 g
Starting system	Strangler
Fast idle gap (at throttle plate)	1.10 mm ± 0.05
Constant CO circuit:	
Air intake aperture	500
Fuel calibration	30
Air calibration	130
CO content at idling	1 to 2%

Carburettor type and data (commencing 7-Series)

Engine type	**1442 cc 1 double choke carb, modified power**
Engine code	6 YD 2
From engine No.	68 200 021
Compression ratio	9.5:1
Carburettor make:	Weber
Type	36 DCNV
Ref No	5
Choke tube	27
Secondary venturi	4.5 short
Main jet	130 ± 2.5
Air correction jet	185 ± 5
Emulsion tube	F 36
Idling fuel jet	40
Idling air jet	145 ± 5
Progression holes	90-90-90-100
Pump injector	50
Float needle	1.75
Float weight	20 g
Float level	52 ± 0.25
Starting system	Strangler
Fast idle gap	0.40 to 0.45 mm
Vacuum kick opening	5.0 to 5.5 mm
Mechanical opening	8.0 to 8.5 mm
CO content at idling	1% to 2%

Carburettor types and data (commencing 8-Series)

Engine type	**1294 cc single carb.**	**1294 cc single carb. low compression**	**1294 cc single carb. modified power**
Engine code	6 G 1	6 G 1A	6 G 1D
Compression ratio	9.5:1	8.8:1	8.8:1
Carburettor make:	Weber	Weber	Solex
Type	32 IBSA	32 IBSA	32 IBSA 5A
Ref No	9	11	110A
Choke tube	26	26	23
Secondary venturi	4.5	4.5	2

	1294 cc single carb	1294 cc single carb low compression	1294 cc single carb modified power
Main jet	137 ± 2.5	137 ± 2.5	112.5 ± 2.5
Air correction jet	160 ± 5	170 ± 5	180 ± 5
Emulsion tube	F6	F83	E7
Econostat fuel jet	120	120	40
Econostat air jet	50 to 200	200	–
Idling fuel jet	40 ± 2.5	42.5 ± 2.5	41.5 ± 2.5 with damper
Idling air jet	185 ± 5	135 ± 5	180
Progression holes	100-110-70	80-110-70-80	4 x 0.6
Pump injector	45	45	45
Pump bypass	40	40	–
Pump stroke	–	–	3 mm
Float needle	1.5	1.5	1.5
Float weight	11 g	11 g	5.7 g
Float level	6 mm ± 2.5	6 mm ± 2.5	17.5 mm ± 2
Starting system	Strangler	Strangler	Strangler
Fast idle gap	1.20 mm ± 0.05	1.20 mm ± 0.05	1.00 mm ± 0.05
Mechanical opening	5.25 mm	5.25 mm	–
Starter cam	–	–	No 4
Constant CO circuit:			
Air intake aperture	–	–	580
Fuel calibration	–	–	35
Air calibration	–	–	150
CO content at idling	1 to 2%	1 to 2%	1 to 2%

Carburettor type and data (commencing 9-Series)

Engine type	**1294 cc double choke carb.**
Engine code	6 G 2
Compression ratio	9.5:1
Carburettor make:	Weber
Type	36 DCNVH
Ref No	1
Choke tube	28
Secondary venturi	4.5 short
Main jet	$125 \pm {}^{2}_{3}$
Air correction jet	175 ± 5
Emulsion tube	F 36
Economy jet, fuel	50 ± 10
Idling jet	$40 \pm {}^{2}_{3}$
Idling air jet	120 ± 5
Progression	70-70-90-100
Pump injector	50 long
Pump stroke	No 2 cam 42
Pump capacity	1.4 to 2.1 cc
Float needle	1.75
Float weight	14.5 g
Float level	42.5 ± 0.25 m
Starting device	By flap
Vacuum opening	5 to 5.5 mm
Mechanical opening	8 to 8.5 mm

Engine type	**1441 cc double choke carb. auto transmission**
Engine code	6 Y 2
Compression ratio	9.5:1
Carburettor make	Weber
Type	36 DCA 2
Choke tube	28
Secondary venturi	4.5
Main jet	$127 \pm {}^{3}_{2}$
Air correction jet	170 ± 5
Emulsion tube	F 46
Pneumatic enrichment	50 ± 10
Idling fuel jet	37 to 42
Idling air jet	145 ± 5
Progression holes	3 x 90 - 100
Pump injector	40 long
Pump stroke	No 2
Pump cam	42
Pump bypass	40
Float needle	1.75
Float weight	14.5 g
Float level	42.5 ± 0.25 (without gasket)
Fast idle gap	0.50 to 0.55 mm
Vacuum kick opening	4.5 to 5 mm
Modulated vacuum kick opening	7.5 to 8 mm

Engine type	1592 cc, double choke carb. auto transmission
Engine code	6 J 2
Compression ratio	9.35:1
Carburettor make:	Weber
Type	36 DCA
Choke tube	29
Secondary venturi	4.5
Main jet	132 ± $\frac{3}{2}$
Air correction jet	175 ± 5
Emulsion tube	F46
Pneumatic enrichment	40 ± 10
Idling fuel jet	37 to 45
Idling air jet	140 ± 5
Progression holes	3 x 90 - 100
Pump injector	40 long
Pump stroke	No 2
Pump cam	42
Pump bypass	40
Float needle	1.75
Float weight	14.5 g
Float level	41 mm ± 0.25 (without gasket)
Automatic choke:	
Throttle opening, cocked	0.50 to 0.60 mm
Vacuum opening, choke	4.0 to 4.5 mm
Modulated opening, choke	6.0 to 7.0 mm

Idling speeds (commencing 9-Series)

1294 cc (code 6 G 2)	900 rpm
1442 cc (code 6 Y 2 automatic)	950 rpm
1592 cc (code 6 J 2 automatic)	950 rpm

Carburettor type and data (commencing A-Series 1979-1980)

Engine type	1294 cc high compression	1294 cc power variation
Engine code	6 G1	6 G1VP
Compression ratio	9.5:1	8.8:1
Carburettor make:	Solex	Solex
Type	32 BISA 7	32 BISA 7
Ref No	117	128
Choke tube	27	23
Main jet	135 ± 2.5	117.5 ± 2.5
Air correction jet	170 ± 5	175 ± 5
Emulsion tube	EC	EC
Econostat fuel	50 ± 5	50 ± 5
Idling fuel jet	39 to 45	39 to 45
Idling air jet	130	130
Pump injector	40	40
Float needle	1.5	1.5
Constant CO:		
Fuel calibration	35	35
Emulsion aerator	160	150

Engine type	1442 cc single choke	1592 cc single choke
Engine code	Y 1	6 J 1
Compression ratio	9.5:1	9.35:1
Carburettor make:	Solex	Solex
Type	32 BISA 7	32 BISA 7
Ref No	121	132
Choke tube	27	30
Main jet	135 ± 2.5	150 ± 5
Air correction jet	155 ± 5	170 ± 10
Emulsion tube	EC	E2
Econostat fuel	55 ± 10	55 ± 10
Idling fuel jet	36 to 42	48 ± 3
Idling air jet	130	130 to 190
Pump injector	40	40
Float needle	1.5	1.5
Constant CO:		
Fuel calibration	35	35
Emulsion aerator	140	160

Engine type	1294 cc high compression	1294 cc double choke
Engine code	6 G1	6 Y 2
Compression ratio	9.5:1	9.5:1

Carburettor make:	Weber or Bressel	Weber or Bressel
Type	32 IBSH	36 DCNVH
Ref No	3	5
Choke tube	26	28
Main jet	$135 \pm {}^{2}_{3}$	$127 \pm {}^{3}_{2}$
Air correction jet	175 ± 5	170 ± 5
Emulsion tube	F6	F46
Fuel enricher	130	–
Idling fuel jet	40 to 45	40 to 42
Idling air jet	200 ± 5	160 ± 5
Progression	100-100-70	3 x 90-100
Pump injector	45	50 long
Float needle	150	175
Float weight	11 g	14.5 g
Float level	6 mm ± 0.25	42.5 mm ± 0.25
Fast idle gap	105 ± 5	40 ± 5
Vacuum enricher	55 ± 10	50 ± 10

Engine type	**1442 cc power variation**	**1442 cc low compression**
Engine code	6 Y 2VP	6 Y 2BTC
Compression ratio	8.8:1	8.8:1
Carburettor make:	Weber	Weber or Bressel
Type	36 DCNVH	36 DCNVH
Ref No	6	10
Choke tube	27	28 with bar
Main jet	$125 \pm {}^{2}_{3}$	$127 \pm {}^{3}_{2}$
Air correction jet	185 ± 5	180 ± 5
Emulsion tube	F46	F46
Idling fuel jet	40 to 45	40 to 45
Idling air jet	125 ± 5	120 ± 5
Progression	70-70-90-100	2 x 70-70-100
Pump injector	50 long	50 long
Float needle	175	175
Float weight	14.5 g	14.5 g
Float level	42.5 mm ± 0.25	42.5 mm ± 0.25
Fast idle gap	40 ± 5	40 ± 5
Vacuum enricher	65 ± 10	50 ± 10

Idling speeds (commencing A-Series)

1294 cc (code 6 G 1)	850 rpm
1442 cc (code 6 Y 1 and 6 Y 2)	900 rpm
1592 cc (code 6 J 1 manual)	900 rpm
1592 cc (code 6 J 2 automatic)	950 rpm

Carburettor type and data (commencing B-Series 1980-1981)

Engine type	**1592 cc**
Engine code	6 J 2
Compression ratio	9.35:1
Carburettor make:	Weber or Bressel
Type	36 DCNVH
Ref No	7
Choke tube	29 with bar
Main jet	135 ± 5
Air correction jet	180 ± 10
Emulsion tube	F46
Fuel enricher	50 ± 10
Idling fuel jet	40 to 45
Idling air jet	135 ± 10
Progression	80-75-85-90
Pump injector	40 long
Float needle	175
Float weight	14.5 g
Float level	42 mm ± 0.25
Vacuum opening	4.25 to 0.25
Modulated opening	8.0 to 8.5 mm

Idling speeds (commencing B-Series)

1294 cc (code 6 G 1)	850 rpm
1442 cc (code 6 Y 1 and 6 Y 2)	900 rpm
1592 cc (code 6 J 1 manual)	900 rpm
1592 cc (code 6 Y 2 automatic and manual)	950 rpm

Carburettor type and data (commencing C-Series 1981-1982)

Engine type	1294 cc
Engine code	6GIE
Compression ratio	9.5:1
Carburettor make:	Solex
Type	32 B1SA8
Ref No	143
Choke tube	26
Secondary venturi	2
Main jet	130 ± 5
Air correction jet	185 ± 5
Emulsion tube	EC
Fuel enricher	60 ± 10
Econostat, fuel	50 ± 10
Econostat, air	300
Idling fuel jet	45 ± 10
Progression slot	4.5 x 0.6
Pump injector	40 ± 5
Pump stroke	3 mm
Float needle	1.5
Float weight	5.7 g

Engine type	1592 cc double choke automatic transmission	1592 cc double choke manual transmission
Engine code	6J2A	6J2A
Compression ratio	9.35:1	9.35:1
Carburettor make:	Weber	Weber or Bressel
Type	36 DCA	36 DCNVA
Ref No	4	17
Choke tube	29	29
Secondary venturi	4.5	4.5
Main jet	150 ± 5	145 ± 5
Air correction jet	165 ± 15	165 ± 15
Emulsion tube	F46	F54
Fuel enricher	30 (idling)	–
Idling fuel jet	40 to 45	40 to 45
Idling air jet	165 ± 15	165 ± 15
Progression	90-80-90-90	90-80-105-110
Pump injector	40	45 ± 5
Pump stroke/cam	2-42	2-42
Pump bypass	40	40
Float needle	1.75	1.75
Float weight	14.5 g	20 g
Float level	42 mm ± 0.25 mm	52 mm ± 0.25 mm
Vacuum opening	4 to 4.5 mm	5 to 5.5 mm
Mechanical opening	8 to 8.5 mm	8 to 9 mm
Modulated opening	6 to 7 mm	–
Cam setting	6.5 mm	–

Idling speeds (commencing C-Series)

1294 cc (code 6G1E)	600 to 700 rpm
1592 cc (code 6J2A – automatic transmission)	950 rpm
1592 cc (code 6J2A – manual transmission)	900 rpm

Carburettor type and data (commencing D-Series 1982-1983)

Engine type	1592 cc double choke Manual transmission	1592 cc Economy
Engine code	6J2A	6J1A
Compression ratio	9.35:1	9.35:1
Carburettor make:	Weber or Bressel	Solex
Type	32 DCNVH	35 BISA 8
Ref No	12	345
Choke tube	29	30
Secondary venturi	4.5	2
Main jet	150 ± 5	150 ± 5
Air correction jet	175 ± 15	170 ± 15
Emulsion tube	F46	E2
Fuel enricher	–	60 ± 10
Econostat, fuel	–	60 ± 10
Econostat, air	–	300
Idling fuel jet	40 to 45	45 to 50
Idling air jet	165 ± 15	130 ± 15
Progression	90-80-90-90	–
Pump injector	40 to 50	40 ± 5

	Manual transmission 1592 cc double choke	1592 cc Economy
Pump stroke/cam	2-42	–
Pump bypass	40	–
Float needle	1.75	1.5
Float weight	14.5 g	5.7 g
Float level	42 mm ± 0.25	–
Vacuum opening	6 mm ± 0.25	–
Mechanical opening	8 to 9 mm	
CO content at idling	2%	

Idling speeds (commencing D-Series)

1592 cc (code 6J2A – manual)	900 rpm
1592 cc (code 6J2A – automatic)	950 rpm
1592 cc (code 6J1A9)	650 rpm

Carburettor type and data (Minx, Rapier 1985-on)

Carburettor make:	Weber or Bressel
Type	36 DCNVH
Ref No	12/103
Choke tube	29
Secondary venturi	4.5
Main jet	145 ± 5
Air correction jet	175 ± 15
Emulsion tube	F46
Fuel enricher	40 ± 10
Idling fuel jet	40/45
Idling air jet	165 ± 15
Progression	90-80-125-90
Pump injector	40 to 50
Pump stroke/cam	2-42
Pump bypass	40
Float needle	1.75
Float weight	14.5 g
Float level	42.5 ± 0.25
Vacuum kick opening	6.25 mm ± 0.25
Modulated opening	7.0 to 8.0 mm

Ignition system

Distributor (commencing 9-Series)

Type	'Hall effect' – electronic
Make	Chrysler, Bosch or Ducellier

Ignition coil (commencing 9-Series)

Primary resistance at 20°C (70°F)	1.4 to 1.6 ohms
Secondary resistance at 20°C (70°F)	8000 to 12 000 ohms

Spark plugs

1294 cc with 1 single choke carburettor:	
7 and 8-Series	Champion N7YC, Marchal GT34-5HA or equivalent
Commencing 9-Series	Champion N9Y, Marchal HT34-5HA, Chrysler A75P, C Bosch W175T 301 or equivalent
Commencing D-Series	Champion N9YC or equivalent
1294 cc with 2 twin choke carburettors:	
7 and 8-series	Champion N7YC, Marchal GT34-5HA or equivalent
Commencing 9-Series	Champion N9YC, Marchal GT34-5HA, Chrysler A75P Bosch W175T 401 or equivalent
Commencing D-Series	Champion N9YC or equivalent
1294 cc with 2 twin choke carburettors:	
7 and 8-series	Champion N7YC, Marchal GT34-2H or equivalent
1294 cc with 1 twin choke carburettor:	
Commencing 9-Series	Champion N9YC, Marchal GT34-5HA, Bosch W175T301 or equivalent
1442 cc – all models:	
Commencing 7-Series	Champion N9YC, Marchal GT34-5HA, Bosch W175T301 or equivalent
1592 cc – all models up to July 1984:	
Commencing 9-Series	Champion N9YC, Marchal GT34-5HA, Bosch W175T301 or equivalent
Commencing D-Series	Champion N9YC or equivalent
1592 cc, July 1984 on	Champion N9YC, Bosch W7DC, Peugeot CP10 or equivalent

Spark plug gap – all models	0.024 to 0.028 in (0.6 to 0.7 mm)

Ignition timing (stroboscopic at idling speed – vacuum disconnected)

1294 cc (single carb, low compression, code 6G1A, commencing 8-Series)	10 to 12° BTDC
1294 cc (single carb, modified power, code 6G1D, commencing 8-Series)	8 to 10° BTDC
1294 cc (single carb, codes 6G1 and 6G1BTC, commencing 9-Series)	8 to 10° BTDC
1294 cc (power variation, code 6G1VP, commencing 9-Series)	10 to 12° BTDC
1294 cc (power variation, code 6G1BP, commencing A-Series)	8 to 10° BTDC
1294 cc (single carb, code 6G2, commencing 9-Series)	10 to 12° BTDC
1294 cc (high compression, code 6G1E, commmencing C-Series)	2° BTDC
1294 cc (high compression, code 6G1E, commencing D-Series)	2 to 4° BTDC
1442 cc (single carb, code 6Y2, commencing 9-Series)	12 to 14° BTDC
1442 cc (power steering, code 6Y2VP, commencing 9-Series)	10 to 12° BTDC
1442 cc (low compression, code 6Y2 BTC, commencing A-Series)	10 to 12° BTDC
1442 cc (single carb, code 6Y1, commencing A-Series)	8 to 10° BTDC
1592 cc (single carb, code 6J2, commencing 9-Series	10 to 12° BTDC
1592 cc (single carb, code 6J1, commencing A-Series)	10 to 12° BTDC
1592 cc (high compression 6J1A)	4° BTDC
1592 cc (single carb, code 6J2A, commencing C-Series)	10 to 12° BTDC

Clutch

Type

1294 cc, commencing 7-Series	180 DBR or 190 Borg and Beck
1442 cc, commencing 7-Series	180 DBR, 190 DBR or 190 Borg and Beck
1592 cc, single choke carburettor	180 DBR, 190 DBR or 190 Borg and Beck
1592 cc, twin choke carburettor	200 DBR

Manual gearbox and final drive

Gear ratios (four-speed gearbox type AT)

4th (commencing 9-Series, code 6Y1, 1442 cc, single carb)	1.04:1

Gearbox type (commencing A-Series)

Four-speed	AC 429
Five-speed	AC 428

Gear ratios (AC gearboxes)

1st	3.166:1
2nd	1.833:1
3rd	1.25:1
4th	0.939:1
5th	0.767:1
Reverse	3.153:1
Final drive ratio	4.214:1

Lubricant capacity (AC gearboxes)

Total:	
AC 429	2.1 pints (1.2 litres)
AC 428	2.3 pints (1.3 litres)
Refill after draining:	
AC 429	1.8 pints (1.0 litre)
AC 428	1.9 pints (1.1 litres)

Lubricant type (AC gearboxes) .... SAE 90 EP gear oil

Gearbox type (commencing D-Series)

Four-speed	BE1/4
Five-speed	BE1/5

Gear ratios (BE1 gearboxes)

1st	3.308:1
2nd	1.833:1
3rd	1.280:1
4th	0.969:1
5th	0.757:1
Reverse	3.333:1
Final drive ratio	4.18:1

Lubricant capacity (BE1 gearboxes)

Refill after draining	3.5 pints (2.0 litres)

Lubricant type (BE1 gearboxes) .... SAE 15W/40 multigrade engine oil

Torque wrench settings (AC428 and AC429 gearbox)

	lbf ft	kgf m
Bearing retainer plate	16	2.2
Drain and filler plugs	30	4.1
End cover	21	2.9
Gearbox casing	21	2.9
Input shaft nut (AC428)	210	29
Input shaft bolt (AC429)	104	14.4
Output shaft nut	148	20.4
Reverse switch	10	1.4
Selector forks and dogs	21	2.9
Support bracket	18	2.5
Selector shaft threaded bush	88	12.1
Cross-shaft support bracket	33	4.6
Relay lever locking bolt	25	3.5
Relay lever locknut	18	2.5
Selector cross-shaft bracket	18	2.5
Clutch housing	33	4,6
Clutch slave cylinder	16	2.2

Torque wrench settings – final drive (AC428 and AC429)

	lbf ft	kgf m
Drain plug	30	4.1
Bearing retainer bolts	21	2.9
Crownwheel-to-carrier bolts	66	9.1
Extension housing bolts	21	2.9
Cover nuts:		
8 mm studs	21	2.9
10 mm studs	37	5.1

Torque wrench settings (BE1/4 and BE1/5 gearbox)

	lbf ft	kgf m
End cover	9	1.2
Input and output shaft nuts	40	5.5
Bearing retainer bolts	11	1.5
Selector rod lockplate	11	1.5
Casing retaining bolts	9	1.2
Reverse idler gear spindle bolt	15	2.0
Gear selector shaft support	11	1.5
Reverse fork	15	2.0
Breather	11	1.5
Reverse lamp switch	18	2.4
Gearbox drain plug	7	0.9
Final drive drain plug	22	3.0
Speedometer pinion adaptor	9	1.2
Extension housing bolts	11	1.5
Final drive casing:		
10 mm diameter bolts	30	4.1
7 mm diameter bolts	9	1.2
Clutch release bearing guide tube	9	1.2

Automatic transmission

General

Type	Torqueflite A415, three-speed
Ratios:	
1st	2.48:1
2nd	1.48:1
3rd	1.00:1
Reverse	2.10:1
Torque converter ratio	Variable between 2:1 and 1:1
Torque converter diameter	9.49 in (241 mm)
Fluid capacity (drain and refill)	5.6 pints (3.2 litres)

Shift speeds (1592 cc engine with 3.0:1 final drive)

Throttle position	Selector position	Gearchange	mph	km/h
Minimum	D	1-2	8-15	13-24
Minimum	D	2-3	11-21	19-34
Wide open	D	1-2	38-48	61-77
Wide open	D	2-3	60-72	
Kickdown	D	3-2	60-34	96-55
Kickdown	D	3-1	35-3	56-5
Closed	D-2	3-2	68	110
Only move the selector lever at speeds below 68 mph (110 km/h)				
Closed	D-1	2-1	37-28	60-45
Only move the selector lever at speeds below 68 mph (110 km/h)				

Torque wrench settings

	lbf ft	kgf m
Torque converter housing shield	9	1.24
Driveplate-to-crankshaft	40	5.52
Driveplate-to-torque converter	40	5.52
Inhibitor switch	7	0.97
LH engine mounting-to-support	16	2.21
LH engine support-to-torque converter housing	22	3.04
Kickdown band adjuster locknut	35	4.83
Oil cooler connector	30	4.14
Oil cooler pipes	16	2.21
Rear engine mounting rubber-to-crossmember	16	2.21
Rear engine mounting-to-support	9	1.24
Rear engine support-to-block	16	2.21
RH engine mounting	37	5.11
Selector cable support bracket bolt	11	1.5

Final drive (automatic transmission)

General

Gear ratio	3.0:1
Fluid capacity (drain and refill)	1.8 pints (1.05 litre)

Torque wrench settings

	lbf ft	kgf m
Cover-to-housing	14	1.9
LH bearing retainer	21	2.9
RH extension housing	21	2.9
Level and drain plugs	24	3.3
Speedometer cable housing	9	1.2

Front suspension and driveshafts

Heavy duty suspension torsion bar identification

Right-hand bar	White and red paint
Left-hand bar	White and blue paint

Torque wrench settings

	lbf ft	kgf m
Driveshaft hub nut (automatic transmission)	145	20.0
Roadwheel bolts	48	6.6

Steering

Front wheel alignment

Front wheel alignment	2.0 mm toe-in to 2.0 mm toe-out

Tyres (Minx and Rapier)

Size	165/82SR13
Pressure (cold):	
Front	26 lbf/in² (1.8 bar)
Rear	26 lbf/in² (1.8 bar)

Add 4 lbf/in² (0.28 bar) for full load or high speed conditions

Torque wrench settings (power-assisted steering)

	lbf ft	kgf m
Pump-to-bracket	18	2.48
Pump bracket-to-engine mounting	15	2.07
Oil inlet pipe	26	3.59
Oil return pipe	30	4.14
Oil return pipe-to-regulator valve	26	3.59

Rear suspension

Coil spring free length

Heavy duty suspension	11.772 in (299 mm)
Camber (non-adjustable)	0° 30′ to 1° 30′ negative
Toe-in (non-adjustable)	0 to 2.0 mm

Braking system

Master cylinder

Type (commencing 9-Series)	Split circuit, incorporating low fluid level indicator
Internal diameter:	
Commencing 9-Series	0.811 in (20.6 mm)
Commencing May 1980	0.866 in (22 mm)

Front brakes (commencing May 1980)

Minimum thickness of disc after resurfacing	0.44 in (11.25 mm)
Minimum thickness of pad and backing plate	0.28 in (7 mm)

Rear brakes (commencing May 1980)

Wheel cylinder inner diameter	0.866 in (22 mm)

Brake servo (commencing May 1980)

Manufacturer	Teves or Bendix

1 *Brake servo unit*
2 *Brake master cylinder reservoir filler cap*
3 *Engine oil filler cap*
4 *Windscreen wiper motor*
5 *Carburettor*
6 *Battery*
7 *Thermostat housing*
8 *Ignition coil*
9 *Cooling system expansion bottle*
10 *Radiator filler cap*
11 *Fuel filter*
12 *Diagnostic socket*
13 *Fuel pump*
14 *Distributor*
15 *Power steering pump reservoir*
16 *Windscreen washer reservoir*
17 *Ignition control unit*

Engine compartment of 1984 Alpine LS with 1592 cc engine. Air cleaner removed for clarity

1 Engine oil drain plug
2 Gearbox oil drain plug (Type BE1/5 transmission)
3 Final drive oil drain plug (Type BE1/5 transmission)
4 Front shock absorber
5 Lower suspension arm
6 Anti-roll bar
7 Torsion bar
8 Gearchange rods
9 Exhaust system
10 Oil filter
11 Brake caliper

Front underbody view of 1984 Alpine LS

1 Handbrake cable
2 Exhaust system
3 Brake equaliser valve
4 Rear brake hose
5 Rear suspension arm
6 Fuel tank
7 Rear anti-roll bar
8 Rear suspension crossmember

Rear underbody view of 1984 Alpine LS

Brake equalising valve

Spring length (with full fuel tank and 176 lb/80 kg at rear of load compartment):	
Standard suspension	6.5 to 6.6 in (165 to 168 mm)
Heavy duty suspension	6.45 to 6.55 in (164 to 167 mm)

Electrical system

Battery capacity (cold climates)

Battery capacity (cold climates)	48 amp hour at 20 hour discharge rate

General dimensions and weights (1984-on)

Overall length

Alpine and Solara	4318 mm (170 in)
Minx and Rapier:	
Saloon	4393 mm (173 in)
Hatchback	4318 mm (170 in)

Kerb weight (all models)

With manual transmission	1040 kg (2292 lb)
With automatic transmission	1090 kg (2402 lb)

Trailer weights

Unbraked:	
1294 cc	520 kg (1146 lb)
1592 cc manual	520 kg (1146 lb)
1592 cc auto	545 kg (1201 lb)
Braked:	
1294 cc	750 kg (1653 lb)
1592 cc manual	1000 kg (2204 lb)
1592 cc auto	1100 kg (2425 lb)
Roof rack load	75 kg (165 lb)

3 Engine

1592 cc engine – description

1 The 1592 cc engine fitted to 1600 models is very similar to the 1294 cc and 1442 cc engines and the majority of procedures are identical to those given in Chapter 1.
2 The piston stroke is identical to that for the 1442 cc engine, and the additional capacity has been achieved by increasing the bore diameter.

1592 cc engine and automatic transmission – removal and refitting

3 Disconnect the battery terminal leads (negative first).
4 Drain the cooling system as described in Chapter 2.
5 Remove the bonnet as described in Chapter 12.
6 Disconnect and remove the air intake hoses, air cleaner assembly, and carburettor intake adaptor.
7 Disconnect the cooling and heating hoses from the engine, and remove the oil level dipstick.
8 Unclip the oil cooler pipes from the transmission and disconnect the vacuum hose from the brake servo.
9 Disconnect the throttle cables from the carburettor and torque converter housing, and detach the selector cable and bracket.
10 Remove the fuel tank filler cap, then disconnect and plug the inlet pipe at the fuel pump.
11 Disconnect and remove the following electrical leads:

(a) *Oil pressure and temperature sender unit leads*
(b) *Battery earth lead at the engine*
(c) *Coil HT and LT leads*
(d) *Coil-to-control unit earth lead*
(e) *Distributor cap and leads, and rotor*

12 Unbolt the power steering pump, remove the drivebelt, unclip the fuel filter, and place the steering pump on the right-hand wing valance.
13 Remove the hub caps and slacken the wheel nuts and driveshaft nuts.
14 Jack up the front of the car and support it on axle stands.
15 Working beneath the car, remove the front shock absorbers and fit the suspension retaining rods (see Chapter 1, Section 4).
16 Remove the front wheels, then remove the alternator cover. Remove the bolt and withdraw the speedometer driven gear from the transmission.
17 Remove the bottom hose from the engine. Disconnect the electrical leads from the alternator, starter, and inhibitor switch.
18 Place a receptacle beneath the transmission, then detach and plug the oil cooler pipes.
19 *With the suspension retaining rods in position,* unscrew the lower suspension balljoint nuts on each side and disconnect the joints using a separator tool.
20 Working on each side in turn, remove the driveshaft nut, pull the hub/disc assembly out and up until it is clear of the driveshaft, then support the assembly on an axle stand. Pull each driveshaft from the transmission.
21 Remove the exhaust pipes from the manifold flange.
22 Remove the engine rear support bracket and mounting, and the oil filter canister.
23 Take the weight of the engine/transmission with a suitable hoist, using the lifting eyes on the power steering pump bracket and the left-hand engine mounting. Depending on the equipment used, it may prove beneficial to lower the car to the ground first.
24 Remove the nuts and bolts from the mounting brackets and lower the engine/transmission to the ground. Make sure that the distributor clears the radiator and front panel.
25 Remove the hoist and withdraw the engine/transmission unit from under the car.
26 Refitting is a reversal of the removal procedure, but the following additional points should be noted:

(a) *Use a suitable thread locking liquid when fitting the rear support-to-cylinder block securing bolts*
(b) *Do not fully tighten the engine mountings until the unit has been rocked side-to-side and front-to-rear; this will settle the mountings. Make sure that the rear mounting is positioned centrally as shown in Fig. 1.35, Chapter 1*
(c) *Tighten the suspension balljoint nuts and driveshaft nuts to the specified torques*
(d) *Adjust the selector and downshift cables as described in Section 8*
(e) *Fill the transmission with 5.6 pints (3.2 litres) of the specified fluid. Refill the cooling system as described in Chapter 2. Allow the engine to idle for a least 2 minutes, then check the automatic transmission fluid level as described in Section 8*

1592 cc engine and automatic transmission – separation and reassembly

27 With the engine/transmission unit out of the car, remove the starter motor and the dirt cover from the torque converter housing.
28 Mark the torque converter and driveplate in relation to each other, then remove the securing bolts.
29 Unbolt the transmission from the cylinder block, noting the location of the brackets, then withdraw the transmission from the engine.
30 Reassembly is a reversal of the separation procedure, but make sure that the marks on the torque converter and driveplate are correctly aligned. If the torque converter has been renewed, first position the engine with Nos 1 and 4 pistons at TDC, then rotate the torque converter until the timing slot on its periphery is aligned with the O-mark in the timing aperture. Always use a thread locking liquid when fitting the torque converter securing bolts.

Driveplate (automatic transmission) – removal and refitting

31 With the transmission separated from the engine, hold the crankshaft pulley bolt with a spanner and unbolt the driveplate and support plate from the crankshaft.
32 Refitting is a reversal of removal, but make sure that the mating faces are clean and free from burrs. Coat the securing bolt threads with a suitable locking liquid, and tighten the bolts to the specified torque.

1592 cc engine (manual transmission) – removal and refitting

33 The procedure is identical to that described in Chapter 1 for the 1294 cc and 1442 cc engines with manual transmission.

Crankshaft spigot bearing – description and renewal

34 The spigot bearing shown in Chapter 1 is applicable to models up to and including Series 8. Commencing with 9-Series models, the bearing is replaced by a bush.
35 To remove the bush, fill the recess with thick grease, then using a close-fitting dowel rod and a hammer, force the bush out (see Fig. 13.2).
36 Clean the recess, then drive in the new bush until the outer edge is flush with the chamfer inner edge.

Timing cover oil seal – renewal with engine in car

37 If an internal leg extractor is available, follow the procedure given in paragraphs 38 to 43 inclusive. If the extractor is not available the timing cover must be removed and the procedure given in paragraphs 44 to 56 should be followed.

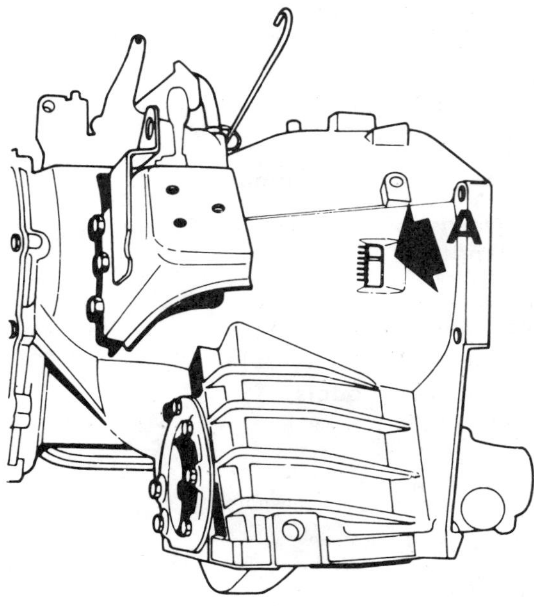

Fig. 13.1 Torque converter timing mark (A) aligned prior to reassembly of engine to transmission (Sec 3)

38 Turn the front wheel on full right-hand lock and remove the dirt cover from the right-hand side.
39 Loosen the alternator mounting and adjustment bolts and remove the drivebelt.
40 Engage top gear, apply the handbrake, and unscrew the crankshaft pulley bolt. Remove the washer and pulley.
41 Insert a 0.5 in (12 mm) diameter bolt into the crankshaft and use this as a platform for the centre screw of the extractor. Remove the oil seal.
42 Smear the new oil seal lip with engine oil, then drive the oil seal into the timing cover with a suitable tube until it is flush.
43 Make sure that the crankshaft pulley is not grooved on the oil seal contact surface, and renew if necessary. The remaining refitting procedure is a reversal of removal, but tighten the pulley bolt to the specified torque and adjust the drivebelt tension as described in Chapter 2.

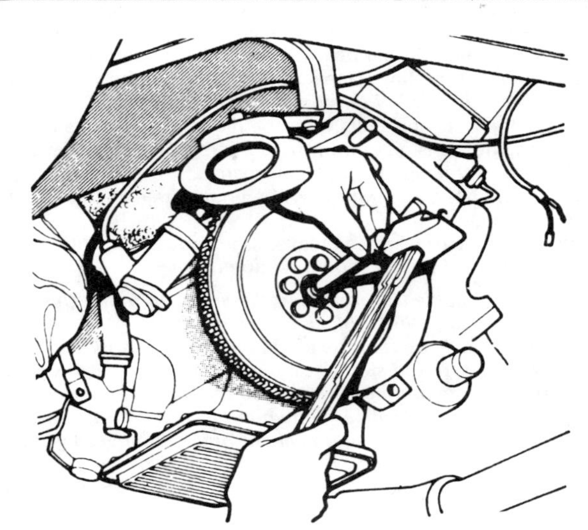

Fig. 13.2 Removing the crankshaft spigot bush (Sec 3)

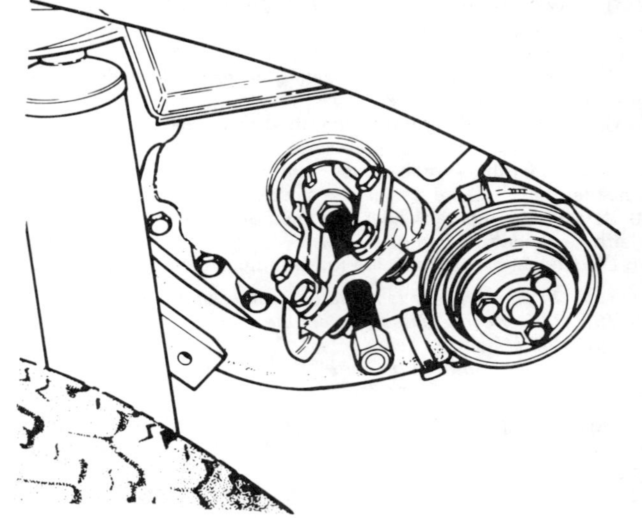

Fig. 13.3 Using an extractor to remove the timing cover oil seal (Sec 3)

44 If the timing cover is to be removed, first disconnect the battery terminal leads.
45 Drain the cooling system as described in Chapter 2.
46 Drain the oil from the engine, and remove the starter motor (Chapter 11).
47 Remove the right-hand dirt cover and remove the alternator (Chapter 11) and the mounting bracket.
48 Disconnect the hose from the timing cover.
49 Remove the water elbow from the sump housing.
50 Remove the sump pan and filter, then unbolt the sump case and withdraw it together with the water pump.
51 Insert a block of wood between a crankshaft web and the crankcase, then unscrew the pulley bolt and remove the washer and pulley. Remove the block of wood.
52 Support the engine with a trolley jack, and remove the engine side mountings, noting the location of shims.
53 Unscrew the securing nuts and bolts, lever the engine approximately 0.25 in (6 mm) to the left, and withdraw the timing cover.
54 Support the timing cover and drive out the oil seal with a suitable drift.
55 Smear the lip of the new oil seal with engine oil, drive the oil seal into the timing cover until flush using a suitable tube. (Fig. 13.4).
56 Check the crankshaft pulley for grooves and renew it if necessary. The remaining refitting procedure is a reversal of removal, but install new gaskets and coat the threads of the sump case-to-crankcase bolts with locking liquid. Tighten all nuts and bolts to the specified torque. Tension the alternator drivebelt and refill the cooling system as described in Chapter 2. Refill the engine with the correct quantity of oil.

Camshaft – removal and refitting with engine in car

57 Remove the timing cover as described in paragraphs 44 to 53 inclusive.
58 Remove the timing chain and camshaft sprocket as described in Chapter 1, Section 12.
59 Remove the cylinder head and cam followers as described in Chapter 1, Section 8.
60 Remove the distributor as described in Chapter 4, Section 3.
61 Remove the camshaft as described in Chapter 1, Section 13. It is not necessary to remove the oil pump or driveshaft.
62 Refitting is a reversal of removal, with reference to the relevant sections of Chapters 1 and 4 and paragraph 56 of this Section. It is recommended that the circlip is fitted to the distributor/oil pump driveshaft *before* fitting the sump, to prevent unnecessary work should the circlip be dropped into the engine.

Connecting rod/piston assemblies (1592 cc engine) – description

63 When refitting the piston and connecting rods to the cylinder block, the notches in the skirts of numbers 2 and 4 pistons must face towards the timing cover end. The notches in the skirts of numbers 1 and 3 pistons must face towards the flywheel end. The lubrication slots on all the connecting rods must face the camshaft side of the engine.

Cylinder head – revised tightening procedure

64 Cylinder head retaining bolts with longer threads (50 mm instead of 28 mm) have been progressively introduced on all engines.
65 When refitting a cylinder head using the longer bolts first lubricate the threads and underside of the bolt heads with engine oil then screw in all the bolts finger tight.
66 Refer to Fig. 1.9 in Chapter 1 and pretighten the bolts to 37 lbf ft (5.1 kgf m) in the sequence shown.
67 In the same sequence, tighten the bolts again to a final torque of 52 lbf ft (7.1 kgf m).
68 On completion of reassembly run the engine with the car stationary until the radiator cooling fan operates. Stop the engine and allow it to cool for at least six hours.
69 Again working in the sequence shown in Fig. 1.9, slacken each bolt half a turn and retighten it to 52 lbf ft (7.1 kgf m), before moving on to the next bolt. Adjust the valve clearances again after retightening.
70 It is not necessary to carry out any further retightening of the cylinder head, but the valve clearances should be checked and, if necessary, readjusted after 1000 miles (1500 km).

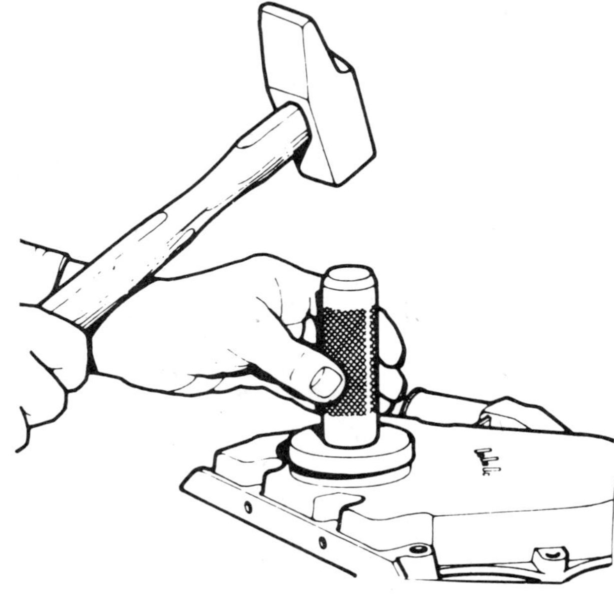

Fig. 13.4 Fitting the timing cover oil seal (Sec 3)

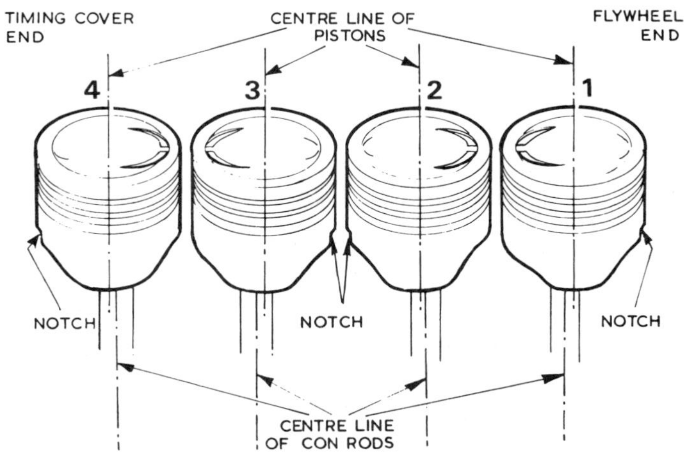

Fig. 13.5 Piston skirt notch location on 1592 cc engine (Sec 3)

Timing chain and sprockets – modifications

71 From approximately May 1982 the tooth profile of the camshaft and crankshaft sprockets has been modified as shown in Fig. 13.6.
72 The old and new type sprockets are not interchangeable and if new parts are to be fitted, it is essential to renew the complete set; i.e. camshaft sprocket, crankshaft sprocket and timing chain.
73 Concurrent with the introduction of new sprockets is the addition of a synthetic rubber damping ring to the camshaft sprocket.
74 The ring locates into a groove in the sprocket and its purpose is to reduce timing gear and chain noise.
75 It is recommended that the damping ring be renewed whenever the timing gear components are being worked on. The ring cannot be fitted to earlier camshaft sprocket assemblies.

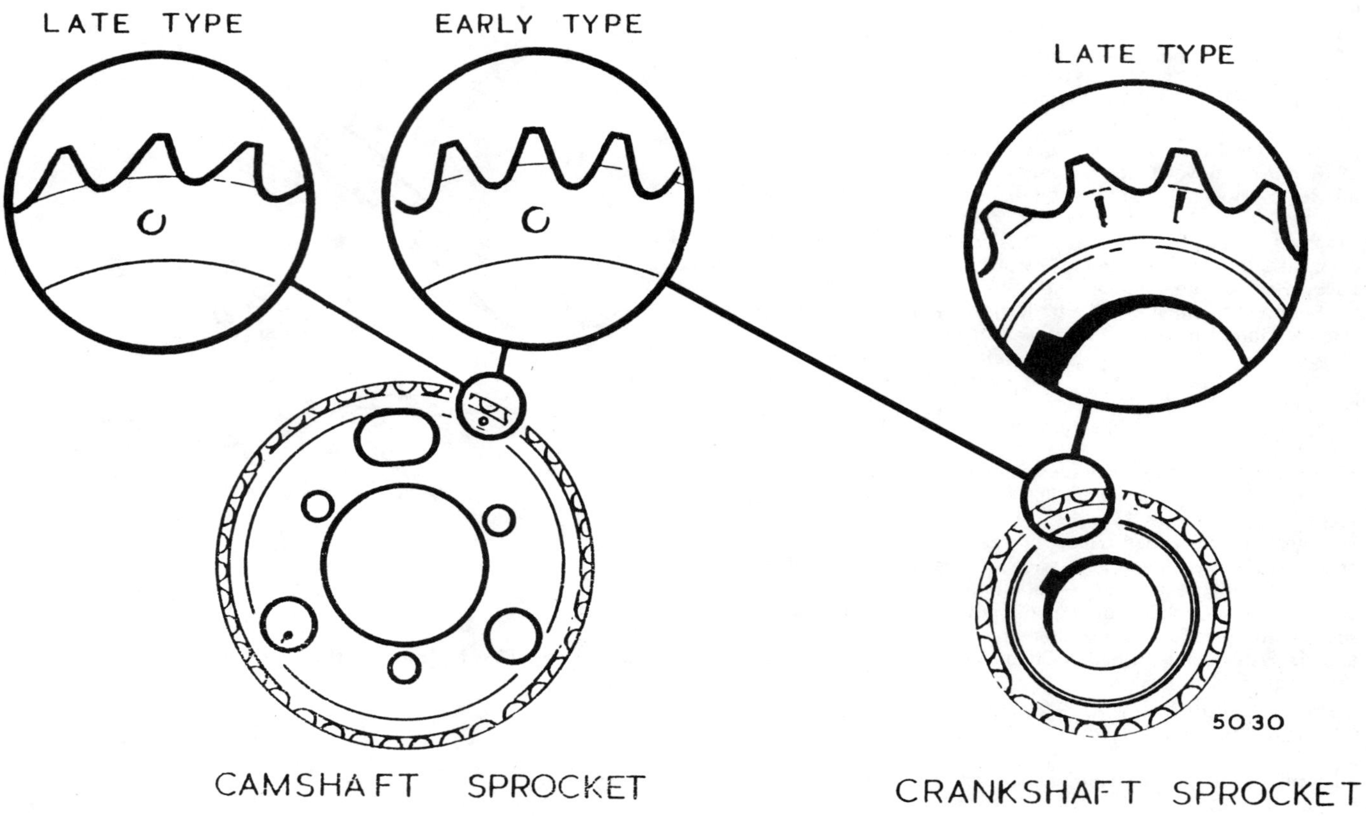

LATE TYPE EARLY TYPE LATE TYPE

CAMSHAFT SPROCKET CRANKSHAFT SPROCKET

5030

Fig. 13.6 Modified camshaft and crankshaft sprocket details (Sec 3)

Camshaft and cam followers – modifications

76 As from the following engine numbers, a modified camshaft, followers and oil pump drive pinion gear have been fitted.

Engine type	Engine number
G1 1294 cc	*F7500413*
Y2 1442 cc	*F8050280*
J1 1592 cc	*F9600767*
J2 1592 cc	*F9201285*
J2 Auto 1592 cc	*F9000250*

77 Refer to Figs. 13.7, 13.8 and 13.9 for details of component modifications.

78 The modified and earlier components must not be intermixed, but a complete repair kit containing camshaft, followers, drive gear and thrust washer is available.

79 Revised valve clearances apply to engines fitted with the modified parts:

Inlet – 0.25 mm (0.010 in) cold
Exhaust – 0.25 mm (0.010 in) cold

Engine oil and filter

80 Later models are equipped with a dipstick which incorporates a low oil level sensor. Refer to Section 12 of this Supplement for details.

81 The oil level should still be checked at weekly intervals in the following way.

82 Check the level cold, preferably first thing in the morning. Withdraw the dipstick, wipe it clean, re-insert it and withdraw it for a second time. The oil level should be up to or near the high mark on the dipstick. Top up if necessary through the oil filler cap, remembering that 1 litre (1.75 pints) is required to raise the oil level from the low to high mark.

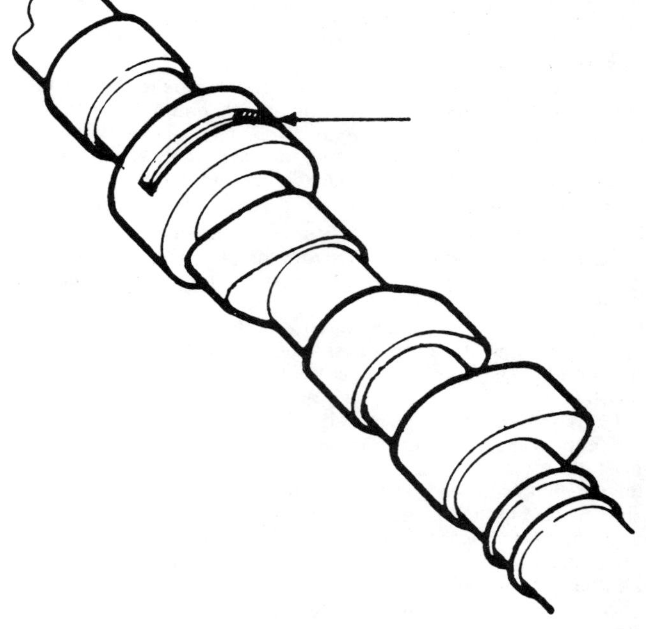

Fig. 13.7 Later type camshaft (Sec 3)

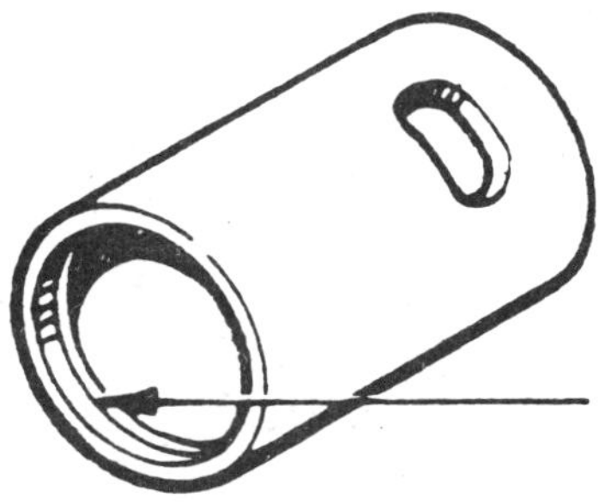

Fig. 13.8 Later type cam follower (Sec 3)

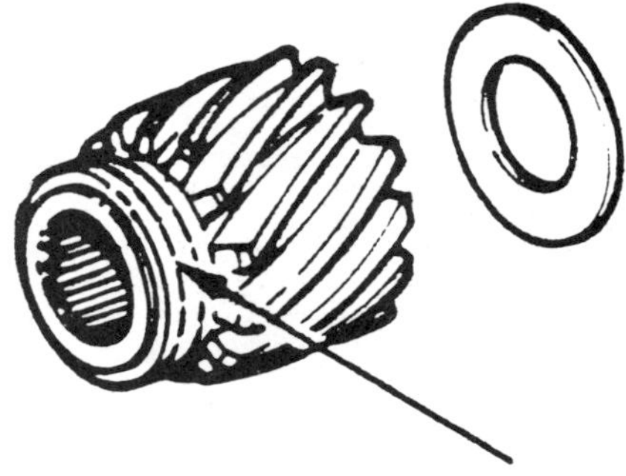

Fig. 13.9 Later type oil pump drive gear (Sec 3)

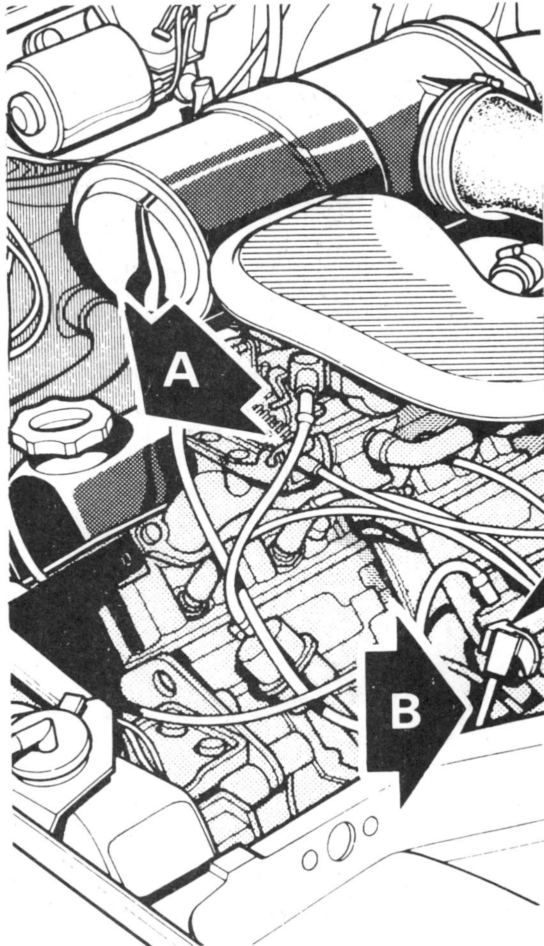

Fig. 13.10 Engine oil filler cap (A) and dipstick (B) (Sec 3)

83 To change the engine oil, drain it hot. Place a suitable container under the sump pan, remove the oil filler cap and unscrew the socket-headed drain plug.
84 While the oil is draining, unscrew the oil filter using a suitable removal tool. If one is not available, drive a large screwdriver through the filter casing and use it as a lever to unscrew it.
85 Refit the drain plug. Smear the rubber sealing ring of the new filter with engine oil and screw it into position using hand pressure only.
86 Fill the engine with the correct grade and quantity of oil, and

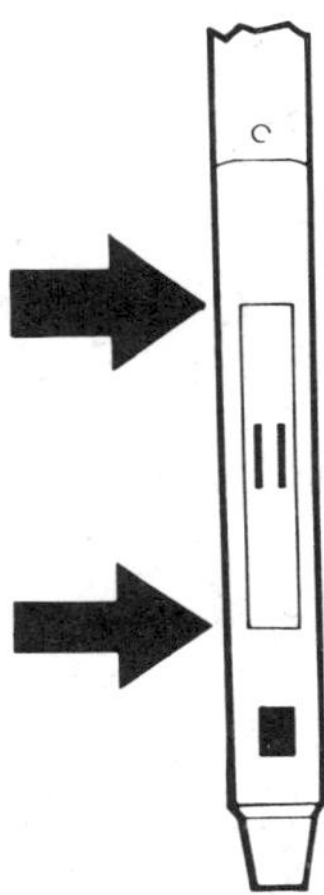

Fig. 13.11 Oil level range on sensor type dipstick (Sec 3)

then start the engine and run it for two or three minutes. The oil warning light will take a few seconds to extinguish. This is normal and is due to the new filter casing having to fill with oil.
87 Switch off the engine, wait a few minutes and then check the oil level and top up to the high mark on the dipstick if necessary.

4 Cooling system

Draining and filling

1 Commencing with Series 9 models the thermostat no longer incorporates a bleed hole, but a bleed plug is fitted to the water outer elbow.
2 When draining the cooling system the plug should be removed to allow all the coolant to drain.
3 When filling the system, pour in coolant until it flows from the bleed plug, then refit the plug and follow the procedure given in Chapter 2.
4 The sealed type of thermostat must never be fitted to an outlet elbow without a bleed plug.

Radiator (automatic transmission models) – removal and refitting

5 A fluid cooler is incorporated in the radiator on automatic transmission models, and therefore it is necessary to drain the fluid and disconnect the hoses when removing the radiator.
6 To drain the fluid refer to Section 8 of this Supplement.
7 Keep the fluid pipes uppermost during the removal procedure,

then drain the remaining fluid into a container. Plug the fluid inlet and outlet to prevent the entry of water and dirt.
8 Refitting is a reversal of removal, but refill the transmission with fluid as described in Section 8.

5 Fuel and exhaust systems

Air cleaner – general

1 A modified type air cleaner is fitted to later models equipped with the 1592 cc engine. Removal and refitting of the unit and renewal of the element are as follows.
2 To remove the air cleaner, undo the four nuts and recover the washers securing the unit to the carburettor (photo).
3 Lift the assembly up, disconnect the air intake and vacuum hoses then remove the air cleaner from the engine (photos).
4 To renew the element, release the clips and lift off the air cleaner bottom cover (photo).
5 Undo the eight screws and withdraw the element retainer (photos).
6 Withdraw the element from the air cleaner body (photo).
7 Refitting is the reverse sequence of removal. Set the temperature control on the side of the unit to the summer or winter position accordingly (photo).

Tamperproofing – description

8 Carburettors for 7-Series models onwards have the suffix A on their type numbers. This indicates that certain devices are fitted to the carburettor to prevent indiscriminate adjustment.
9 On Solex carburettors tamperproof caps are fitted over both the mixture and throttle stop screws, and once in position no adjustment is possible. On Weber and Bressel carburettors the mixture

5.2 Air cleaner-to-carburettor fixing nuts

5.3A Air cleaner cold air intake hose

5.3B Air cleaner warm air intake hose

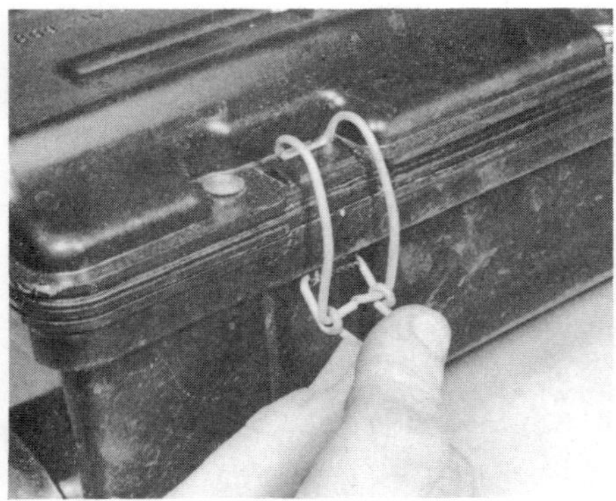
5.4 Air cleaner cover clip

5.5A Unscrewing element retainer screw

5.5B Air cleaner element retainer

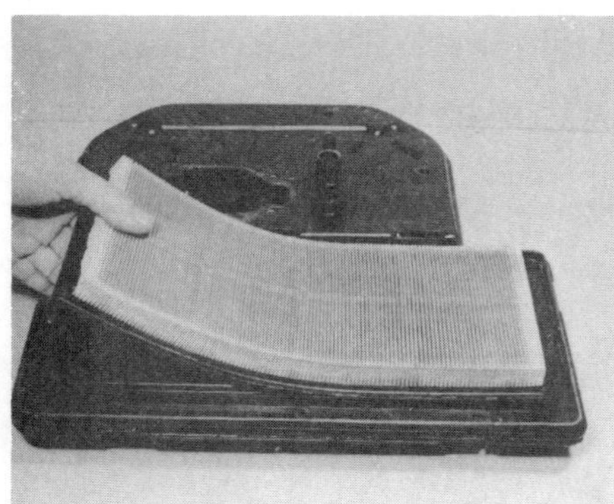
5.6 Air cleaner element

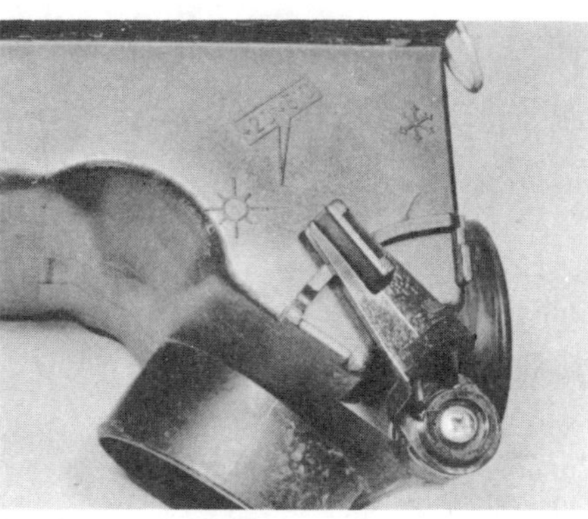
5.7 Temperature control air cleaner body

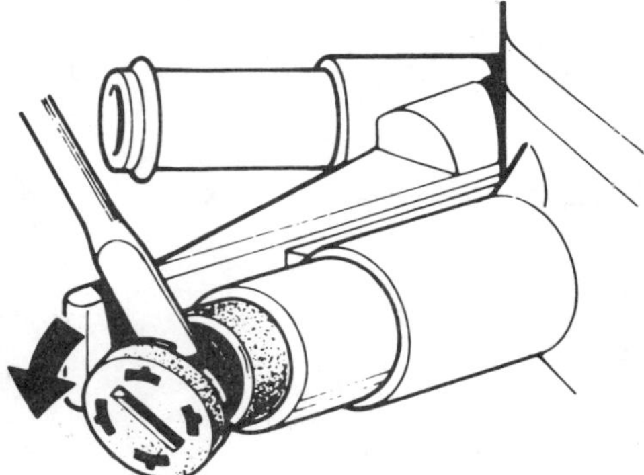
Fig. 13.12 Removing tamperproof cap from idle mixture screw on Solex carburettor (Sec 5)

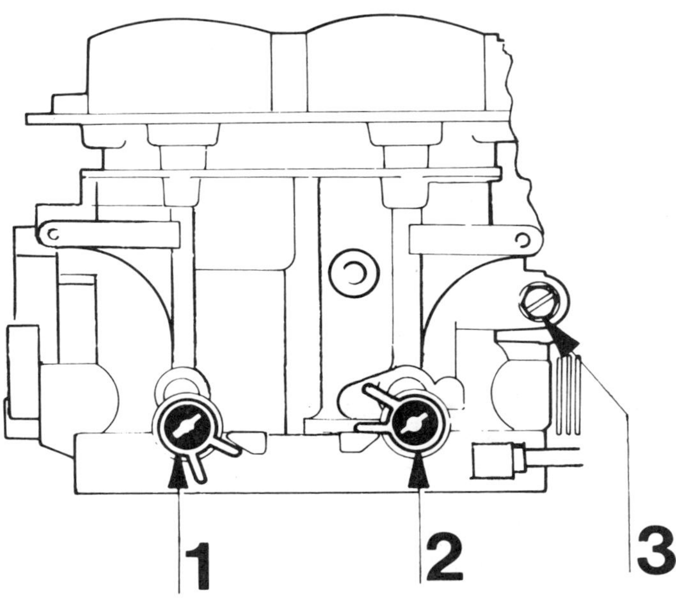

Fig. 13.13 Limited travel type mixture screws (1 and 2) and throttle stop screw (3) on Weber 36 DCA carburettor (Sec 5)

screws are fitted with tamperproof caps which allow 300° of adjustment. After the car has covered 1000 miles (1600 km), the dealer fits a tamperproof cap to the throttle stop screw which allows 90° of adjustment.
10 Both types of tamperproof cap can be removed if necessary to make adjustments, but this involves the destruction of the cap. In some territories this action is illegal and therefore current legislation in this respect should be checked first.

Tamperproof caps – removal and renewal
11 On Solex carburettors the mixture screw cap is removed by first prising the end off with a screwdriver.
12 Unscrew and remove the mixture screw, noting the number of turns necessary, then separate the screw and spring from the steel sleeve. Discard the plastic sleeve.
13 Reassemble the screw, spring, retainer and sleeve, and insert the mixture screw to its original position in the carburettor.
14 To fit a new cap, press it onto the first position on the mixture screw. After making the adjustment press the cap to its second free-wheeling position on the mixture screw.
15 To remove the throttle stop screw cap on Solex carburettors, break it off with a pair of pliers.
16 To fit a new cap, engage the legs with the bracket and push it firmly into position. The tab of the old cap will be displaced when the new cap is being installed.
17 On Weber and Bressel carburettors the caps are removed by crushing with a pair of pliers, but care should be taken not to bend the adjustment screws.
18 After adjusting the mixture, the caps must be fitted so that it is not possible to further enrich the mixture, ie after pushing the cap on the screw, it must not be possible to turn it anti-clockwise.
19 To fit a new cap to the throttle stop adjustment screw, first adjust the idling speed as described in Chapter 3.
20 Turn the throttle stop screw outward (anti-clockwise) until the idling speed drops by 50 rpm, then push the cap onto the screw so that it is not possible to turn it anti-clockwise. Now turn the screw clockwise to regain the correct idling speed.

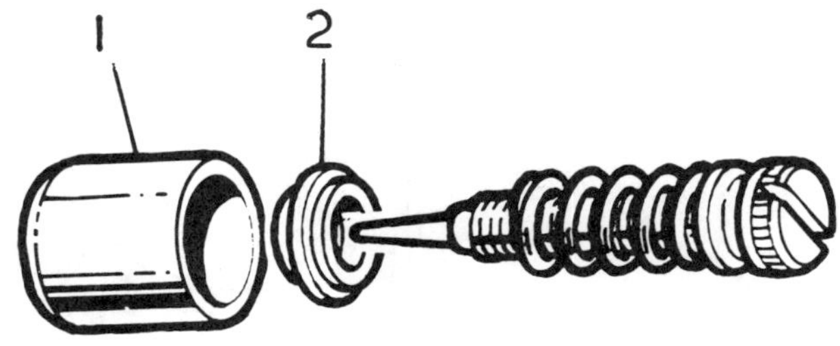

Fig. 13.14 Tamperproof cap components on Solex carburettor (Sec 5)

1 Steel sleeve *2 Retainer*

Weber 36 DCNV carburettor – fast idle adjustment
21 Turn the choke lever to the position shown in Fig. 13.16.
22 Loosen the locknut and position the adjustment screw so that it just touches the fast idle cam.
23 Tighten the locknut and release the choke lever.

Weber DCNVH carburettor – description
24 The Weber DCNVH carburettor is very similar to the DCNVA, the only difference being smaller main jets and a vacuum-operated enrichening device. The arrangement produces greater economy at part throttle engine speeds.
25 Refer to Fig. 13.17. Under part throttle and cruising conditions, vacuum on the engine side of the throttle valve will be sufficient to pull the diaphragm and pushrod from the ball valve, and the enrichening circuit will be inoperative. Under full throttle and high speed conditions the vacuum is reduced, and the diaphragm and pushrod will force the ball from its seat. The enrichening circuit is opened to give the required mixture correction for these conditions.

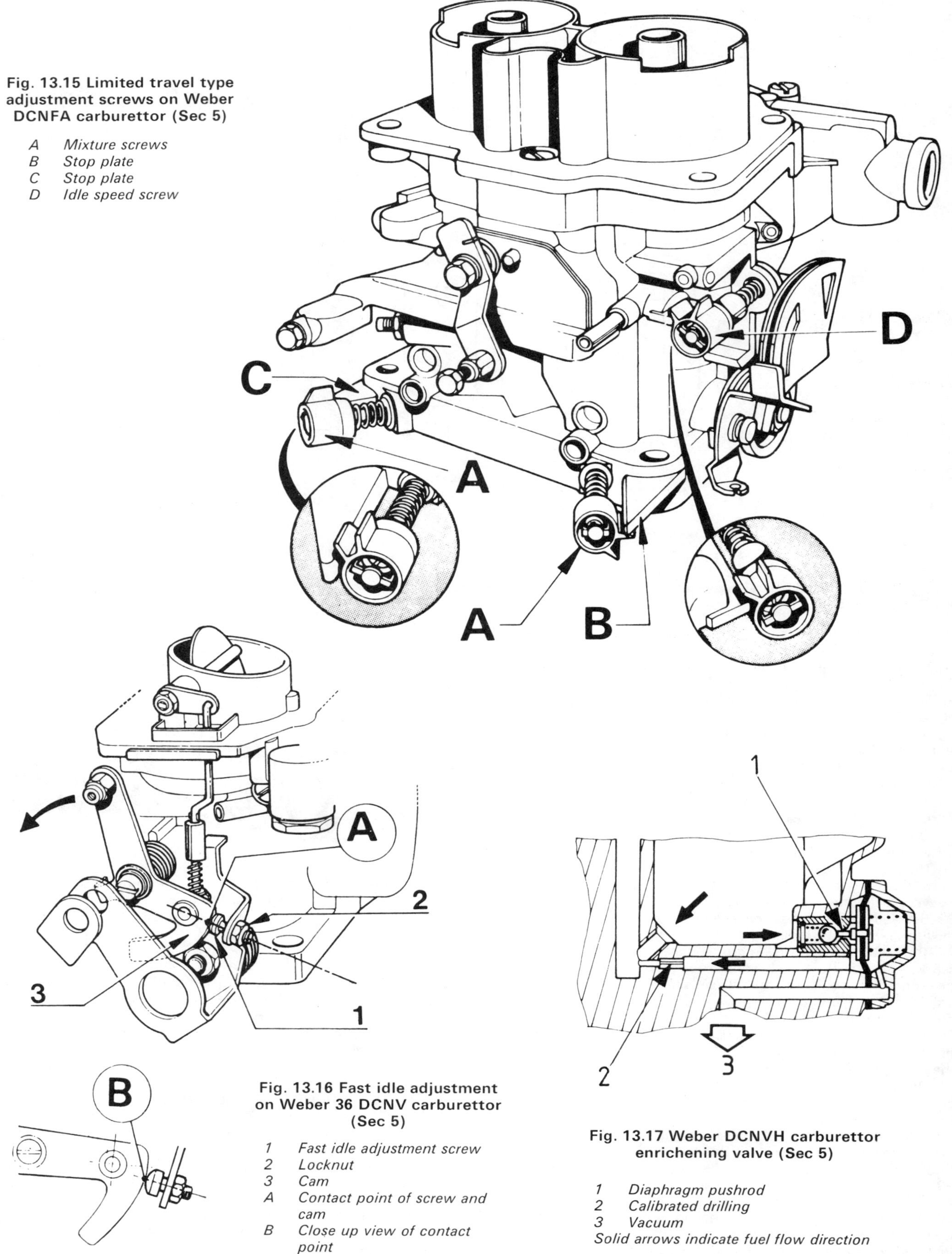

Fig. 13.15 Limited travel type adjustment screws on Weber DCNFA carburettor (Sec 5)

A Mixture screws
B Stop plate
C Stop plate
D Idle speed screw

Fig. 13.16 Fast idle adjustment on Weber 36 DCNV carburettor (Sec 5)

1 Fast idle adjustment screw
2 Locknut
3 Cam
A Contact point of screw and cam
B Close up view of contact point

Fig. 13.17 Weber DCNVH carburettor enrichening valve (Sec 5)

1 Diaphragm pushrod
2 Calibrated drilling
3 Vacuum
Solid arrows indicate fuel flow direction

Weber 36 DCA carburettor – description

26 The Weber 36 DCA carburettor is similar to the DCNVH type described in paragraphs 24 and 25, but in addition it is fitted with a coolant-heated automatic choke.

27 The automatic choke operates as follows. When the engine is cold a thermostatic spring inside the automatic choke holds the choke butterflies closed. As the engine warms up, the coolant circulating through the choke cover heats the spring, and the butterflies are progessively opened.

28 Unlike many similar automatic chokes, the DCA arrangement does not have to be set before starting from cold. When the engine is stopped, a spring-loaded pushrod opens the throttle valve to release the fast idle screw from the choke cam, thus allowing the automatic choke to adjust to the engine temperature. As soon as the engine is started, inlet manifold vacuum withdraws the spring-loaded pushrod and allows the screw to rest on the fast idle cam.

Weber 36 DCA carburettor – idling adjustment

29 With the engine at normal operating temperature, connect a tachometer and adjust the throttle stop screw so that the engine runs at the specified idling speed.

30 Adjust the mixture screws alternately until the highest engine speed is achieved.

31 Readjust the throttle stop screw to regain the specified idling speed, then turn one mixture screw clockwise to reduce the idling speed by 25 rpm. Similarly, turn the other mixture screw clockwise to reduce the idling speed by a further 25 rpm.

32 Turn the throttle stop screw clockwise to increase the idling speed to the specified rpm.

33 Switch off the engine and remove the tachometer.

Weber 36 DCA carburettor – removal and refitting

34 Remove the air cleaner and adaptor as described in Chapter 3 or earlier in this Chapter.

35 Have ready two suitable plugs (old spark plugs may be used), then disconnect the two hoses from the automatic choke cover and plug them immediately. Identify each hose for location.

36 Disconnect the vacuum advance pipe, fuel supply pipe, and throttle cable from the carburettor.

37 Unscrew the four retaining nuts and withdraw the carburettor from the inlet manifold.

38 Refitting is a reversal of removal, but use a new flange gasket and adjust the throttle cable as described in paragraphs 39 to 40 inclusive. After running the engine, check and if necessary top up the cooling system level.

Weber 36 DCA carburettor – throttle cable adjustment

39 A cambox is fitted between the carburettor and accelerator pedal and therefore a special procedure must be followed. The cambox interconnects the throttle cable, accelerator pedal cable, throttle valve (downshift) cable, and cruise control servo cable.

40 Adjust the accelerator pedal cable first as follows. Remove the cambox cover and insert a 6 mm diameter twist drill through the side hole and through the cams. (The drill must be 6 mm – 1/4 in will not fit).

41 Loosen the cable locknut, then turn the adjuster outwards to remove any free play until the drill is lightly pinched.

42 Tighten the locknut and remove the twist drill.

43 The throttle cable must be adjusted next. Loosen the cable locknut at the carburettor end, and remove the air cleaner and adaptor as described in Chapter 3.

44 Hold the choke butterflies fully open, and push the fast idle screw fully in towards the automatic choke against the tension of the pushrod spring.

45 Release the choke while still depressing the fast idle screw. Adjust the cable position until it has a maximum free play of 0.079 in (2 mm) and the stop of the accelerator pedal and throttle cable cams in the cambox are in light contact.

46 Tighten the locknut and refit the air cleaner.

47 Since the cambox interconnects four cables, it is recommended that the throttle valve (downshift) cable and cruise control servo cable are also adjusted in that order as described in Sections 8 and 12.

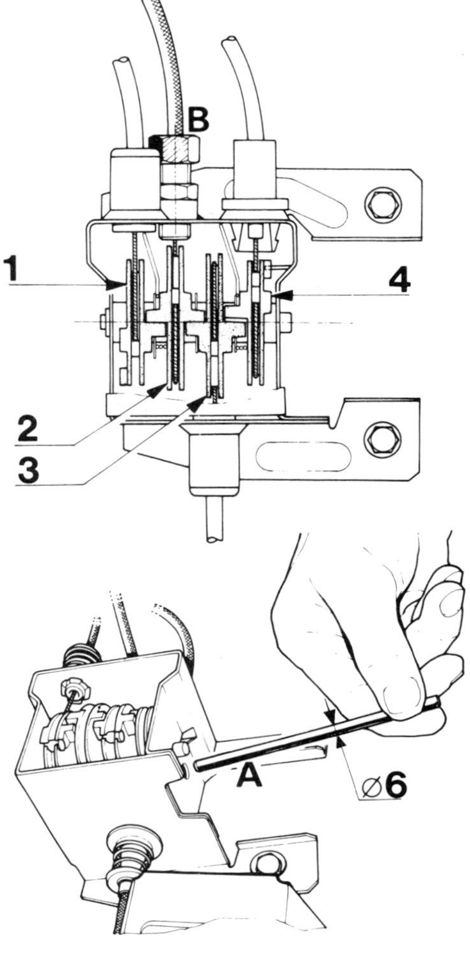

Fig. 13.18 Control cable cambox on automatic transmission models (Sec 5)

1	*Throttle valve cam*	4	*Cruise control cable cam*
2	*Accelerator cable cam*	A	*6.0 mm diameter twist drill*
3	*Throttle cable cam*	B	*Cable adjuster*

Weber 36 DCA carburettor – float level adjustment

48 The procedure is identical to that described in Chapter 3, but Refer to Fig. 13.20 for the dimension checking point on the float.

Weber 36 DCA carburettor automatic choke – removal, refitting and adjustment

49 Remove the carburettor as described earlier.

50 Remove the three retaining screws and withdraw the cover and thermostatic spring. Remove the plastic disc, noting its position.

51 Disconnect the operating link from the choke spindle by removing the circlip.

52 Remove the retaining screws and withdraw the automatic choke.

53 If necessary, unscrew the retaining screws and lift off the diaphragm cover and spring. Examine the diaphragm and if it requires renewal, cut off the inner lips of the diaphragm bush and withdraw the diaphragm and pushrod.

54 Make sure that the mating surfaces are clean, then insert the new diaphragm and pushrod. Push the centre of the diaphragm to force the plastic bush into position.

55 Refit the spring and cover and tighten the retaining screws in diagonal sequence.

56 To refit the automatic choke, locate a new O-ring over the vacuum supply tube, position the assembly on the carburettor, and tighten the retaining screws.

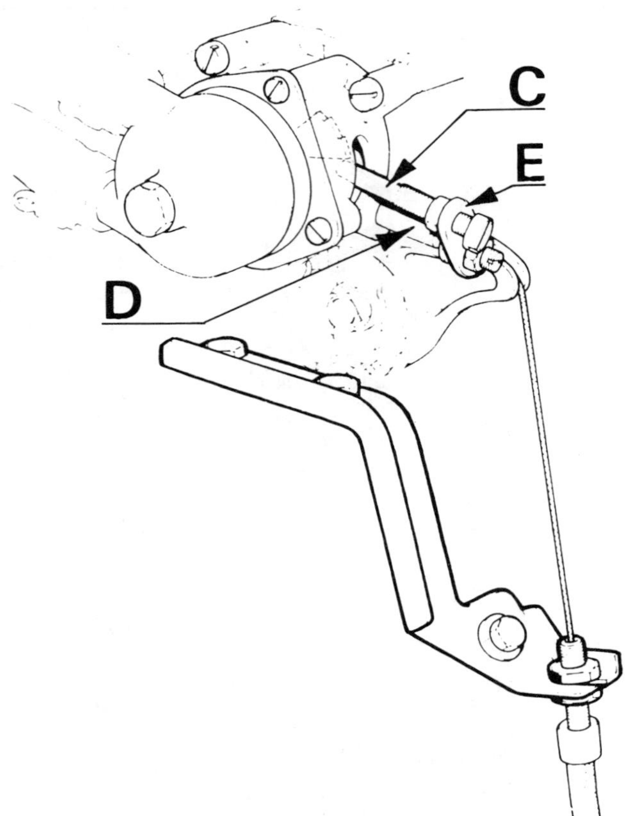

Fig. 13.19 Throttle cable and fast idle components on Weber 36 DCA carburettor (Sec 5)

C Fast idle screw
D Vacuum diaphragm pushrod
E Throttle lever

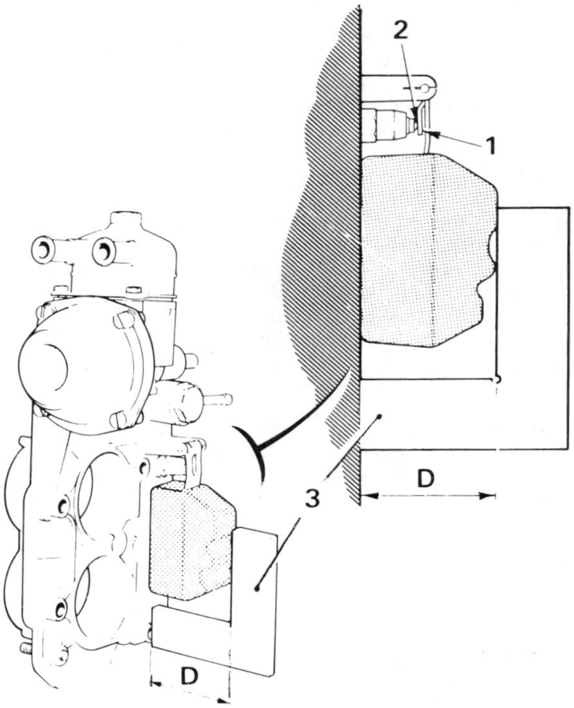

Fig. 13.20 Weber 36 DCA carburettor float checking diagram (Sec 5)

1 Tab
2 Fuel inlet needle valve ball
3 Checking gauge
D Float setting dimension

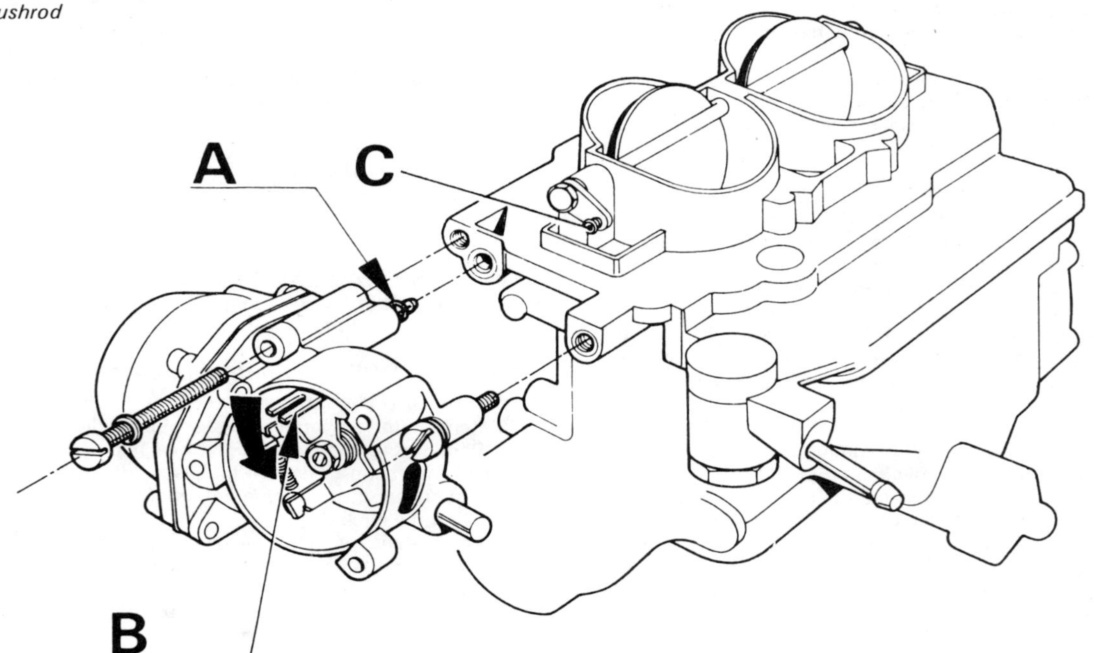

Fig. 13.21 Weber 36 DCA carburettor automatic choke (Sec 5)

A O-ring seal *B Fork* *C Choke operating lever*

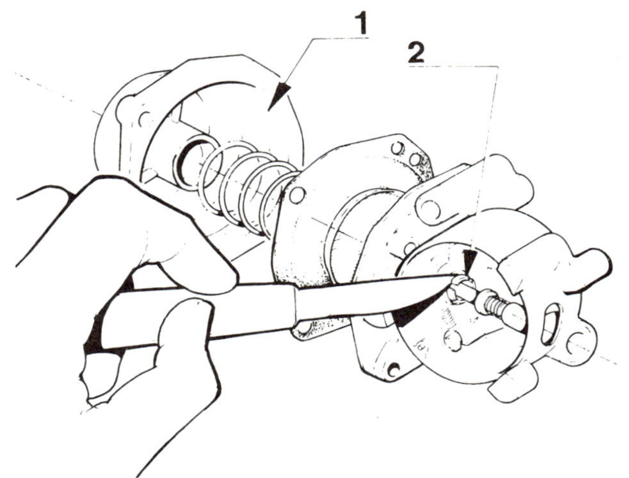

Fig. 13.22 Removing the diaphragm from the Weber 36 DCA carburettor automatic choke (Sec 5)

1 Cover *2 Bush retaining lips*

57 Connect the operating link and refit the circlip.
58 To adjust the vacuum pushrod throttle opening, first fully depress the pushrod and release it slowly to the stop screw. Open the throttle and check that the pushrod remains stationary. If it moves out, polish and lubricate the contact faces of the stop screw and pushrod until the operation is correct.
59 Hold the choke butterflies fully open, then using feeler blades, check that the throttle butterflies are opened by between 0.020 and 0.024 in (0.50 and 0.60 mm). If necessary, loosen the locknut and reposition the stop screw, then tighten the locknut.
60 To adjust the fast idle screw, release the choke and push the stop screw until the fast idle screw abuts the high position on the fast idle cam.
61 While still depressing the stop screw, invert the carburettor and check the position of the throttle butterflies in relation to the progression holes. On 1442 cc engines the butterflies should be between positions A and B in Fig. 13.24 on 1592 cc engines they should be between positions B and C.
62 If necessary, loosen the locknut and reposition the fast idle screw, then tighten the locknut.
63 To adjust the choke butterfly opening, release the throttle, then position the fast idle screw on the second step of the fast idle cam.
64 Using a twist drill check that the choke butterflies are opened by (C) between 0.236 and 0.256 in (6 and 6.5 mm).
65 If necessary bend the lug (A in Fig. 13.25).
66 Check that the operating link operates freely through its full movement.
67 To adjust the vacuum diaphragm openings, fully open the throttle butterflies and depress the vacuum pushrod. Turn the choke operating fork anti-clockwise to fully compress the vacuum pushrod spring, then using a twist drill check that the choke butterflies are opened between 0.157 and 0.177 in (4.0 and 4.5 mm). With the operating fork released the choke butterflies should be opened between 0.236 and 0.276 in (6.0 and 7.0 mm).
68 If adjustment is necessary, bend the tang (B in Fig. 13.26).
69 Refit the plastic disc, hook the thermostat spring on the fork, and refit the cover with the retaining screws finger tight.
70 Turn the choke cover in both directions to check that the choke butterflies open and close smoothly.
71 With ambient temperature at approximately 20°C (68°F), turn the cover anti-clockwise until the choke butterflies just close. Mark the housing opposite the cover alignment mark.
72 Turn the cover clockwise until the choke butterflies are just fully open and make a second mark on the housing.
73 Position the cover alignment mark midway between the two housing marks and tighten the retaining screws.
74 Refit the carburettor to the engine as described earlier.

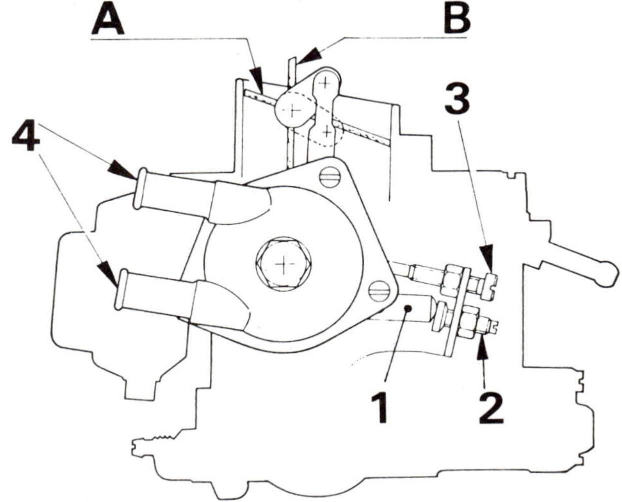

Fig. 13.23 Automatic choke component on the Weber 36 DCA carburettor (Sec 5)

1 Vacuum diaphragm pushrod
2 Stop screw
3 Fast idle screw
4 Coolant hose connections
A Choke valve plate closed
B Choke valve plate open

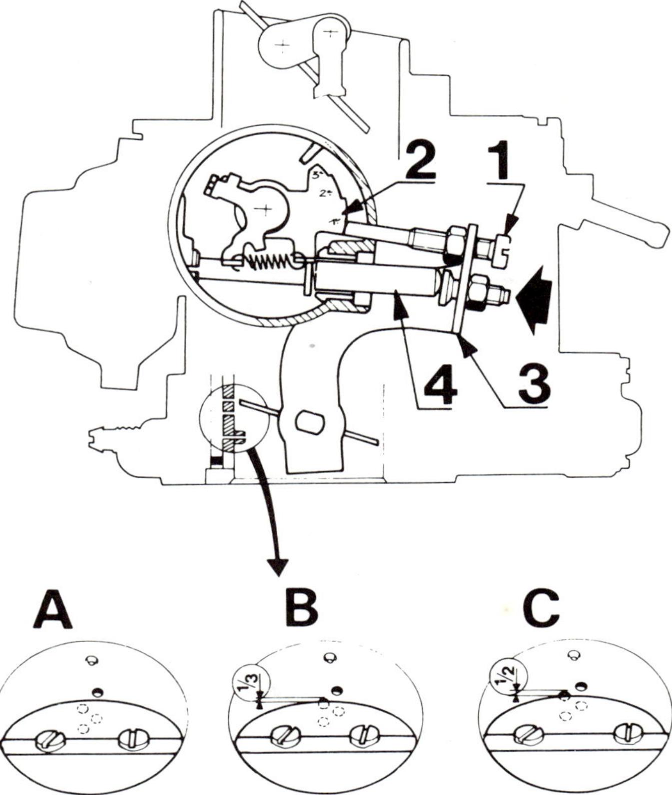

Fig. 13.24 Fast idle screw adjustment on the Weber 36 DCA carburettor (Sec 5)

1 Fast idle screw
2 Cam
3 Throttle lever
4 Vacuum diaphragm pushrod
A First hole uncovered
B Second hole one third uncovered
C Second hole half uncovered

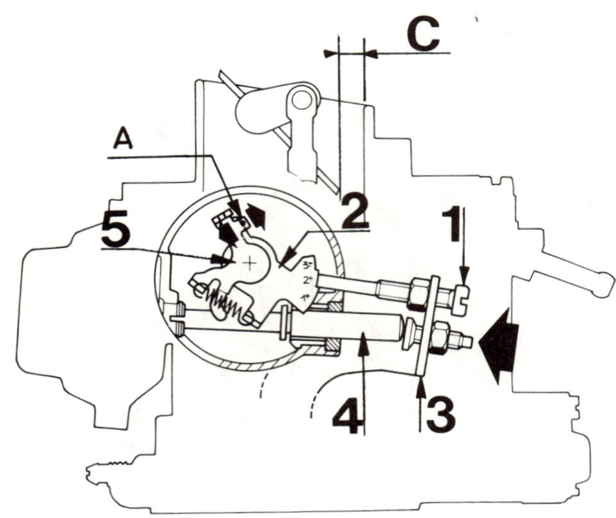

Fig. 13.25 Choke valve plate opening adjustment on Weber 36 DCA carburettor (Sec 5)

1 *Fast idle screw*
2 *Cam*
3 *Throttle lever*
4 *Vacuum diaphragm pushrod*
5 *Fork lever*
A *Adjustment lug*
C *Choke valve plate opening*

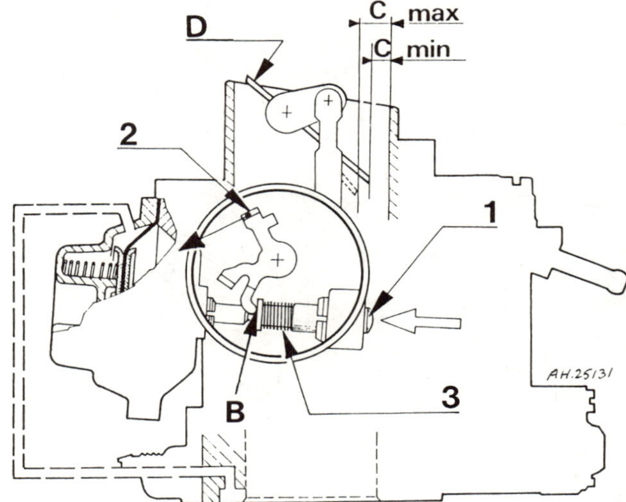

Fig. 13.26 Choke vacuum diaphragm opening adjustment on Weber 36 DCA carburettor (Sec 5)

1 *Vacuum diaphragm pushrod*
2 *Fork lever*
3 *Spring*
B *Adjustment tab*
C *Choke valve plate opening*
D *Choke valve plate*

Weber 36 DCA carburettor venturi heater

75 In order to eliminate stalling after cold starting on automatic transmission models which is caused by icing, electric heater elements are fitted to the carburettor on 1985 models.

76 The elements may be fitted to earlier carburettors, provided their bodies are cast to accept them, by carrying out the following operations.

77 Disconnect the battery and remove the air cleaner from the carburettor.

78 Drain the cooling system.

79 Screw the thermal switch supplied in the fitting kit into the cylinder head coolant outlet.

80 Fit the heating elements and their brackets to the carburettor.

81 Connect the relay (Fig. 13.28) and wiring harness with wires running as shown in the wiring diagram (Fig. 13.29).

82 Connect the large-section red wire with red insulation to a positive feed from the fuse box.

83 Fit the connector to the relay.

84 Connect the black wire to earth.

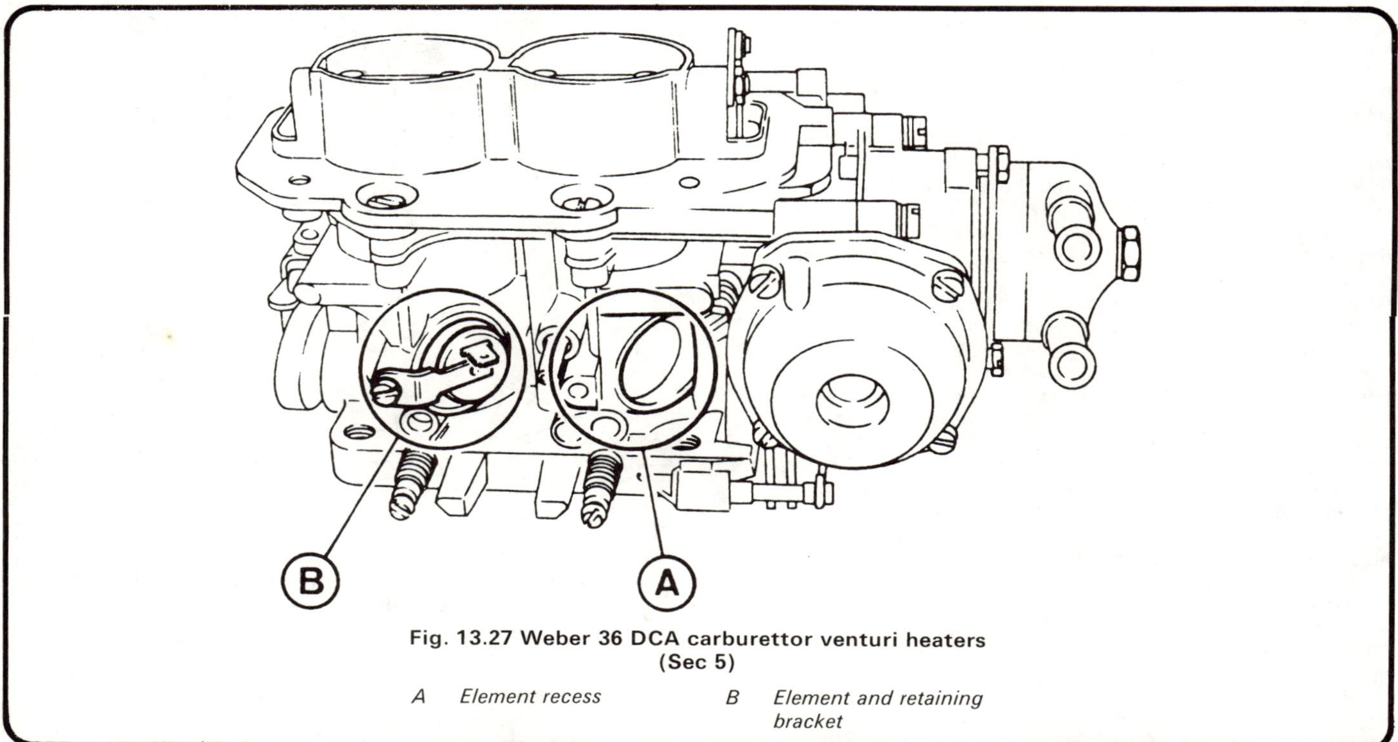

Fig. 13.27 Weber 36 DCA carburettor venturi heaters (Sec 5)

A *Element recess*
B *Element and retaining bracket*

Fig. 13.28 Relay (R) Weber 36 DCA carburettor venturi heater (Sec 5)

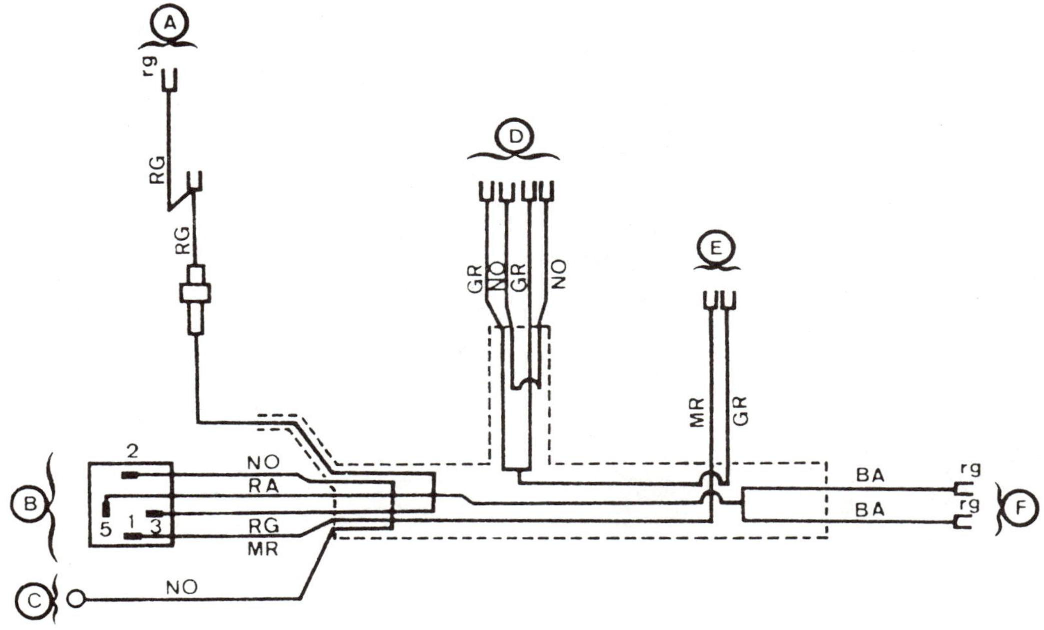

Fig. 13.29 Wiring diagram for Weber 36 DCA carburettor venturi heater (Sec 5)

A *Large section red wire with red insulator to positive feed from fuse box*
B *Relay and connector*
C *Black earth wire*
D *Grey two-way male and female connectors to main wiring harness*
E *Grey and brown wires to thermal switch*
F *White wires with red insulators to carburettor heater elements*

Wiring colour code

BA White
GR Grey
MR Brown
NO Black
RG Red

85 Fit the grey male and female two-way connectors to the main wiring harness.
86 Connect the grey and brown wires to the thermal switch.
87 Connect the two white wires with red insulators to the carburettor heater elements.
88 Reconnect the battery, refit the air cleaner and fill the cooling system.

Accelerator and choke cables – all models except 1294 cc (1984-on)

89 The method of attachment of the accelerator and choke cables fitted to later vehicles with Weber or Bressel carburettors and manual transmission are shown in the accompanying photos.
90 The accelerator inner cable end fitting locates in a slot in the carburettor throttle linkage lever. The cable end can be slipped out or refitted by holding the linkage fully open by hand then sliding the cable sideways (photo). The outer cable can then be withdrawn from its support bracket.
91 Cable adjustment is by means of two knurled plastic nuts on the outer cable (photo). The cable should be adjusted to provide a small amount of inner cable free play with the throttle closed.
92 The choke cable is secured to the choke control linkage on the carburettor by a screw and to the support bracket by a clamp. Slacken the screws to remove the cable (photo).

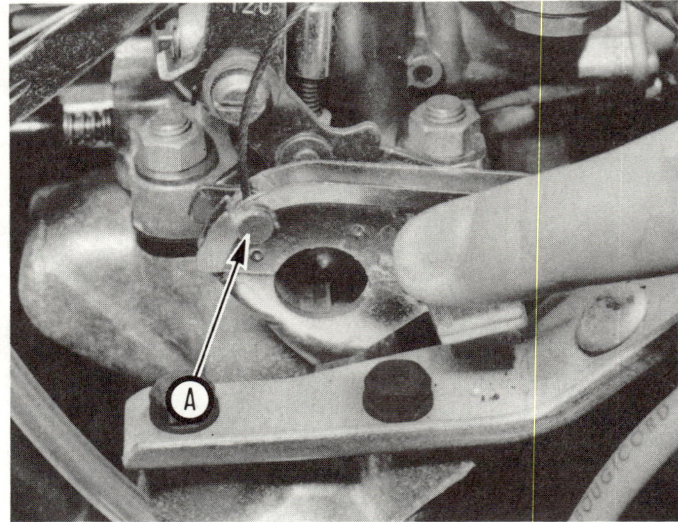

5.90 Accelerator cable end fitting (A)

5.91 Accelerator cable adjusting units (arrowed)

5.92 Choke cable clamp and linkage retaining screws (arrowed)

93 The cable should be adjusted so that the choke flap closes fully when the choke knob is pulled fully out and opens fully when the knob is pushed in.

Accelerator and choke cables – 1294 cc (1984 on)

94 The linkage is as shown in Fig. 13.30, the accelerator cable being retained at the carburettor bracket by a clip in a groove. Adjust the clip to give slight free play in the cable, but ensure that full throttle is obtainable.
95 It is important that the accelerator cable is routed under the brake master cylinder as shown in Fig. 13.31.

Fuel tank gauge transmitter unit – modification

96 The fuel gauge transmitter unit fitted to later models is manufactured in plastic and has a brass tag connector to earth the unit to the tank.
97 When fitting this type of transmitter unit to the tank, first locate the rubber seal so that its lug is positioned in line with the brass connector. This is necessary to prevent the seal being wedged between the connector and the tank during fitting, which would in turn prevent the earth circuit from being made.

Exhaust system – description

98 Commencing with 8 Series models, the flexible section of the exhaust pipe is discontinued and a spring-tensioned spherical balljoint is fitted.
99 To separate the joints compress the springs with a pair of grips, unhook them, and pull the joints apart.
100 Thoroughly clean the joints before assembling them and smear the contact surfaces with a suitable lubricant.

Exhaust system – vibration

101 If the special balljoints in the front downpipe of the exhaust system seize a certain amount of vibration may be transmitted at engine speeds of 2000 to 3000 rpm when in gear.
102 To eradicate this, it is necessary to partially disconnect the exhaust system just sufficiently to allow the balljoints to move through maximum travel. when the exhaust system is detached it is most important that no attempt is made to pull the ends of the wire clips to unhook the pipe.
103 Use a pair of grips with suitable wide jaws to compress the spring whilst pushing on the inner end of the clips (towards each other) so that one clip may be unhooked from the pipe. Repeat this procedure with the opposing pipe, then spray the spherical joint

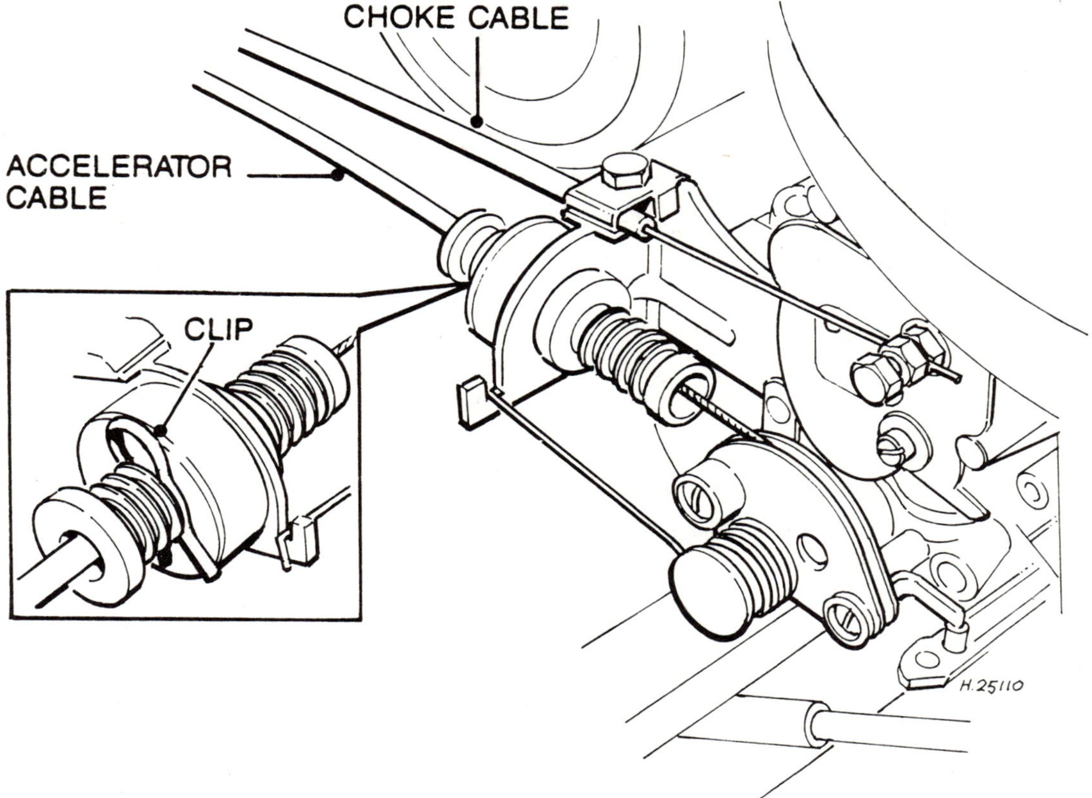

Fig. 13.30 Accelerator and choke cables on later 1294 cc engines (Sec 5)

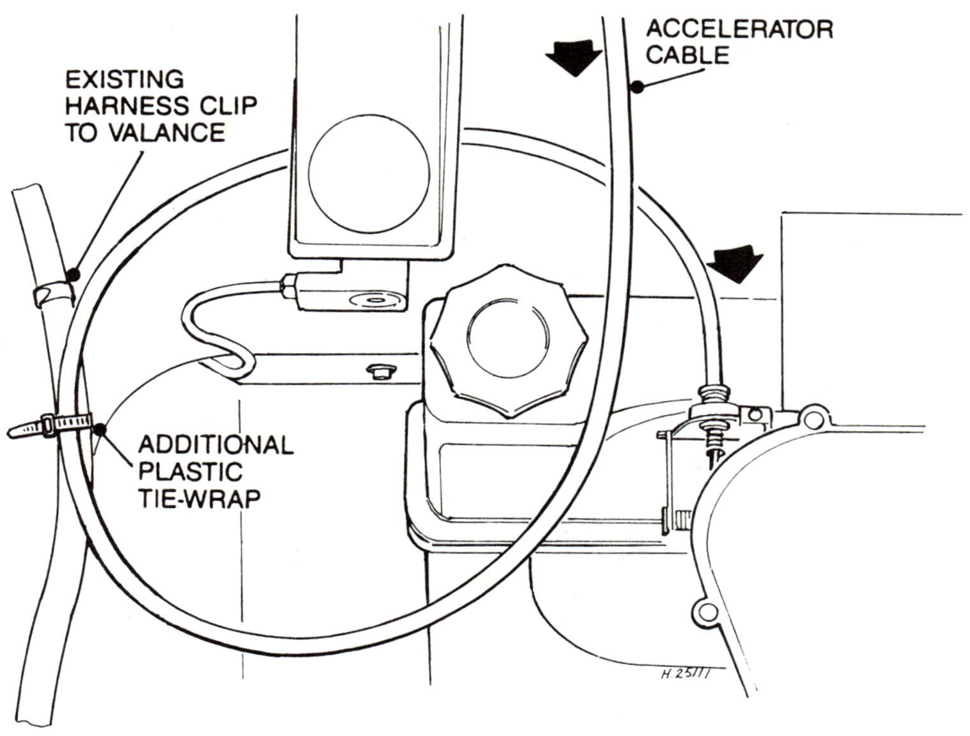

Fig 13.31 Accelerator cable routing on 1294 cc models (Sec 5)

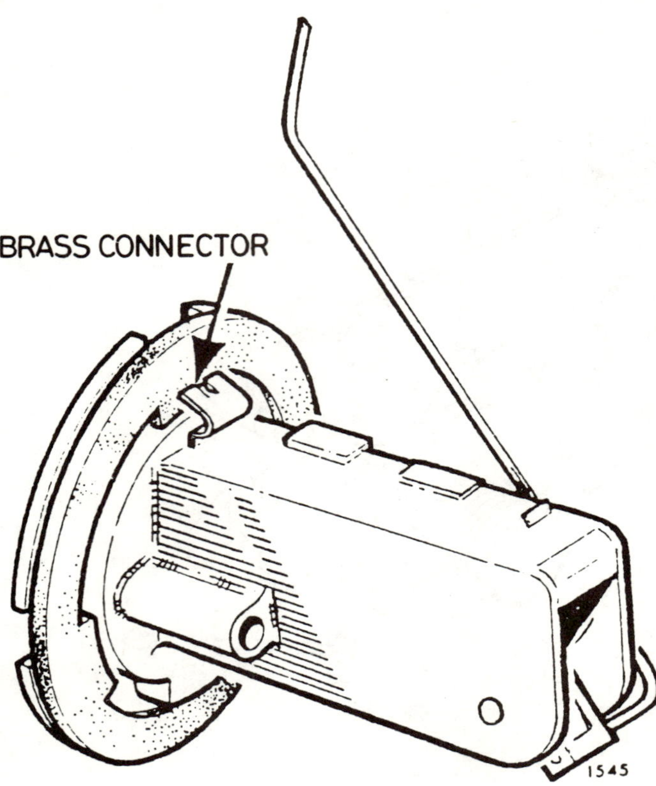

Fig. 13.32 Later type fuel tank sender unit (Sec 5)

faces with some anti-seize solution ensuring that the spray is worked fully into the joints.

104 Reassemble the exhaust system in the reverse order of removal. Note that a small amount of vibration is normal when the car is in neutral.

6 Ignition system

Hall effect ignition system – description

1 All models from 9 Series onward are fitted with the Hall effect ignition system. This operates in a similar manner to the system described in Chapter 4, but the distributor-mounted components are different.

2 Instead of the reluctor and pick-up, the Hall effect system incorporates a permanent magnet, a detector/amplifier, and four vanes. When a vane is masking the detector/amplifier no voltage is induced in the detector, and under these conditions the control unit passes current through the low tension windings of the coil.

3 Rotation of the distributor will uncover the detector and cause it to be influenced by the magnetic field of the permanent magnet. The 'Hall effect' induces a small voltage in the detector plate which is then amplified and triggers the control unit to interrupt the low tension current in the coil.

4 The control unit in the Hall effect system incorporates a circuit which switches off the low tension circuit if the time between consecutive signals exceeds 1.5 seconds. The coil and internal circuits are therefore protected if the ignition is left switched on inadvertently.

5 Distributors of Bosch, Chrysler or Ducellier manufacture may be used in the Hall effect system, but all are essentially the same in their design and construction.

Hall effect distributor – servicing

6 Servicing is identical to that described in Chapter 4, except that on the Chrysler distributor the cap is retained by screws. Additionally the distributor should be lubricated every 5000 miles (8000 km) as follows.

7 Remove the distributor cap and pull off the rotor arm (taking care not to bend the metal vanes on the Chrysler distributor).

8 Apply two or three drops of engine oil to the felt in the recess on top of the distributor shaft.

9 Refit the rotor and cap.

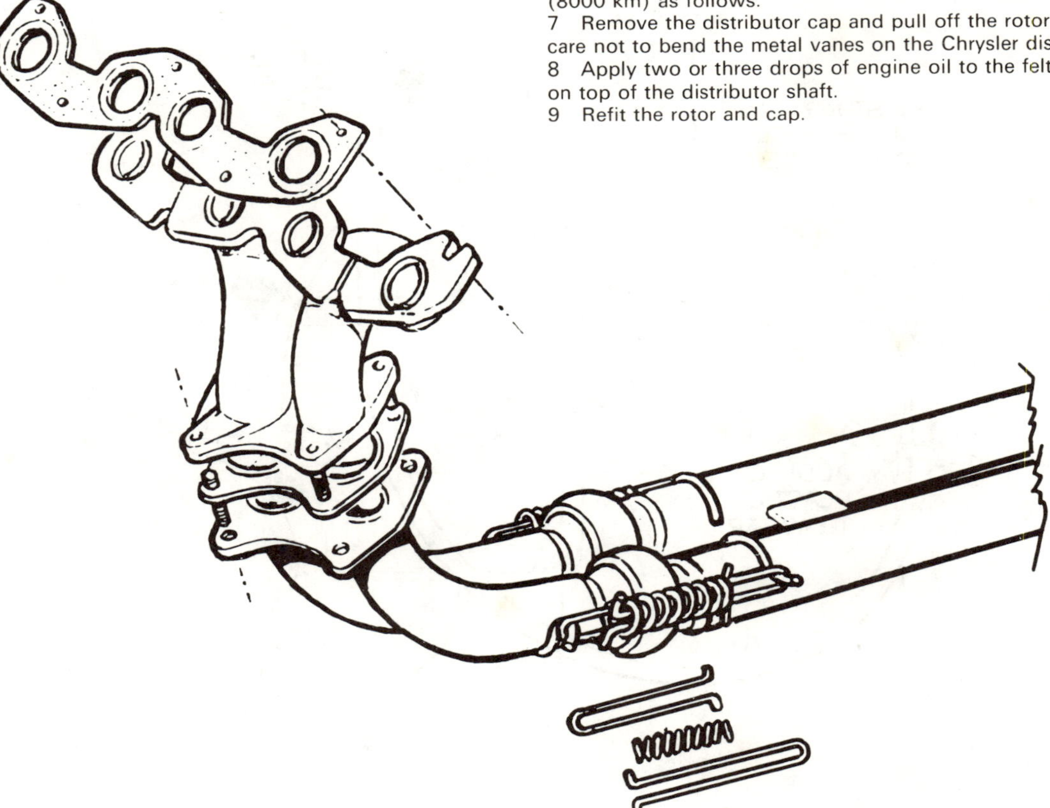
Fig. 13.33 Later type exhaust downpipe connections (Sec 5)

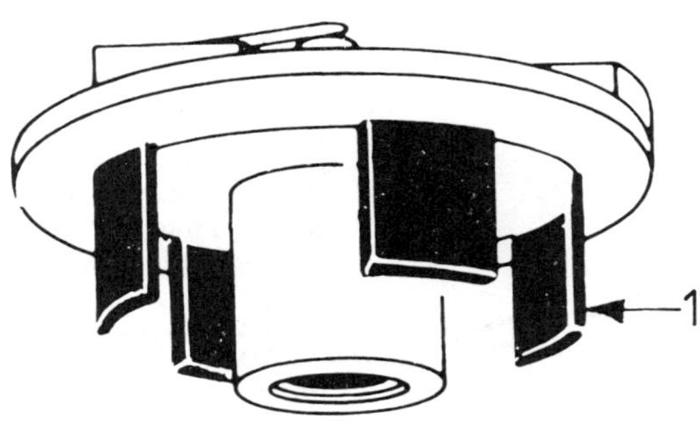

Fig. 13.34 Chrysler type Hall effect distributor rotor and vane (1) (Sec 6)

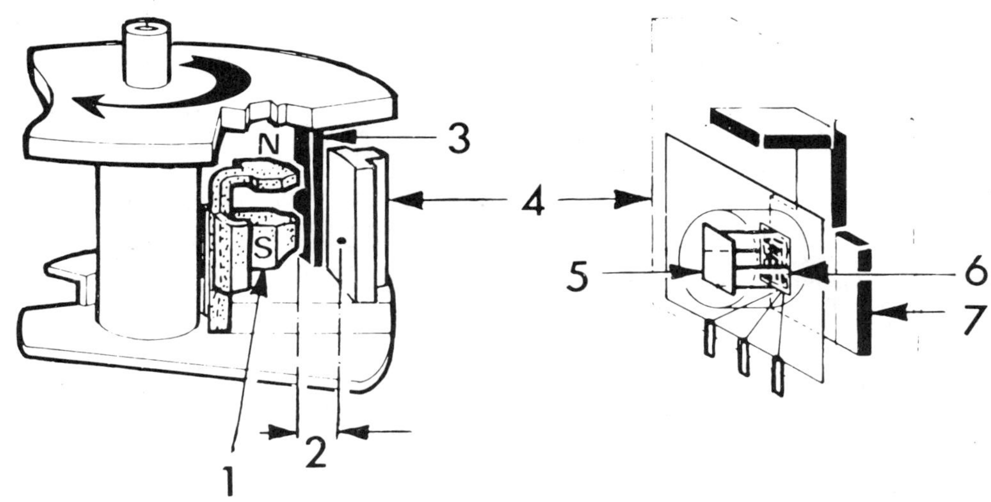

Fig. 13.35 Hall effect ignition system pick-up assembly (Sec 6)

1	*Magnet*	*3*	*Vane*	*5*	*Detector plate*	*7*	*Armature*
2	*Air gap*	*4*	*Detector/amplifier*	*6*	*Printed circuit amplifier*		

Hall effect distributor – timing to the engine

10 Remove the No 1 (flywheel end) spark plug, place a finger over the hole, and rotate the crankshaft until compression is felt.
11 Continue turning the crankshaft until the timing mark on the flywheel/torque converter is aligned with the specified timing mark on the clutch housing.
12 With the distributor removed and the cap withdrawn, turn the rotor to face the direction of the No 1 segment of the cap. The **trailing** edge of the No 1 vane must be in line with the magnet as shown in Figs. 13.36 and 13.37.
13 Keep the vane in the correct position, insert the distributor and engage the driving dog, and tighten the clamp screw. Refit the cap.
14 Finally, check the ignition timing with a strobe lamp as described in Chapter 4.

Spark plugs

15 Refer to the Specifications at the beginning of this Supplement for spark plug recommendations for later models.
16 It may be more convenient to remove the air cleaner or ducts for better access to the spark plugs.
17 Check the fitted sequence of the HT leads and identify them if necessary with tape or other means (No 1 at the flywheel end of the engine).
18 Pull the HT leads from the spark plugs by gripping the rubber connectors, not the leads.
19 Brush or vacuum any dirt from around the spark plug recesses, then unscrew and remove the spark plugs.

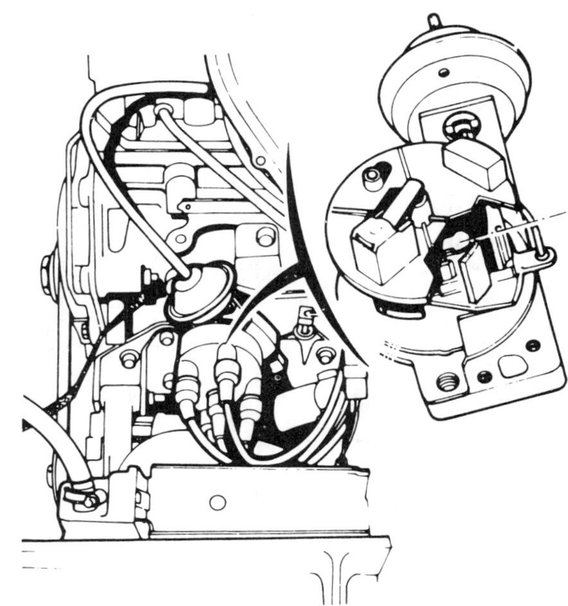

Fig. 13.36 Vane timing position on the Chrysler distributor (Sec 6)

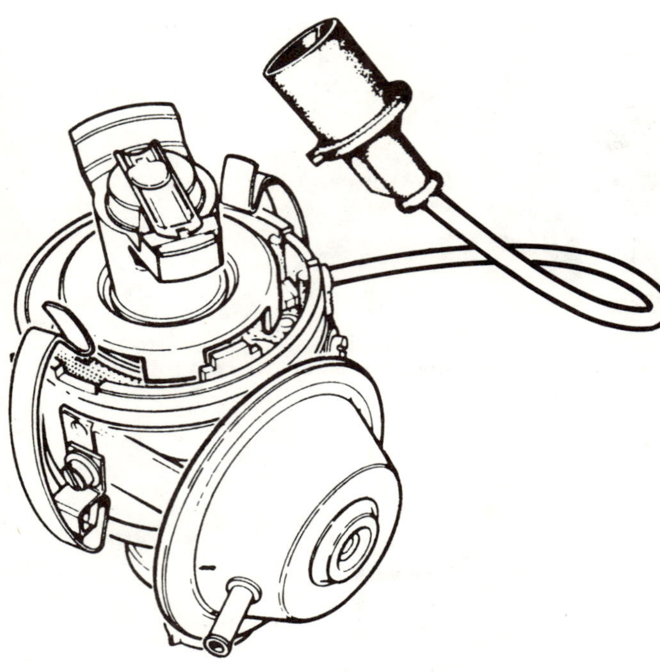

Fig. 13.37 Vane timing position on the Bosch distributor (Sec 6)

20 Remove the rubber dust excluding seals from the plugs, renewing them if they are deformed.
21 Refer to Chapter 4, Section 7 for plug conditions and cleaning details.
22 Clean out the plug recesses without allowing grit to drop into the cylinder bore. Fit the rubber seals to the plugs.
23 Apply a smear of grease to the spark plug threads and screw them into place using hand pressure, then finally tighten to the specified torque. Do not overtighten.
24 Reconnect the HT leads in their correct order.

7 Manual transmission

Gearbox types – general

1 Since their introduction, Alpine and Solara models have been equipped with five different manual gearbox types.

2 The original unit fitted was the type AT gearbox available in four-speed configuration only. This gearbox is covered in detail in Chapter 6 and in later Sections of this Chapter.

3 Models commencing A-series were fitted with the type AC gearbox in four-speed configuration (AC429) and in five-speed configuration (AC428).

4 Models commencing D-series are fitted with the BE1 type gearbox in either four-speed (BE1/4) or five-speed (BE1/5) configuration.

AT gearbox – routine maintenance

5 A filler/level plug and drain plug are fitted to both the gearbox and final drive housing. Check the oil level cold, by removing both filler/level plugs. Top up if necessary until oil begins to trickle out of the filler/level plug holes. Drain hot, by removing the filler/level plugs and the drain plugs (see *Routine Maintenance* at the front of this Manual).

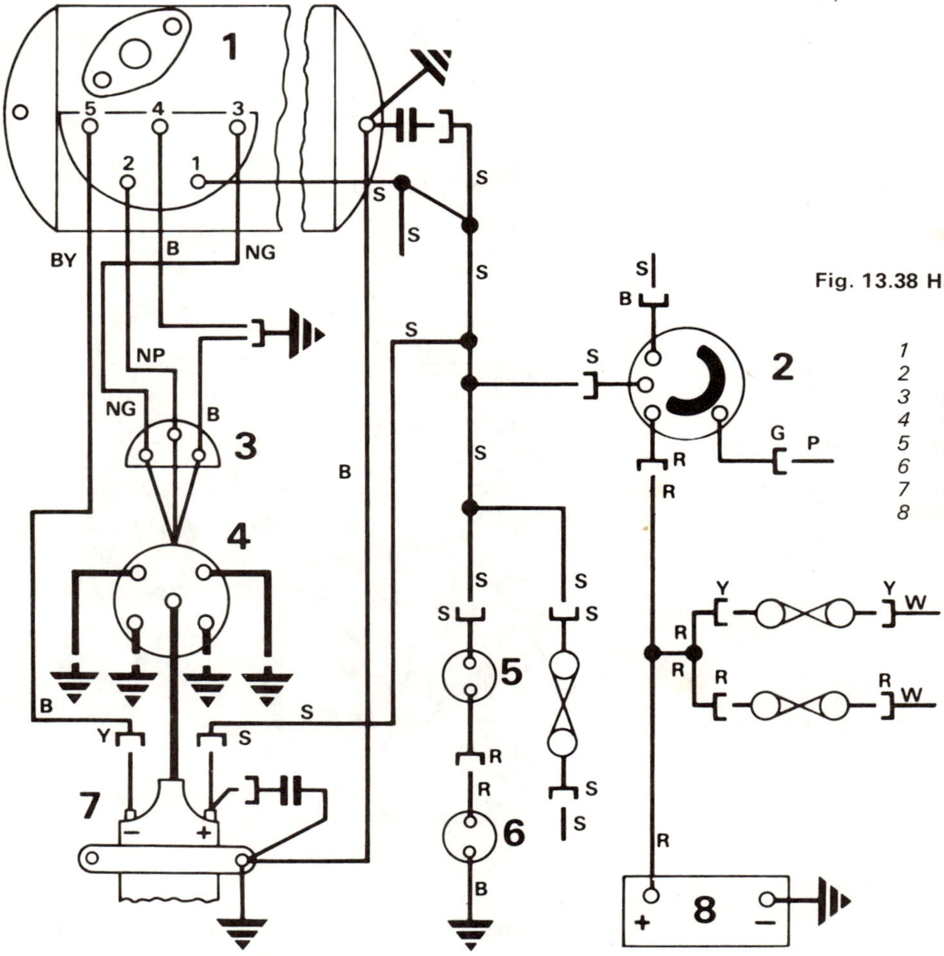

Fig. 13.38 Hall effect ignition system wiring diagram (Sec 6)

1 *Control unit*
2 *Ignition switch*
3 *3-pin connector, distributor*
4 *Distributor*
5 *Sensor, engine cooling fan*
6 *Motor, engine cooling fan*
7 *Ignition coil*
8 *Battery*

Colour code

B Black
G Green
N Brown
P Purple
R Red
S Slate
W White
Y Yellow

AC gearbox – routine maintenance

6 On the AC428 and AC429 gearboxes drain plugs are provided on the gearbox and final drive housings although the same lubricating oil is used for both assemblies and is able to migrate from one unit to the other.

7 The oil level is checked by means of a dipstick located in the final drive housing and replenished through the breather hole in the gearbox.

8 To top up or refill the gearbox clean the area around the breather and unscrew it using a 14 mm spanner. Fill or top up the gearbox until the level is between the high and low marks on the dipstick. The lubricant type and capacity are given in the Specifications. The difference between the high and low marks on the dipstick is 0.25 pint (0.15 litre). Refit the breather securely after filling. Drain the gearbox and final drive oil hot after with drawing the dipstick and unscrewing the drain plugs.

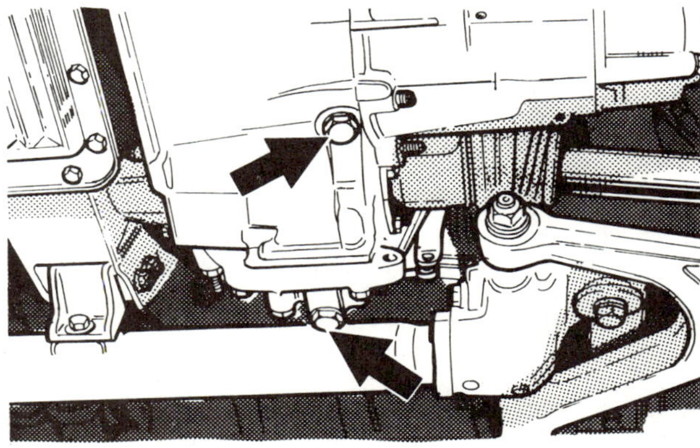

Fig. 13.39 Drain plugs on AC type manual gearbox (Sec 7)

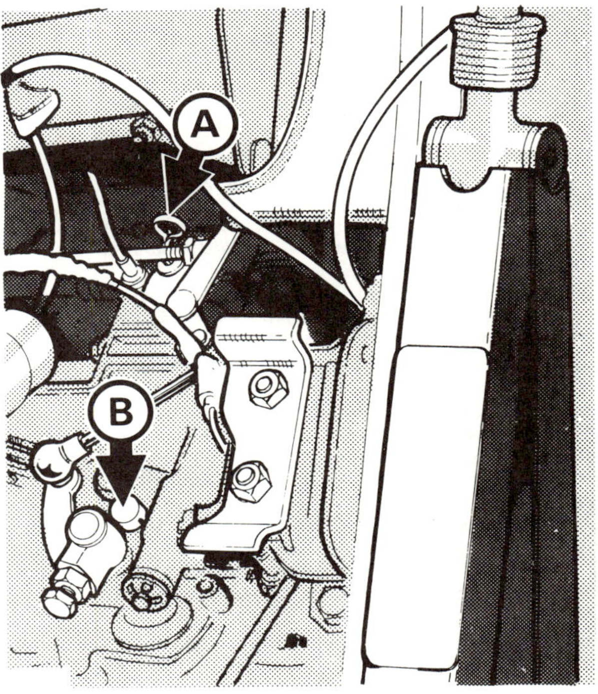

Fig. 13.40 Dipstick (A) and breather (B) on AC type gearbox (Sec 7)

BE1 gearbox – routine maintenance

9 On the BE1 type gearboxes no provision is made for oil level checking owing to the design of the casings and this is therefore not a service requirement.

10 When renewing the oil, remove the filler and drain plugs (Fig. 13.41). When refilling, only pour in the exact quantity of specified lubricant due to the lack of level checking facility.

Gearbox (type AT) – dismantling and reassembly

11 As from October 1977 the reverse selector shaft in the top cover incorporates grooves instead of flats, and a plunger is fitted instead of a detent ball. With this exception the repair procedure is identical to that described in Chapter 6.

12 As from January 1978 the gear lever is supported in rigid upper and lower bush housings, instead of a rigid lower housing and spring-tensioned upper housing as previously fitted. The gear lever must be supported firmly in the housings, at the same time allowing free movement. If necessary, a 0.02 in (0.5 mm) shim can be fitted between the housings.

13 As from May 1978 longer baulk ring springs are fitted to the synchromesh units on 2nd, 3rd and 4th speeds. The new springs are identified by their radius of 1.272 in (32.3 mm) as against the 1.173 in (29.8 mm) radius of the previous springs. Note that the larger springs must never be fitted to 1st speed, otherwise difficult engagement will occur.

Final drive (type AT gearbox) – removal and refitting

14 The instructions for removing the final drive unit in Chapter 6 include removing the gearbox separately, then removing the final drive/clutch bellhousing. However, it may be found easier to remove the complete final drive and gearbox assembly as one unit, then separate the two sub-assemblies on the bench. The following paragraphs describe this method.

15 Disconnect the battery negative lead.

16 Detach the inlet pipe from the fuel pump, unclip it, and pull it through the engine support bracket.

17 Disconnect the gearshift link rod and the reversing light switch leads.

18 Unscrew the upper clutch housing-to-cylinder block bolts and the engine support bracket-to-mounting pad bolts.

19 Loosen the left-hand front driveshaft nut and roadwheel bolts.

20 Jack up the front of the car and support it on axle stands.

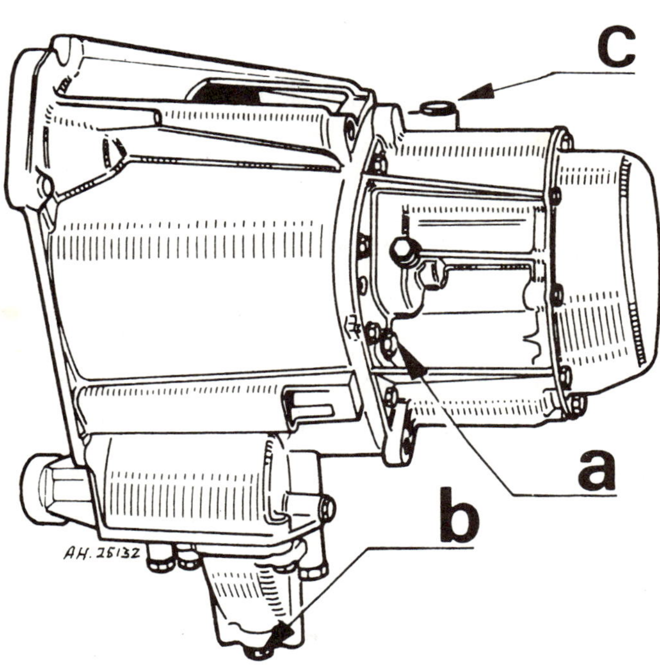

Fig. 13.41 Gearbox drain plug (a), final drive drain plug (b) and filler plug (c) on BE1 type gearbox (Sec 7)

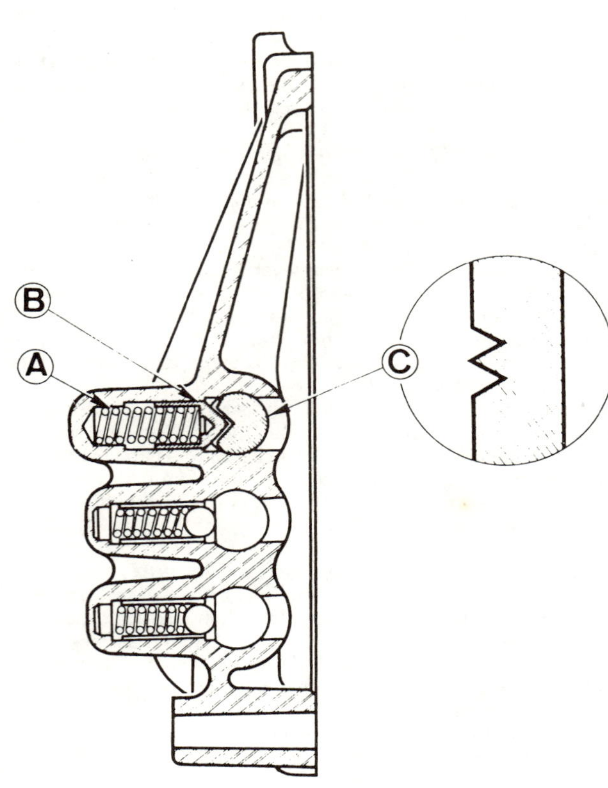

Fig. 13.42 Sectional view of selector shafts fitted to 1977 and later type AT gearbox (Sec 7)

A Detent spring
B Detent plunger
C Reverse selector shaft

21 Drain the oil from the final drive unit.
22 Remove the left-hand and right-hand driveshafts as described in Chapter 7.
23 Unscrew the securing bolts and remove the clutch slave cylinder without disconnecting the hose.
24 Remove the engine mounting bracket and spacer.
25 Disconnect the spring (if fitted), then remove the gearchange relay lever and spacer.
26 Unscrew the bolt and withdraw the speedometer cable and housing, then remove the driven gear.
27 Remove the cover and starter motor as described in Chapter 11.
28 Support the final drive/gearbox assembly, unscrew the remaining securing bolts and withdraw the complete unit from the engine. Take care not to allow the weight of the unit to bear on the input shaft while it is engaged with the clutch.
29 Unbolt the gearbox from the final drive/clutch housing, discard the O-rings, and recover the pinion shims.
30 Commence refitting by connecting the final drive/clutch housing to the gearbox as described in Chapter 6, Section 2.
31 Insert a guide stud into the upper hole in the cylinder block and refit the dirt cover.
32 Engage a gear, then locate the unit to the engine and insert two securing bolts. Remove the guide stud, insert the remaining bolts and tighten them to the specified torque.
33 Refit the clutch slave cylinder, starter motor, and speedometer cable.
34 Refit the engine mounting and the gearshift support and link rod. Adjust the link rod as described in Chapter 6 and in paragraph 38 of this Section.
35 Refit the driveshafts as described in Chapter 7.
36 Reconnect the fuel pipe and reversing light switch leads.
37 Refill the gearbox/final drive with the specified amount of oil.

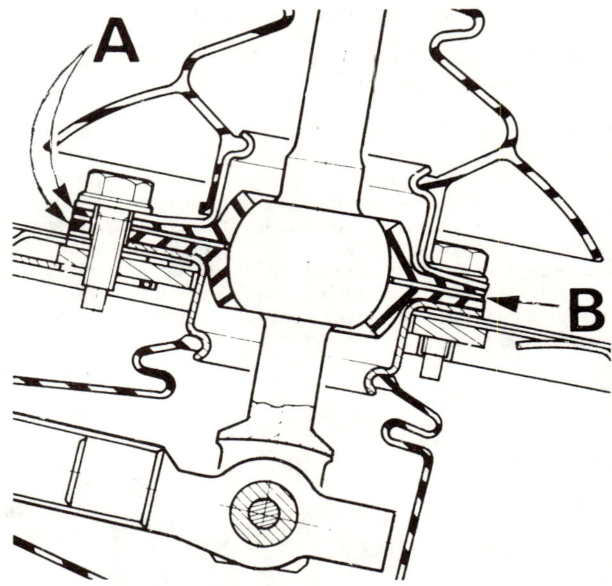

Fig. 13.43 Sectional view of gear lever support on 1978 and later type AT gearbox (Sec 7)

A Bush housings
B Shim

Gearchange mechanism (type AT gearbox) – adjustment

38 As from vehicle identification number S6 MDI 6P 521943 (RHD) or 447738 (LHD), the adjustment dimensions for the lateral and transverse rods have been increased by 0.118 in (3 mm). The old and new dimensions are shown in Figs. 13.44 and 13.45.

Gearbox (type AC428 and AC429) – removal and refitting

39 Disconnect the battery negative lead.
40 Detach the fuel supply pipe from the gearbox.
41 Disconnect the supply wires from the reversing light switch and unclip them from the left-hand engine support.
42 Unbolt the coil bracket and battery earth lead, and place them to one side.
43 Attach a hoist to the thermostat housing, jack up the front of the car and support it on axle stands, then take the weight of the left-hand side of the engine.
44 Remove the starter motor upper retaining bolt, then unscrew the two nuts and three bolts and withdraw the left-hand engine mounting.
45 Unscrew the locknut and bolt from the relay lever and move the lever towards the front of the car. Remove the three securing bolts from the gearbox.
46 Loosen both front hub nuts, then remove both front roadwheels.
47 Raise the engine slightly, then remove the front shock absorbers and fit the suspension retaining rods (see Chapter 7).
48 Working under the left-hand wing, unbolt the mounting bracket from the gearbox and recover the three detent springs and gasket.
49 Remove the drain plugs and drain the gearbox oil into a suitable container. Refit the plugs.
50 Disconnect the gear selector rods at the gearbox, then remove the two nuts securing the cross-shaft to the final drive. Raise the relay lever and withdraw the cross-shaft from the gearbox.
51 Using a extractor, release the left-hand suspension lower balljoint.
52 Unscrew the hub nut and pull the brake disc caliper, and hub outwards until the outer end of the driveshaft is released. Support the hub in its raised position with a length of wood.
53 Using a tapered drift or lever, release the driveshaft from the final drive and remove it.
54 Repeat the procedure given in paragraphs 51 to 53 for the right-hand driveshaft.

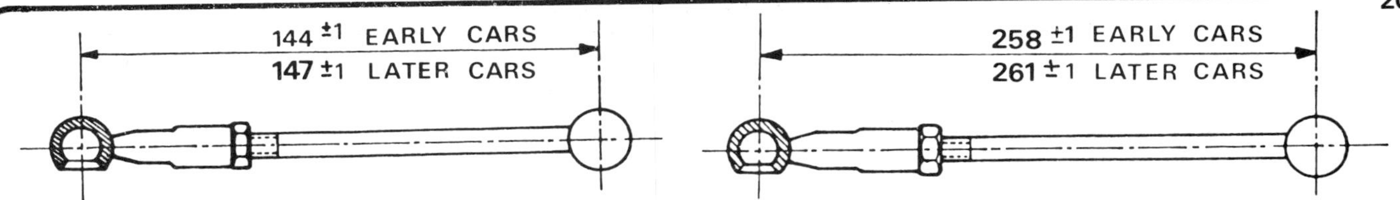

Fig. 13.44 Gearchange lateral rod adjustment dimension in mm – type AT gearbox (Sec 7)

Fig. 13.45 Gearchange transverse rod adjustment dimension in mm – type AT gearbox (Sec 7)

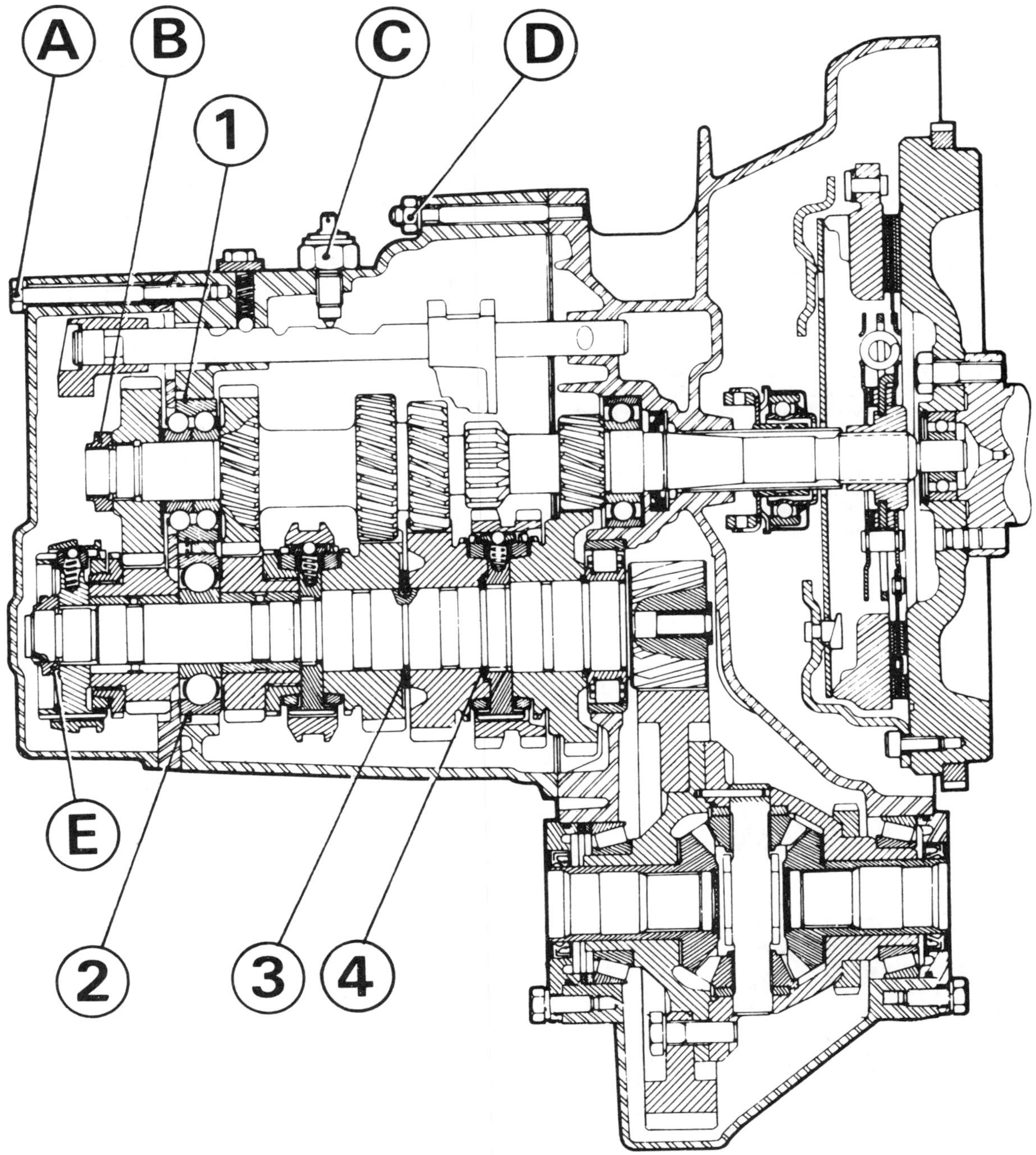

Fig. 13.46 Sectional view of type AC428 gearbox (Sec 7)

1 Bearing, input shaft
2 Bearing, output shaft
3 Thrust washers, 2nd speed gear
4 Circlip, 1st/2nd synchromesh gear
A End cover bolt
B Input shaft nut
C Reverse switch
D Gearbox casing nut
E Output shaft nut

55 Unscrew the bolt and remove the speedometer cable.
56 Unbolt the clutch slave cylinder, leaving the hydraulic hose connected. Place the slave cylinder to one side.
57 Remove the starter motor, with reference to Chapter 11 if necessary.
58 Support the gearbox/final drive assembly on a transmission stand or trolley jack.
59 Remove the remaining clutch housing securing bolts, then withdraw the gearbox from the engine. Take care not to allow the weight of the unit to bear on the input shaft while it is engaged with the clutch.
60 Refitting is a reversal of removal, but not the following additional points:

- (a) *Thread suitable studs into the cylinder block to ensure correct alignment of the engine and gearbox.*
- (b) *Make sure that the battery earth lead is fitted under the coil bracket rear bolt.*
- (c) *Retain the detent balls and springs with masking tape during the refitting procedure.*

Gearbox (type AC428) – dismantling, overhaul and reassembly

61 Clean the exterior of the gearbox with paraffin and wipe dry.
62 Unscrew the socket screw, then unbolt the end cover from the gearbox and remove the gasket.
63 Turn the selector shaft to engage any gear but 5th, then remove the bolt from the 5th gear selector fork, and move the synchro sleeve to engage the 5th gear. The input and output shafts are now locked, but it will be necessary to fit a clamp to the 5th speed synchro unit (Fig. 13.47) to prevent the unit from springing apart.
64 Bend back the peening, then unscrew the input and output shaft nuts using the special socket obtainable from tool hire agents.
65 Pull off the 5th speed synchro and selector fork, then remove the baulk ring, 5th gear, and bush from the output shaft.
66 Remove the 5th driving gear from the input shaft.
67 Unscrew the socket screws and remove the bearing retainer plate.
68 Extract the input and output shaft bearings sufficiently to remove the circlips.
69 Unscrew the reversing light switch. With the clutch housing downwards, remove the nuts, then tap the casing free and withdraw it upwards from the clutch housing. Remove the gasket.
70 Unscrew the 5th selector dog locking screw. Move all the selectors to neutral, then withdraw the 5th/reverse selector rod while pulling against the interlock plunger. Recover the selector dog.
71 Remove the reverse fork and pivot and recover the plunger and spring.
72 Unbolt the reverse gear relay lever and support.
73 Remove the reverse idler gear spindle and the reverse idler gear.
74 Remove the interlock plunger from between the 3rd/4th and 5th/reverse selector rod apertures.
75 Unscrew the locking bolts from the 1st/2nd and 3rd/4th selector forks and from the 3rd/4th selector dog.
76 Remove the selector rods and the interlock plunger from between the apertures. Remove the selector forks and dog.

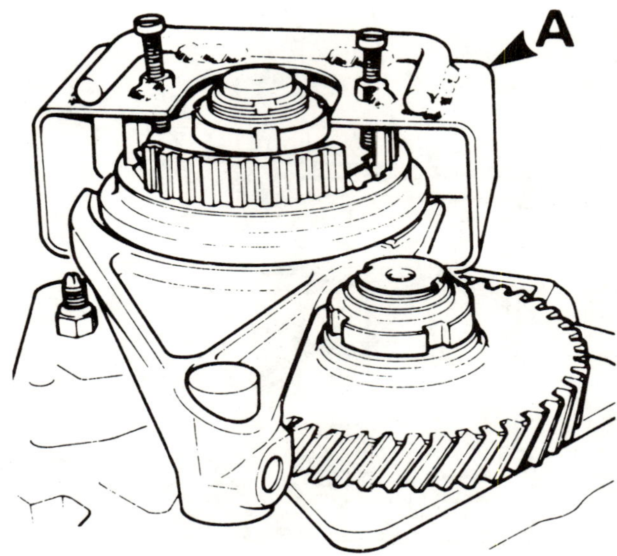

Fig. 13.47 5th speed synchro unit clamp (A) on type AC428 gearbox (Sec 7)

Fig. 13.48 Reverse selector components on type AC428 gearbox (Sec 7)

A	*Reverse gear relay lever*	*C*	*Reverse idler gear spindle*
B	*Relay lever support*	*D*	*Reverse idler gear*

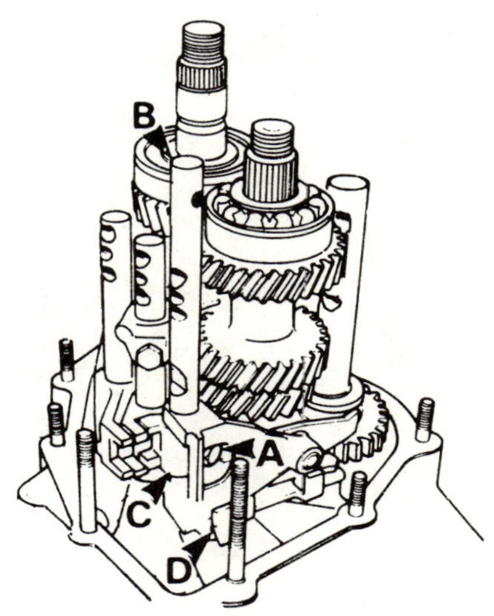

Fig. 13.49 5th/reverse selector components on type AC428 gearbox (Sec 7)

A	*Lockscrew on selector dog*	*C*	*Selector dog*
B	*5th/reverse selector rod*	*D*	*Plunger and spring*

77 Withdraw the input and output shafts together from the clutch housing.
78 To dismantle the input shaft, first note the fitted position of the bearing, then use a puller to remove it from the 5th gear end. Similarly remove the bearing from the clutch end of the shaft.
79 Before dismantling the output shaft, note that the baulk rings for 3rd and 4th speed gears and 1st and 2nd speed gears are identical, but must be kept separate for refitting to their respective gears.
80 Using a puller, remove the ball-bearing from the output shaft, then remove the 4th speed gear, bush, baulk ring, 3rd/4th synchro unit, baulk ring, and 3rd speed gear.
81 Remove the ring and withdraw the two half thrust washers, followed by the 2nd speed gear and baulk ring.
82 Remove the circlip retaining the 1st/2nd synchro unit, using old feeler blades to prevent the circlip damaging the shaft.
83 Withdraw the 1st/2nd synchro unit, baulk ring, and 1st speed gear.
84 Remove the circlip retaining the clutch and bearing, again using old feeler blades to prevent damage to the shaft.
85 Using a puller, remove the bearing from the shaft.
86 Clean all components in paraffin, wipe dry, and examine them for wear and damage. Check the gear wheels for chipping of the teeth and the bearings for excessive wear. If the condition of the synchromesh units is suspect, dismantle them as follows. Mark the hub and sleeve in relation to each other, then wrap the assembly in cloth and push the hub from the sleeve. Recover the rollers, springs and pins (5th speed synchro) or balls, springs and plates (1st/2nd and 3rd/4th speed synchro).
87 Check the synchro unit components for wear, particularly the internal teeth. Check the ends of the selector forks for wear. Check the gearbox casing for cracks, in particular around the bearing apertures and bolt holes.
88 Using a screwdriver, prise the oil seal from the clutch housing. If the clutch release bearing guide tube is worn, drive it from the clutch housing using a suitable drift. Coat the end of the new tube with a liquid locking agent and drive it into the clutch housing so that it projects 1.6 in (40.5 mm). tap the new oil seal into the housing using a suitable metal tube.
89 If the selector shaft in the gearbox casing requires renewal, remove the threaded bush followed by the selector shaft and spring. Remove the O-ring and copper washer. Refitting is a reversal of removal, but dip the O-ring in oil first.
90 Commence refitting by assembling the 5th speed synchro unit. Place the 5th speed gear on the bench with the cone upwards, then fit the baulk ring and sliding sleeve. Insert the springs and pins, then depress the springs and locate the rollers in the grooves. With the retaining plate in position, lift the sleeve until the rollers engage the detent.
91 To assemble either the 1st/2nd or 3rd/4th synchro units, place the sliding sleeve on the bench with the selector fork groove at the bottom. With the hub shoulder uppermost, insert the springs, plates and balls. Depress the balls and press the hub into the sleeve.
92 Drive the clutch end bearings onto the output shaft with a suitable metal tube, and secure it by fitting the circlip, using old feeler blades to prevent damage to the shaft. Note that the larger face of the circlip must face away from the bearing.
93 Fit the 1st speed gear, baulk ring and 1st/2nd synchro unit. Secure with the circlip, taking care not to damage the shaft.
94 With the 1st/2nd synchro hub pushed towards the clutch end bearing, use a feeler blade to check that the clearance of the circlip in its groove is not more than 0.002 in (0.05 mm). If it is, fit a thicker circlip as necessary.

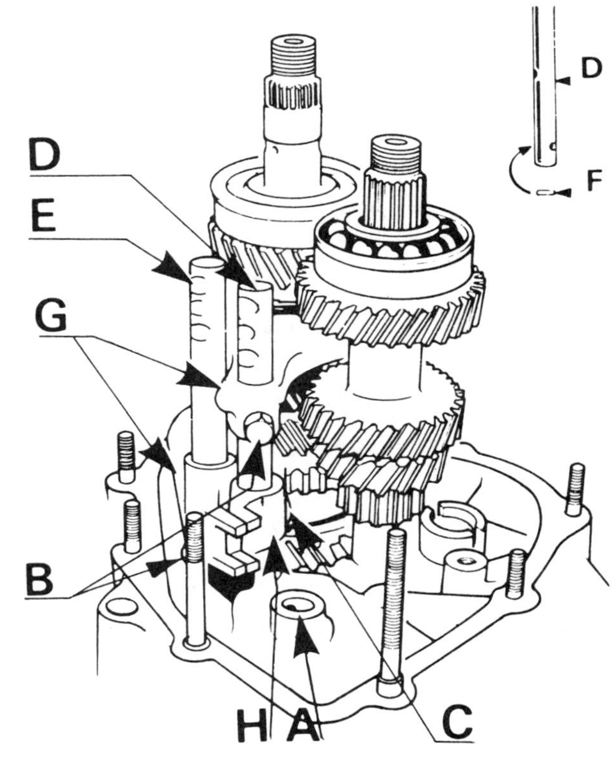

Fig. 13.50 1st/2nd and 3rd/4th selector components on type AC428 gearbox (Sec 7)

A 5th/reverse rod aperture
B Locking bolt, 3rd/4th selector fork
C Locking bolt, 3rd/4th selector dog
D 3rd/4th selector rod
E 1st/2nd selector rod
F Interlock pin
G Selector fork
H 3rd/4th selector dog

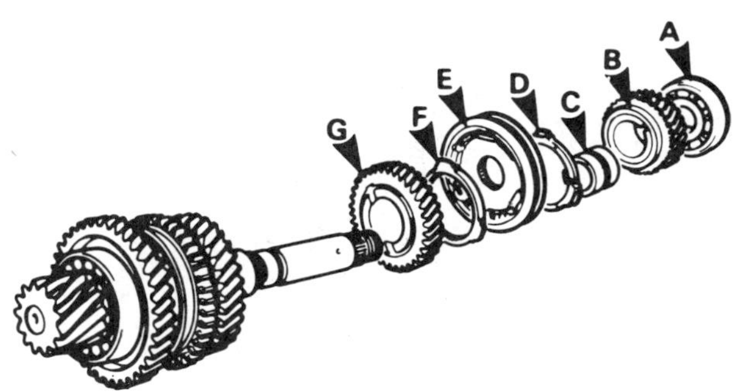

Fig. 13.51 Output shaft components on type AC428 gearbox (Sec 7)

A Bearing
B 4th speed gear
C Bush, 4th speed
D Baulk ring, 4th speed
E Hub and sleeve assembly
F Baulk ring, 3rd speed
G 3rd speed gear

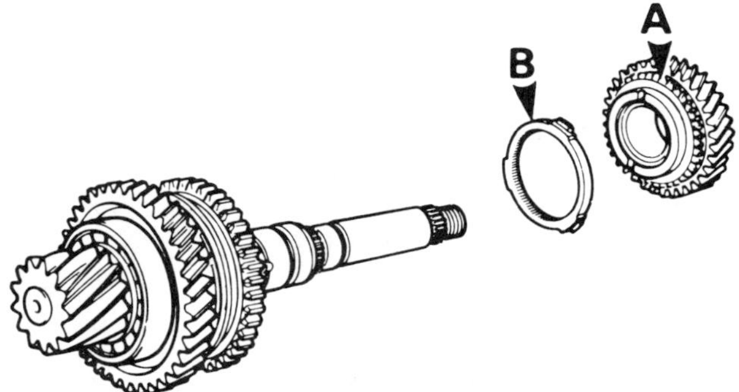

Fig. 13.52 2nd speed gear (A) and baulk ring (B) on type AC428 gearbox (Sec 7)

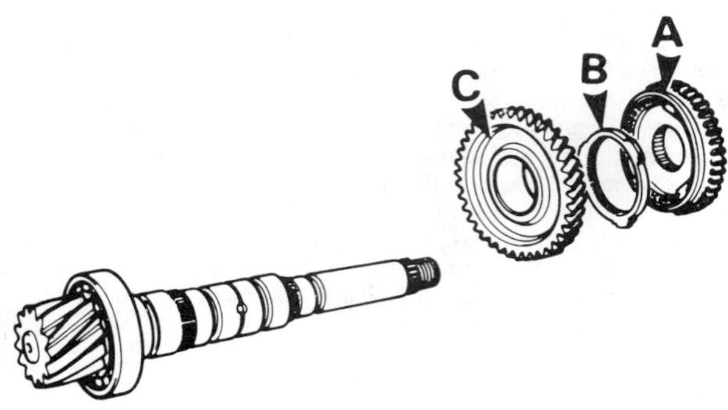

Fig. 13.53 1st/2nd synchro unit (A), baulk ring (B) and 1st speed gear (C) on type AC428 gearbox (Sec 7)

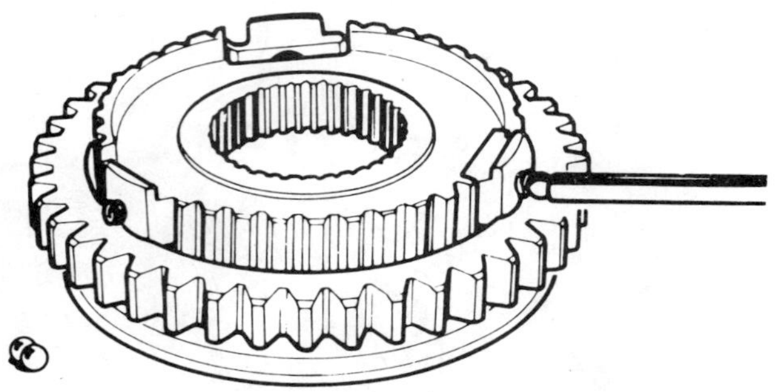

Fig. 13.55 Assembling 1st/2nd synchro unit on type AC428 gearbox (Sec 7)

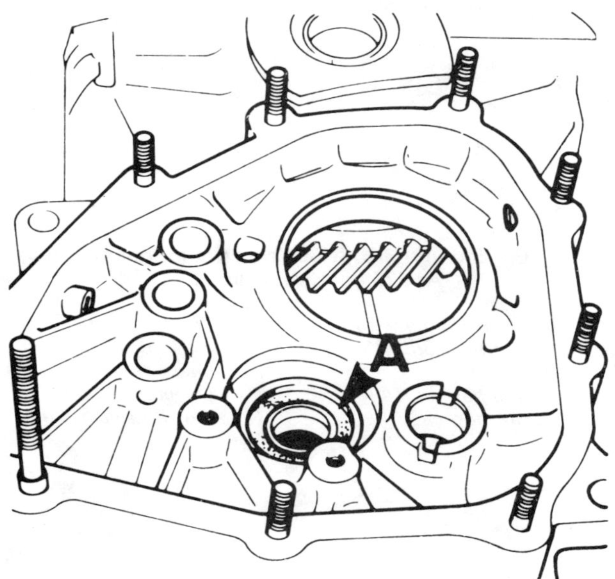

Fig. 13.56 Location of clutch housing oil seal (A) on type AC428 gearbox (Sec 7)

95 Fit the baulk ring and 2nd speed gear.
96 Fit the half thrust washers, then push the gear towards the synchro unit and use a feeler blade to check that the clearance of the washers in the groove is not more than 0.002 in (0.05 mm). If it is, fit a pair of thicker thrust washers.
97 Locate the retaining ring over the thrust washers.
98 Fit the 3rd speed gear, followed by the baulk ring, 3rd/4th synchro unit, baulk ring, 4th speed bush, and 4th speed gear.

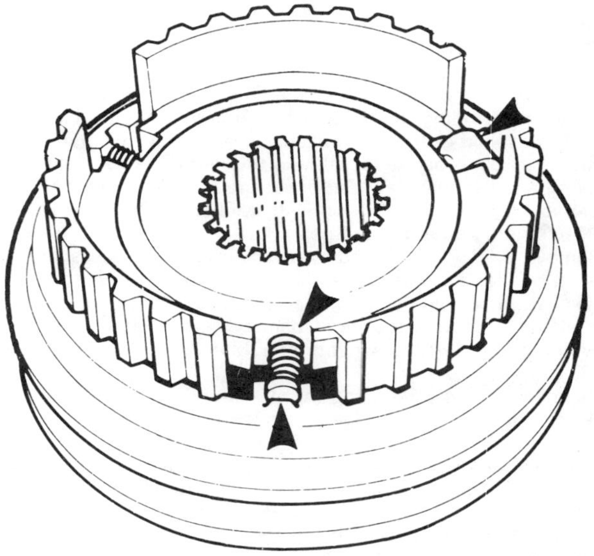

Fig. 13.54 5th speed synchro unit showing roller and spring on type AC428 gearbox (Sec 7)

99 If the gearbox housing or shaft bearings have been renewed, the position of the bearings must be checked with the circlips fitted. The outer faces of the outer tracks must be between 0.002 in (0.05 mm) below the face of the casing and 0.0008 in (0.02 mm) above the face, giving a maximum preload of 0.0008 in (0.02 mm). If necessary the circlip thickness should be changed.
100 Place the clutch housing face down on the bench. Mesh the input and output shafts together and lower them into the housing.
101 Fit the 1st/2nd and 3rd/4th selector forks to the synchromesh sleeves with the locking bolt holes towards the clutch housing.
102 Insert the 1st/2nd selector rod through the fork and into the housing, fit the locking bolt and washer and tighten the bolt to the specified torque.
103 Insert the interlock plunger between the 1st/2nd and 3rd/4th selector rod bores.
104 Grease the interlock pin and locate it in the 3rd/4th selector rod, then, with the 3rd/4th wand in position, insert the selector rod through the fork and dog and into the housing. Fit the locking bolts and washers and tighten the bolts to the specified torque.
105 Insert the interlock plunger between the 3rd/4th and 5th/reverse selector rod bores.
106 check that the 1st/2nd and 3rd/4th dogs are in neutral.
107 Locate a new O-ring on the reverse idler gear spindle. Check that the locating pin is in position, then fit the reverse idler gear and spindle into the clutch housing.
108 Fit the reverse selector fork support bracket and tighten the retaining bolts. Fit the reverse selector fork to the support bracket and reverse gear, then fit the spring and plunger.
109 Locate the 5th/reverse selector dog and insert the 5th/reverse selector rod. Fit the locking bolt and washer and tighten the bolt to the specified torque.
110 Locate a new gasket on the clutch housing and retain it with a little grease.
111 Check that all the gears are in neutral, then lower the gearbox casing over the shafts and engage the selector finger with the selector forks and dogs. Make sure that the selector shaft can be depressed and raised freely.
112 Fit the casing retaining washers and nuts and tighten them evenly.
113 Fit the circlips to the input and output shaft bearings.
114 Fit the bearing retaining plate. Apply a liquid locking agent to the threads of the socket screws, then insert and tighten them to the specified torque.
115 Fit the input shaft bearing outer half.
116 Fit the 5th speed driving gear to the input shaft, and the 5th speed driven gear to the output shaft.

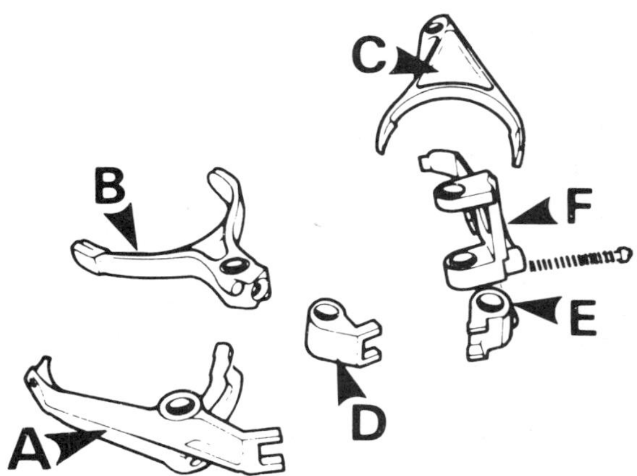

Fig. 13.57 Selector fork identification – type AC428 gearbox (Sec 7)

A 1st/2nd selector fork
B 3rd/4th selector fork
C 5th speed gear selector fork
D 3rd/4th selector dog
E 5th reverse selector dog
F Reverse relay lever

117 Fit the bush to the output shaft, followed by the baulk ring, 5th speed synchro unit, and retainer plate.

118 Engage 5th gear and turn the selector shaft to engage any other gear. The input and output shafts are now locked.

119 Fit the nuts to the shafts, tighten them to the specified torque, then peen them to lock.

120 With the selector shaft in neutral, pull the 5th speed sliding sleeve to neutral and remove the clamp.

121 Fit the locking bolt and washer to the 5th speed selector fork and tighten the bolt.

122 Slacken the 4th speed stop screw locknut and turn the screw outwards several turns. With 4th speed fully engaged, turn the screw inwards until it just contacts the sliding sleeve, then release the selector shaft and turn the screw inwards one complete turn. Tighten the locknut.

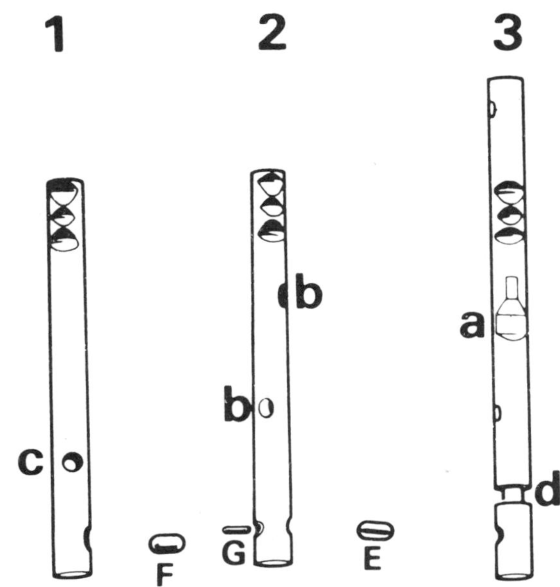

Fig. 13.58 Selector rod identification – type AC428 gearbox (Sec 7)

1 1st/2nd selector rod
2 3rd/4th selector rod
3 5th/reverse selector rod
a Flat groove
b Tapped holes
c Tapped hole
d Circular groove
E Interlock plunger
F Interlock plunger
G Interlock pin

123 Locate a new gasket on the gearbox casing and retain it with a little grease.

124 Fit the cover and tighen the socket screw and bolts to the specified torque.

125 With the gearbox on its side, slacken the 3rd and 5th speed stop screw locknuts and unscrew the screws several turns. Adjust the stop screws using the procedure described in paragraph 122.

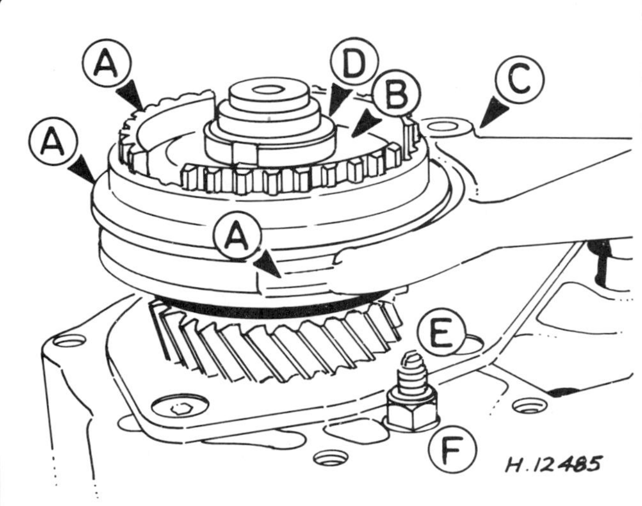

Fig. 13.59 5th speed synchro and 4th speed stop screw location – type AC428 gearbox (Sec 7)

A Synchromesh unit and fork assembly
B Retainer plate
C Securing nut (early type)
D Securing nut (early type)
E Stop screw, 4th gear
F Locknut

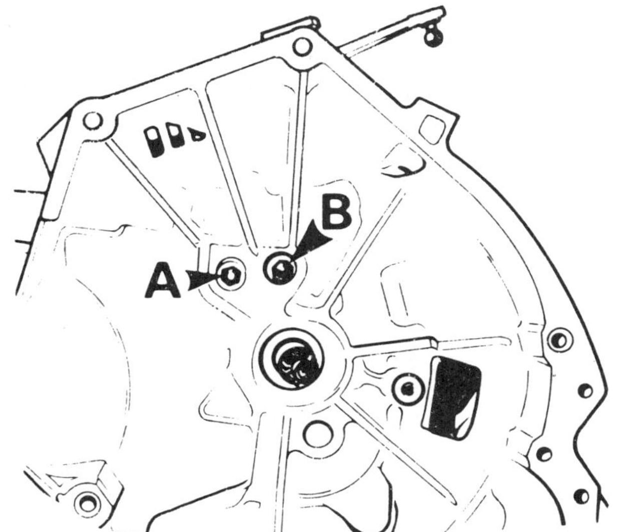

Fig. 13.60 3rd speed (A) and 5th speed (B) adjustment stop screws on type AC428 gearbox (Sec 7)

Gearbox (type AC429) – dismantling, overhaul and reassembly

126 The four-speed gearbox type AC429 fitted to later models bears a strong resemblance to the five-speed gearbox described above. The main difference, apart of course from the absence of 5th gears and associated components, is that the gearbox end cover locates the input and output shaft outer bearings.
127 Removal and refitting procedures are as given for the AC428 gearbox, paragraphs 39 to 60.
128 Dismantling, overhaul and reassembly procedures are broadly as given for the five-speed gearbox, ignoring any references to 5th gear components. Further differences are noted in the following paragraphs.
129 **End cover.** As noted above, the end cover locates the input and output shaft bearings. Therefore a bearing retainer plate is not fitted, and there is no gasket between the cover and the gearbox case.
130 **Reverse idler spindle.** The reverse idler spindle is secured by a stop plate and bolt instead of by a locating pin.
131 **Input shaft securing bolt.** A bolt (35 mm across flats) is used to secure the input shaft components, instead of the nut used on the five-speed gearbox shaft.
132 **Output shaft securing nut.** The output shaft nut has a *left-hand thread.* Special tool No 270349 00 is specified for slackening or tightening this nut; note that if this or a similar tool is used, the effective length of the torque wrench used in conjunction with the tool will be increased and a proportionately lower setting must be used on the torque wrench.
133 **Stop screw adjustment.** Adjustment of the 3rd and 4th speed stop screws is as described in paragraph 122. Note however that the 4th speed stop screw is located on the outside of the gearbox end cover.
134 **Gearshift mechanism adjustment.** Adjustment is as described in paragraphs 148 to 153, except that the length of the longitudinal rod (paragraph 150) should be 28.07 to 28.15 in (713 to 715 mm).

Final drive (type AC428 and AC429) – reassembly and adjustment

135 Whilst dismantling of the later type final drive follows closely the procedure described in Chapter 6, the reassembly and adjustment procedures differ somewhat.
136 Fit the assembled crownwheel carrier, with outer bearing tracks to the housing. Smear the mating face of the cover with sealant and fit it to the housing, tightening the nuts to the specified torque.
137 Fit the extension housing, **without sealant,** and tighten the securing bolts to the specified torque.
138 If any components have been renewed, the bearing preload must be adjusted as follows.
139 Fit the thickest selective washer available (5.65 mm) to the outer track of the bearing on the crownwheel side. Fit the bearing retainer plate, **without** the oil seal or O-ring.
140 With the bearing retainer plate uppermost, tighten the retaining bolts progressively to the specified torque. Rotate the crownwheel and strike the housing with a wooden or plastic hammer whilst doing this to settle the bearings.
141 Remove the retainer plate and the selective washer.
142 Using a dial gauge, measure the distance from the face of the final drive housing to the bearing outer track. Call this dimension A.
143 Similarly measure the height of the inner face of the bearing thrust plate above the retainer flange. Call this dimension B.
144 Subtract dimension B from A and add the specified preload (0.15 mm) to determine the thickness of selective washer required.

Example:

Dimension A	*14.145 mm*
Minus dimension B	*– 9.60 mm*
	4.85 mm
Add preload	*+ 0.15 mm*
Required thickness	*5.00 mm*

145 Selective washers are available in thicknesses of 4.25 to 5.65 mm in steps of 0.05 mm.
146 Smear the choosen selective washer with grease and fit it against the bearing outer track. Fit a new oil and O-ring in the bearing retainer, then fit the retainer, **without sealant,** and tighten the bolts to the specified torque.
147 Refit the speedometer pinion assembly (if removed), using a new O-ring, and secure with the locking bolt. Reassembly of the final drive is now complete.

Gearchange mechanism (type AC428 and AC429) – adjustment

148 Check that the length of the gearchange lower transverse rod between balljoint centres is between 7.9 and 8.0 (201 and 203 mm). If not, adjust it to this dimension.
149 Make sure that the relay lever and mechanism is secured, and that the gear lever assumes a vertical position in neutral.
150 Check that the length of the longitudinal rod is between 27.90 and 27.95 in (708 and 710 mm) for AC428 gearboxes and between 28.07 and 28.15 in (713 and 715 mm) for AC429 gearboxes. also check that the length of the upper transverse rod is between 9.65 and 9.72 in (245 and 247 mm) between centres (photo), and the lower transverse rod is between 7.90 and 8.0 in (201 and 203 mm). Adjust the rods as necessary.
151 With the gear lever in 1st, 3rd or 5th, it should be approximately vertical when viewed from the side. If not, slacken the gear lever housing retaining bolts and move the housing backward or forward as necessary. Tighten the bolts after making the adjustment.
152 If reverse gear can be engaged without depressing the gear lever, loosen the inhibitor bracket bolts (photo), then move the gear lever fully to the right, then rearwards slightly to take up the free play, and finally turn the gear lever fully clockwise. Hold the gear

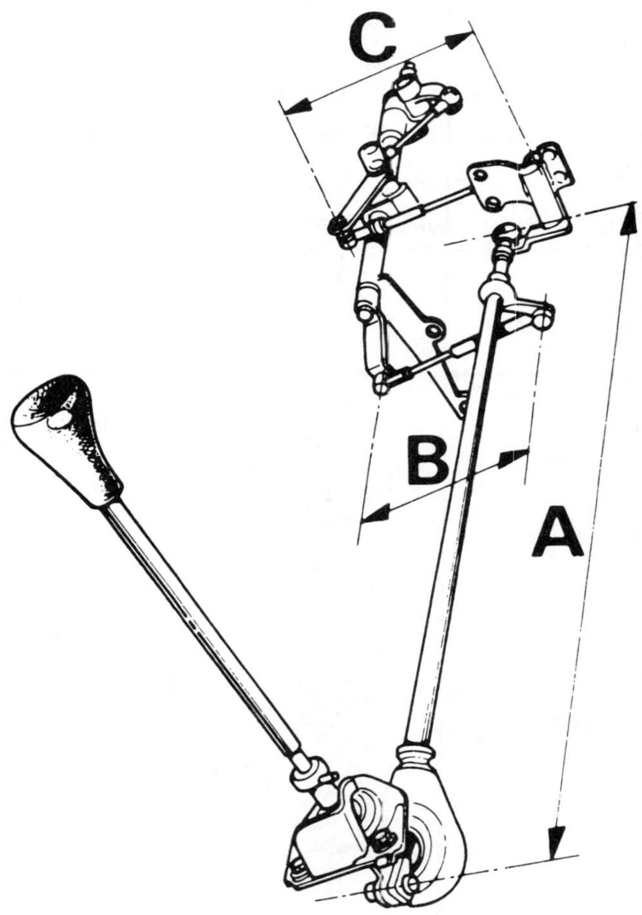

Fig. 13.61 Gearchange rod setting diagram for type AC gearbox (Sec 7)

A Longitudinal rod
B Transverse lower rod
C Transverse upper rod
For dimensions see paragraph 150

7.150 Gearchange upper transverse rod (arrowed) on AC428 gearbox

7.152 Gear lever and reverse inhibitor bracket on AC429 gearbox

lever in this position, then adjust the bracket so that it just contacts the stop pin and tighten the bolts.
153 If reverse gear can still be engaged without depressing the gear lever, renew the bracket and stop pin.

Gearbox (type BE1) – removal and refitting

154 Remove and refitting procedures are essentially the same as for the AC428 and AC429 gearboxes described earlier in this Section with the exception of the gearchange mechanism attachments. On the BE1 gearbox separate the ball sockets on the control rods from the balljoints on the levers as necessary to release the linkage from the gearbox. Refit the linkage using the reverse of the removal procedure.

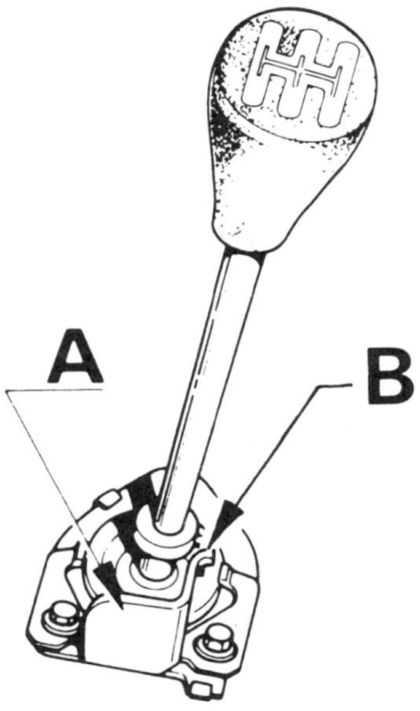

Fig. 13.62 Reverse inhibitor bracket adjustment on type AC gearbox (Sec 7)

A Bracket *B Stop pin*

Gearbox (type BE1) – dismantling, overhaul and reassembly

Dismantling into major assemblies (type BE1/5)

155 Remove the gearbox from the car as described earlier in this Section.
156 Remove the eight bolts and washers which secure the end cover. Remove the cover (photo).
157 Make alignment marks between the 5th gear synchro hub and its sliding sleeve.
158 Engage 5th gear, then drive out the 5 mm roll pin which secures 5th gear selector fork to the selector rod (photo).
159 Hold 5th gear selector fork in the engaged position and return the gear selector to neutral so that the selector rod moves through the fork.
160 Engage any other gear to lock up the shafts, then unscrew and remove the 28 mm nut from the end of the input shaft. If the nut is staked in position it may be necessary to relieve the staking (photo).
161 Remove 5th gear synchro hub, sliding sleeve and selector fork from the input shaft. Be prepared for the ejection of the detent ball from the selector fork.
162 Refit the 5th gear sliding sleeve and hub and engage 5th gear

7.156 Removing gearbox end cover

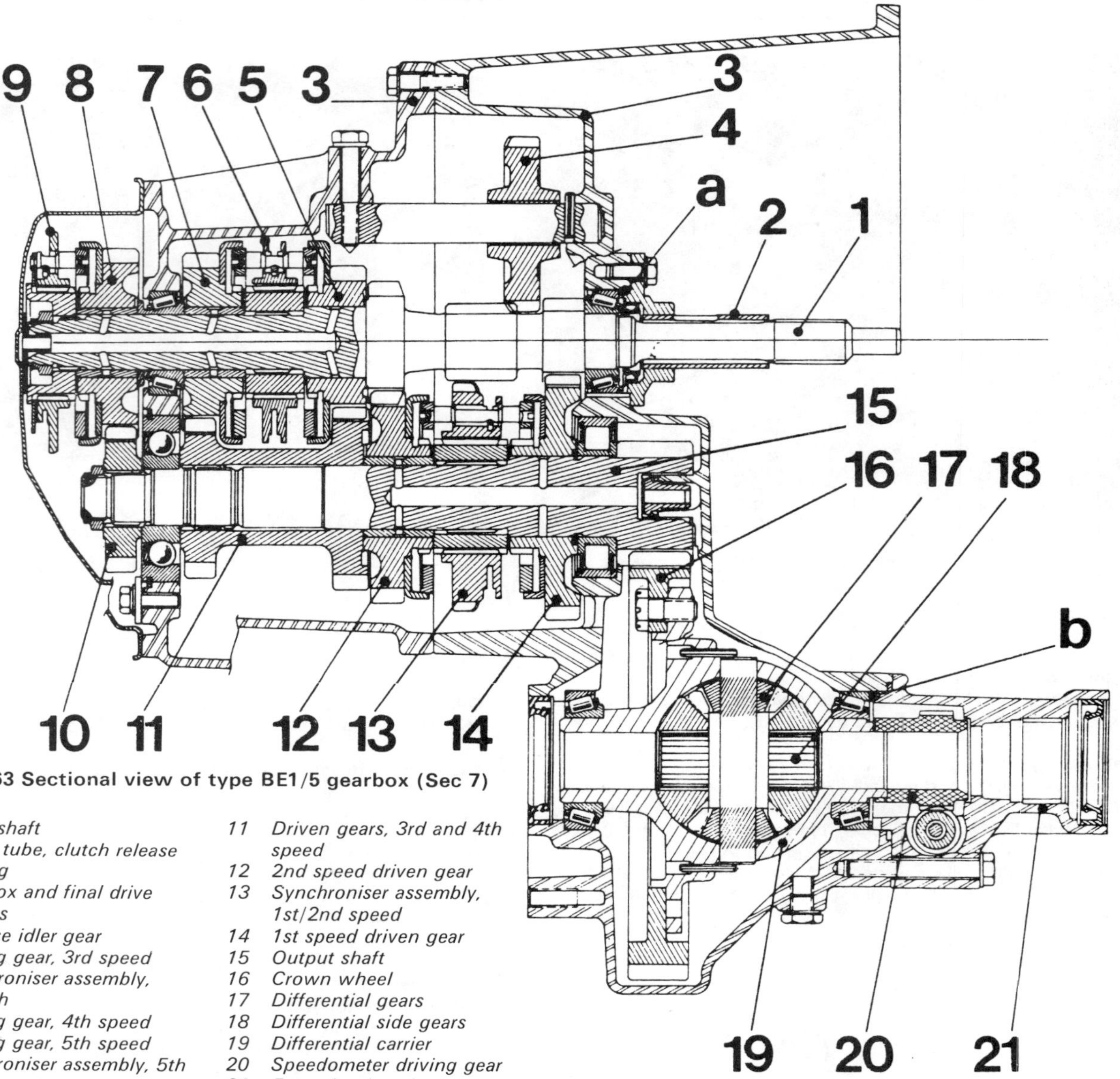

Fig. 13.63 Sectional view of type BE1/5 gearbox (Sec 7)

1 *Input shaft*
2 *Guide tube, clutch release bearing*
3 *Gearbox and final drive casings*
4 *Reverse idler gear*
5 *Driving gear, 3rd speed*
6 *Synchroniser assembly, 3rd/4th*
7 *Driving gear, 4th speed*
8 *Driving gear, 5th speed*
9 *Synchroniser assembly, 5th speed*
10 *Driven gear, 5th speed*
11 *Driven gears, 3rd and 4th speed*
12 *2nd speed driven gear*
13 *Synchroniser assembly, 1st/2nd speed*
14 *1st speed driven gear*
15 *Output shaft*
16 *Crown wheel*
17 *Differential gears*
18 *Differential side gears*
19 *Differential carrier*
20 *Speedometer driving gear*
21 *Extension housing*
a and b Adjustment shims

7.158 5th gear selector fork roll pin (arrowed)

7.160 Unscrewing input shaft nut

7.162 Unscrewing output shaft nut

7.166 Unscrewing selector rod lockplate bolt

7.169 Gear selector shaft retaining circlip

again. Relieve the staking from the output shaft nut and remove the nut (photo). Remove the sliding sleeve and hub again.
163 Remove from the input shaft the 5th gear, its bush and the spacer.
164 Remove the two bolts and washers which secure the output shaft rear bearing.
165 Remove the circlip from the output shaft rear bearing by prising up its ends. The circlip should be renewed anyway, so do not be afraid of breaking it. Raise the output shaft if the circlip is jammed in its groove.
166 Extract its securing bolt and remove the selector rod lockplate (photo).
167 Remove the bolt which retains the reverse idler gear spindle.
168 Remove the 13 bolts and washers which secure the end casing to the main casing. Withdraw the end casing: it is located by dowels, and may need striking with a wooden or plastic mallet to free it. Do not use a metal hammer, nor lever in between the joint faces. Note the location of the clutch cable bracket.
169 Remove the selector arm and spring from the gear selector shaft. Remove the circlip and washer, push the shaft in and recover the O-ring (photo).
170 Drive out the roll pins which secure the selector finger and the interlock bracket to the selector shaft.
171 Inspect the cover which protects the end of the selector shaft. If it is tapered, proceed to the next paragraph. If it is cylindrical, use pliers or a self-gripping wrench to extract it, then press the gear selector shaft towards the cover so that the circlip and washer can be extracted from the end of the shaft. These components are not fitted to earlier models; there are associated changes to the shaft

3
4
5
6

Fig. 13.64 5th speed synchro components BE1/5 gearbox (Sec 7)

3 5th speed synchro hub
4 5th speed gear
5 Bush
6 Spacer

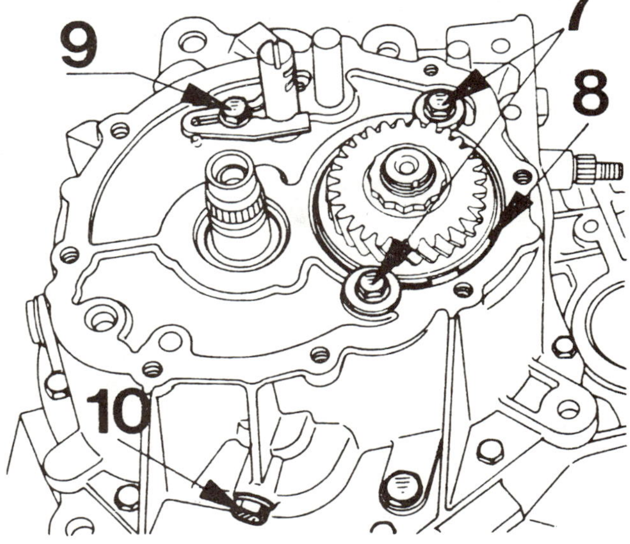

Fig. 13.65 Type BE1/5 gearbox end casing (Sec 7)

7 Output shaft bearing retaining bolts
8 Output shaft circlip
9 Selector rod lockplate retaining bolt
10 Reverse idler gear spindle retaining bolt

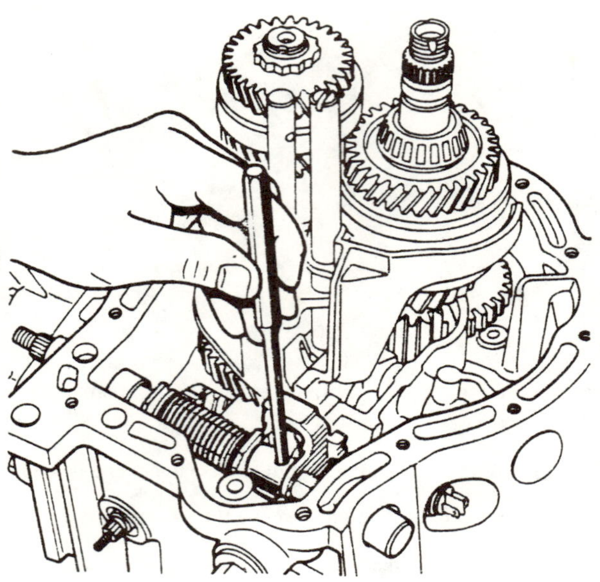

Fig. 13.66 Removing selector finger/interlock bracket-to-shaft roll pin – type BE1/5 gearbox (Sec 7)

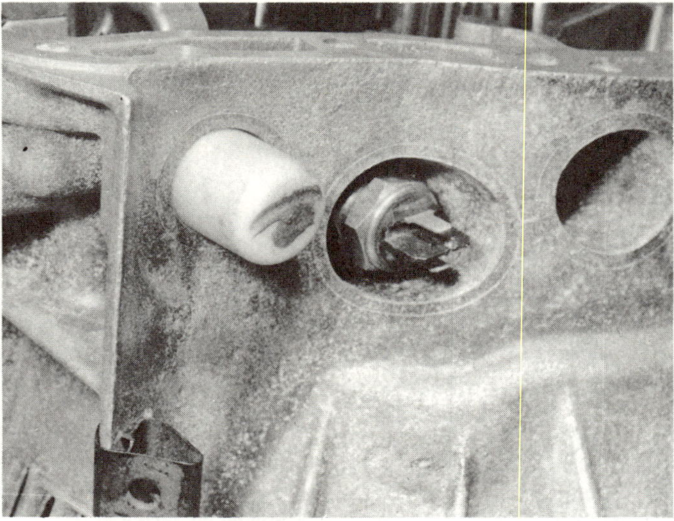

7.171A Selector shaft cover

7.171B Selector shaft displaced for circlip removal

7.173 Removing reverse idler spindle

itself, the main casing and 5th gear selector fork (photos).

172 Pull the selector shaft out of the gearbox. From inside the gearbox recover, as they are freed from the shaft, the selector finger, the interlock bracket, the spring and its cup washers. Notice which may round the washers are fitted.

173 Screw the reverse idler spindle retaining bolt back into the spindle and use it as a lever to extract the spindle (photo). Remove the reverse idler gear itself.

174 Remove the swarf-collecting magnet from the casing (photo).

175 Carefully lift out the two geartrains with their shafts, the selector forks and the selector rods.

176 Remove the spring support bracket from inside the main casing.

177 If not already removed, drive out the selector shaft end cover, using a drift of diameter no greater than 14 mm.

178 Extract the lubrication jet, using a wire hook.

179 Unscrew and remove the reversing lamp switch.

180 Remove the nut and washer which secure the reverse selector fork spindle. Remove the spindle and the selector fork. Recover the detent plunger and spring (photos).

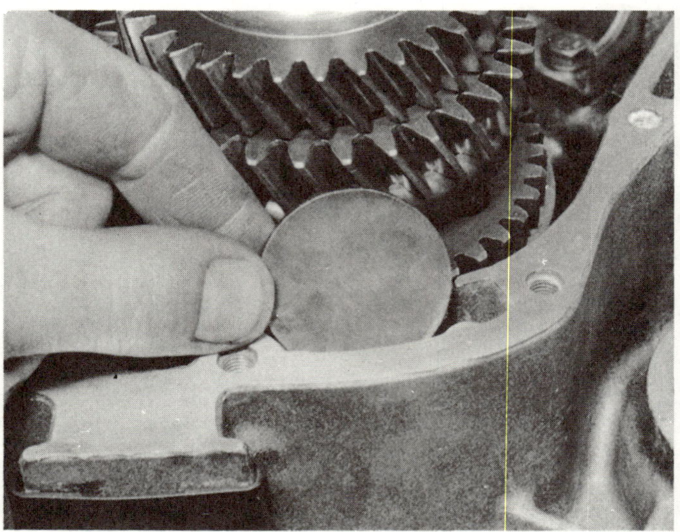

7.174 Swarf collecting magnet

Fig. 13.67 Removing geartrains – type BE1/5 gearbox (Sec 7)

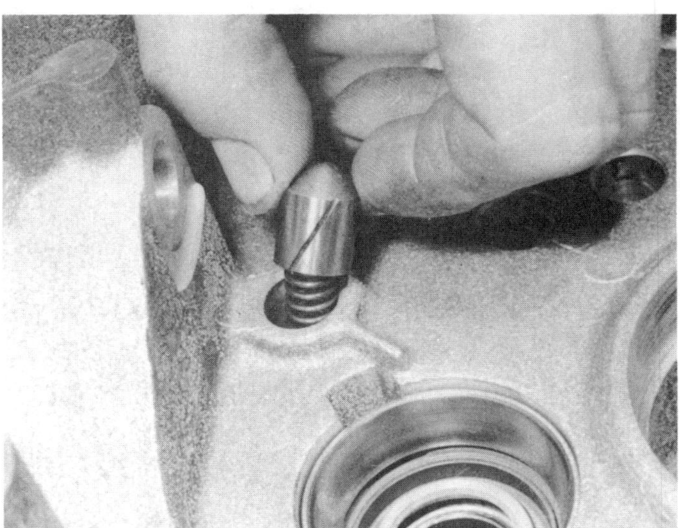

7.180B Reverse detent plunger and spring

181 Unscrew and remove the breather from the main casing (photo).
182 Turning to the clutch housing, remove the clutch release bearing (if not aleady done). Pull off the release fork.
183 Unbolt and remove the release bearing guide tube.

7.180A Reverse selector fork

184 From behind the tube remove the preload shim and the outer track of the input shaft front bearing.
185 To remove the final drive unit, first unbolt and remove the speedometer pinion and its adaptor (photo).
186 Unbolt and remove the extension housing. Recover the speedometer driving gear and the bearing preload shim (photos).
187 Unbolt the final drive half housing. Remove the half housing and final drive unit. Note the location of the gearchange pivot bracket.
188 Identify the final drive bearing outer tracks: if they are to be re-used they must be refitted on the same sides.
189 Remove the selector lever from the main casing. It is retained by a circlip and a washer.
190 If it is wished to remove the clutch release lever balljoint, do so with a slide hammer having a suitable claw. (A new gearbox will not necessarily be fitted with a balljoint).
191 The gearbox is now dismantled into its major assemblies.

Dismantling into major assemblies (type BE1/4)
192 As mentioned previously, the four-speed and five-speed gearboxes differ only in respect of the 5th gear and associated components. The only difference in dismantling procedure is in the method used to slacken the input and output shaft nuts.
193 The nuts can be slackened with the gearbox in the vehicle. To do so, engage a gear and apply the handbrake. Remove the end cover and slacken the nuts on the input and output shafts. (The input shaft nut is combined with an oil thrower). Refit the cover and proceed to remove the gearbox.
194 With the gearbox out of the vehicle, the best way to lock the shafts is to engage a gear and immobilise the input shaft with a clutch disc to which a metal bar has been welded (Fig. 13.68) it is unwise to attempt to grip the input shaft splines with any other tool, as damage may be caused.
195 With the input and output shaft nuts slackened, proceed as described for the five-speed box, making appropriate allowances.

Examination and renovation
196 Refer to Chapter 6, Section 4 for a general guide to examination of gearbox components.
197 Circlips, roll pins, gaskets, oil seals and locking devices should all be renewed as a matter of course. Prise out the old oil seal from the clutch release bearing guide tube, but do not fit the new seal until so instructed during reassembly.
198 If new input shaft or differential bearings are to be fitted, a selection of preload shims will be required. Read through the relevant procedure before starting work.

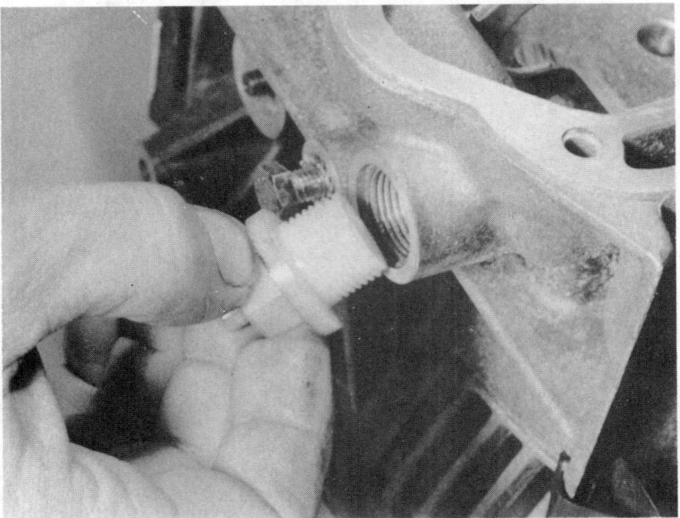
7.181 Removing the breather

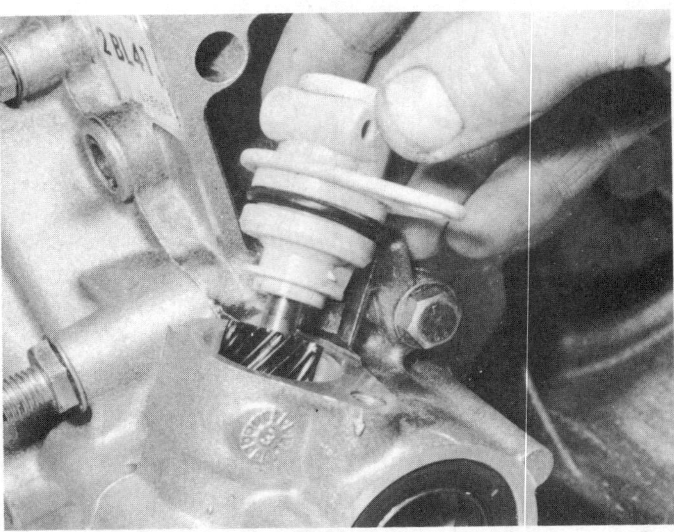
7.185 Removing speedometer pinion

7.186A Removing the final drive extension housing

7.186B Speedometer drivegear

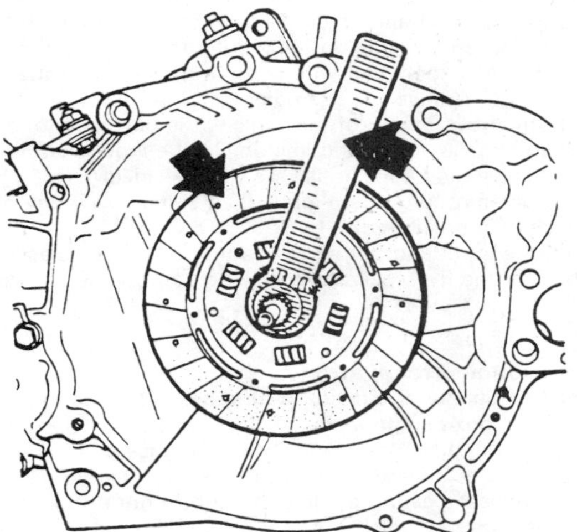
Fig. 13.68 Method of locking input shaft using special tool – type BE1/4 gearbox (Sec 7)

Input shaft – dismantling and reassembly

199 Remove the 3rd and 4th gear components from the input shaft by supporting the assembly under the 3rd gear and pressing or driving the shaft through. Protect the end of the shaft. Once the rear bearing is free, the outer components can be removed from the shaft in order: 4th gear and its bush, 3rd/4th synchro sleeve and hub and 3rd gear (photos).

200 Mark the synchro sleeve and hub relative to each other to show which side faces 4th gear.

201 Remove the front bearing from the shaft, preferably with a press or a bearing puller. As a last resort it may be possible to support the bearing and drive the shaft through: be sure to protect the end of the shaft if this is done.

202 Once the input shaft bearings have been removed, they must be renewed. Press the rear bearing outer track from the end casing and press in the new track, making sure it enters squarely.

203 Before commencing reassembly, make sure that the input shaft is free from burrs and wear marks. Lubricate all parts as they are fitted.

204 Fit a new front bearing to the shaft, using a suitable tube to press or drive it home.

205 Fit 3rd gear, 3rd/4th synchro hub and sleeve, 4th gear and its bush. Take care not to get 3rd and 4th gears mixed up, they are similar in appearance (4th gear has more teeth). If the original

7.199A Input shaft bearing

7.199B Removing 4th speed gear

7.199C Removing 4th speed gear bush

7.199D 3rd/4th synchro sleeve

7.199E 3rd/4th synchro-hub

7.199F 3rd speed gear

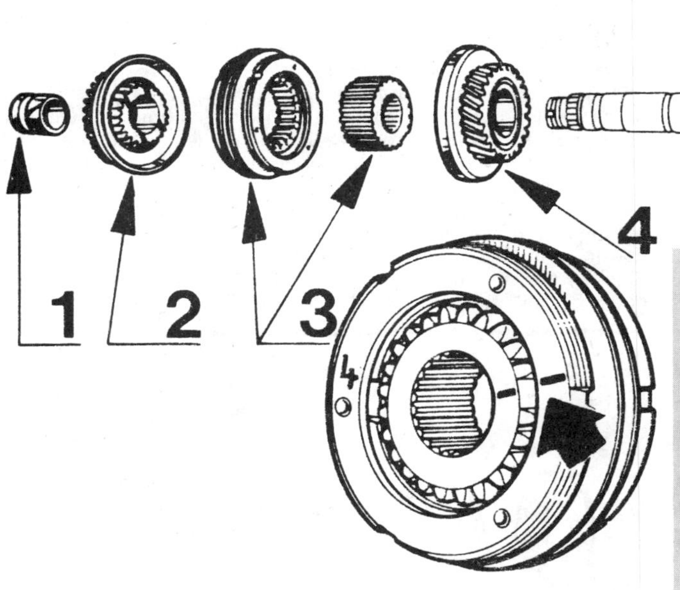

Fig. 13.69 Input shaft components – type BE1/5 gearbox (Sec 7)

1 4th gear bush
2 4th gear
3 3rd/4th synchro sleeve and hub
4 3rd gear
Arrow indicates mating marks

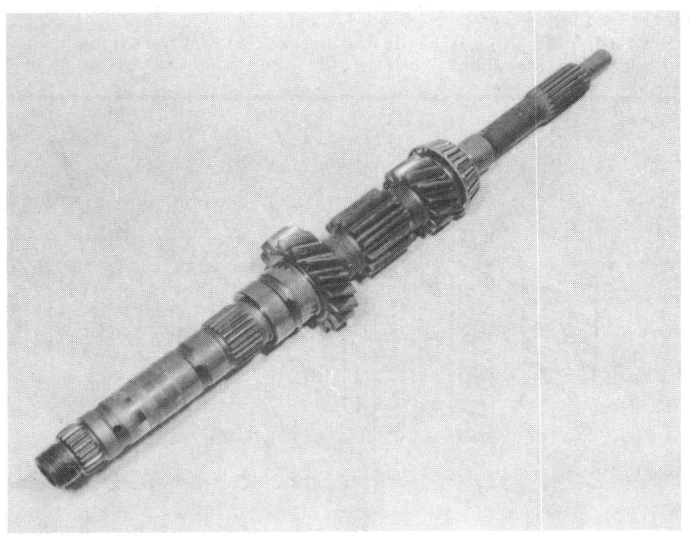
7.199G Input shaft stripped, except for front bearing

synchro components are being refitted, observe the mating marks made during dismantling.
206 Fit a new bearing to the shaft, again using a piece of tubing.
207 The input shaft is now reassembled.

Output shaft – dismantling and reassembly

208 Remove 5th gear (when applicable) and the rear bearing from the output shaft. Use a puller or bearing extractor if they are a tight fit on the shaft (photo).
209 Remove 3rd/4th gear assembly, 2nd gear and its bush (photos).
210 Make alignment marks between the 1st/2nd synchro hub and sleeve, then remove them from the shaft (photos).
211 Remove 1st gear and the half washer (early models) or needle thrust bearing and circlip (later models) (photos).
212 Press or drive the shaft out of the pinion end bearing, protecting the end of the shaft.
213 Before commencing reassembly, make sure that the shaft is free from burrs or wear marks. Lubricate all parts as they are fitted.
214 Fit the pinion end bearing to the shaft, using a piece of tube to drive or press it home. On later models, fit a new circlip.
215 Fit the half washers above the bearing, using a smear of grease to hold them in position. On later models, fit the needle thrust bearing instead.
216 Refit 1st gear, taking care not to dislodge the half washers (when fitted).
217 Refit the 1st/2nd synchro unit, observing the mating marks made when dismantling. The chamfer on the external teeth must face toward 1st gear.
218 Fit 2nd gear and its bush.
219 Fit the 3rd/4th gear assembly, making sure it is the right way round.
220 Fit the rear bearing, with the circlip groove nearest the tail of the shaft.
221 Fit the 5th gear, when applicable, with its boss towards the bearing. On early four-speed models fit the washer; on later models fit the spacer.
222 Fit a new nut to the output shaft but do not tighten it yet. Assembly of the output shaft is now complete.

Selector mechanism – dismantling and reassembly

223 One of the unusual features of this gearbox is that the detent balls are located in the forks (photo).
224 Rotate the 1st/2nd and 3rd/4th selector rod to disengage the detent slots from the balls, then remove the rod from the forks.
225 Where applicable, remove 5th gear selector rod from the 1st/2nd fork.
226 Examine the forks and rods for wear and damage and renew as necessary.
227 Commence reassembly by inserting 5th gear selector rod into the 1st/2nd fork.
228 Offer the 3rd/4th fork to the 1st/2nd fork so that their holes and selector fingers align.
229 Insert the 1st/2nd and 3rd/4th selector rod, positioning the locking slot as shown (Fig. 13.72). Bring all the selector finger slots into line to position the selectors in neutral (photo).

Differential – dismantling and reassembly

230 Unbolt the crownwheel from the differential housing.
231 Remove the side gears by pushing them round inside the housing until they can be removed (photo).
232 Drive out the roll pins which secure the differential gear spindle.
233 Remove the spindle, the differential gears and their washers (photo).
234 Use a press or bearing extractor to remove the bearings.
235 Examine all parts for wear and damage, and renew as necessary. Lubricate all parts as they are assembled.
236 Fit the bearings, using a piece of tube to press or drive them home.
237 Fit the spindle with the differential gears and thrust washers. Secure the spindle with new roll pins, which should be driven in until they are centrally located in their holes (photo). Note that on later models, modified differential gears and thrust washers are used, see Fig. 13.74.
238 Fit the side gears, one at a time, and work them into their proper positions. Retain them in this position using tool 8.0317M or equivalent inserted from the crownwheel side (photo).
Note: *As from February 1985, a ring is fitted to centralise the differential gears. The use of the special retaining tool 80317M is no longer required where this ring is fitted (Fig. 13.75).*
239 Fit the crownwheel with its chamfer towards the differential housing. Secure with the bolts, tighten them in diagonal sequence to the specified torque.

Reassembly of major units (type BE1/5)

240 Commence reassembly by fitting the selector lever into the main casing. Make sure that the locating dowel is in position in the final drive housing mating face.
241 Apply jointing compound to the mating face, then fit the differential assembly with its bearing tracks (photo).
242 Fit the final drive half housing and the extension housing, but only tighten their securing bolts finger tight at this stage.
243 Fit a new oil seal, lips well greased, to the other side of the final drive housing from the extension.
244 Remove the extension housing, fit a preload shim 2.2 mm thick to the bearing outer track and refit the extension housing (without its O-ring). Rotate the crownwheel while tightening the extension housing bolts until the crownwheel just starts to drag. This operation seats the bearing.
245 Remove the extension housing and the preload shim. With an accurate depth gauge, measure the distance from the final drive housing joint face to the bearing outer track. Call this dimension A. Similarly measure the protrusion of the spigot on the extension housing above the joint face. Call this dimension B (photos).

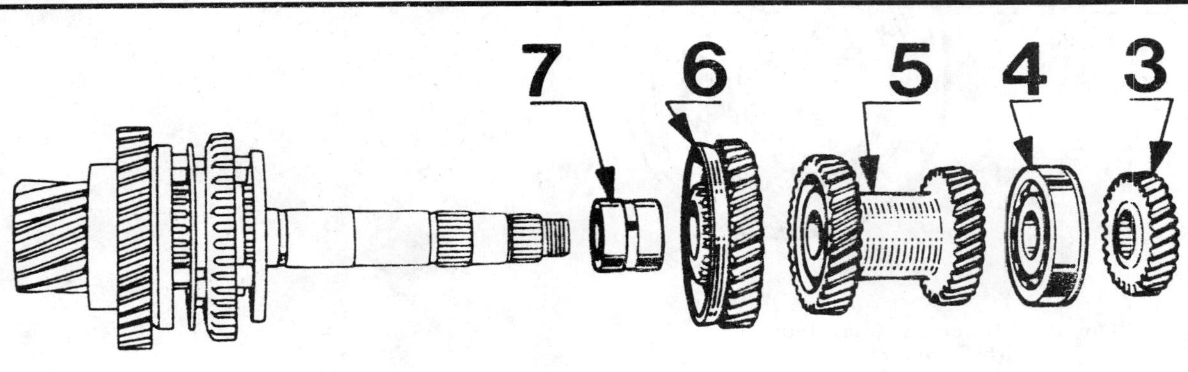

Fig. 13.70 Output shaft components – type BE1/5 gearbox (Sec 7)

3 5th gear
4 Rear bearing
5 3rd/4th gear assembly
6 2nd gear
7 2nd gear bush

7.208 Output shaft rear bearing

7.209A 3rd/4th speed gear assembly on output shaft

7.209B 2nd speed gear on output shaft

7.209C 2nd speed gear bush

7.210A 1st/2nd synchro sleeve

7.210B 1st/2nd synchro-hub

7.211A Output shaft 1st speed gear

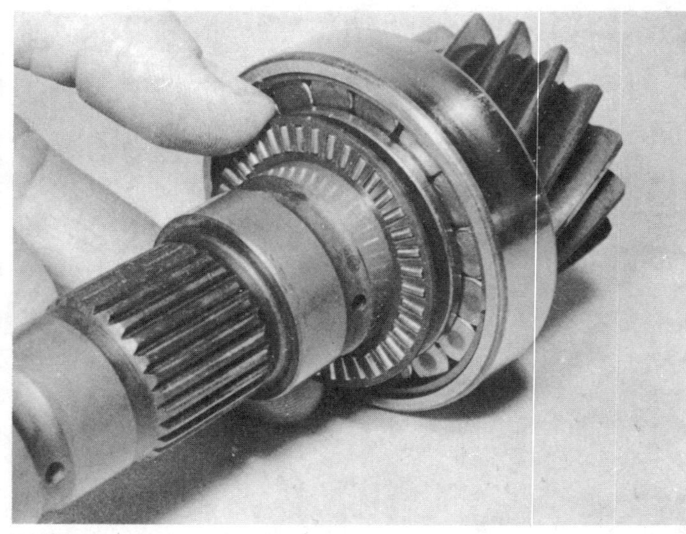
7.211B Needle thrust bearing

7.211C Output shaft bearing circlip (arrowed)

7.223 Selector fork with captive detent ball and spring

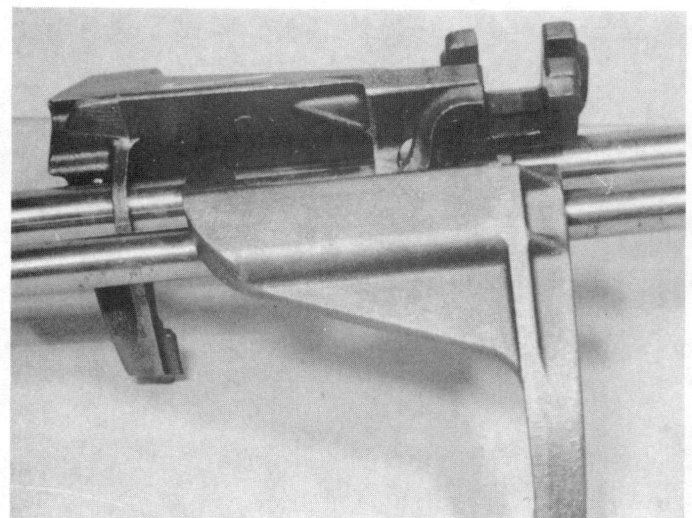
7.229 Selector forks and rods in neutral position

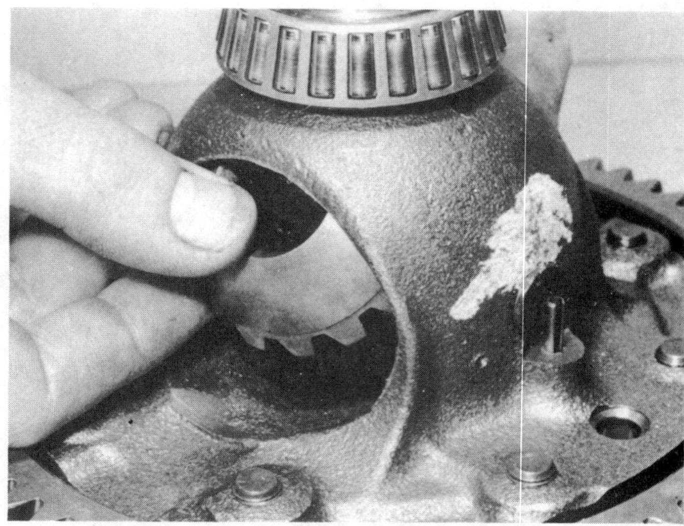
7.231 Removing differential side gear

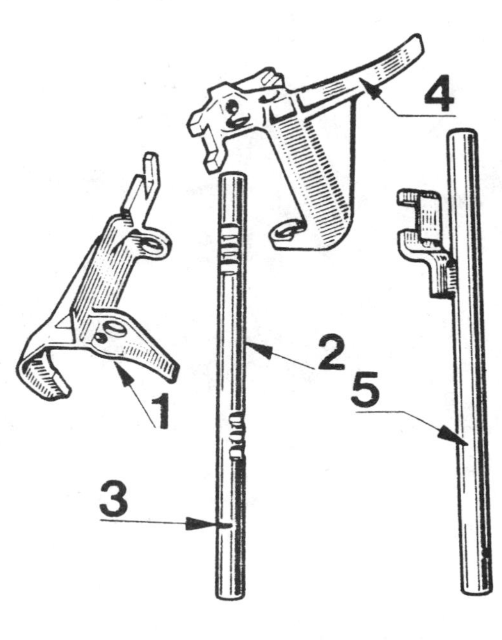

Fig. 13.71 Selector fork and rod identification – type BE1/5 gearbox (Sec 7)

1 3rd/4th fork
2 1st/2nd and 3rd/4th selector rod
3 Locking slot
4 1st/2nd fork
5 5th selector rod

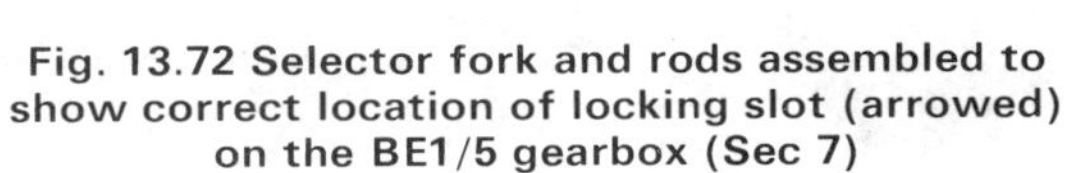

Fig. 13.72 Selector fork and rods assembled to show correct location of locking slot (arrowed) on the BE1/5 gearbox (Sec 7)

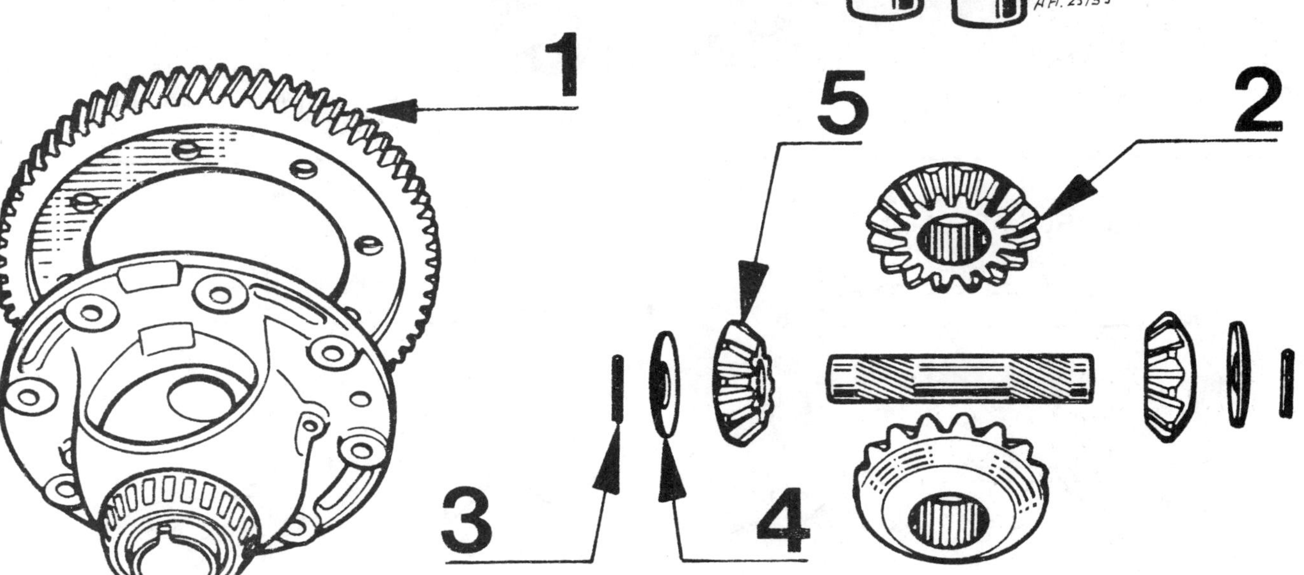

Fig. 13.73 Final drive/differential components – type BE1/5 gearbox (Sec 7)

1 Crownwheel
2 Side gear
3 Roll pin
4 Washer
5 Planet gear

7.233 Differential spindle and gears

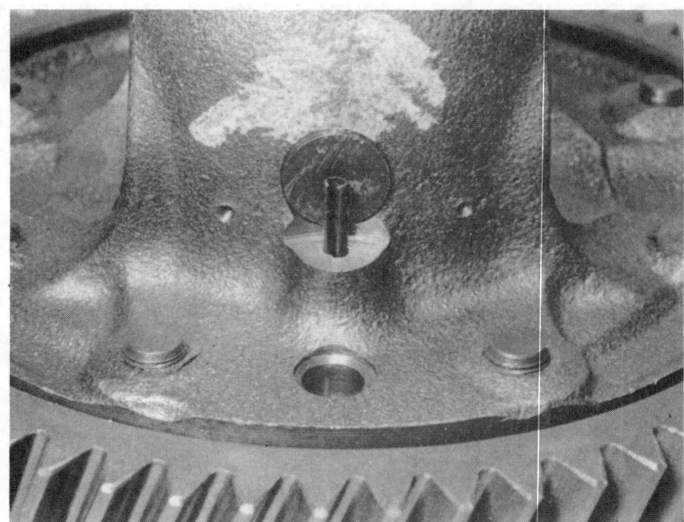

7.237 Differential spindle roll pin correctly fitted

7.238 Differential side gear retaining tool in position

7.241 Differential correctly located. Note housing alignment dowel (arrowed)

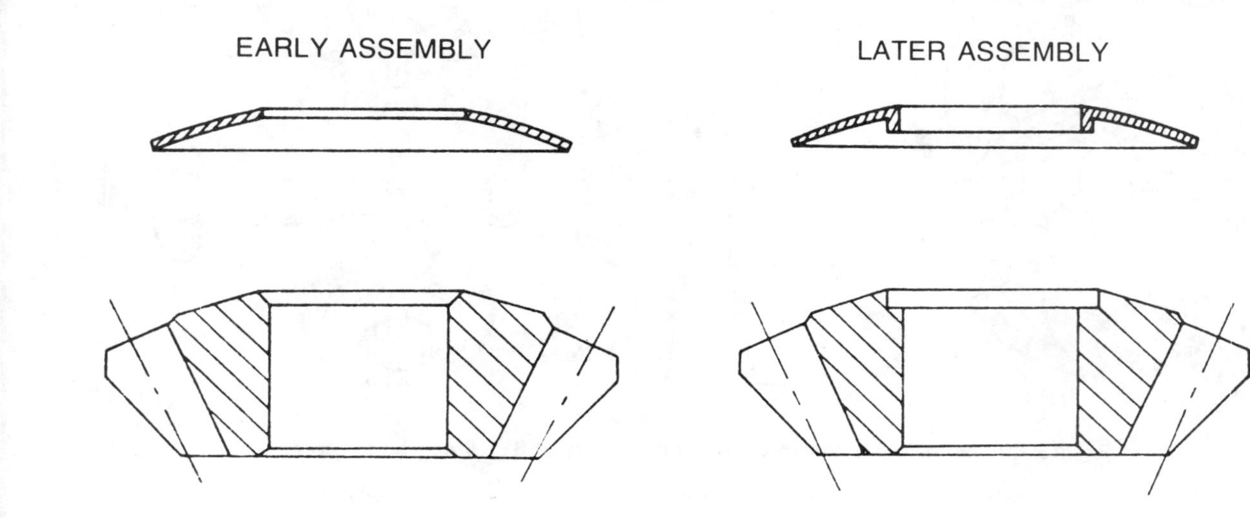

Fig. 13.74 Two types of differential gears and thrust washers on type BE1 gearbox (Sec 7)

7.245A Measuring from the joint face to the bearing outer track

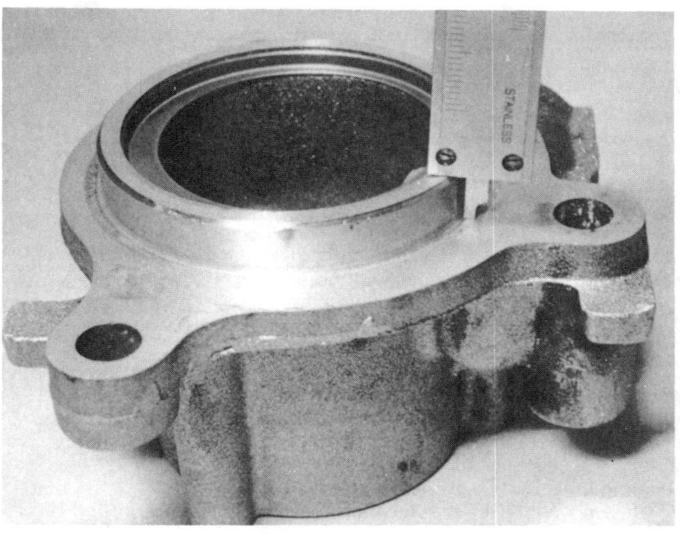

7.245B Measuring spigot protrusion

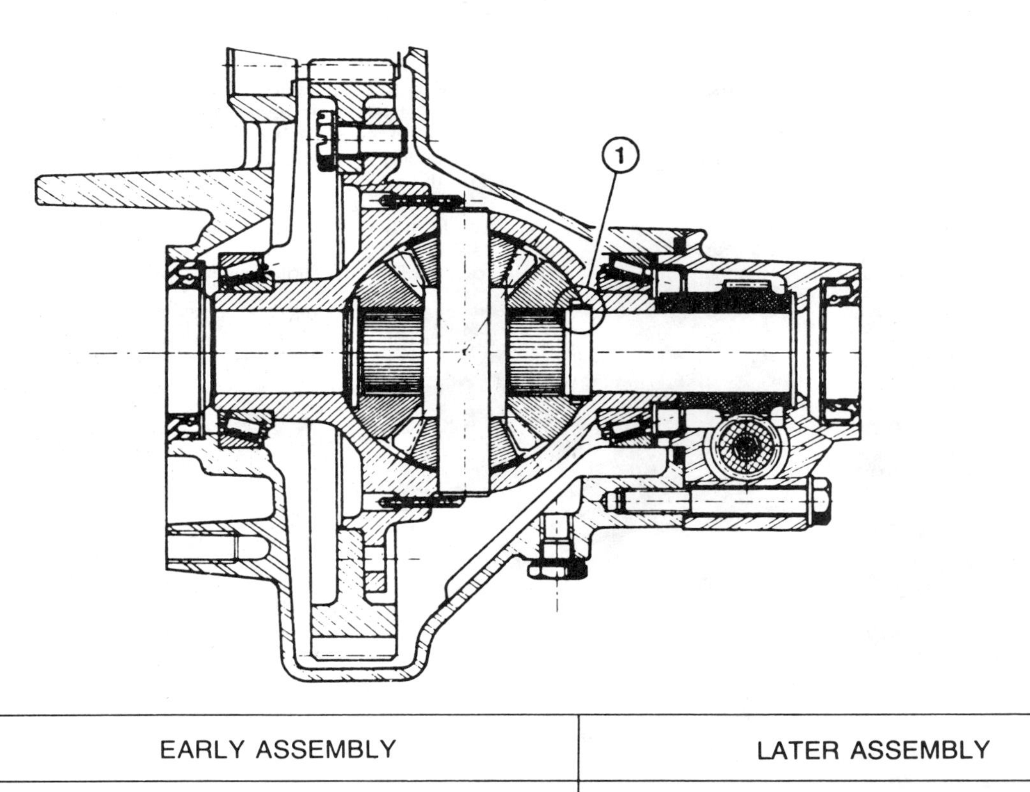

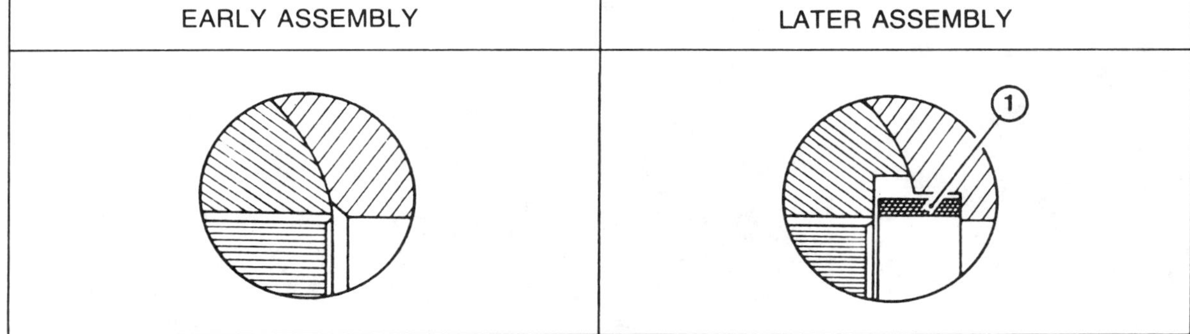

Fig. 13.75 Differential with later type side gear centralising ring (1) on BE1 gearbox (Sec 7)

246 The thickness S of preload shim required is determined by this formula:

S = (A – B) + 0.10 mm

The extra 0.10 mm is the preload factor for the bearings. Shims are available in thicknesses of 1.1 to 2.2 in steps of 0.1 mm.
247 Tighten the final drive half housing securing bolts to the specified torque.
248 Fit the preload shim just determined, the speedometer driving gear and the extension housing with a new O-ring (photo). Tighten the securing bolts to the specified torque.
249 Fit and secure the speedometer pinion and its adaptor.
250 Fit a new oil seal, lips well greased, into the extension housing.
251 Fit a new gear selector shaft oil seal in the main casing.
252 From the clutch housing side, fit the clutch release bearing guide tube. Do not use a gasket under the guide tube flange, and only tighten the bolts finger tight. Invert the casing and fit a preload spacer (any size) and the input shaft bearing outer track (photos).
253 Fit the gear selector shaft spring bracket and tighten its securing bolts to the specified torque.
254 If removed, fit the two locating dowels in the main casing mating face.
255 Fit and tighten the breather.
256 Refit the lubrication jet (photo).
257 Fit the reverse detent spring and plunger. Depress the plunger and fit the reverse selector fork and its spindle. Tighten the spindle securing nut to the specified torque.
258 Fit the reversing lamp switch using a new copper washer. Tighten it to the specified torque (photo).
259 Assemble the geartrains and the selector forks and rods. Offer the whole assembly to the gearcase (photo).
260 Fit the reverse idler spindle and gear, with the chamfer towards the rear of the gearbox. Make sure the pin in the shaft is correctly located.
261 Refit the swarf-collecting magnet.
262 Insert the spring and washers into the bracket (photo).
263 Enter the selector shaft into the casing, passing it through the compressed spring and washers inside the casing. Also engage the shaft with the selector finger and the interlock bracket. It may be helpful to keep the finger and the bracket together with a short length of rod (maximum diameter 14 mm) which can be withdrawn as the selector shaft enters (photo).
264 Make sure that the flat on the shaft and the roll pin hole are correctly orientated (Fig. 13.76). Secure the selector finger and the interlock bracket with two new roll pins. The slots in the roll pins should be 80° away from each other and in line with the longitudinal axis of the shaft (photo).
265 On later models, fit the washer and a new circlip to the cover end of the shaft.
266 On all models, refit the selector shaft cover if it was removed.
267 To the lever end of the selector shaft fit a new O-ring, a washer and a new circlip.
268 Apply jointing compound to the main casing/end casing mating face. Fit the end casing, making sure that the input and output shafts and the selector rod pass through their respective holes. Fit the thirteen securing bolts and tighten them progressively to the specified torque; remember to fit the clutch cable bracket (photo).
269 Fit the reverse idler spindle bolt, using a new washer. Tighten the bolt to the specified torque.
270 Fit the drain plugs, using new washers, and tighten them to the specified torque.
271 Fit the selector rod lockplate. Secure it with its bolt and washer, tightening the bolt to the specified torque.
272 Fit the output shaft bearing circlip, making sure it is properly located in the groove.
273 Fit the output shaft bearing retaining washers and bolts. Tighten the bolts to the specified torque.
274 Fit the spacer (shoulder towards the bearing), 5th gear bush and 5th gear to the input shaft. Also fit the sliding sleeve and hub, but not the selector fork (photos).
275 Lock up the geartrains by engaging 5th gear with the sliding sleeve and any other gear with the selector shaft. Fit the output shaft nut and tighten it to the specified torque, then lock it by staking its skirt into the groove.
276 Remove the 5th gear sliding sleeve and hub, then refit them

7.248 Fitting the preload shim for the differential bearings

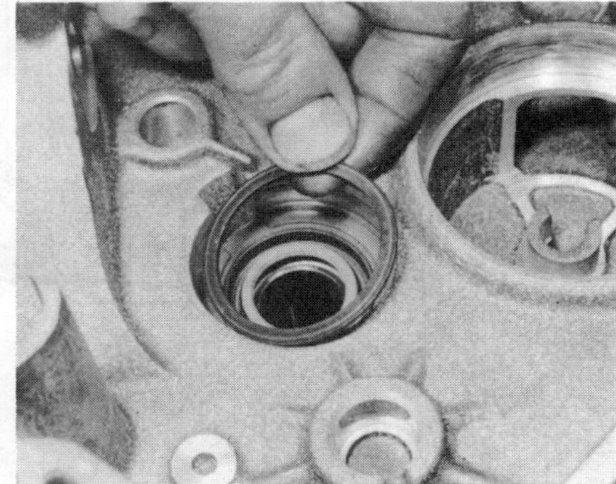
7.252A Input shaft preload spacer

7.252B Input shaft front bearing outer track

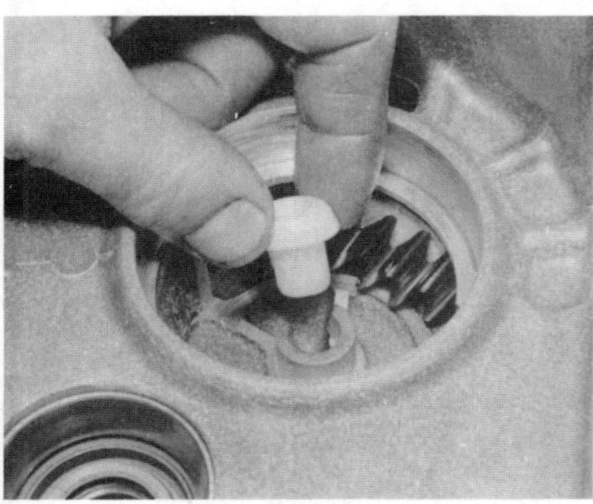
7.256 Fitting the lubrication jet

7.258 Reversing lamp switch

7.259 Fitting geartrains meshed together

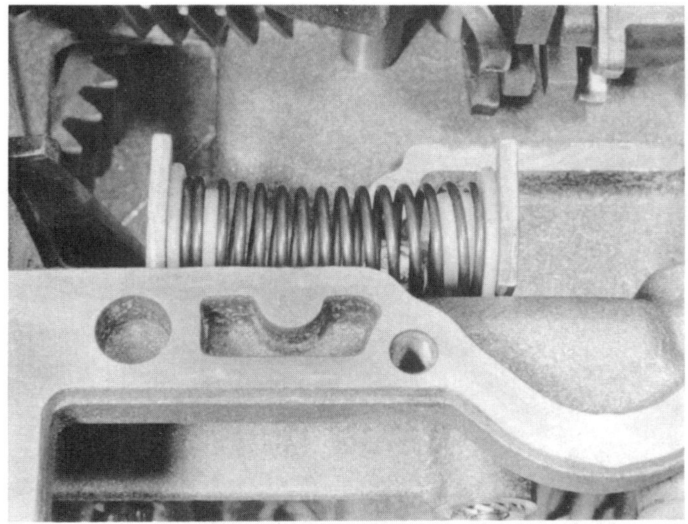

7.262 Selector shaft spring and washers inside bracket

7.263 Fitting the selector shaft

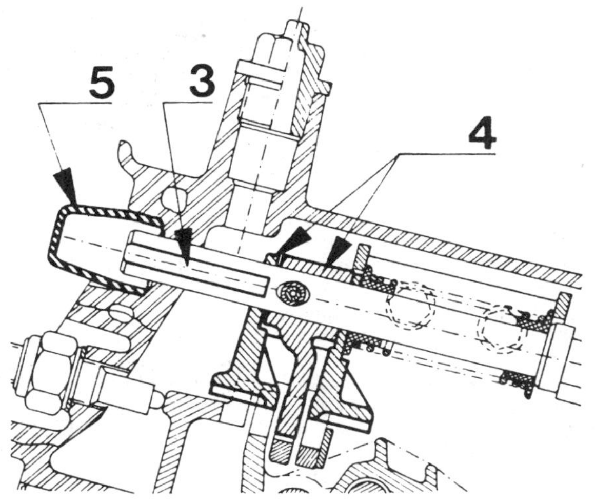

Fig. 13.76 Selector shaft and finger correctly orientated on BE1 gearbox (Sec 7)

3 Flat on selector shaft
4 Selector finger and interlock bracket
5 Selector shaft cover

7.264 Selector finger/interlock bracket roll pins

7.268 Clutch cable bracket

7.274A 5th speed gear spacer

7.274B Input shaft 5th speed gear bush

7.276A 5th speed selector fork detent ball and spring

7.276B Detent ball and spring retaining roll pin

with the selector fork. If the original components are being refitted, observe the mating marks made when dismantling. As the fork is being lowered into position, insert the detent ball into its hole. Alternatively, extract the rollpin and insert the detent ball and spring from the other end (photos).

277 Engage two gears again, then fit the input shaft nut and tighten it to the specified torque. Lock the nut by staking.

278 Secure 5th gear selector fork to its rod with a new roll pin.

279 Coat the mating faces with jointing compound, then refit the rear cover. Use thread locking compound on the securing bolts and tighten them to the specified torque.

280 Turn to the clutch housing and remove the release bearing guide tube. If a new release lever balljoint is to be fitted, do so now: put thread locking compound on its splines and drive it in.

281 Refit the clutch release bearing guide tube with a preload spacer 2.4 mm thick and without a gasket. Insert the retaining bolts and tighten them progressively, at the same time rotating the input shaft. Stop tightening when the shaft starts to drag: the bearings are then correctly seated.

282 Remove the guide tube and shim. Using a depth gauge, accurately measure the distance from the bearing outer track to the joint face on the casing. Call this dimension C. Similarly measure the protrusion of the spigot on the guide tube flange above the joint face. Call this dimension D (photos).

7.282A Measuring from joint face to bearing outer track

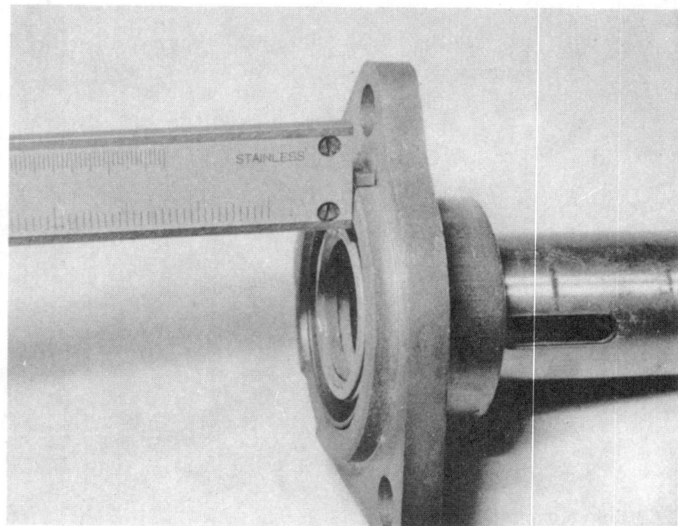

7.282B Measuring spigot protrusion

283 The thickness T of preload shim required is given by the formula:

T = (C – D) + 0.03 mm

The extra 0.03 mm is to provide bearing endfloat, and allows for the thickness of the gasket which will be fitted. Shims are available in thicknesses from 0.7 to 2.4 mm in steps of 0.1 mm.
284 Fit a new oil seal, lips well greased, to the guide tube.
285 Fit the preload shim (of calculated thickness), a new gasket and the guide tube (photo). Secure with the bolts and tighten them to the specified torque.
286 Refit the clutch release fork and release bearing.
287 If not already done, refit the gearchange levers, making sure that they are in the correct position. Also refit the clutch bellcrank (if removed) and the gearchange pivot bracket (photos).
288 Reassembly of the transmission is now complete. Do not refill it with oil until the driveshafts have been engaged.

Reassembly of major units (type BE1/4)
289 The only difference in reassembling the four-speed box, apart from the obvious absence of 5th gear components, lies in the method of locking the geartrains when tightening the input and output shaft nuts.
290 Refer to paragraphs 193 or 194 and use one of the methods described there. Remember to stake the nuts after tightening.

Gearchange mechanism (type BE1) – adjustment
291 Place the gear lever in neutral then lightly move it from side to side noting that there is a small amount of free play before spring resistance is felt.
292 Take up the play in the left-hand direction until the spring resistance can just be felt.
293 Align the end of a ruler with the 1st/2nd gear mark on the gear lever knob as shown in Fig. 13.77.
294 Move the gear lever against spring resistance as far as it will go to the left without moving the ruler and note the amount by which it moves. This should be 1.30 to 1.45 in (33 to 37 mm).
295 If the noted lever travel is incorrect, release the gear lever boot and extract the retaining clip at the base of the lever.
296 Lift up the gear lever eccentric so that it is clear of the splines and turn it through the number of splines necessary to obtain the correct travel dimension. Note that one spline equals approximately 1 mm of gear lever movement.
297 Repeat the checks described above and when correct refit the retaining clip and gear lever boot.

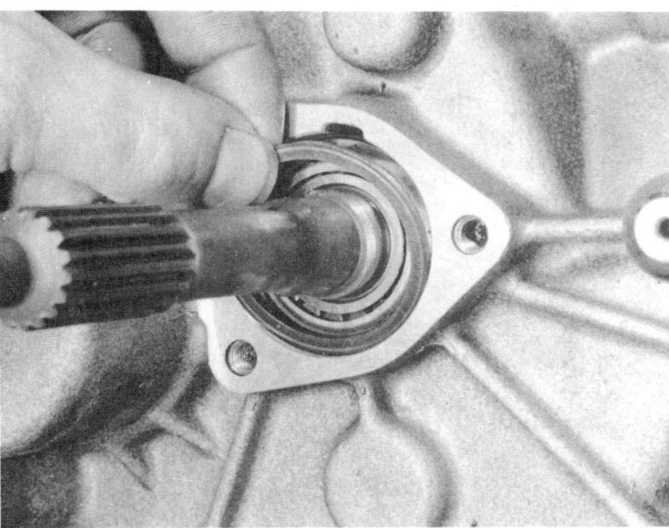
7.285 Input shaft preload shim

7.287A Gearchange lever coil spring

7.287B Refitting clutch bellcrank

7.287C Gearchange rod pivot bracket, note lock washers

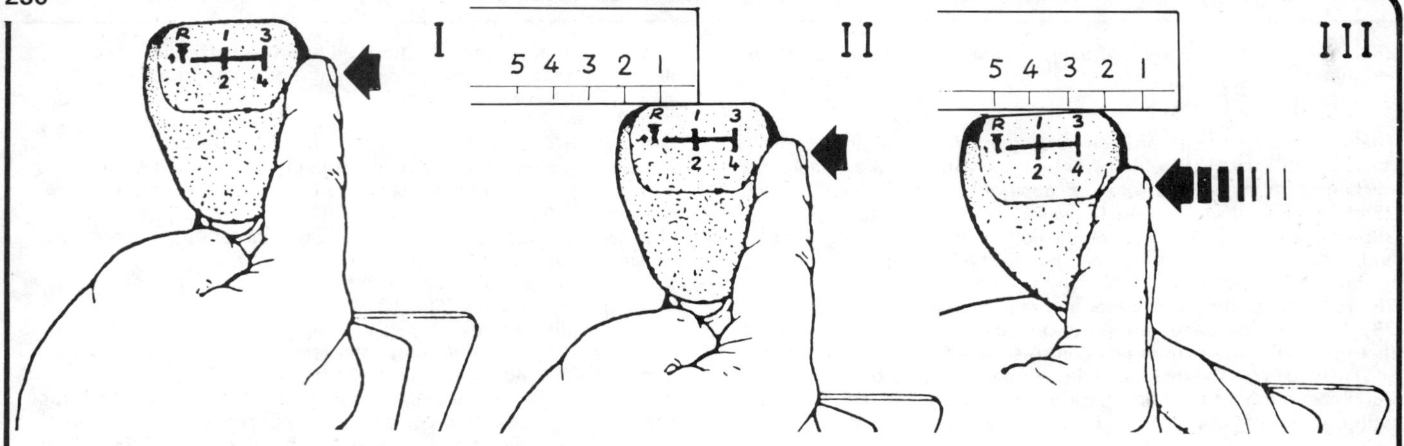

Fig. 13.77 Checking gearlever travel – type BE1 gearbox (Sec 7)

I Move lever to left to take up play
II Align a ruler with the 1st/2nd mark on the knob
III Measure lever travel by moving it to the left against spring resistance

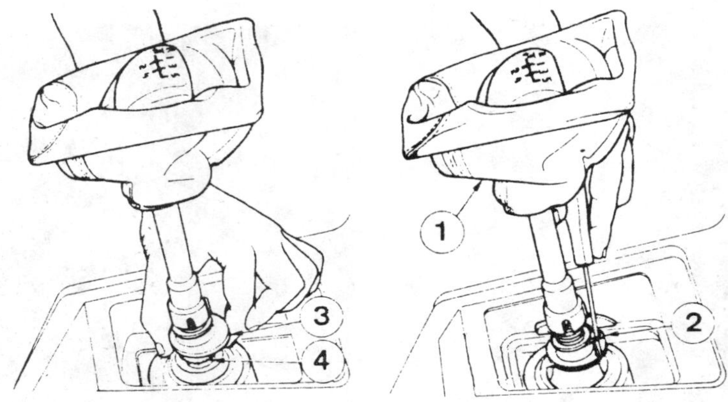

Fig. 13.78 Gear lever travel adjustment items on BE1 gearbox (Sec 7)

1 Gaiter 2 Clip 3 Eccentric 4 Splines

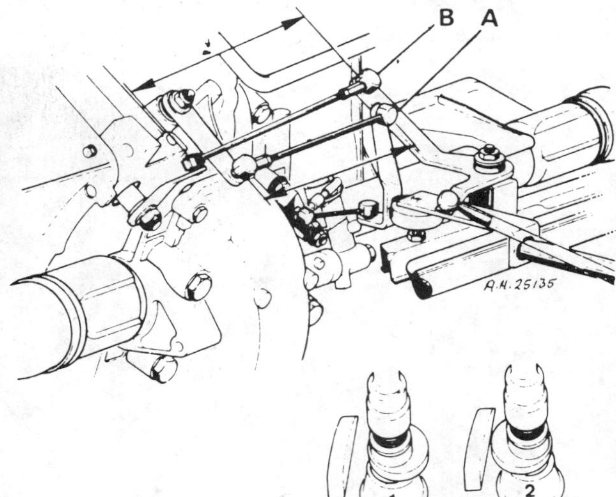

Fig. 13.79 Gearchange mechanism adjustment for type BE1 gearbox (Sec 7)

1 Eccentric in minimum position
2 Eccentric in maximum position
A Selection link
B Engagement link

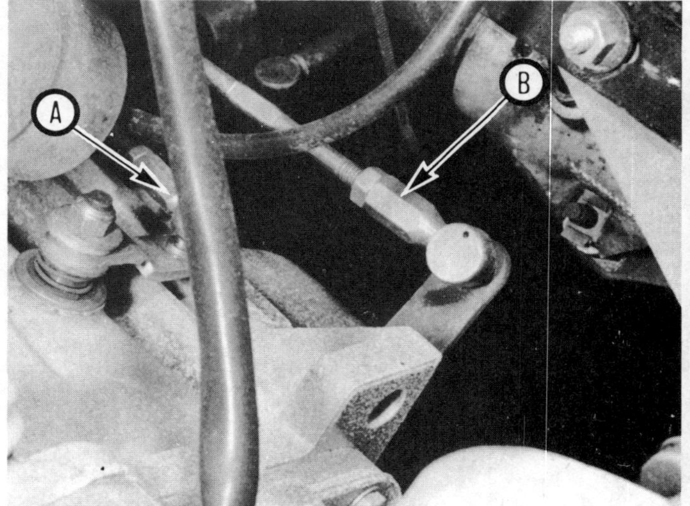

7.298 Gearchange selection link (A) and engagement link (B) on BE1 transmission

298 If the correct movement cannot be obtained even with the eccentric in its maximum or minimum position, alter the length of the selection link (A of Fig. 13.79) as follows (photo):

(1) Lever eccentric in minimum position – lengthen the link by 0.23 in (6 mm)
(2) Lever eccentric in maximum position – shorten the link by 0.23 in (6 mm)

(The initial datum length of the selection link (A) is 5.83 in or 148 mm).
299 Also check that the engagement link (B) of Fig. 13.79 is set to 8.84 in (224.5 mm) between centres.

8 Automatic transmission and final drive

General description

1 The automatic transmission fitted to certain models is the Torqueflite A415. It is fully automatic and incorporates a fluid-filled torque converter and a planetary gear unit.
2 Three forward speeds and one reverse are provided, and a kickdown facility is provided for rapid acceleration when a change to a lower speed range is required.
3 Due to the complexity of the automatic transmission there are only a few jobs which can be carried out by the home mechanic. If a fault develops in the unit which is not covered by the following procedure, it is imperative that the car is taken to a main agent who will have the necessary equipment to test the transmission while in position in the car.

Routine maintenance

4 Every 10 000 miles (16 000 km) or 10 months, whichever occurs first, the fluid level should be checked. To do this, run the engine until normal operating temperature has been reached.
5 With the engine idling, move the selector lever through each position, pausing five seconds at each detent. Return the selector lever to the 'N' position and leave the engine idling.
6 Wipe the dipstick tube which is located on the left-hand side of the engine compartment.
7 Withdraw the dipstick, wipe it with a lint-free rag, reinsert it and withdraw it once more.
8 Note the fluid level on the dipstick which should be just above the 'MIN' mark on the dipstick. If the car has just been driven for 10 miles (16 km) or more the level should be near, but not above the 'MAX' mark.
9 If topping up is necessary, add the specified type of automatic transmission fluid, through the dipstick tube, until the level is correct.
10 On later models a dipstick with different markings is fitted (Fig. 13.81). With this type of dipstick the checking procedure is the same except that the level is checked with the engine switched off. If the car has not been driven, the level should be in the 'FROID' zone, either side of the reference mark. If the car has been driven for 10 miles (16 km) or more the level should be in the 'CHAUD' zone, either side of the reference mark.
11 It is also necessary to check the fluid level in the final drive housing at the same service interval. The final drive assembly uses the same specification of transmission fluid as the main unit, but occupies a separate housing and is checked and filled by means of a filler plug on the side of the unit (the upper plug if two are fitted). The fluid level must be maintaned up to the filler plug orifice.

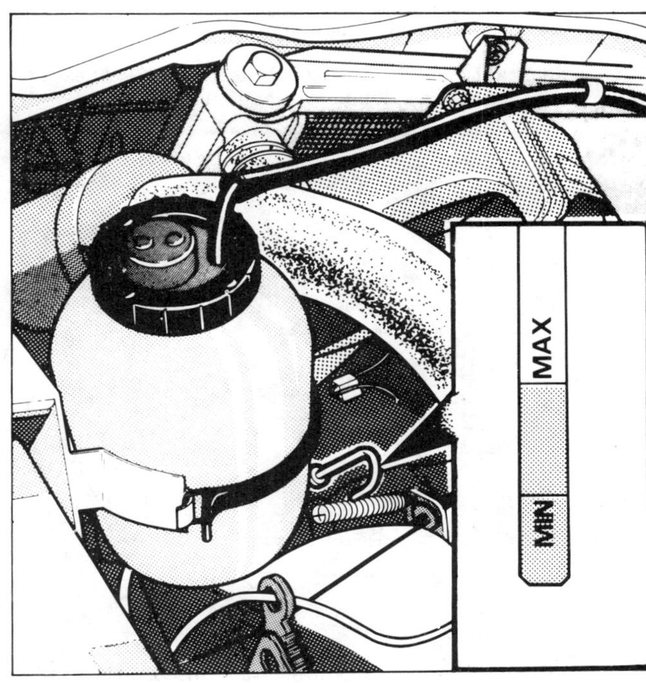

Fig. 13.80 Automatic transmission fluid dipstick location and early type dipstick markings (Sec 8)

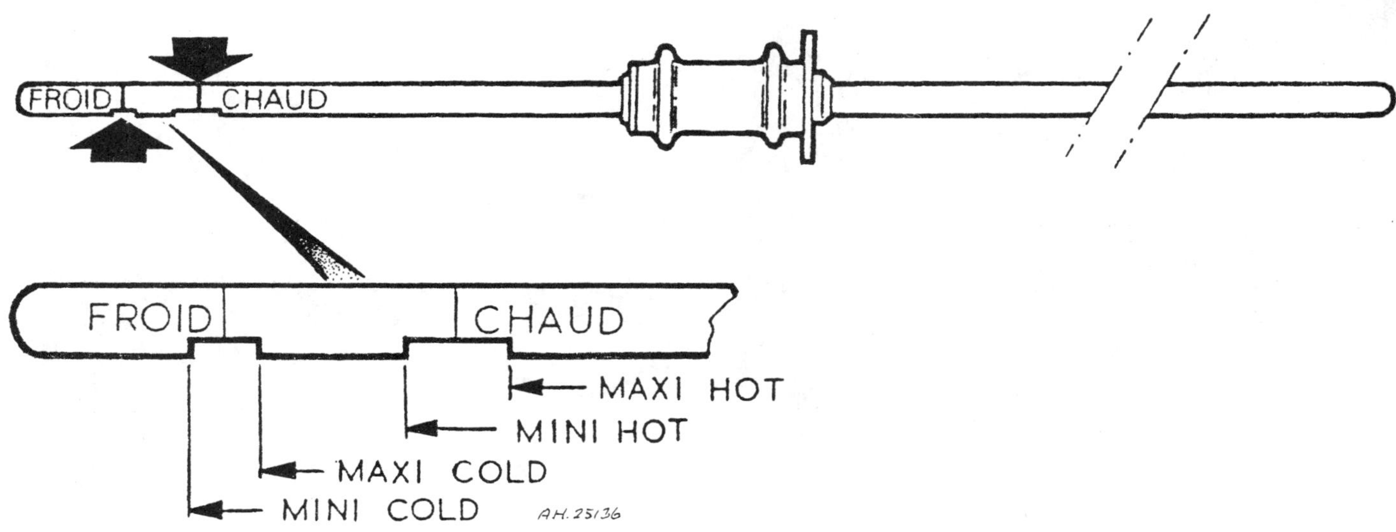

Fig. 13.81 Later type automatic transmission fluid dipstick (Sec 8)

12 If the car is used mainly for short journeys, trailer towing, or more than 50% operation in heavy traffic during hot weather, the fluid in the transmission should be renewed at 30 000 mile (48 000 km) intervals. To do this first jack up the front of the car and support it on axle stands.
13 Clean the transmission sump and place a suitable container beneath it. Loosen but do not remove the sump bolts, then tap the sump to break the seal. When the fluid has drained, move the bolts and withdraw the sump. Take care if the vehicle has just been run, the transmission fluid may be very hot.
14 Remove the screws and withdraw the filter.
15 Detach the magnet from the valve block, clean it, and refit it. If the filter is saturated with sediment renew it, otherwise thoroughly clean it with clean transmission fluid. Clean and wipe dry the sump and transmission housing.
16 Refit the filter and sump, using sealing compound to seal the sump flange.
17 Lower the car to the ground and pour 5.6 pints (3.2 litres) of the recommended transmission fluid through the filler tube.
18 Allow the engine to idle for two or three minutes, then check and top up the fluid level as described earlier.
19 After renewing the fluid, the front (kickdown) band should be adjusted as described in paragraphs 21 to 23.

Automatic transmission – removal and refitting

20 The automatic transmission can be removed leaving the engine in the car, but as the method requires the use of a transmission stand, it is recommended that the engine and transmission are removed together as described in Section 3.

Automatic transmission – front band adjustment

21 The front (kickdown) band adjuster is located on the upper front of the transmission. First fully loosen the locknut and check that the screw turns freely in the casing.
22 Using a torque wrench, tighten the screw to 6 lbf ft (0.8 kgf m), then back it off 2.5 turns.
23 Tighten the locknut to the specified torque.

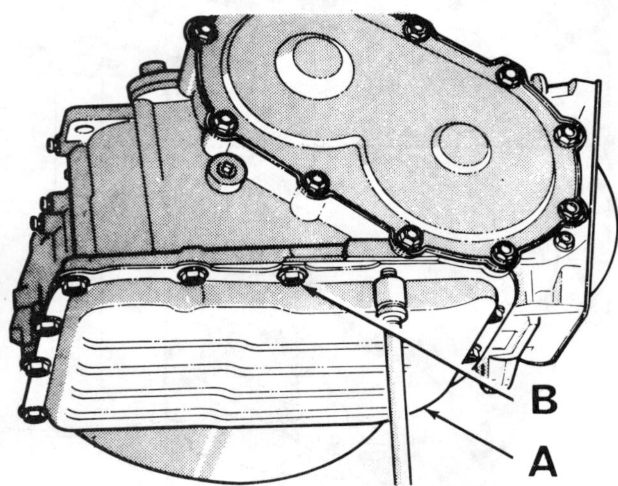

Fig. 13.82 Removing automatic transmission sump (A) and retaining bolts (B) (Sec 8)

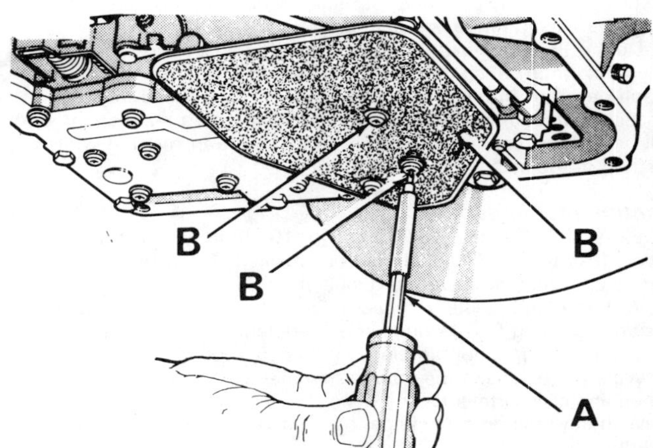

Fig. 13.83 Removing automatic transmission filter (Sec 8)

A Screwdriver *B Screws*

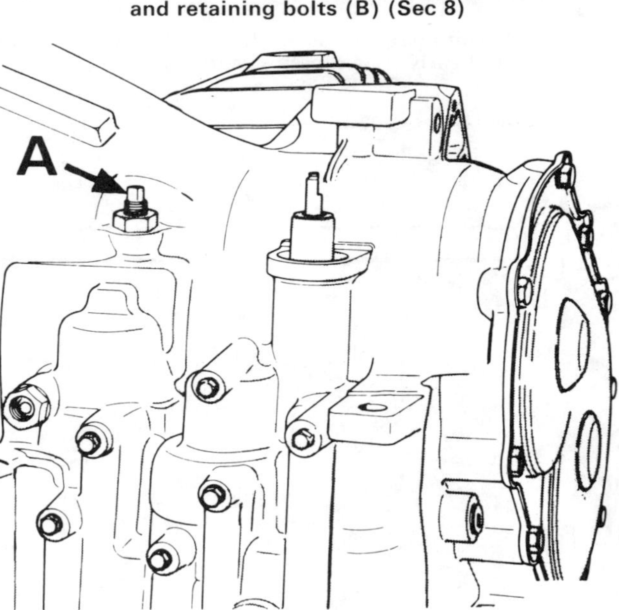

Fig. 13.84 Automatic transmission front band adjuster (A) (Sec 8)

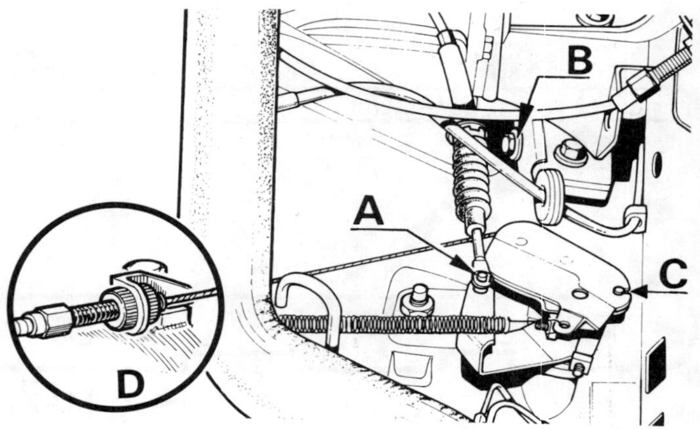

Fig. 13.85 Automatic transmission selector and throttle valve cables (Sec 8)

A Selector cable clevis pin
B Selector cable support bolt
C Kickdown cable nipple
D Kickdown cable adjuster

Automatic transmission – selector cable adjustment

24 Loosen the selector cable support bracket bolt and make sure that the bracket is free to move within the elongated hole.
25 Move the selector lever to position P. Working in the engine compartment, move the selector lever on the transmission fully forward to the P position.
26 Tighten the support bracket bolt to the specified torque.
27 Check that the engine will only start in positions N and P, and that the reversing lights only operate in position R.

Throttle valve (downshift) cable – adjustment

28 Loosen the locknuts at the cable support bracket on the transmission.
29 Remove the cambox cover plate (see Fig. 13.18).
30 With the throttle valve lever held against its internal stop and the stops of the throttle valve and accelerator pedal cams in contact, adjust the cable until the free play has just been eliminated.
31 Tighten the locknuts and refit the cambox cover plate.

Final drive (automatic transmission) – description

32 The automatic transmission housing and final drive housing are integral, although a double-lipped oil seal separates the two compartments because the lubricant levels are different.
33 The automatic transmission output shaft is connected by helical gears to the transfer shaft which is integral with the final drive pinion.

Differential unit (automatic transmission) – removal and refitting

34 Disconnect the battery terminals leads.
35 Remove the front shock absorbers and fit the retaining rods (see Chapter 1, Section 4).
36 Loosen both front driveshaft nuts and the left-hand wheel bolts.
37 Remove the fuel tank filler cap, then disconnect the inlet pipe at the fuel pump. Disconnect the fuel sensor (if fitted), and pull the pipe and grommet from the transmission lifting eye.
38 With a hoist connected to the lifting eye, take the weight of the transmission. Unbolt the left-hand engine mounting.
39 Jack up the front of the car and support it on axle stands.
40 Remove the speedometer adaptor from the right-hand side of the extension housing.
41 Remove the drain plug and drain the final drive oil into a container.
42 Unscrew the nut and use a separator tool to disconnect the lower suspension balljoint on the right-hand side.
43 Unscrew the hub nut, then pivot the roadwheel and hub outward until it clears the driveshaft. Remove the right-hand driveshaft, then loosely reassemble the lower balljoint.
44 Remove the left-hand wheel, and withdraw the left-hand driveshaft using the same procedure.
45 Disconnect the inhibitor switch leads.
46 Detach the exhaust pipe from the manifold and remove the rear engine mounting.
47 Unscrew the bolts from the left-hand side differential bearing retainer. Fabricate a tool similar to that shown in Fig. 13.89 and use it to rotate the retainer and break the seal.
48 Withdraw the retainer (see Fig. 13.90). Unbolt the rear cover from the final drive housing.
49 Support the differential unit and unscrew the bolts from the right-hand extension housing. Turn the housing to break the seal, then withdraw it.
50 Lower the left-hand side of the engine, and withdraw the differential unit from under the suspension crossmember.
51 Refitting is a reversal of removal, but the following additional points should be noted:

(a) *The crownwheel and pinion are matched and must only be renewed together*
(b) *Apply a bead of silicone sealant to the mating surfaces of the bearing retainer, extension housing, and final drive casing before reassembling. On early models a joint was fitted to the final drive cover; this should be discarded and silicone sealant used instead*
(c) *Where new components have been fitted the bearing preload must be adjusted as described in paragraphs 52 to 57.*
(d) *Refill the final drive with the specified quantity of lubricant*

Differential unit (automatic transmission) – bearing preload adjustment

52 The differential unit bearings must be finally assembled with a preload of 0.006 to 0.011 in (0.15 to 0.29 mm). To achieve this, first remove the outer track from the left-hand bearing retainer using an extractor.
53 Remove the shim and measure its thickness, then select a shim 0.019 in (0.48 mm) thinner and fit this to the retainer followed by the outer track.
54 Assemble the final drive using a silicone sealant on the right-hand extension housing; leave the left-hand housing dry.
55 Tighten the bolts to the specified torque, checking at the same time that the differential rotates freely. If it seizes, a thinner shim should be selected.
56 Using a dial gauge, record the total side-to-side movement of the crownwheel. Add the desired preload to this dimension plus the thickness of the shim fitted. Subtract 0.002 in (0.05 mm) for the sealant, and the resultant is the thickness of the shim required. Example:

Shim fitted	*0.061 in*
Desired preload	*0.009 in*
Dial gauge reading	*0.006 in*
Total	*0.076 in*
Minus sealant	*0.002 in*
Shim required	*0.074 in*

57 Remove the existing shim and fit the correct shim, then assemble the bearing retainer and cover using a silicone sealant.

9 Front suspension, driveshafts and hubs

Front hub bearings – description

1 As from September 1977, the front hub bearings incorporate integral inner and outer oil seals instead of the separate seals used earlier. The bearings are also retained in position by circlips instead of the threaded sleeve used previously.
2 The modification affects the stub axles, hubs, driveshafts and brake discs. These components are not interchangeable with the earlier type, except for the modified brake disc which may be fitted to earlier models.

Driveshafts – description

3 The modified driveshaft mentioned in paragraph 2 can be identified by the length of the outer splined end which is 3.327 to 3.406 in (84.5 to 86.5 mm) to the locating shoulder, compared with 3.189 to 3.268 in (81.0 to 83.0 mm) on the earlier type.
4 Further modification has been made to the driveshaft as follows. In April 1978 a rubber sealing washer was fitted to protect the hub bearing from dirt (see Fig. 13.92). Make sure when fitting this washer that it is pushed fully onto the driveshaft. In June 1978 the washer was replaced by a steel deflector and two springs were fitted in the inner constant velocity joint.
5 On automatic transmission models, stops are fitted to the right-hand inner CV joint. When dismantling the joint, the stops must be bent with a screwdriver (see Fig. 13.93) in order to release the rollers. Tap the stops back during reassembly.
6 Commencing February 1985 on Minx and Rapier models, a solid driveshaft is fitted to the right-hand side instead of the previously used tubular shaft. Later production models are fitted with solid driveshafts on both sides of the vehicle.
7 When renewing a driveshaft; a tubular type may be replaced with a solid shaft as they are interchangeable, and one of either type may be fitted on opposite sides of the same vehicle.

Front hub bearings (from September 1977) – removal and refitting

8 The procedure is identical to that described in Chapter 7, but once removed, the bearings cannot be refitted because it is impossible to relocate the oil seals. New bearings must therefore be fitted.

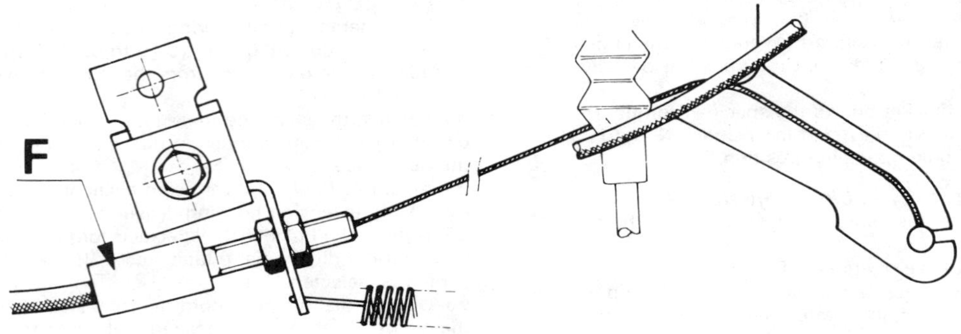

Fig. 13.86 Automatic transmission throttle valve cable locknuts and adjuster sleeve (F) (Sec 8)

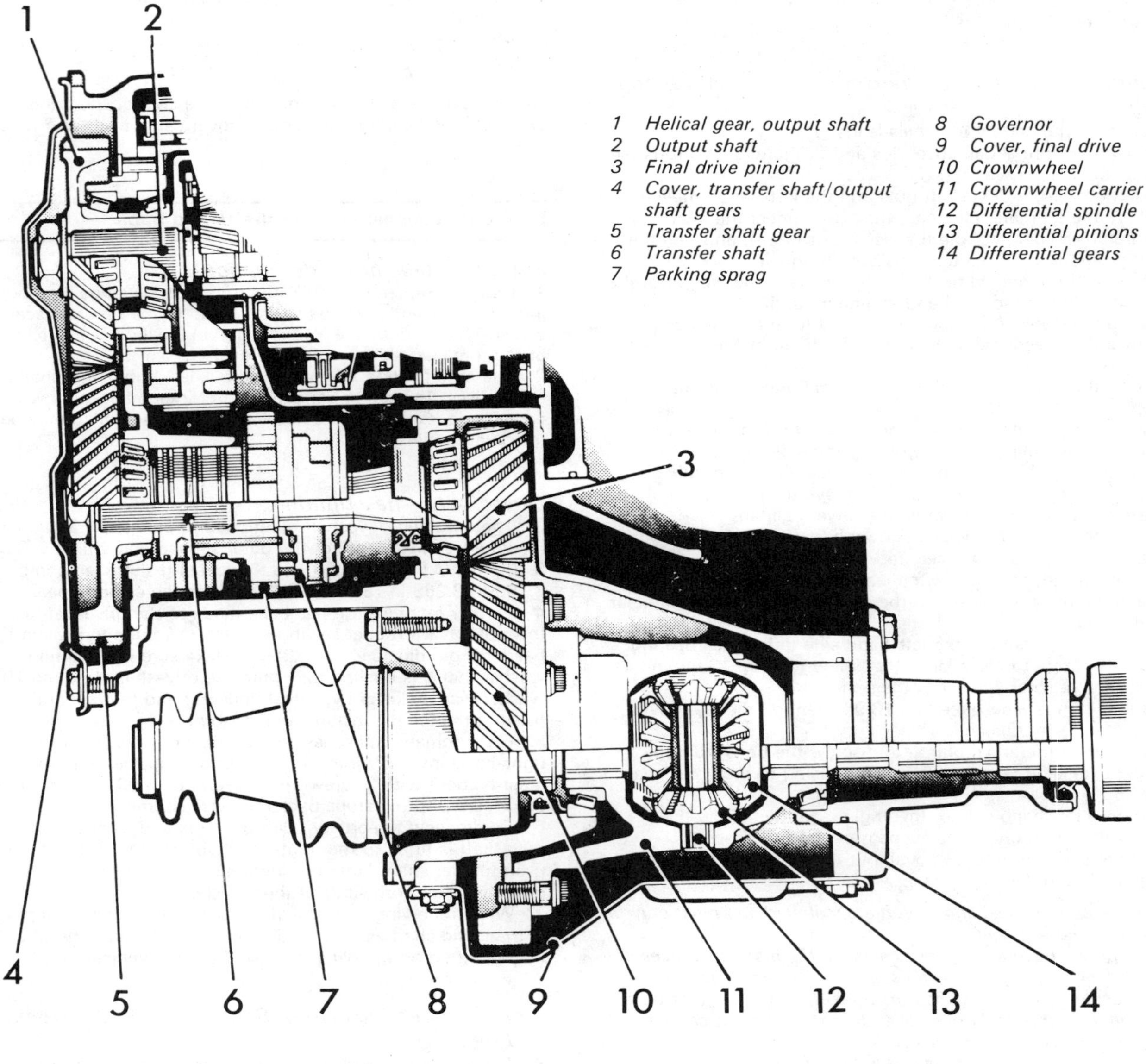

1 Helical gear, output shaft
2 Output shaft
3 Final drive pinion
4 Cover, transfer shaft/output shaft gears
5 Transfer shaft gear
6 Transfer shaft
7 Parking sprag
8 Governor
9 Cover, final drive
10 Crownwheel
11 Crownwheel carrier
12 Differential spindle
13 Differential pinions
14 Differential gears

Fig. 13.87 Cutaway view of automatic transmission final drive (Sec 8)

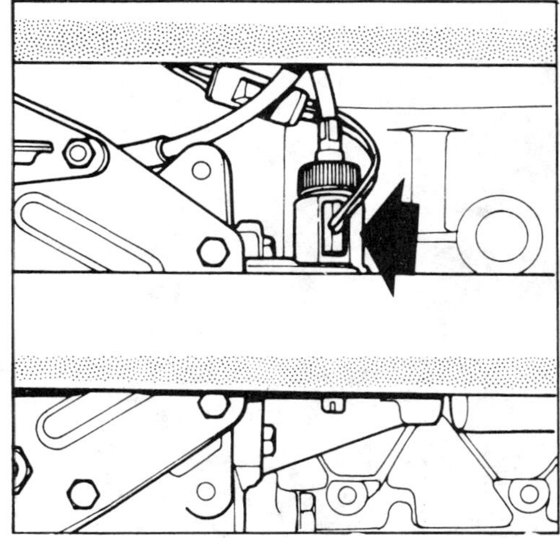

Fig. 13.88 Speedometer drive adaptor (arrowed) on automatic transmission (Sec 8)

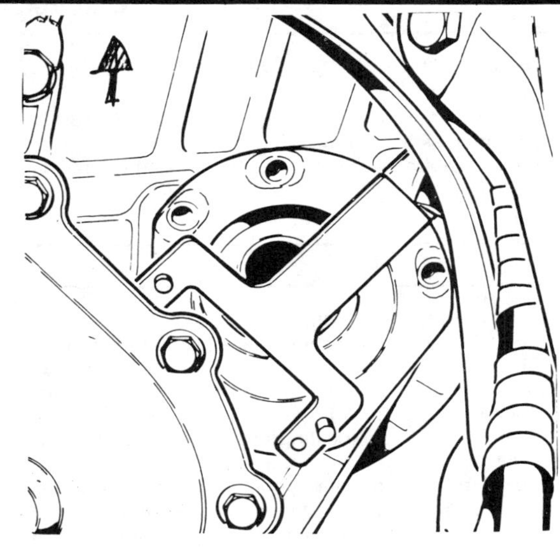

Fig. 13.89 Tool for rotating the automatic transmission differential bearing retainer (Sec 8)

Fig. 13.90 Removing the differential bearing retainer from the left-hand side of the automatic transmission (Sec 8)

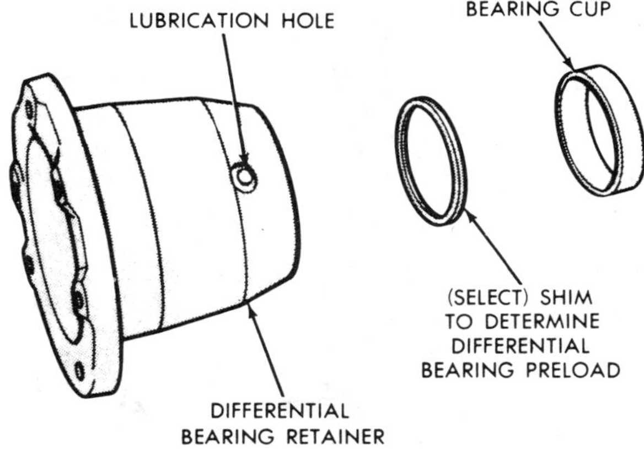

Fig. 13.91 Left-hand differential bearing retainer components on automatic transmission (Sec 8)

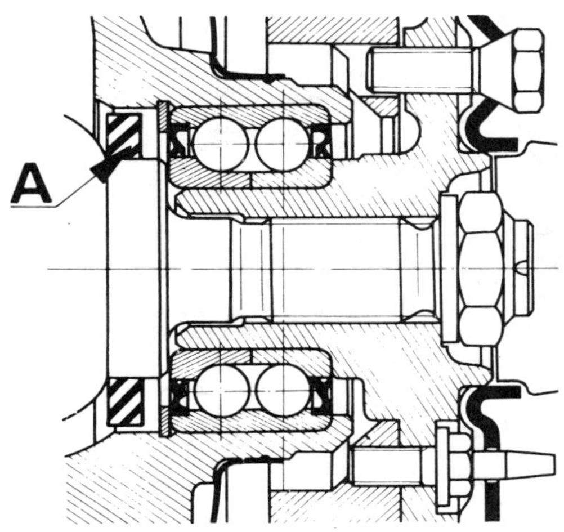

Fig. 13.92 Sectional view of later type front hub bearings with integral oil seals (Sec 9)

A Sealing washer

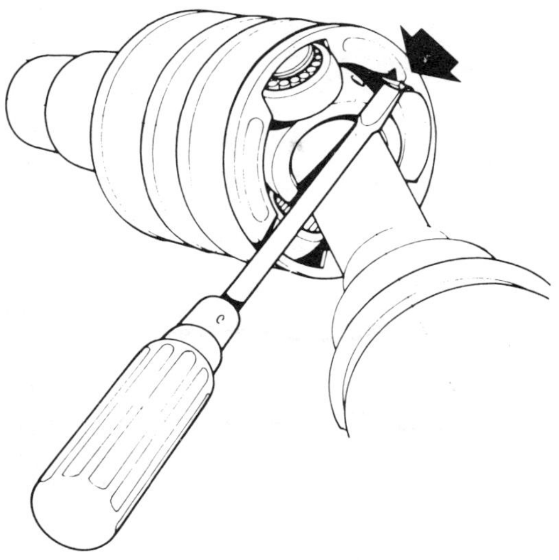

Fig. 13.93 Bending up CV joint roller stops (arrowed) – automatic transmission (Sec 9)

Front suspension (commencing May 1980) – modifications

9 As from the introduction of the Solara, and from May 1980 for the Alpine, the front crossmember tube diameter has been increased. At the same time the subframe rear mounting has been modified (photos).

10 Steering

Steering gear (manual) – description

1 As from May 1978 two modifications have been made to the steering gear. The damper lockwasher is no longer fitted and there is no groove in the adjusting plug. Also the O-ring on the pinion has been replaced by an oil seal. Both types of pinion are easily identified; the O-ring type has a locating groove, whereas the oil seal type has a flat bearing surface for the seal.

2 The tie-rod inner balljoint on later models is locked to the rack by peening instead of by the locknut fitted to earlier models. Because of this, tie-rods which are removed should not be re-used.

3 After tightening the modified balljoint, use a round-nosed punch to peen the metal flange onto the flats of the rack.

9.9A Solara front subframe rear mounting

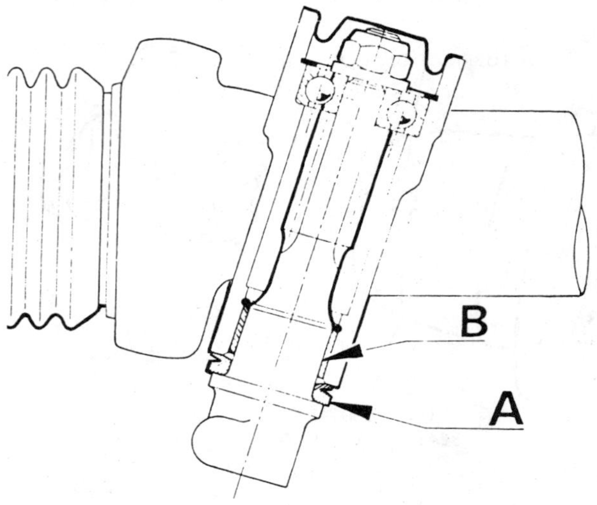

Fig. 13.94 Sectional view of later type manual steering gear (Sec 10)

A Oil seal *B White metal bush*

9.9B Solara front subframe-to-crossmember tube bolts

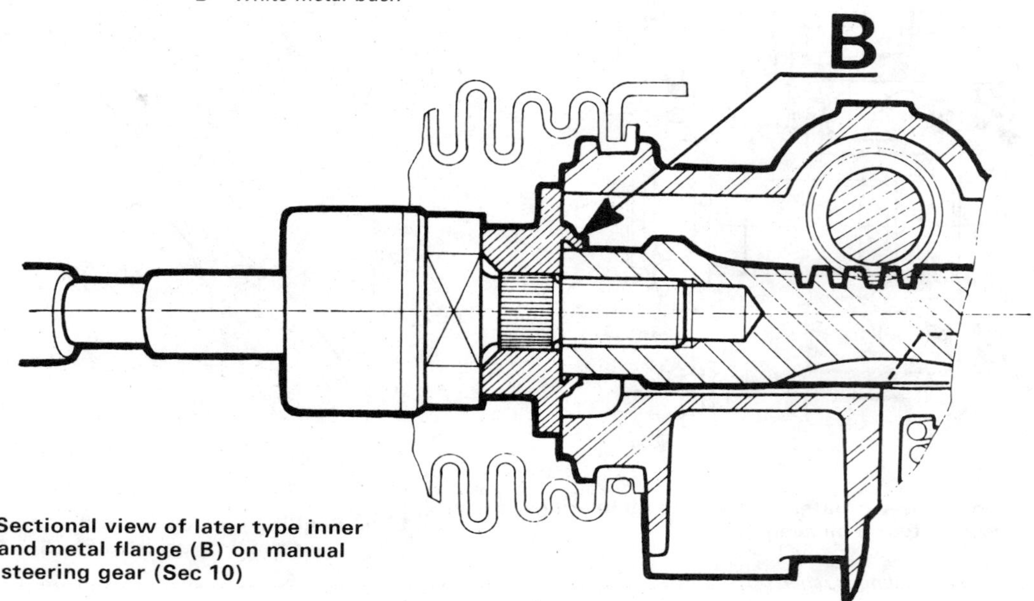

Fig. 13.95 Sectional view of later type inner balljoint and metal flange (B) on manual steering gear (Sec 10)

Steering gear (power-assisted) – description

4 Power-assisted steering became available on Alpine models in 1980, and is also fitted to most Solara, and all Minx and Rapier models.

5 The rack and pinion steering gear is very similar in design to the manual type but relies on power assistance provided by a hydraulic pump driven by a belt from the crankshaft pulley.

Steering (power-assisted) – maintenance

6 Check the fluid level in the pump reservoir regularly after a run when the fluid is hot.

7 With the vehicle standing on a level surface, clean around the filler cap and remove it. Wipe the attached dipstick clean, insert it and then withdraw it for the second time. The fluid should be up to the hot mark. Top up if necessary with the specified type of fluid.

8 Check the tension and condition of the drivebelt regularly. Tension the drivebelt as described in the next sub-section.

Pump drivebelt (power-assisted steering) – adjustment

9 With the pump carrier mounting bolts loose, check that the drivebelt is fully engaged with the pump and crankshaft pulleys.

10 Insert a lever through the pulley guard, and raise the pump until there is 0.5 in (13 mm) of free movement halfway along the run of the belt.

11 Tighten the bolts and recheck the tension.

12 Run the engine for two or three minutes, then switch off and recheck the tension.

13 Never pull on the reservoir to adjust the position of the pump. A new drivebelt should be retensioned after 600 miles (1000 km).

Steering gear (power-assisted) – removal and refitting

14 Working inside the car, set the roadwheels in the straight-ahead position, and remove the pinch-bolt which secures the lower steering column to the pinion shaft.

15 Disconnect the battery negative terminal.

16 Detach the automatic selector lever cable from the transmission and bracket.

17 Unclip the pump supply and return hoses. Jack up the front of the car and support it on axle stands.

18 Disconnect the front exhaust pipe from the manifold.

19 Place a suitable container under the steering gear, then disconnect the return pipe, followed by the feed pipe, from the steering gear. Allow the fluid to drain.

20 Release the selector cable. Disconnect the tie-rod balljoints from the steering arms using a separator tool.

21 Unscrew the mounting bolts and remove the shim packs, noting their location and quantity.

22 Detach the crossmember and rubber block.

23 Unbolt the engine mounting bracket and push it to the left-hand side as far as possible.

24 Raise the steering gear into the left-hand wing valance aperture, then tilt it down, passing the right-hand track rod between the floor and anti-roll bar.

25 Withdraw the steering gear from the car.

26 Refitting is a reversal of removal, but the following additional points should be noted:

(a) *Due to the restricted access, the fluid pipe unions must be entered before the steering gear is finally positioned, and fully tightened later*

(b) *Apply a locking liquid to the engine mounting bolts before inserting them*

(c) *Install a new exhaust flange gasket*

(d) *Fill the pump reservoir with fluid, then refit the cap and turn the steering wheel from lock to lock several times. Top up the reservoir until it is half full, then start the engine and again turn the steering wheel from lock to lock several times. When all the air has been purged from the circuit, run the car until the fluid is hot, and top up the level to the FULL mark on the dipstick*

(e) *Check the pipes and unions for leakage*

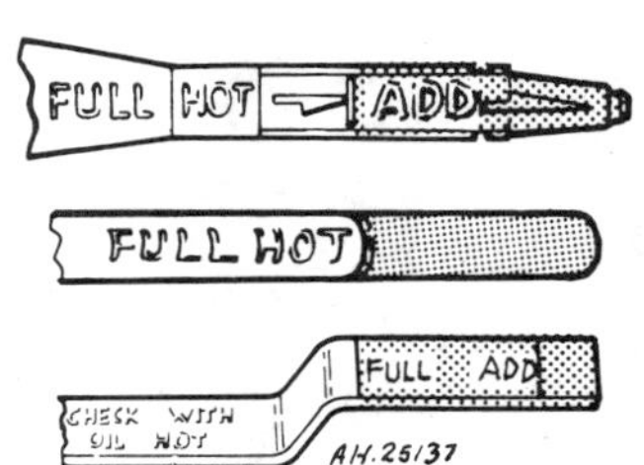

Fig. 13.96 Typical power steering fluid dipsticks (Sec 10)

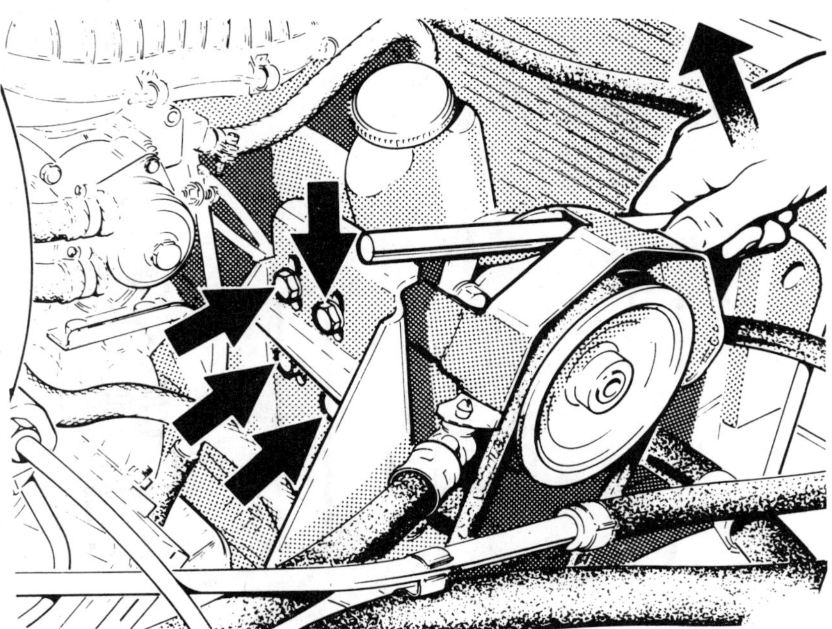

Fig. 13.97 Adjusting power steering pump drivebelt tension. Direction for tensioning and carrier bolts arrowed (Sec 10)

Hydraulic pump (power-assisted steering) – removal and refitting

27 Disconnect the battery negative terminal.
28 Loosen the mounting bolts, lower the pump, and remove the drivebelt.
29 Cover the distributor, alternator, and the alternator drivebelt with polythene sheeting.
30 Loosen the feed pipe union. Unscrew the nut and withdraw the petrol filter from the pump carrier.
31 Remove the nuts securing the pump to the carrier.
32 Loosen the hose clip at the base of the reservoir.
33 Remove the pump from the carrier, noting the location of any spacers. Remove the filler cap and empty the fluid into a suitable receptacle.
34 Disconnect the feed and return pipes and withdraw the pump. Drain any remaining fluid and plug all pipes and apertures.
35 Unbolt the belt guard from the pump.
36 Refitting is a reversal of removal, but tension the drivebelt as described in paragraphs 9 to 13, and refill the system with fluid as described in paragraph 26.

Power steering pipes – modifications

37 A modified end flare on the power steering high pressure pipes (C in Fig. 13.100) has been introduced on models from approximately June 1983 to overcome leakage problems associated with the earlier single end flare pipes (A).
38 If leakage if experienced with the earlier type of pipe a special insert is available (B), which should be fitted into the pump adaptor before fitting the pipe.

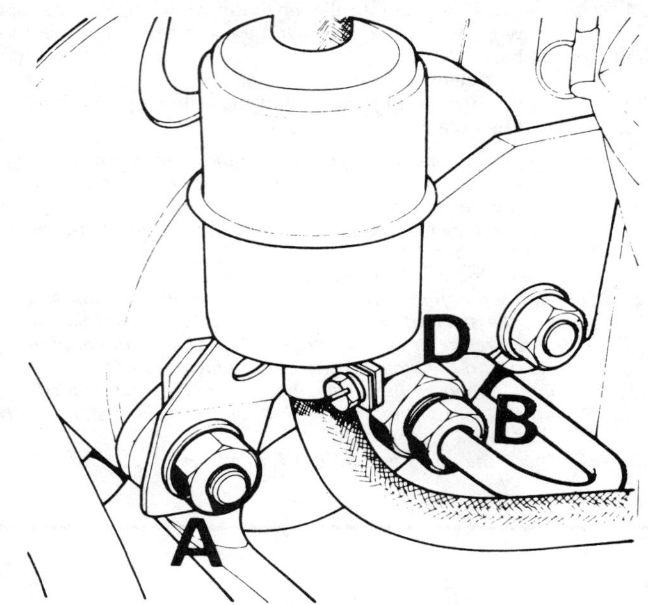

Fig. 13.98 Rear view of power steering pump (Sec 10)

A Fuel filter mounting bracket
B Power steering feed pipe union
D Power steering regulator valve cap

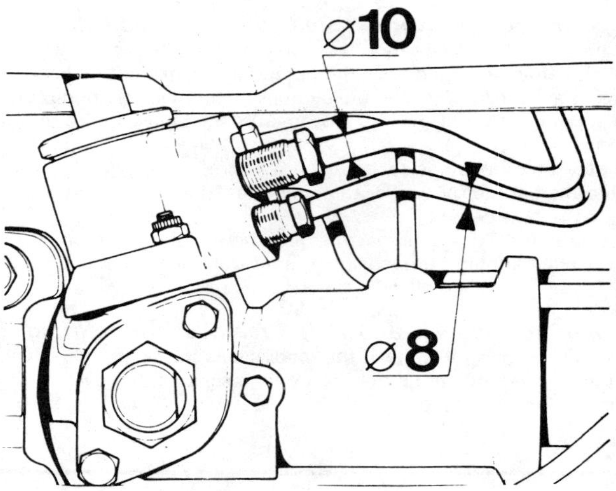

Fig. 13.99 Location of power steering fluid pipes (Sec 10)

Feed pipe – 8.0 mm diameter
Return pipe – 10.0 mm diameter

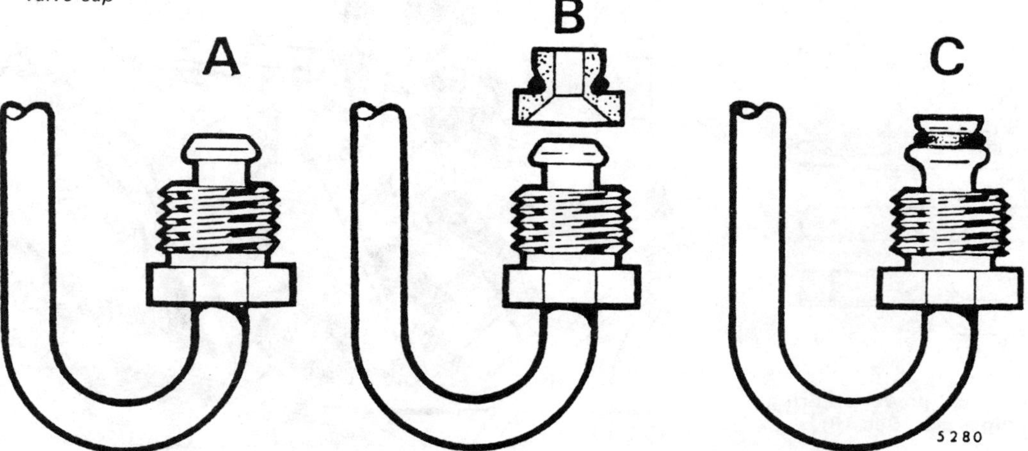

Fig. 13.100 Power steering pipeline flares and unions (Sec 10)

A Early type B Early type with insert C Modified type

Wheels and tyres

39 Wheels and tyres should give no real problems in use provided that a close eye is kept on them with regard to excessive wear or damage. To this end, the following points should be noted.

40 Ensure that tyre pressures are checked regularly and maintained correctly. Checking should be carried out with the tyres cold and not immediately after the vehicle has been in use. If the pressures are checked with the tyres hot, an apparently high reading will be obtained owing to heat expansion. Under no circumstances should an attempt be made to reduce the pressures to the quoted cold reading in this instance, or effective underinflation will result.

41 Underinflation will cause overheating of the tyre owing to excessive flexing of the casing, and the tread will not sit correctly on the road surface. This will cause a consequent loss of adhesion and excessive wear, not to mention the danger of sudden tyre failure due to heat build-up.

42 Overinflation will cause rapid wear of the centre part of the tyre tread coupled with reduced adhesion, harsher ride, and the danger of shock damage occurring in the tyre casing.

43 Regularly check the tyres for damage in the form of cuts or bulges, especially in the sidewalls. Remove any nails or stones embedded in the tread before they penetrate the tyre to cause deflation. If removal of a nail *does* reveal that the tyre has been punctured, refit the nail so that its point of penetration is marked. Then immediately change the wheel and have the tyre repaired by a tyre dealer. Do *not* drive on a tyre in such a condition. In many cases a puncture can be simply repaired by the use of an inner tube of the correct size and type. If in any doubt as to the possible consequences of any damage found, consult your local tyre dealer for advice.

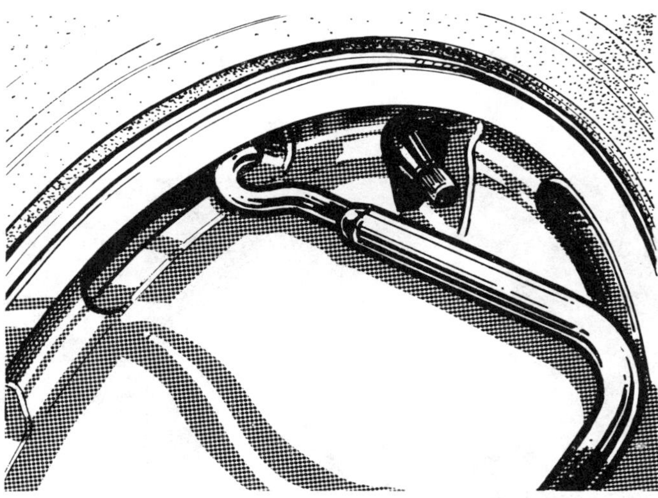

Fig. 13.101 Removing roadwheel trim (Sec 10)

44 Periodically remove the wheels and clean any dirt or mud from the inside and outside surfaces. Examine the wheel rims for signs of rusting, corrosion or other damage. Light alloy wheels are easily damaged by 'kerbing' whilst parking, and similarly steel wheels may become dented or buckled. Renewal of the wheel is very often the only course of remedial action possible.

45 The balance of each wheel and tyre assembly should be maintained to avoid excessive wear, not only to the tyres but also to the steering and suspension components. Wheel imbalance is normally signified by vibration through the vehicle's bodyshell, although in many cases it is particularly noticeable through the steering wheel. Conversely, it should be noted that wear or damage in suspension or steering components may cause excessive tyre wear. Out-of-round or out-of-true tyres, damaged wheels and wheel bearing wear/maladjustment also fall into this category. Balancing will not usually cure vibration caused by such wear.

46 Wheel balancing may be carried out with the wheel either on or off the vehicle. If balanced on the vehicle, ensure that the wheel-to-hub relationship is marked in some way prior to subsequent wheel removal so that it may be refitted in its original position.

47 General tyre wear is influenced to a large degree by driving style – harsh braking and acceleration or fast cornering will all produce more rapid tyre wear. Interchanging of tyres may result in more even wear, but this should only be carried out where there is no mix of tyre types on the vehicle. However, it is worth bearing in mind that if this is completely effective, the added expense of replacing a complete set of tyres simultaneously is incurred, which may prove financially restrictive for many owners.

48 Front tyres may wear unevenly as a result of wheel misalignment. The front wheels should always be correctly aligned according to the settings specified by the vehicle manufacturer.

49 Legal restrictions apply to the mixing of tyre types on a vehicle. Basically this means that a vehicle must not have tyres of differing construction on the same axle. Although it is not recommended to mix tyre types between front axle and rear axle, the only legally permissible combination is crossply at the front and radial at the rear. When mixing radial ply tyres,textile braced radials must always go on the front axle, with steel braced radials at the rear. An obvious disadvantage of such mixing is the necessity to carry two spare tyres to avoid contravening the law in the event of a puncture.

50 In the UK, the Motor Vehicles Construction and Use Regulations apply to many aspects of tyre fitting and usage. It is suggested that a copy of these regulations is obtained from your local police if in doubt as to the current legal requirements with regard to tyre condition, minimum tread depth, etc.

Roadwheel trim (Minx and Rapier) – removal and refitting

51 The hooked end of the car jack handle is used to release the wheel trim clips on these cars.

52 Insert the tol into the first slot in the trim anti-clockwise from the tyre valve. Pull sharply to disengage the first retaining clip.

53 Release the remaining clips in a similar way.

54 Refit by aligning the hole in the trim with the tyre valve and pushing the trim fully home to fully engage the retaining clips with their retaining ring.

11 Braking system

General description

1 A brake pad wear warning system is fitted to all models commencing with 8-Series. A wear contact is incorporated in the lining and this completes the circuit to the warning lamp on the instrument panel. The wear contact is fitted to each inner brake pad.

2 Before removing the pads on models so equipped the spring clip on the side of the caliper must be removed and the pad lead disconnected. When refitting, the lead must be located beneath the spring clip as shown in Fig. 13.103. The remainder of the pad renewal procedure is as given in Chapter 10.

3 From January 1978 the DBA rear wheel cylinders incorporate a steel cup under the rubber seal. This prevents the seals collapsing when the brake circuit is initially filled, using a vacuum process, by the manufacturers.

4 Commencing with 9-Series models, a low fluid level indicator is fitted instead of the pressure differential switch (photo).

Disc brake pads (up to May 1980) DBA Bendix – removal and refitting

5 Loosen the roadwheel bolts, jacking up the front of the car and support it on axle stands. Remove the roadwheel.

6 Using a stiff wire brush, clean away all dirt and grit from around the front of the caliper.

7 Extract the two small spring clips (Fig. 13.104) then using a hammer and small punch, drive out the sliding keys that locate the caliper yoke in the bracket.

8 Push the caliper yoke downwards then tip it forward at the top to disengage it from the bracket. Lift away the caliper, with hydraulic hose still attached and suspend it from a convenient place under the wheel arch. Do not let it hang from the hydraulic hose.

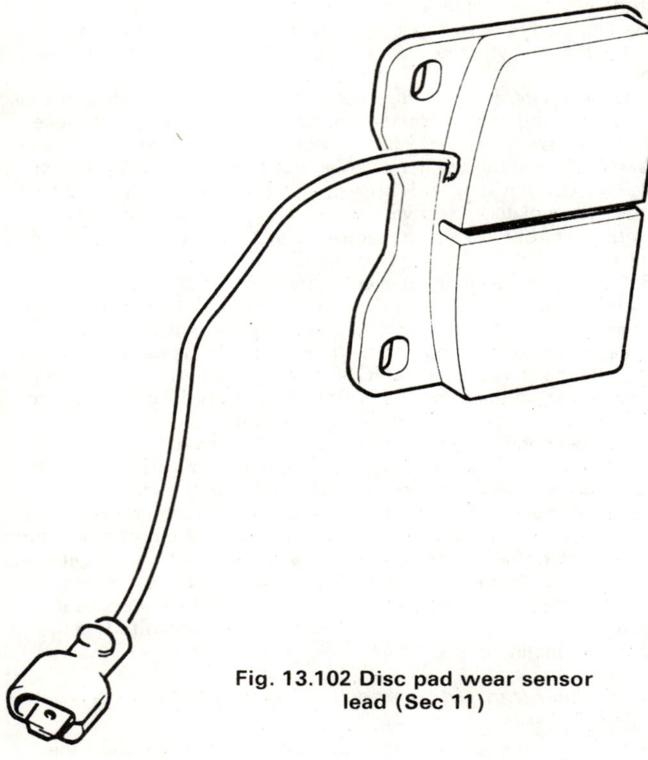

Fig. 13.102 Disc pad wear sensor lead (Sec 11)

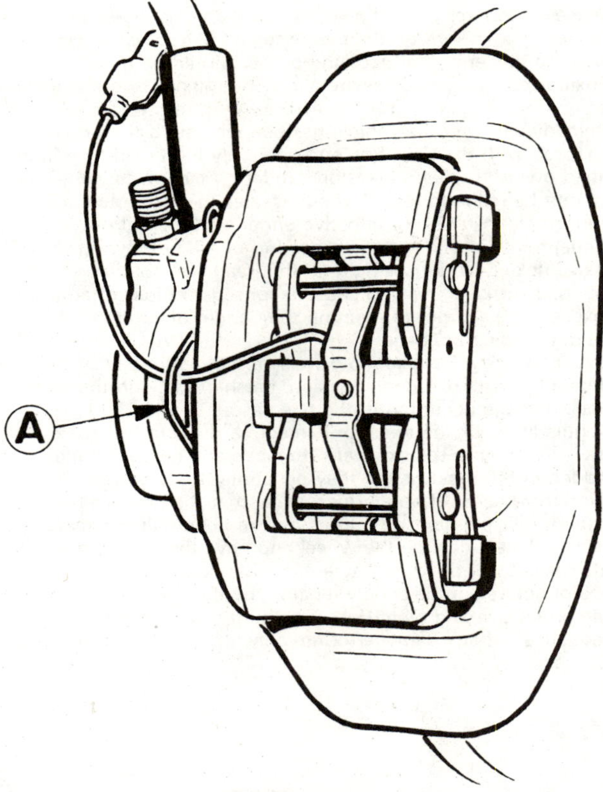

Fig. 13.103 Disc pad wear sensor lead correctly routed (Sec 11)

A Spring clip

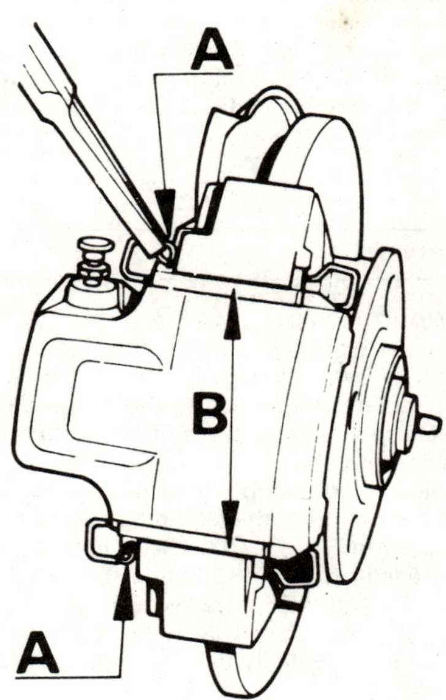

Fig. 13.104 DBA Bendix type brake caliper – pre-May 1980 (Sec 11)

A Spring clip *B Sliding keys*

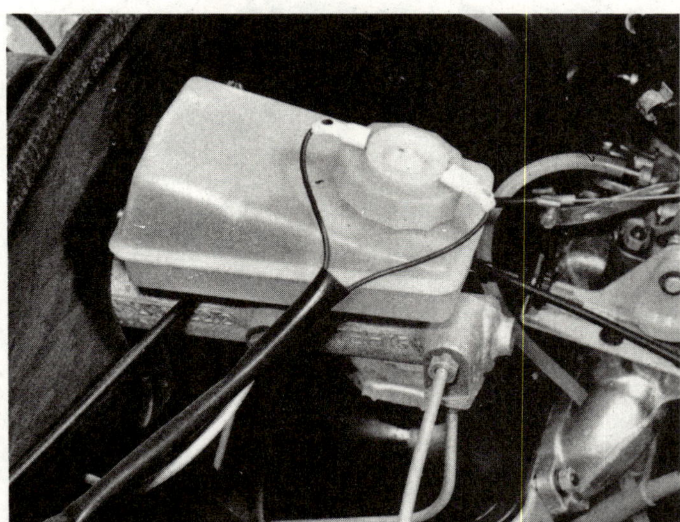

11.4 Brake master cylinder fluid reservoir with low level switch

9 Remove the upper and lower steady springs, noting their fitted position then lift out the two brake pads.
10 The cleaning and checking procedures are similar to those given in Chapter 10 with reference to the Specifications at the beginning of this Supplement.
11 If new pads are being fitted, first push the piston into the caliper cylinder using a suitable clamp and at the same time making sure that the brake fluid does not overflow from the reservoir.
12 Place the steady springs on the pads then fit the pads to the bracket. If the pads have a retaining stop fitted at one end of their backing plate, this must be at the bottom when the pads are fitted (Fig. 13.105).
13 Refit the yoke to the bracket and insert the lower sliding key. Lever the yoke down using a screwdriver and insert the upper key. Secure the keys in place using new spring clips.
14 Depress the brake pedal several times to bring the pads into contact with the disc then check the brake fluid lever in the master cylinder reservoir.

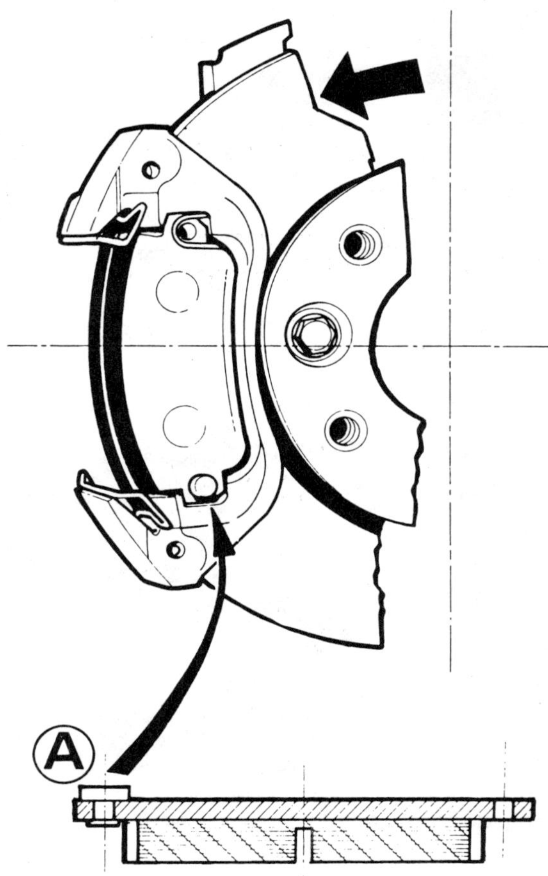

Fig. 13.105 Pad retaining stop (A) – DBA Bendix brakes, pre-May 1980 (Sec 11)

15 Refit the roadwheel and lower the car to the ground.

Disc brake pads (commencing May 1980) Teves – removal and refitting

16 Loosen the roadwheel bolts. Jack up the front of the car and support it on axle stands. Remove the roadwheel.
17 Disconnect the pad wear warning light lead and remove it from the clip (photo).
18 Press off the retaining spring from the caliper (photo).
19 Remove the sealing caps, then unscrew and remove the socket screws from the inside of the caliper using an Allen key (photos).
20 Withdraw the caliper together with the piston side pad, then prise the pad from the piston (photos).
21 Support the caliper on a stand without straining the hose.
22 Remove the outer pad from the fixing support (photo).
23 The cleaning and checking procedures are similar to those given in Chapter 10, with reference to the Specifications at the beginning of this Supplement.
24 If new pads are being fitted, first push the piston into the caliper cylinder, at the same time making sure that the brake fluid does not spill from the reservoir.
25 Fit the outer pad to the fixed support and the inner pad to the piston.
26 Locate the caliper to the disc. Apply a liquid agent to the threads of the socket screws, then insert them and tighten them to the specified torque.
27 Check that the caliper moves freely on the fixed support, then refit the sealing plugs and reconnect the pad wear warning light lead.
28 Fit the retaining spring to the caliper.
29 Refit the roadwheel and lower the car to the ground. Depress

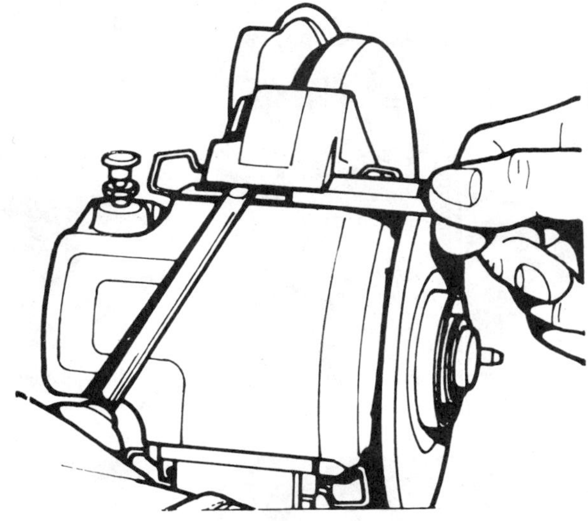

Fig. 13.106 Fitting caliper sliding key – DBA Bendix disc brake, pre-May 1980 (Sec 11)

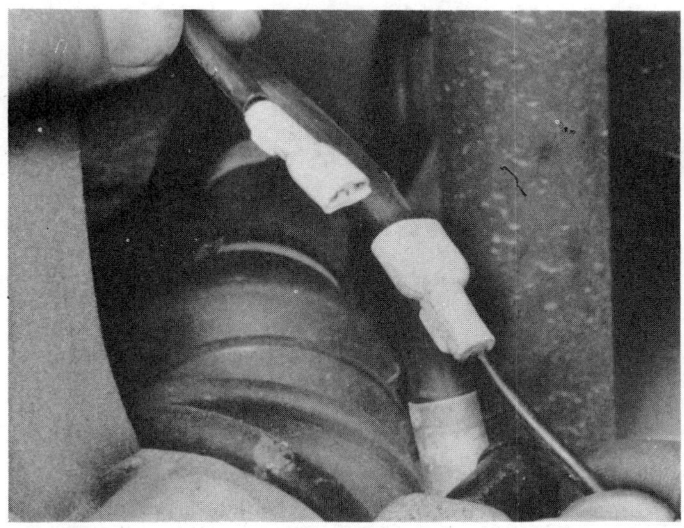

11.17 Disc pad wear sensor lead connector

11.18 Disc pad retaining spring

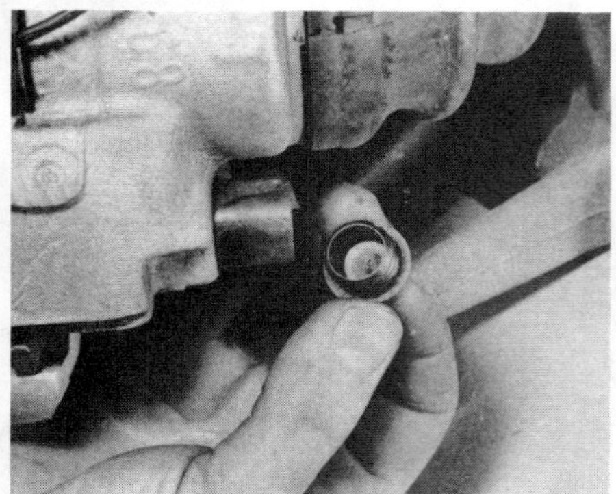
11.19A Caliper screw sealing cap

11.19B Unscrewing caliper socket-headed screw

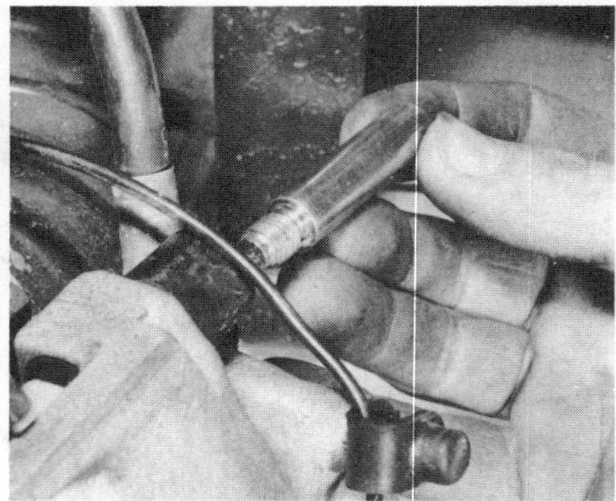
11.19C Removing socket-headed screw

11.20A Withdrawing caliper with inboard disc pad

11.20B Removing inboard pad from piston

11.22 Removing outboard disc pad

the brake pedal several times, then check the level of brake fluid in the reservoir and top up if necessary.

Disc brake pads (commencing May 1980) DBA Bendix – removal and refitting

30 Loosen the roadwheel bolts. Jack up the front of the car and support it on axle stands. Remove the roadwheel.

31 Disconnect the pad wear warning light lead and remove it from the clip.

32 Remove the spring clip from the lower socket screw and remove the screw.

33 Swivel the caliper clear of the lower pin, then withdraw it downwards and support it on a stand without straining the hose.

34 Remove the upper socket screw. Withdraw the disc pads and springs.

35 The cleaning and checking procedures are similar to those given in Chapter 10, with reference to the Specifications at the beginning of this Supplement. Make sure that the socket screw rubbers are not

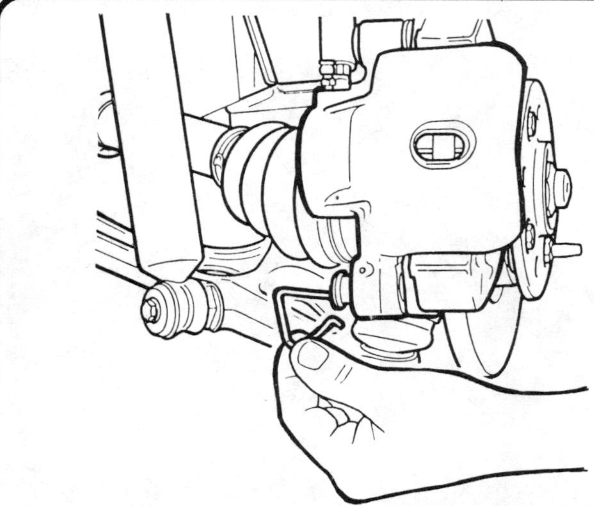
Fig. 13.107 Socket screw spring clip – DBA Bendix disc brake, May 1980-on (Sec 11)

Fig. 13.108 Swivelling brake caliper downwards – DBA Bendix disc brake, May 1980-on (Sec 11)

damaged. Using brake grease, lubricate the fixed support pins, the socket screw apertures in the caliper, and the sliding contact surfaces of the caliper.
36 If new pads are being fitted, first push the piston into the caliper cylinder, at the same time making sure that the brake fluid does not spill from the reservoir.
37 Fit the disc pads and springs, noting that the inner pad has a wide groove and an arrow indicating the forward movement of the disc.
38 Fit the upper socket screw and slide the caliper onto it.
39 Lower the caliper over the lower pin and fit the lower socket screw. Fit the retaining spring and check that the caliper is free to slide on the fixed support.
40 Reconnect the pad wear warning light lead and fit it to the clip.
41 Refit the roadwheel and lower the car to the ground. Depress the brake pedal several times, then check the lever of brake fluid in the reservoir and top up if necessary.

Brake disc (from September 1977) – removal and refitting

42 Refer to Section 9, paragraph 8. Since the front hub must be removed in order to remove the brake disc, it will be necessary to renew the hub bearings at the same time.

Rear brake shoe linings (commencing 9-Series) – inspection

43 The rear brake backplates fitted to 9-Series models onward incorporate an inspection hole at the front. Removal of the rubber plug exposes the primary shoe, and the lining thickness can easily be checked without removing the brake drum.

Rear brake shoes (Teves) – removal and refitting

44 Jack up the rear of the car and support it on axle stands. Remove the roadwheel.
45 Release the handbrake, then remove the securing screws and withdraw the drum. If difficulty is experienced, remove the plug from the backplate. Insert a screwdriver through a brake drum stud hole and disengage the adjuster lever from the automatic adjuster serrated nut. Insert a screwdriver through the backplate and turn the serrated nut to release the brake shoes.
46 With the brake drum removed, turn the serrated nut to its minimum adjustment while depressing the adjuster lever.
47 Unhook the lower return spring, the adjuster link spring and the upper return spring.
48 Using pliers, depress the steady springs and washers, turn the washers through 90° and withdraw them from the steady pins.

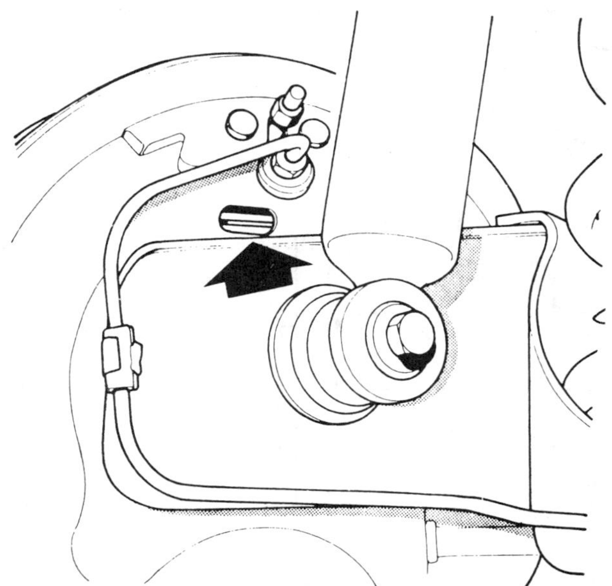

Fig. 13.110 Access hole (arrowed) in Teves type rear brake backplate (Sec 11)

49 Disconnect the handbrake cable from the lever and withdraw both shoes. Place a strong rubber band around the wheel cylinder to retain the pistons.
50 Clean and check the brake shoes and backplate as described in Chapter 10.
51 Fit the adjuster link to the leading (primary) shoe.
52 Fit the trailing (secondary) shoe to the adjuster link, making sure that the longer end of the fork is toward the backplate. Fit the adjuster return spring.
53 Pull the bottom ends of the shoes apart and fit the upper return spring.
54 Lubricate the ends of the shoes with brake grease, then locate them over the hub.

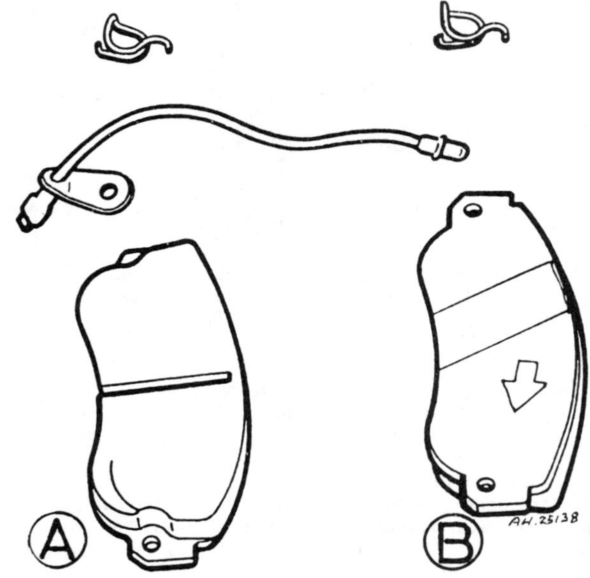

Fig. 13.109 DBA Bendix disc pads – May 1980-on (Sec 11)

A Outboard pad *B Inboard pad*

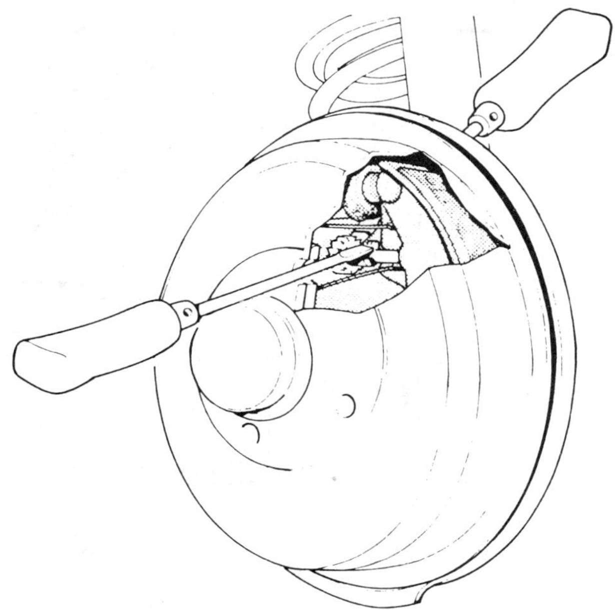

Fig. 13.111 Releasing brake shoes on Teves type rear brake (Sec 11)

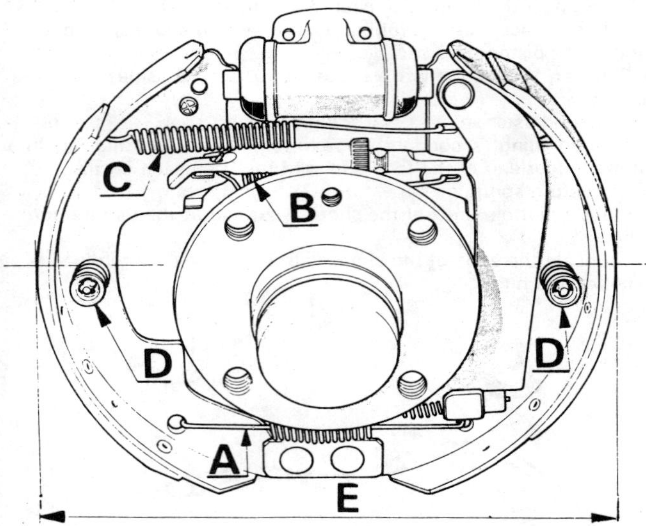

Fig. 13.112 Teves type rear brake assembly (Sec 11)

A Lower return spring
B Link return spring
C Upper return spring
D Steady springs
E Shoe diameter – 8.98 in (228 mm)

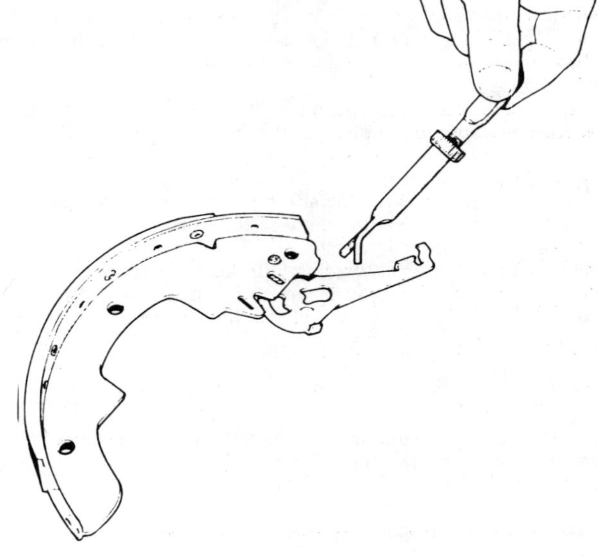

Fig. 13.113 Fitting Teves type rear brake shoe adjuster link (Sec 11)

55 Remove the rubber band. Locate one shoe on the wheel cylinder piston, then lever the other shoe onto the remaining piston.
56 Locate the lower ends of the shoes in the anchor plate. Refit the steady springs and washers.
57 Refit the lower return spring.
58 Push the handbrake lever forward and attach the cable nipple to it. Release the lever, and check that the cable is just taut with the lever against its stop on the trailing shoe. If necessary adjust the cable as described in Chapter 10.
59 Turn the serrated nut on the adjuster to position the shoes so that the brake drum will just slide over them. Make sure that the operating lever engages the serrated nut.
60 Refit the brake drum and roadwheel.
61 Start the engine and depress the brake pedal thirty times to operate the automatic adjuster.
62 Adjust the handbrake if necessary as described in Chapter 10.
63 Lower the car to the ground.

Rear wheel cylinder (Teves) – removal and overhaul

64 The Teves rear wheel cylinder is similar in construction to the Girling type described in Chapter 10, and therefore the same procedures apply.

Low fluid level indicator – testing

65 Switch on the ignition.
66 Unscrew the cap from the master cylinder reservoir and lift it so that the float drops to the bottom of the spindle.
67 Release the handbrake lever and check that the warning lamp on the instrument panel lights. If not, the switch, wiring, or lamp are faulty.
68 Refit the cap and switch off the ignition.

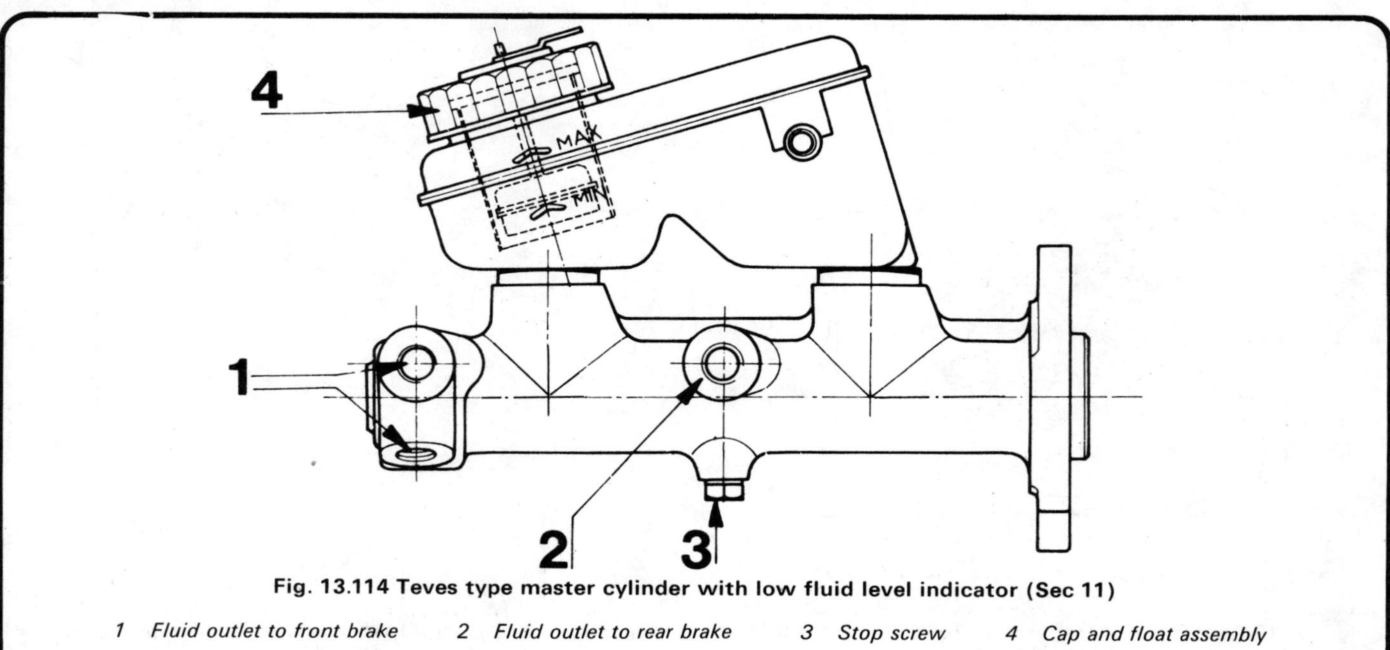

Fig. 13.114 Teves type master cylinder with low fluid level indicator (Sec 11)

1 Fluid outlet to front brake *2 Fluid outlet to rear brake* *3 Stop screw* *4 Cap and float assembly*

Master cylinder (Teves) – dismantling and reassembly
69 The Teves master cylinder fitted to 9-Series models onward is shown in Fig. 13.115. The main difference compared with the earlier type is that no pressure differential switch is fitted.

12 Electrical system

Part A General equipment

Headlamp wash/wipe system – description
1 The headlamp wash/wipe components are shown in Fig. 13.116. The system operates automatically when the windscreen washer is switched on with the sidelamps also operating.

Headlamp wash/wipe wiper motor – removal and refitting
2 Disconnect the battery negative terminal and remove the radiator grille.
3 Disconnect the leads from the headlight unit.
4 Disconnect the leads from the wiper motor, and detach the fluid supply tube.
5 Slide out the retaining plates, then depress the top of the headlamp lens and remove the adjusting knob from its slot.
6 Withdraw the headlamp and wiper assembly. Disconnect the wiper brush and arms.
7 Remove the two screws and the securing nut, and separate the wiper motor from the headlamp.
8 Refitting is a reversal of removal, but adjust the wiper motor so that the brush applies even pressure to the headlamp. Make sure that the hooked end of the spring is located on the side of the plate when refitting the adjusting knob.

Tailgate wash/wipe system – description
9 The tailgate wiper motor is located on the tailgate inner panel; the washer bottle and pump are located within the left-hand quarter panel.

Tailgate wiper and washer jet – adjustment
10 The wiper blade parking position and the washer jet aiming point are shown in Fig. 13.118. The jet is adjusted by inserting a pin in the nozzle and moving it to the desired position.

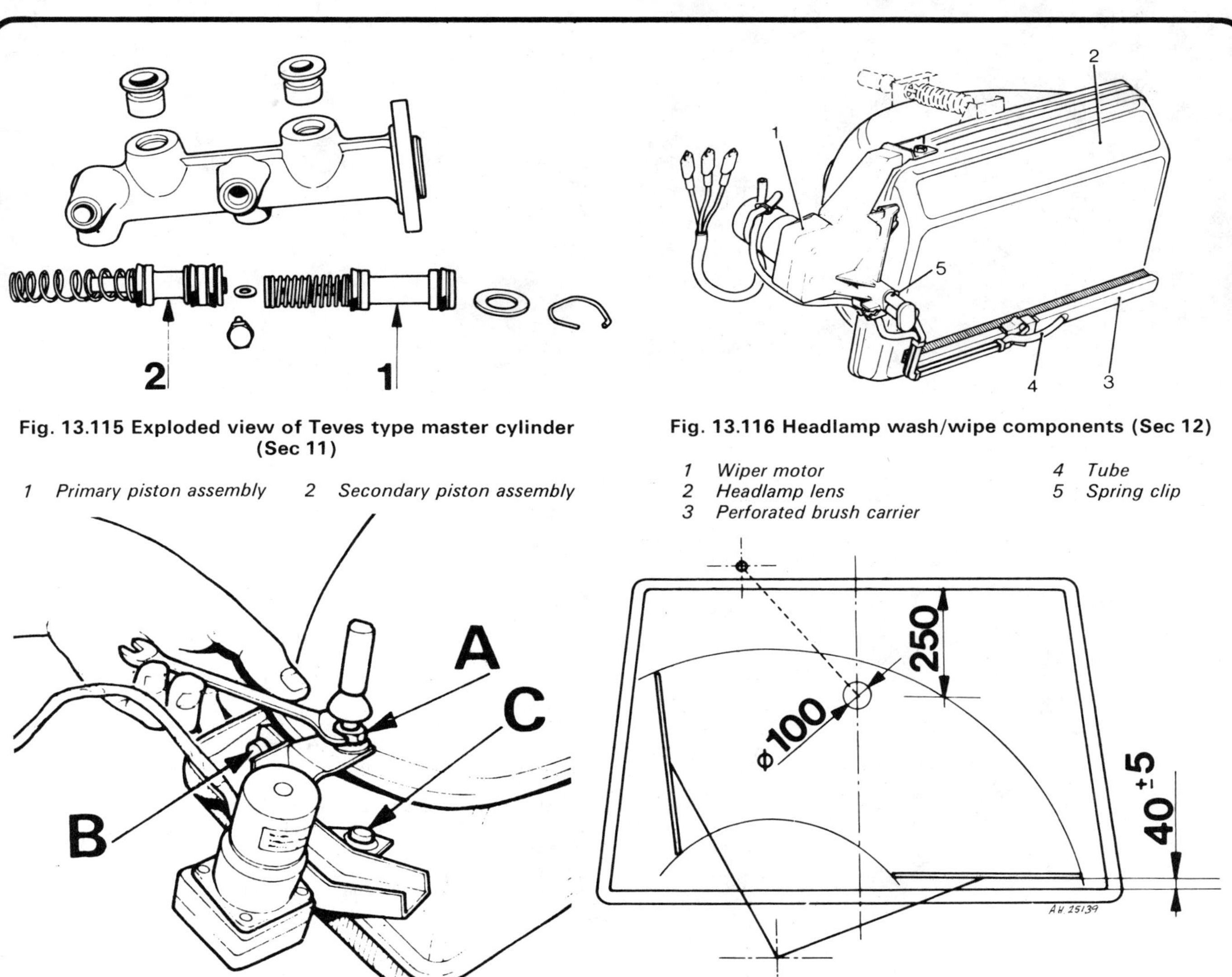

Fig. 13.115 Exploded view of Teves type master cylinder (Sec 11)

1 Primary piston assembly *2 Secondary piston assembly*

Fig. 13.116 Headlamp wash/wipe components (Sec 12)

1 Wiper motor
2 Headlamp lens
3 Perforated brush carrier
4 Tube
5 Spring clip

Fig. 13.117 Wiper motor fixings to headlamp (Sec 12)

A Nut
B Screw
C Screw

Fig. 13.118 Tailgate washer jet striking point and wiper blade parking position. All dimensions in mm (Sec 12)

Tailgate wiper motor – removal and refitting

11 Disconnect the battery negative terminal.
12 Remove the wiper arm and blade.
13 Remove the spindle nut, washer, seal, and spacer.
14 Raise the tailgate and disconnect the motor supply leads.
15 Unscrew the mounting bracket bolts and withdraw the wiper motor. Detach the bracket if necessary.
16 Refitting is a reversal of removal, but assemble the components to the tailgate loosely at first, then tighten the mounting bolts and finally the spindle nut. Adjust the wiper blade as described in paragraph 10.

Tailgate wash/wipe reservoir and pump – removal and fitting

17 Disconnect the battery negative terminal.
18 Remove the rear lamp lens and lamp body.
19 Peel back the trim, unscrew the two screws, and withdraw the reservoir through the lamp aperture.
20 Disconnect the fluid supply tube.
21 To remove the pump, disconnect the supply lead and fluid tube.
22 Unscrew the retaining bolts and withdraw the pump through the lamp aperture.
23 Refitting is a reversal of removal, but make sure that the pump earth lead is located under the mounting bolt. Refill the reservoir and check the operation of the wiper and washer.

Headlight bulb – removal and refitting

24 On later models it is necessary to remove the rubber cover at the rear of the headlight unit to gain access to the bulb (photo).
25 With the wiring plug disconnected, release the retaining clip and withdraw the bulb (photo). Avoid touching the bulb glass with your fingers. If touched, clean the glass with methylated spirit.
26 Refitting is the reverse of removal.

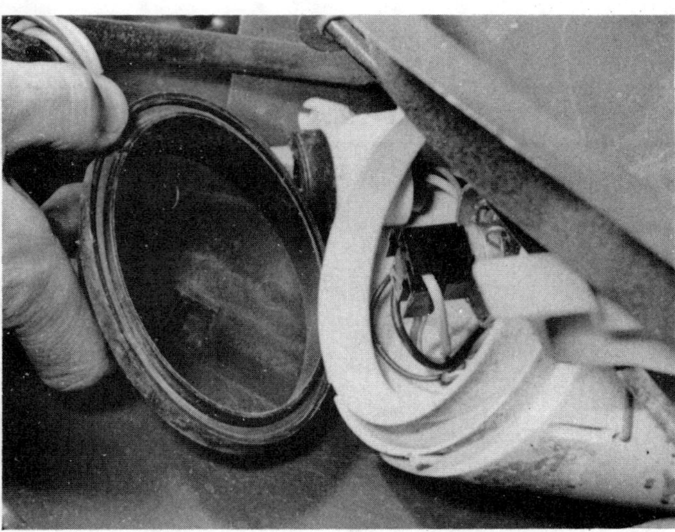
12.24 Headlamp rear rubber cover

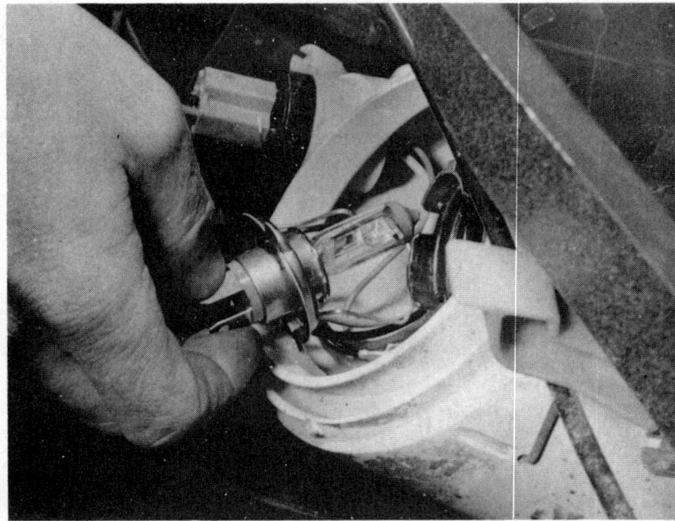
12.25 Removing headlamp bulb

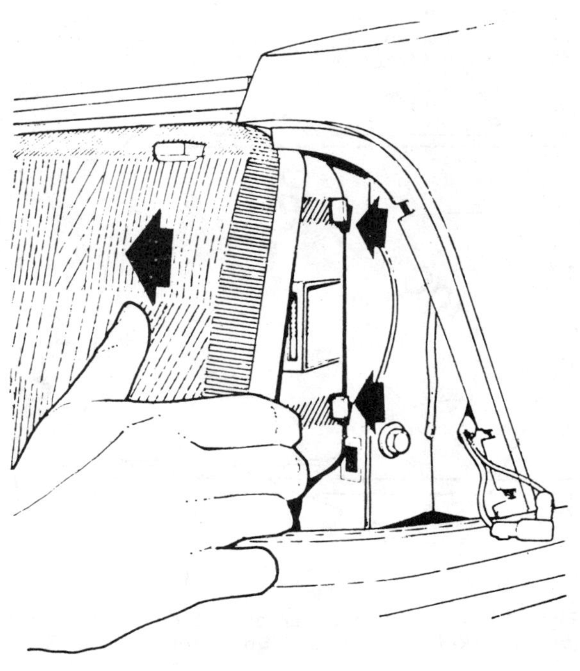
Fig. 13.119 Removing A-Series headlamp (Sec 12)

Headlight unit (commencing A-Series) – removal and refitting

27 Disconnect the battery negative terminal.
28 Remove the front grille. From underneath the wing, remove the nuts securing the front direction indicator lamp, withdraw the lamp and disconnect the wiring.
29 Disconnect the multi-plug, and remove the screws securing the headlamp to the front panel struts.
30 Slide the headlamp towards the grille position and withdraw it.
31 Refitting is a reversal of removal, but do not overtighten the direction indicator lamp nuts, otherwise the headlamp glass may be damaged.

Headlights (commencing A-Series) – alignment

32 A-Series (1979 to 1980) models are equipped with a two-position lever on each headlamp which can be manually adjusted to compensate for load variations in the car. Before making any adjustment to the beam, both levers must be in the 'vide-unladen' position.
33 Two nylon nuts are provided for beam adjustment, the outer nut adjusts vertical movement and the inner nut adjusts lateral movement.

Rear lamp cluster (commencing A-Series) – bulb and lamp renewal

34 Working in the rear compartment, push the spring catch to release the bulbholder (photo).
35 Remove the bayonet type bulbs by pushing and twisting anti-clockwise. Refitting is a reversal of removal.
36 To remove the lamp, disconnect the battery negative terminal and remove the bulbholders.

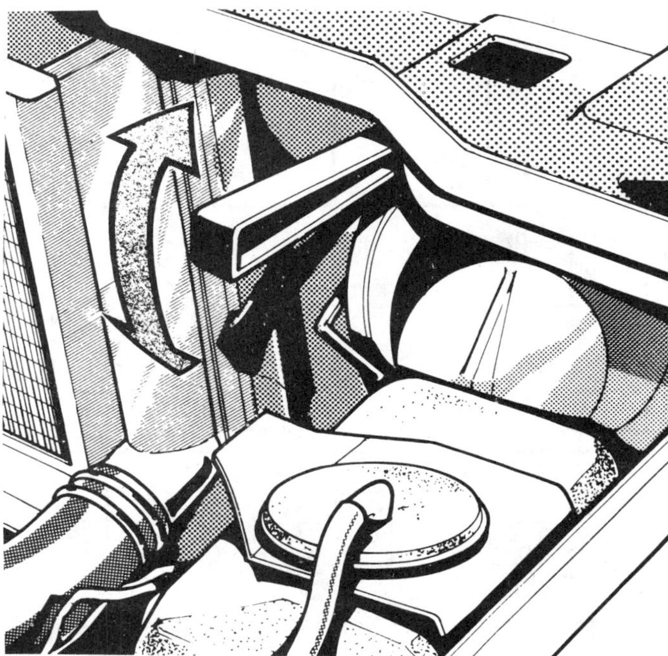

Fig. 13.120 Headlamp adjuster lever on A-Series vehicles (Sec 12)

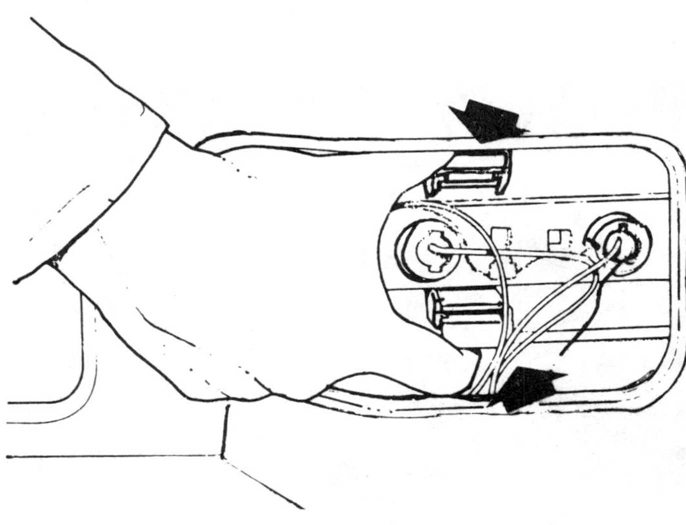

Fig. 13.121 Removing rear lamp cluster on A-Series vehicles. Pressure points (arrowed) (Sec 12)

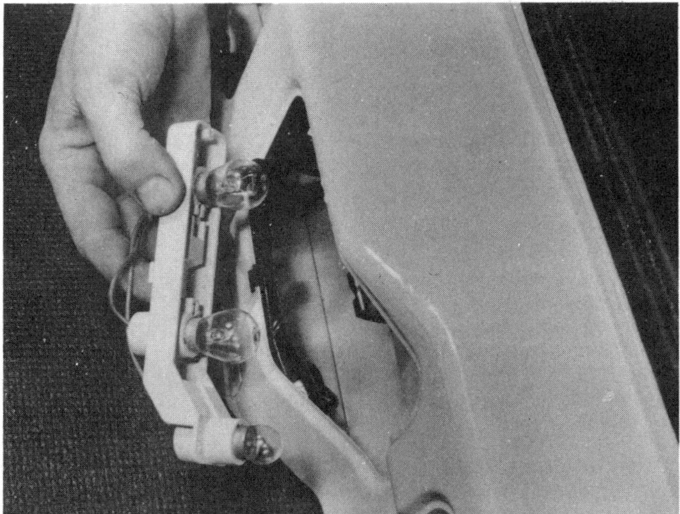

12.34 Removing rear lamp bulbholder from Solara

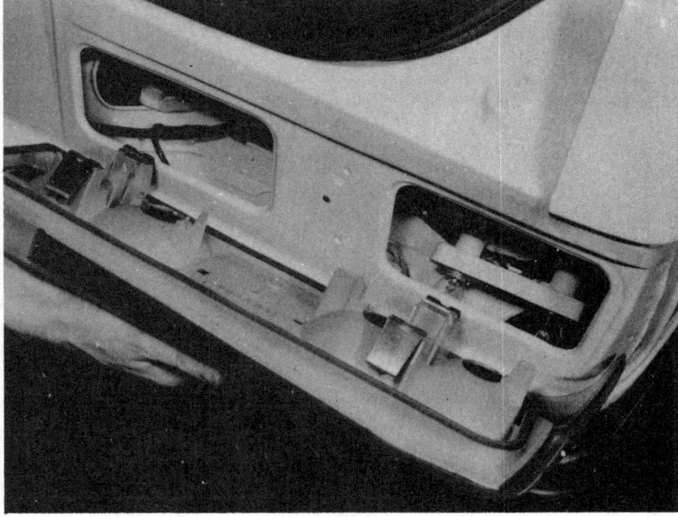

12.37 Removing Solara lens/lamp body

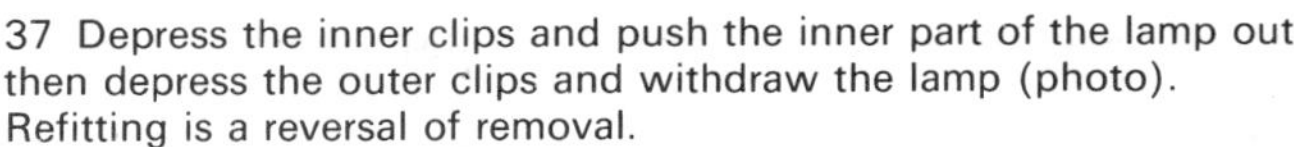

37 Depress the inner clips and push the inner part of the lamp out, then depress the outer clips and withdraw the lamp (photo). Refitting is a reversal of removal.

Fuses (commencing A-Series) description

38 In addition to the fusebox there are three in-line fuses located as follows:

- (a) *1.6 amp fuse for radio/cassette player, located on radio bracket*
- (b) *1.6 amp fuse for electric door locks, located under right-hand facia*
- (c) *5 amp fuse for tailgate wiper motor, located under centre facia*

39 16 amp fuses are fitted to the fusebox for the red/yellow and purple/purple supply leads. The remaining fuses are 10 amp. The circuits protected by the fuses are as follows:

Fuse	Circuit
16 amp	*Heated rear window relay and element*
10 amp	*Electric clock, direction indicator flasher unit, interior lamps, luggage compartment and glovebox lamps, engine oil level indicator control box*
10 amp	*Rear fog lamps*
10 amp	*Side, tail, and number plate lamps, instrument illumination bulbs*
16 amp	*Heater motor, cigarette lighter, headlamp wash/wipe, tailgate wiper and washer pump*
10 amp	*Windscreen wiper motor and washer pump, reverse and stop lamps, fuel flow meter, trip computer, speed control (if applicable)*

Electric window lift – description

40 The early electric window lift system incorporated console-mounted switches, but these have been superseded by door-mounted switches (photo). The circuit diagram for the later system is shown in Fig. 13.122. Removal and refitting procedures for the door glass are identical to those given for manually operated windows in Chapter 12.

12.40 Door-mounted electric window switch

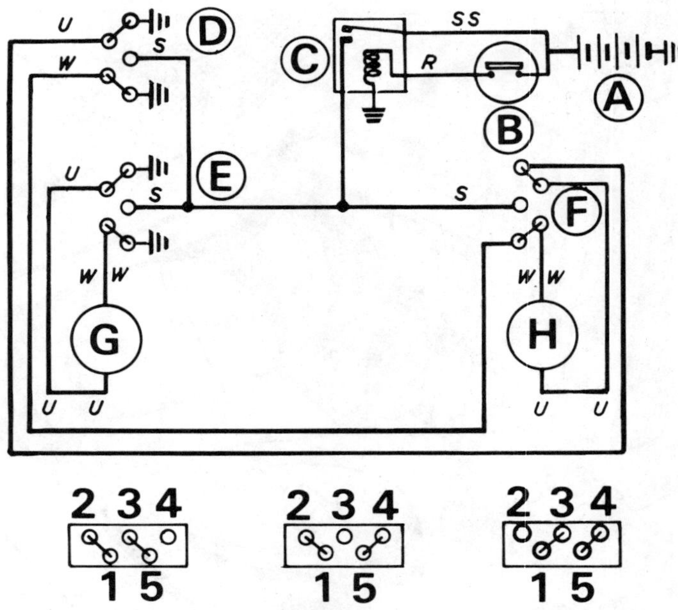

Fig. 13.122 Wiring diagram for power-operated windows with door mounted switches/ LHD shown, RHD similar (Sec 12)

A Battery
B Ignition/starter switch
C Relay
D Right-hand window lift switch on driver's door
E Left-hand window lift switch on driver's door
F Right-hand window lift switch on passenger door
G Left-hand window lift motor
H Right-hand window lift motor

Electric door locks – description

41 The central locking system is fitted to certain models commencing with the A-Series and enables the driver to operate all the door locks from the driver's door (Fig. 13.123).

Electric door lock control box/servo motor – removal and refitting

42 Disconnect the battery negative terminal.
43 Remove the door trim (see Chapter 12).
44 Loosen the control box/servo motor securing nuts, then unscrew the three screws and withdraw the striker backplate.
45 Remove the control box/servo motor-to-door panel nut and washer, and withdraw the lockplate. Do not disconnect the linkage.
46 Detach the control box/servo motor from the lockplate and unclip the harness.
47 Disconnect the multi-plug and connector, detach the connecting link, and withdraw the control box/servo motor.
48 Refitting is a reversal of removal, but check the operation of all the locks on completion.

Cruise control cable – adjustment

49 Loosen the cable locknut at the servo end.
50 Shorten the cable length until the cam moves from its stop in the cambox and there is 0.079 in (2.0 mm) clearance between the stops of cams 4 and 3 (see Fig. 13.18).
51 Turn the adjuster 1.5 turns to lengthen the cable, then tighten the locknut.

Trip computer – description

52 The trip computer incorporates a digital read-out clock in addition to the micro-processor circuitry necessary to provide the driver with information on elapsed time, distance travelled, fuel consumed, average fuel consumption, and average speed. External information is fed to the computer by two units – the fuel flow meter and the speed/distance sensor.

Trip computer – removal and refitting

53 Disconnect the battery negative lead.
54 Remove the ashtray, then remove the screws securing the centre facia switch panel.
55 Remove the heater plate and bracket screws, and pull the switch panel forwards.
56 Remove the springs retaining the trip computer. Push the springs forwards, then lift them from the brackets.
57 Disconnect the supply wire plugs and withdraw the trip computer.
58 Refitting is a reversal of removal.

Trip computer fuel flow meter – removal and refitting

59 Disconnect the battery negative lead.
60 Loosen the clip and disconnect the fuel outlet pipe from the flow meter.
61 Loosen the clip and disconnect the fuel inlet pipe at the filter.
62 Disconnect the two supply wire plugs from the main harness.
63 Remove the two screws and withdraw the flow meter and bracket from the wing valance. Remove the flow meter from the bracket.
64 Refitting is a reversal of removal.

Trip computer speed sensor unit – removal and refitting

65 Jack up the front of the car and support it on axle stands.
66 Disconnect the battery negative lead.
67 On manual transmission models disconnect both speedometer cables and the wiring plug, and unbolt the sensor unit.
68 On automatic transmission models disconnect the speedometer cable and the wiring plug, unbolt the adaptor, then remove the spring clip and pinion and withdraw the sensor unit.
69 Refitting is a reversal of removal, but on the automatic transmission model lubricate the O-ring seals with engine oil.

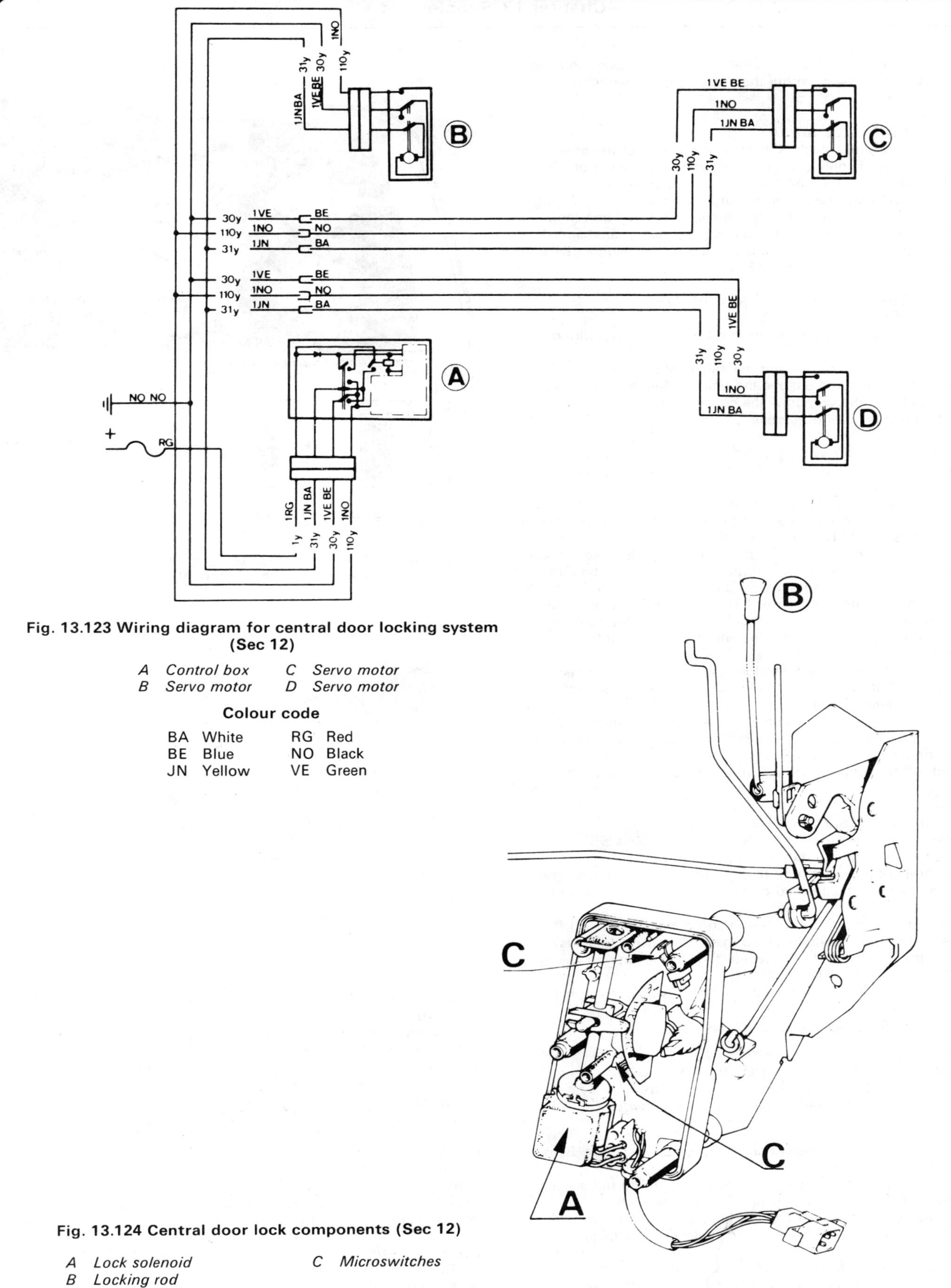

Fig. 13.123 Wiring diagram for central door locking system (Sec 12)

A *Control box*
B *Servo motor*
C *Servo motor*
D *Servo motor*

Colour code

BA	White	RG	Red
BE	Blue	NO	Black
JN	Yellow	VE	Green

Fig. 13.124 Central door lock components (Sec 12)

A *Lock solenoid*
B *Locking rod*
C *Microswitches*

Trip computer – Solara SX

70 The trip computer on some of these models has been modified to incorporate the display within the tachometer. A new control panel is also fitted.

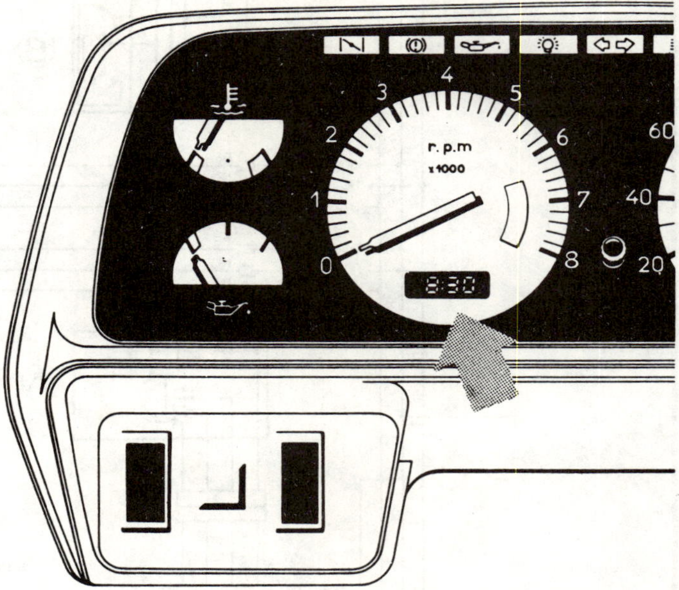

Fig. 13.125 Trip computer display incorporated in tachometer on Solara SX (Sec 12)

Cruise control – description

71 The cruise control switch is located on the end of the direction indicator lever and is only operational at speeds in excess of 30 mph (50 km/h). With the unit switched on it is possible to programme a selected speed into the unit computer by depressing the SET knob at the particular speed required. A servo unit then automatically controls the throttle to maintain the required speed until such time as the brake pedal is applied. The speed can be regained by operating the RESUME switch.

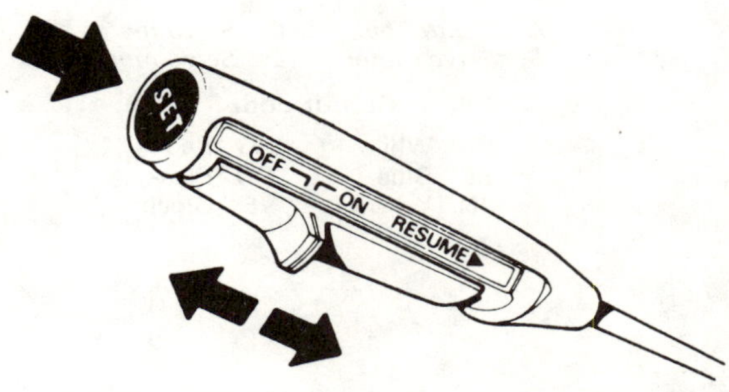

Fig. 13.126 Cruise control switch (Sec 12)

Cruise control unit – removal and refitting

72 Disconnect the battery negative lead.
73 Disconnect the two speedometer cables and the two wiring plugs.
74 Remove the vacuum hose and the cable hairpin clip.
75 Unscrew the nuts and remove the bracket. Remove the mounting nuts and bolts and withdraw the cruise control unit.
76 Refitting is a reversal of removal.

Cruise control servo wiring feed – testing

77 Disconnect the two wiring plugs from the servo unit, then switch on the ignition.
78 With the cruise control switch in the OFF position, connect a voltmeter between earth and each of the wiring plug connections in turn. Zero volts should be indicated (see Fig. 13.127).
79 With the cruise control switch in the ON position make the same check described in paragraph 78. Battery voltage should be indicated on connections 1 and 3, zero voltage on connection 2. If zero voltage is indicated on connection 1 the stop-light may be faulty. The switch incorporates double contacts, and when the footbrake is operated the stop-light contacts close but the cruise control contacts open, and vice versa.
80 With the SET button depressed, repeat the check described in paragraph 78. Battery voltage should be indicated on connections 1 and 2, zero voltage on connection 3.
81 With the cruise control switch in the RESUME position, repeat the check described in paragraph 78. Battery voltage should be indicated on all three connections.
82 Switch off the ignition and reconnect the two wiring plugs to the servo unit.

Cruise control (Rapier models)

83 A limited number of Rapier models have been built with automatic transmission and cruise control.
84 The cruise control system on these models is switched on and off by a separate switch, instead of the stalk switch used on other models.
85 The system servo unit circuit also incorporates a relay.
86 A memory system is employed so that the cruising speed is retained in the memory until the switch (A), or the ignition key, is switched off (Fig. 13.128).
87 The system will not operate at speeds of less than 25 mph (40 kph).
88 The operational sequence is as follows:

> *Switch on the push button switch (A)*
> *When the desired roadspeed is attained, depress the switch (B) (see Fig. 13.129) momentarily. The roadspeed will be maintained until the brake pedal is applied or switch (A) is switched off.*

89 The predetermined roadspeed can be resumed if switch (B) is pulled upwards momentarily.
90 Cruising speed can be increased by accelerating to the desired roadspeed and then depressing switch (B) momentarily.
91 To decrease the set speed without braking, depress and hold switch (B) down until the desired new roadspeed is attained.

Electronic engine oil level indicator – description

92 The electronic engine oil level indicator warns the driver of a low engine oil level whenever the engine is started by making the oil warning light flash on and off. The system incorporates a sensor on the engine dipstick and an electronic control unit beneath the parcel shelf.
93 The dipstick sensor (photo) is a resistor through which a small current is passed. When the resistor is completely immersed in oil, its increase in temperature due to the current will be minimal, but when the oil level is low the resistor temperature (and thus its resistance) will rise, causing an increased voltage which triggers a circuit in the control unit to operate the flashing light.

Electronic oil level indicator dipstick – variations

94 Three different types of electronic dipstick have been used on Minx, Rapier, Alpine and Solara models, each having different level markings. The three types together with a manual dipstick for reference are shown in Fig. 13.131.

Oil level indicator control unit – removal and refitting

95 Disconnect the battery negative lead.
96 Remove the screws and detach the control unit from under the parcel shelf. Disconnect the multi-plug and withdraw the unit.
97 Refitting is a reversal of removal.

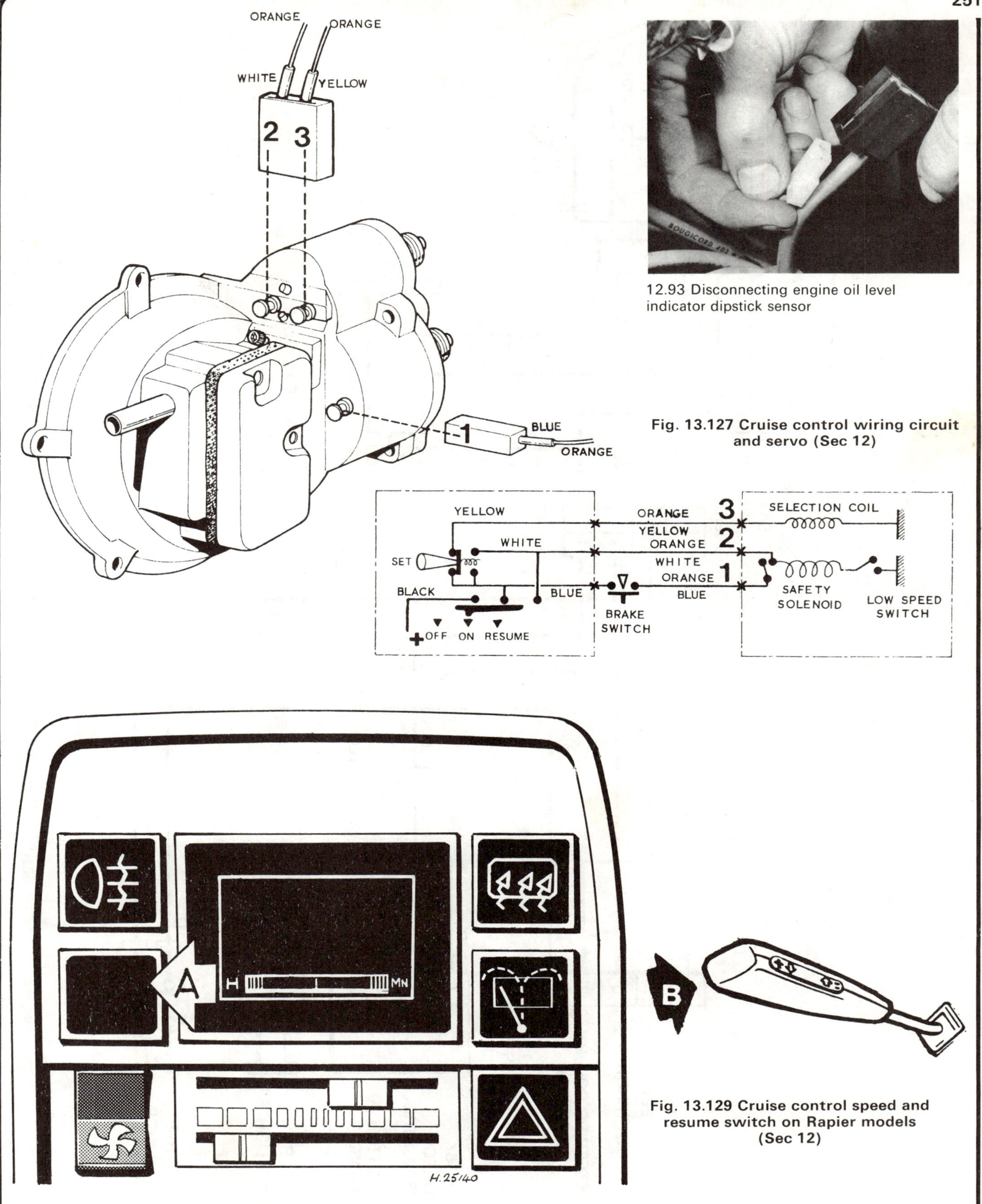

12.93 Disconnecting engine oil level indicator dipstick sensor

Fig. 13.127 Cruise control wiring circuit and servo (Sec 12)

Fig. 13.128 Cruise control on/off switch on Rapier models (Sec 12)

Fig. 13.129 Cruise control speed and resume switch on Rapier models (Sec 12)

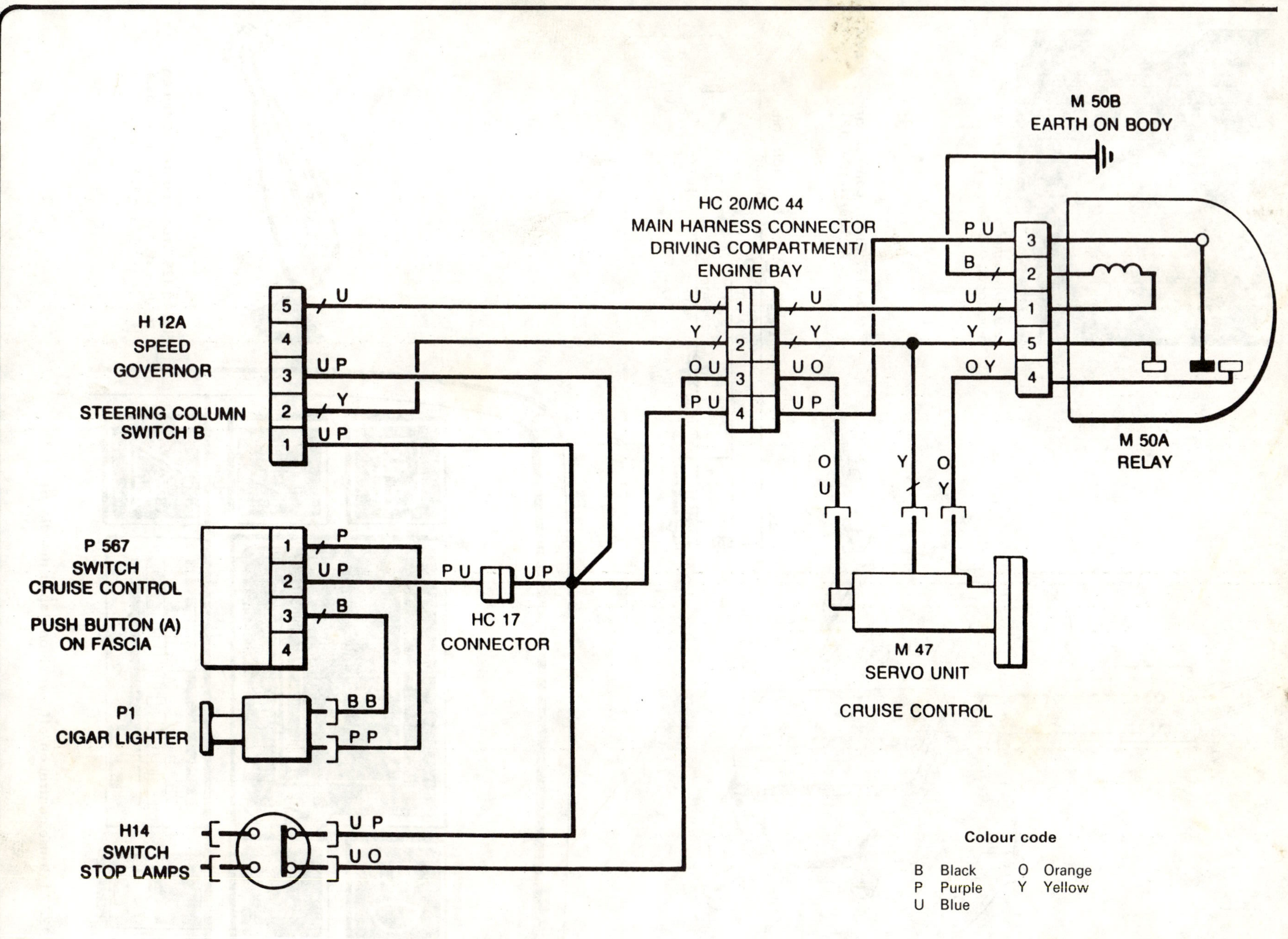

Fig. 13.130 Wiring diagram for 1985 Rapier cruise control system (Sec 12)

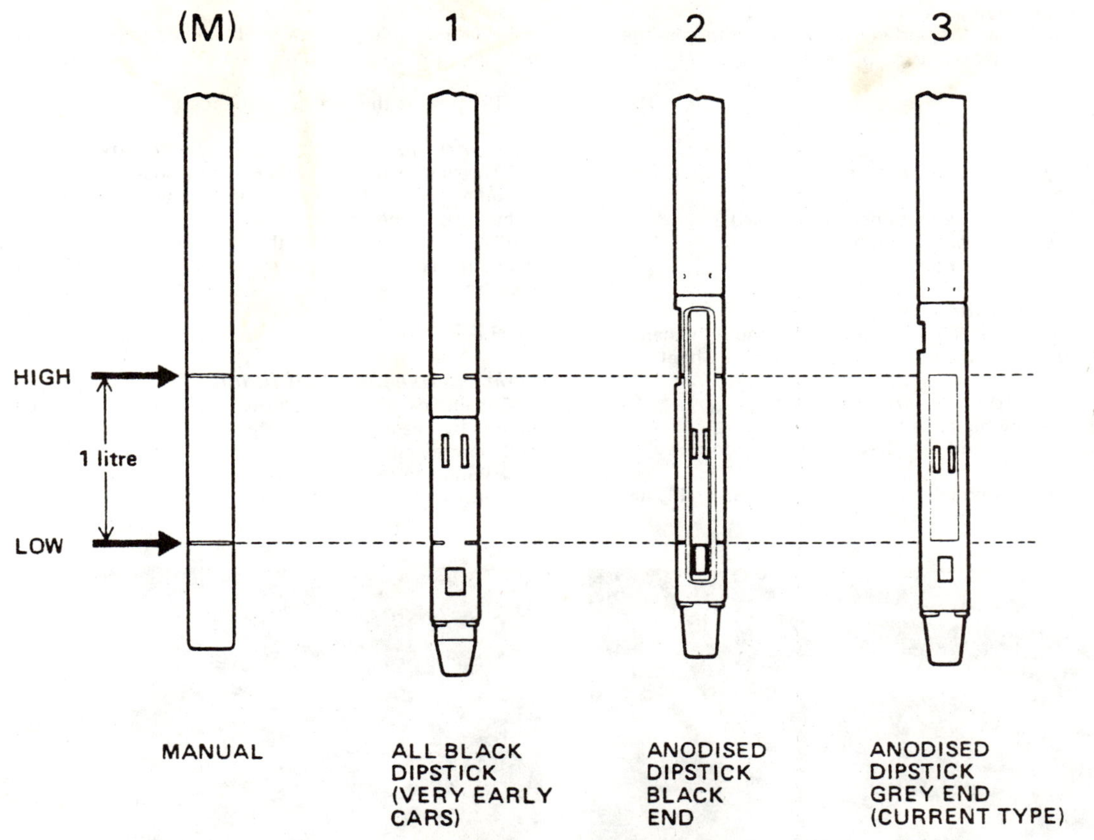

Fig. 13.131 Various types of electronic oil level indicator dipsticks (Sec 12)

Oil level indicator circuit – testing

98 The circuit diagram for the oil level indicator is shown in Fig. 13.132. If a fault develops, the procedure given in the following paragraphs should be carried out.

99 Check that battery voltage is available at the ignition switch, fuse, and control unit.

100 Check for continuity from the control unit multi-plug to the dipstick, and for continuity of the control unit earth wire.

101 Check the insulation and continuity of all related wiring.

102 The system can be tested by withdrawing the dipstick, then starting the engine.

103 The oil warning light should flash on and off as soon as oil pressure has been established.

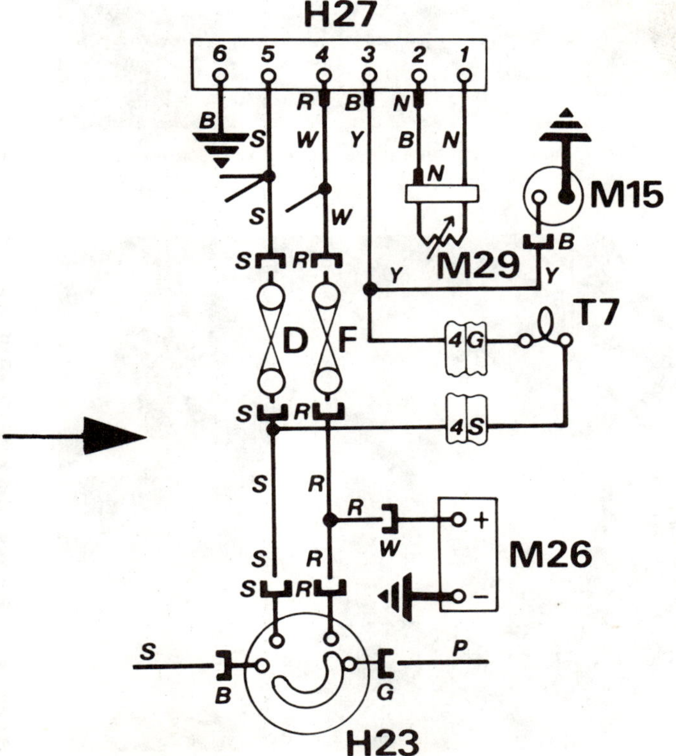

Fig. 13.132 Wiring diagram for electronic oil level indicator (Sec 12)

D	*Fuse, ignition controlled*	*M26*	*Battery*
F	*Fuse, battery supply*	*M29*	*Sensor, electronic oil level*
H23	*Ignition switch*	*T7*	*Low oil level and oil pressure warning light*
H27	*Control unit, electronic oil level*		
M15	*Sender unit, oil warning light*		

Colour code

B	Black	R	Red
G	Green	S	Slate
N	Brown	W	White
P	Purple	Y	Yellow

Radio/cassette player – removal and refitting

104 Disconnect the battery negative lead.
105 Remove the retaining screws and withdraw the radio/cassette player sufficiently to disconnect the supply wiring and aerial lead (photos).
106 Remove the radio/cassette player.
107 Refitting is a reversal of removal.

Boot light (Solara) – removal and refitting

108 Disconnect the battery negative lead.
109 Prise the boot light from the crossmember (photo).
110 Disconnect the supply leads and withdraw the light.
111 Refitting is a reversal of removal.

Horn (Solara) – description

112 The horn on Solara models is located beneath the right-hand wing. Testing procedures are identical to those given in Chapter 11.

Speedometer cable – removal and refitting

113 Disconnect the battery negative terminal.
114 Undo the retaining bolt and withdraw the cable from the gearbox.
115 Remove the instrument panel, as described in Chapter 11, just sufficiently to enable the cable to be disconnected from the rear of the speedometer.
116 Pull the cable through the bulkhead after releasing the grommet, release the cable from its support clips and remove it from the car.
117 Refitting is the reverse sequence of removal.

Direction indicator flasher unit – removal and refitting

118 The flasher unit is plugged into a socket located beneath the instrument panel. The unit is removed by simply pulling it out of its connectors in the socket.
119 The actual location of the unit is dependent on model, but it can be traced by operating the direction indicators and tracking down the unit by sound.

PART B Mobile radio equipment

Aerials – selection and fitting

The choice of aerials is now very wide. It should be realised that the quality has a profound effect on radio performance, and a poor, inefficient aerial can make suppression difficult.

A wing-mounted aerial is regarded as probably the most efficient for signal collection, but a roof aerial is usually better for suppress-

12.105A Extracting radio/cassette player upper fixing screws

12.105B Extracting radio/cassette player lower fixing screws

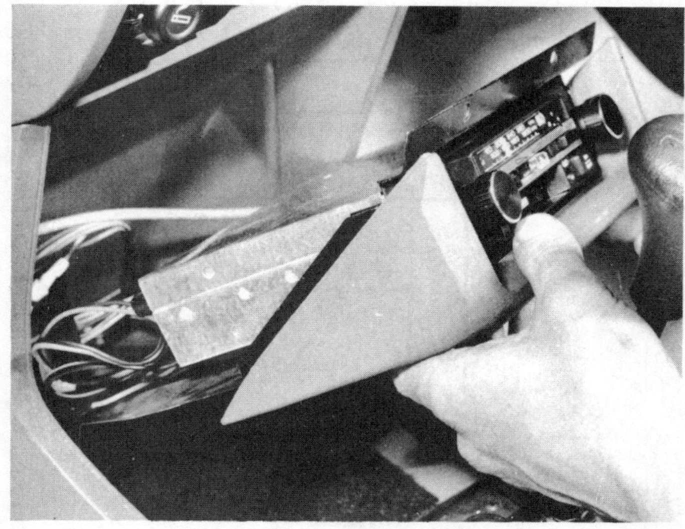

12.105C Withdrawing radio/cassette player

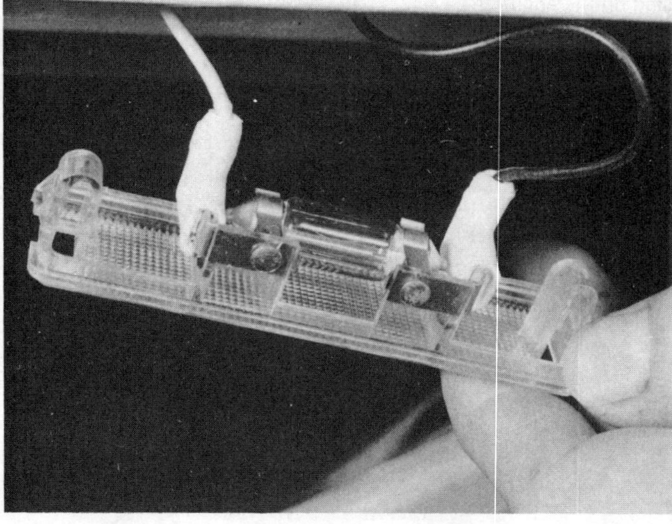

12.109 Boot light

ion purposes because it is away from most interference fields. Stick-on wire aerials are available for attachment to the inside of the windscreen, but are not always free from the interference field of the engine and some accessories.

Motorised automatic aerials rise when the equipment is switched on and retract at switch-off. They require more fitting space and supply leads, and can be a source of trouble.

There is no merit in choosing a very long aerial as, for example, the type about three metres in length which hooks or clips on to the rear of the car, since part of this aerial will inevitably be located in an interference field. For VHF/FM radios the best length of aerial is about one metre. Active aerials have a transistor amplifier mounted at the base and this serves to boost the received signal. The aerial rod is sometimes rather shorter than normal passive types.

A large loss of signal can occur in the aerial feeder cable, especially over the Very High Frequency (VHF) bands. The design of feeder cable is invariably in the co-axial form, ie a centre conductor surrounded by a flexible copper braid forming the outer (earth) conductor. Between the inner and outer conductors is an insulator material which can be in solid or stranded form. Apart from insulation, its purpose is to maintain the correct spacing and concentricity. Loss of signal occurs in this insulator, the loss usually being greater in a poor quality cable. The quality of cable used is reflected in the price of the aerial with the attached feeder cable.

The capacitance of the feeder should be within the range 65 to 75 picofarads (pF) approximately (95 to 100 pF for Japanese and American equipment), otherwise the adjustment of the car radio aerial trimmer may not be possible. An extension cable is necessary for a long run between aerial and receiver. If this adds capacitance in excess of the above limits, a connector containing a series capacitor will be required, or an extension which is labelled as 'capacity-compensated'.

Fitting the aerial will normally involve making a 7/8 in (22 mm) diameter hole in the bodywork, but read the instructions that come with the aerial kit. Once the hole position has been selected, use a centre punch to guide the drill. Use sticky masking tape around the area for this helps with marking out and drill location, and gives protection to the paintwork should the drill slip. Three methods of making the hole are in use:

(a) Use a hole saw in the electric drill. This is, in effect, a circular hacksaw blade wrapped round a former with a centre pilot drill.

(b) Use a tank cutter which also has cutting teeth, but is made to shear the metal by tightening with an Allen key.

(c) The hard way of drilling out the circle is using a small drill, say 1/8 in (3 mm), so that the holes overlap. The centre metal drops out and the hole is finished with round and half-round files.

Whichever method is used, the burr is removed from the body metal and paint removed from the underside. The aerial is fitted tightly ensuring that the earth fixing, usually a serrated washer, ring or clamp, is making a solid connection. *This earth connection is important in reducing interference.* Cover any bare metal with primer paint and topcoat, and follow by underseal if desired.

Aerial feeder cable routing should avoid the engine compartment and areas where stress might occur, eg under the carpet where feet will be located. Roof aerials require that the headlining be pulled back and that a path is available down the door pillar. It is wise to check with the vehicle dealer whether roof aerial fitting is recommended.

Loudspeakers

Speakers should be matched to the output stage of the equipment, particularly as regards the recommended impedance. Power transistors used for driving speakers are sensitive to the loading placed on them.

Before choosing a mounting position for speakers, check whether the vehicle manufacturer has provided a location for them. Generally door-mounted speakers give good stereophonic reproduction, but not all doors are able to accept them. The next best position is the rear parcel shelf, and in this case speaker apertures can be cut into the shelf, or pod units may be mounted.

For door mounting, first remove the trim, which is often held on by 'poppers' or press studs, and then select a suitable gap in the inside door assembly. Check that the speaker would not obstruct glass or winder mechanism by winding the window up and down.

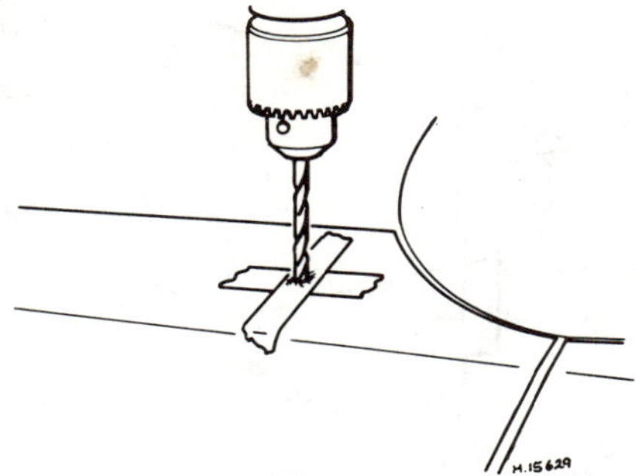

Fig. 13.133 Drilling the bodywork for aerial mounting

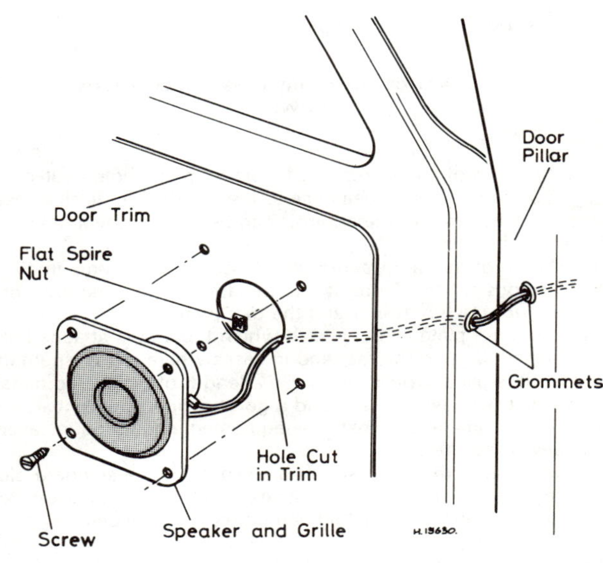

Fig. 13.134 Door-mounted speaker installation

A template is often provided for marking out the trim panel hole, and then the four fixing holes must be drilled through. Mark out with chalk and cut cleanly with a sharp knife or keyhole saw. Speaker leads are then threaded through the door and door pillar, if necessary drilling 10 mm diameter holes. Fit grommets in the holes and connect to the radio or tape unit correctly. Do not omit a waterproofing cover, usually supplied with door speakers. If the speaker has to be fixed into the metal of the door itself, use self-tapping screws, and if the fixing is to the door trim use self-tapping screws and flat spire nuts.

Rear shelf mounting is somewhat simpler but it is necessary to find gaps in the metalwork underneath the parcel shelf. However, remember that the speakers should be as far apart as possible to give a good stereo effect. Pod-mounted speakers can be screwed into position through the parcel shelf material, but it is worth testing for the best position. Sometimes good results are found by reflecting sound off the rear window.

Unit installation

Many vehicles have a dash panel aperture to take a radio/audio unit, a recognised international standard being 189.5 mm x 60 mm. Alternatively a console may be a feature of the car interior design and this, mounted below the dashboard, gives more room. If neither facility is available a unit may be mounted on the underside of the parcel shelf; these are frequently non-metallic and an earth wire from the case to a good earth point is necessary. A three-sided

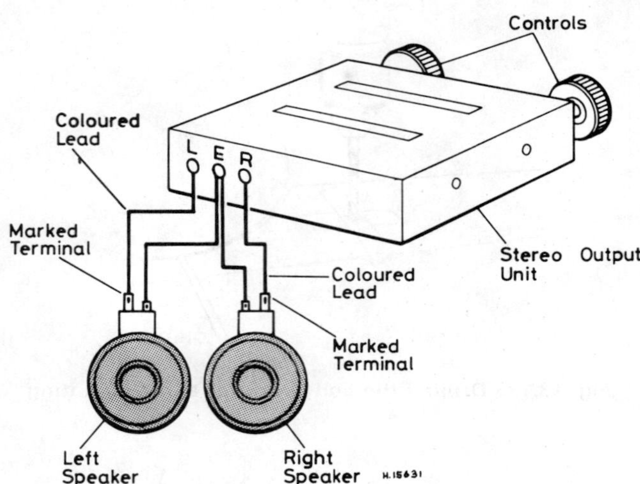

Fig. 13.135 Speaker connections must be correctly made as shown

cover in the form of a cradle is obtainable from car radio dealers and this gives a professional appearance to the installation; in this case choose a position where the controls can be reached by a driver with his seat belt on.

Installation of the radio/audio unit is basically the same in all cases, and consists of offering it into the aperture after removal of the knobs *(not* push buttons) and the trim plate. In some cases a special mounting plate is required to which the unit is attached. It is worthwhile supporting the rear end in cases where sag or strain may occur, and it is usually possible to use a length of perforated metal strip attached between the unit and a good support point nearby. In general it is recommended that tape equipment should be installed at or nearly horizontal.

Connections to the aerial socket are simply by the standard plug terminating the aerial downlead or its extension cable. Speakers for a stereo system must be matched and correctly connected, as outlined previously.

Note: *While all work is carried out on the power side, it is wise to disconnect the battery earth lead.* Before connection is made to the vehicle electrical system, check that the polarity of the unit is correct. Most vehicles use a negative earth system, but radio/audio units often have a reversible plug to convert the set to either + or − earth. *Incorrect connection may cause serious damage.*

The power lead is often permanently connected inside the unit and terminates with one half of an in-line fuse carrier. The other half is fitted with a suitable fuse (3 or 5 amperes) and a wire which should go to a power point in the electrical system. This may be the accessory terminal on the ignition switch, giving the advantage of power feed with ignition or with the ignition key at the 'accessory' position. Power to the unit stops when the ignition key is removed. Alternatively, the lead may be taken to a live point at the fusebox with the consequence of having to remember to switch off at the unit before leaving the vehicle.

Before switching on for initial test, be sure that the speaker connections have been made, for running without load can damage the output transistors. Switch on next and tune through the bands to ensure that all sections are working, and check the tape unit if applicable. The aerial trimmer should be adjusted to give the strongest reception on a weak signal in the medium wave band, at say 200 metres.

Interference

In general, when electric current changes abruptly, unwanted electrical noise is produced. The motor vehicle is filled with electrical devices which change electric current rapidly, the most obvious being the contact breaker.

When the spark plugs operate, the sudden pulse of spark current causes the associated wiring to radiate. Since early radio transmitters used sparks as a basis of operation, it is not surprising that

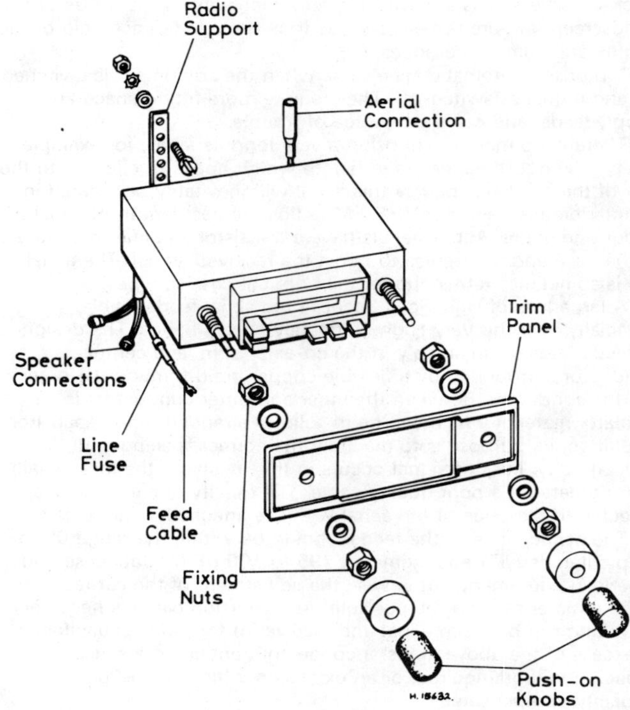

Fig. 13.136 Mounting component details for radio/cassette unit

the car radio will pick up ignition spark noise unless steps are taken to reduce it to acceptable levels.

Interference reaches the car radio in two ways:

(a) by conduction through the wiring.
(b) by radiation to the receiving aerial.

Initial checks presuppose that the bonnet is down and fastened, the radio unit has a good earth connection *(not* through the aerial downlead outer), no fluorescent tubes are working near the car, the aerial trimmer has been adjusted, and the vehicle is in a position to receive radio signals, ie not in a metal-clad building.

Switch on the radio and tune it to the middle of the medium wave (MW) band off-station with the volume (gain) control set fairly high. Switch on the ignition (but do not start the engine) and wait to see if irregular clicks or hash noise occurs. Tapping the facia panel may also produce the effects. If so, this will be due to the

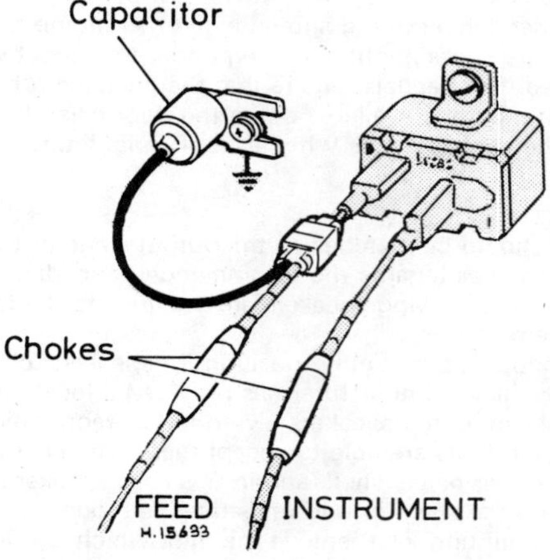

Fig. 13.137 Voltage stabiliser interference suppression

voltage stabiliser, which is an on-off thermal switch to control instrument voltage. It is located usually on the back of the instrument panel, often attached to the speedometer. Correction is by attachment of a capacitor and, if still troublesome, chokes in the supply wires.

Switch on the engine and listen for interference on the MW band. Depending on the type of interference, the indications are as follows.

A harsh crackle that drops out abruptly at low engine speed or when the headlights are switched on is probably due to a voltage regulator.

A whine varying with engine speed is due to the dynamo or alternator. Try temporarily taking off the fan belt – if the noise goes this is confirmation.

Regular ticking or crackle that varies in rate with the engine speed is due to the ignition system. With this trouble in particular and others in general, check to see if the noise is entering the receiver from the wiring or by radiation. To do this, pull out the aerial plug, (preferably shorting out the input socket or connecting a 62 pF capacitor across it). If the noise disappears it is coming in through the aerial and is *radiation noise.* If the noise persists it is reaching the receiver through the wiring and is said to be *line-borne.*

Interference from wipers, washers, heater blowers, turn-indicators, stop lamps, etc is usually taken to the receiver by wiring, and simple treatment using capacitors and possibly chokes will solve the problem. Switch on each one in turn (wet the screen first for running wipers!) and listen for possible interference with the aerial plug in place and again when removed.

Electric petrol pumps are now finding application again and give rise to an irregular clicking, often giving a burst of clicks when the ignition is on but the engine has not yet been started. It is also possible to receive whining or crackling from the pump.

Note that if most of the vehicle accessories are found to be creating interference all together, the probability is that poor aerial earthing is to blame.

Component terminal markings

Throughout the following sub-sections reference will be found to various terminal markings. These will vary depending on the manufacturer of the relevant component. If terminal markings differ from those mentioned, reference should be made to the following table, where the most commonly encountered variations are listed.

Alternator	*Alternator terminal (thick lead)*	*Exciting winding terminal*
DIN/Bosch	B+	DF
Delco Remy	+	EXC
Ducellier	+	EXC
Ford (US)	+	DF
Lucas	+	F
Marelli	+B	F

Ignition coil	*Ignition switch terminal*	*Contact breaker terminal*
DIN/Bosch	15	1
Delco Remy	+	–
Ducellier	BAT	RUP
Ford (US)	B/+	CB/–
Lucas	SW/+	–
Marelli	BAT/+B	D

Voltage regulator	*Voltage input terminal*	*Exciting winding terminal*
DIN/Bosch	B+/D+	DF
Delco Remy	BAT/+	EXC
Ducellier	BOB/BAT	EXC
Ford (US)	BAT	DF
Lucas	+/A	F
Marelli		F

Suppression methods – ignition

Suppressed HT cables are supplied as original equipment by manufacturers and will meet regulations as far as interference to neighbouring equipment is concerned. It is illegal to remove such suppression unless an alternative is provided, and this may take the form of resistive spark plug caps in conjunction with plain copper HT cable. For VHF purposes, these and 'in-line' resistors may not be effective, and resistive HT cable is preferred. Check that suppressed cables are actually fitted by observing cable identity lettering, or measuring with an ohmmeter – the value of each plug lead should be 5000 to 10 000 ohms.

A 1 microfarad capacitor connected from the LT supply side of the ignition coil to a good nearby earth point will complete basic ignition interference treatment. *NEVER fit a capacitor to the coil terminal to the contact breaker – the result would be burnt out points in a short time.*

If ignition noise persists despite the treatment above, the following sequence should be followed:

(a) Check the earthing of the ignition coil; remove paint from fixing clamp.

(b) If this does not work, lift the bonnet. Should there be no change in interference level, this may indicate that the bonnet is not electrically connected to the car body. Use a proprietary braided strap across a bonnet hinge ensuring a first class electrical connection. If, however, lifting the bonnet increases the interference, then fit resistive HT cables of a higher ohms-per-metre value.

(c) If all these measures fail, it is probable that re-radiation from metallic components is taking place. Using a braided strap between metallic points, go round the vehicle systematically – try the following: engine to body, exhaust system to body, front suspension to engine and to body, steering column to body (especially French and Italian cars), gear lever to engine and to body (again especially French and Italian cars), Bowden cable to body, metal

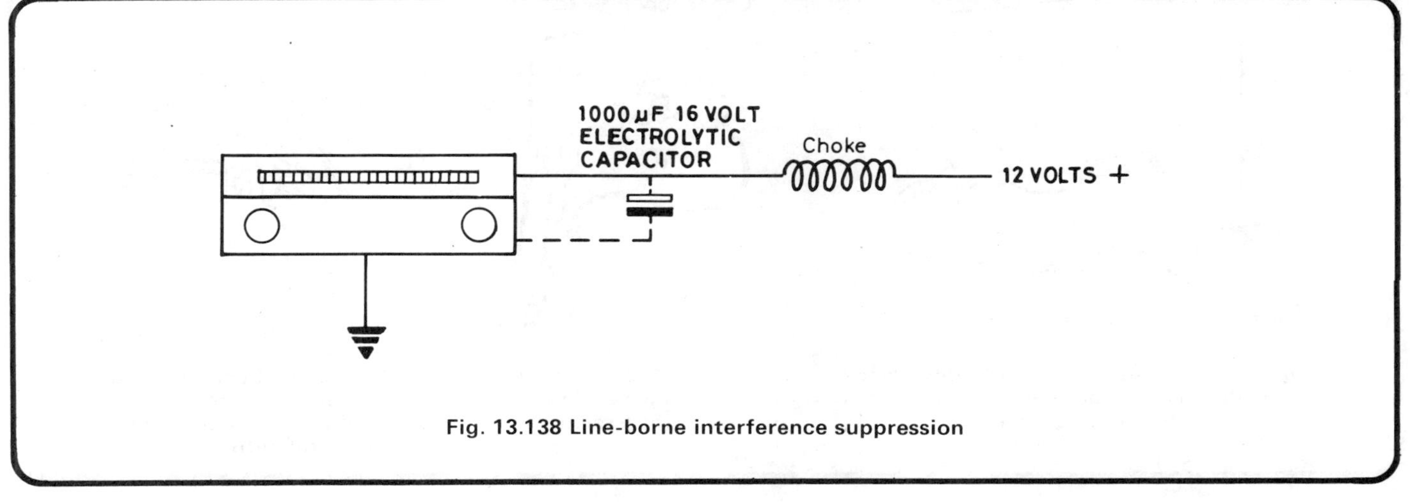

Fig. 13.138 Line-borne interference suppression

parcel shelf to body. When an offending component is located it should be bonded with the strap permanently.

(d) As a next step, the fitting of distributor suppressors to each lead at the distributor end may help.

(e) Beyond this point is involved the possible screening of the distributor and fitting resistive spark plugs, but such advanced treatment is not usually required for vehicles with entertainment equipment.

Electronic ignition systems have built-in suppression components, but this does not relieve the need for using suppressed HT leads. In some cases it is permitted to connect a capacitor on the low tension supply side of the ignition coil, but not in every case. Makers' instructions should be followed carefully, otherwise damage to the ignition semiconductors may result.

Suppression methods – generators

Alternators should be fitted with a 3 microfarad capacitor from the B+ main output terminal (thick cable) to earth. Additional suppression may be obtained by the use of a filter in the supply line to the radio receiver.

It is most important that:

(a) Capacitors are never connected to the field terminals of an alternator.

(b) Alternators must not be run without connection to the battery.

Suppression methods – voltage regulators

Alternator regulators come in three types:

(a) Vibrating contact regulators separate from the alternator. Used extensively on continental vehicles.

(b) Electronic regulators separate from the alternator.

(c) Electronic regulators built-in to the alternator.

In case (a) interference may be generated on the AM and FM (VHF) bands. For some cars a replacement suppressed regulator is available. Filter boxes may be used with non-suppressed regulators. But if not available, then for AM equipment a 2 microfarad or 3 microfarad capacitor may be mounted at the voltage terminal marked D+ or B+ of the regulator. FM bands may be treated by a feed-through capacitor of 2 or 3 microfarad.

Electronic voltage regulators are not always troublesome, but where necessary, a 1 microfarad capacitor from the regulator + terminal will help.

Integral electronic voltage regulators do not normally generate much interference, but when encountered this is in combination with alternator noise. A 1 microfarad or 2 microfarad capacitor from the warning lamp (IND) terminal to earth for Lucas ACR alternators and Femsa, Delco and Bosch equivalents should cure the problem.

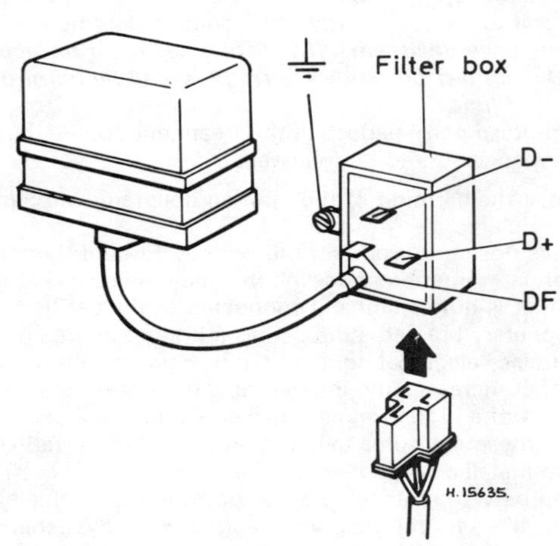

Fig. 13.139 Typical filter box for vibrating contact voltage regulator (alternator equipment)

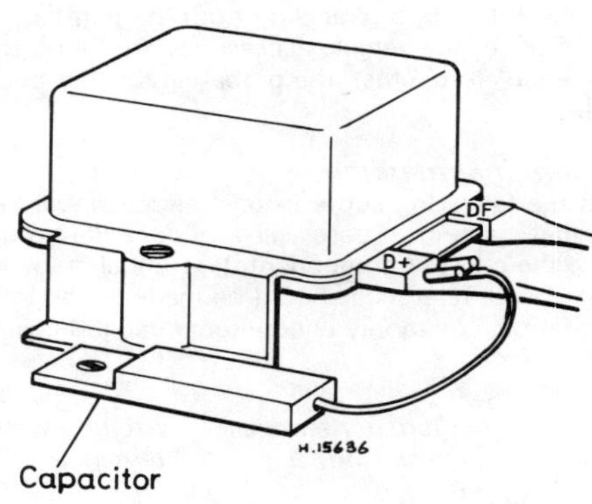

Fig. 13.140 Suppression of AM interference by vibrating contact voltage regulator (alternator equipment)

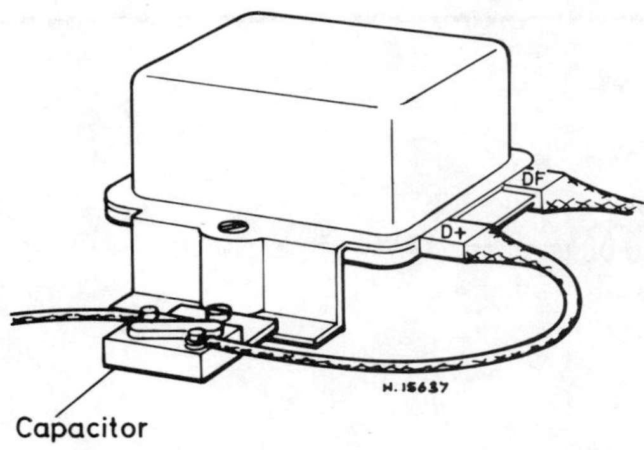

Fig. 13.141 Suppression of FM interference by vibrating contact voltage regulator (alternator equipment)

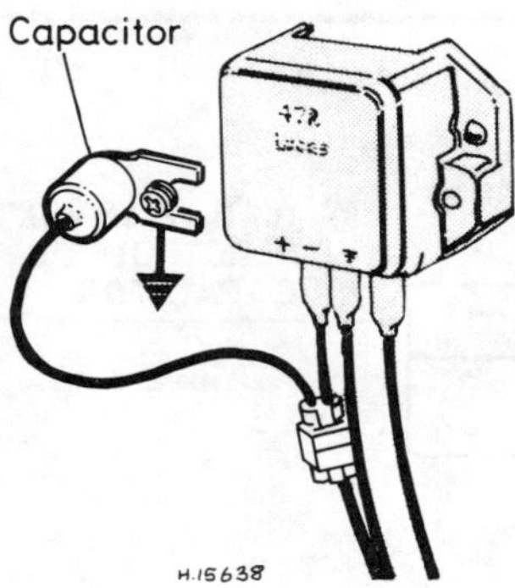

Fig. 13.142 Electronic voltage regulator suppression

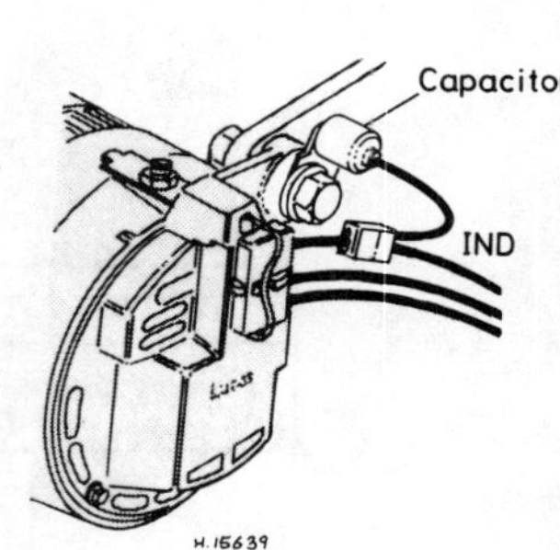

Fig. 13.143 Suppression of interference from electronic voltage regulator when integral with alternator

Suppression methods – other equipment

Wiper motors – Connect the wiper body to earth with a bonding strap. For all motors use a 7 ampere choke assembly inserted in the leads to the motor.

Heater motors – Fit 7 ampere line chokes in both leads, assisted if necessary by a 1 microfarad capacitor to earth from both leads.

Electronic tachometer – The tachometer is a possible source of ignition noise – check by disconnecting at the ignition coil CB terminal. It usually feeds from ignition coil LT pulses at the contact breaker terminal. A 3 ampere line choke should be fitted in the tachometer lead at the coil CB terminal.

Horn – A capacitor and choke combination is effective if the horn is directly connected to the 12 volt supply. The use of a relay is an alternative remedy, as this will reduce the length of the interference-carrying leads.

Electrostatic noise – Characteristics are erratic crackling at the receiver, with disappearance of symptoms in wet weather. Often shocks may be given when touching bodywork. Part of the problem is the build-up of static electricity in non-driven wheels and the acquisition of charge on the body shell. It is possible to fit spring-loaded contacts at the wheels to give good conduction between the rotary wheel parts and the vehicle frame. Changing a tyre sometimes helps – because of tyres' varying resistances. In difficult cases a trailing flex which touches the ground will cure the problem. If this is not acceptable it is worth trying conductive paint on the tyre walls.

Fuel pump – Suppression requires a 1 microfarad capacitor between the supply wire to the pump and a nearby earth point. If this is insufficient a 7 ampere line choke connected in the supply wire near the pump is required.

Fluorescent tubes – Vehicles used for camping/caravanning frequently have fluorescent tube lighting. These tubes require a relatively high voltage for operation and this is provided by an inverter (a form of oscillator) which steps up the vehicle supply voltage. This can give rise to serious interference to radio reception, and the tubes themselves can contribute to this interference by the pulsating nature of the lamp discharge. In such situations it is important to mount the aerial as far away from a fluorescent tube as possible. The interference problem may be alleviated by screening the tube with fine wire turns spaced an inch (25 mm) apart and earthed to the chassis. Suitable chokes should be fitted in both supply wires close to the inverter.

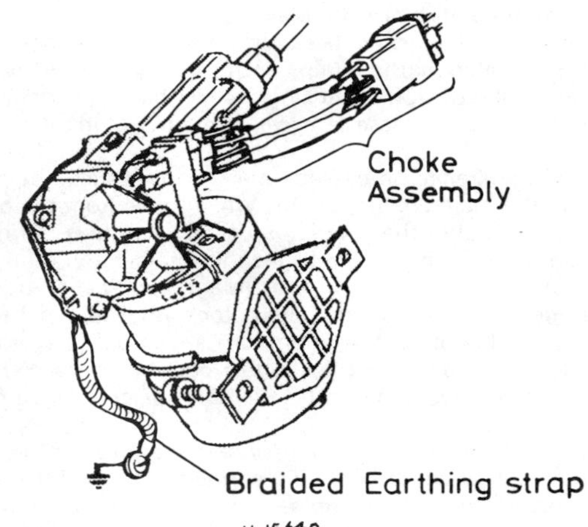

Fig. 13.144 Wiper motor suppression

Radio/cassette case breakthrough

Magnetic radiation from dashboard wiring may be sufficiently intense to break through the metal case of the radio/cassette player. Often this is due to a particular cable routed too close and shows up as ignition interference on AM and cassette play and/or alternator whine on cassette play.

The first point to check is that the clips and/or screws are fixing all parts of the radio/cassette case together properly. Assuming good earthing of the case, see if it is possible to re-route the

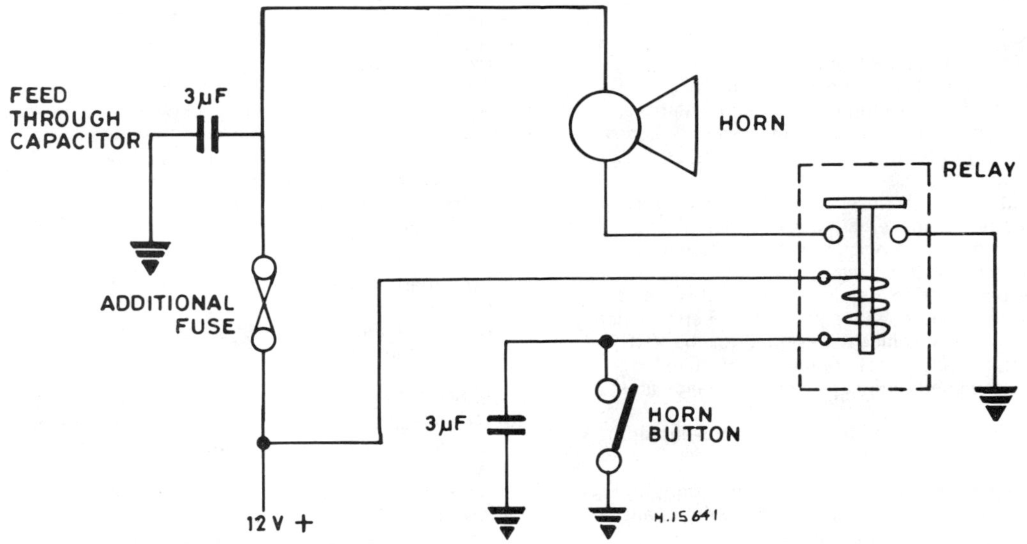

Fig. 13.145 Use of relay to reduce horn interference

offending cable – the chances of this are not good, however, in most cars.

Next release the radio/cassette player and locate it in different positions with temporary leads. If a point of low interference is found, then if possible fix the equipment in that area. This also confirms that local radiation is causing the trouble. If re-location is not feasible, fit the radio/cassette player back in the original position.

Alternator interference on cassette play is now caused by radiation from the main charging cable which goes from the battery to the output terminal of the alternator, usually via the + terminal of the starter motor relay. In some vehicles this cable is routed under the dashboard, so the solution is to provide a direct cable route. Detach the original cable from the alternator output terminal and make up a new cable of at least 6 mm² cross-sectional area to go from alternator to battery with the shortest possible route. *Remember – do not run the engine with the alternator disconnected from the battery.*

Ignition breakthrough on AM and/or cassette play can be a difficult problem. It is worth wrapping earthed foil round the offending cable run near the equipment, or making up a deflector plate well screwed down to a good earth. Another possibility is the use of a suitable relay to switch on the ignition coil. The relay should be mounted close to the ignition coil; with this arrangement the ignition coil primary current is not taken into the dashboard area and does not flow through the ignition switch. A suitable diode should be used since it is possible that at ignition switch-off the output from the warning lamp alternator terminal could hold the relay on.

Connectors for suppression components

Capacitors are usually supplied with tags on the end of the lead, while the capacitor body has a flange with a slot or hole to fit under a nut or screw with washer.

Connections to feed wires are best achieved by self-stripping connectors. These connectors employ a blade which, when squeezed down by pliers, cuts through cable insulation and makes connection to the copper conductors beneath.

Chokes sometimes come with bullet snap-in connectors fitted to the wires, and also with just bare copper wire. With connectors, suitable female cable connectors may be purchased from an auto-accessory shop together with any extra connectors required for the cable ends after being cut for the choke insertion. For chokes with bare wires, similar connectors may be employed together with insulation sleeving as required.

VHF/FM broadcasts

Reception of VHF/FM in an automobile is more prone to problems than the medium and long wavebands. Medium/long wave transmitters are capable of covering considerable distances, but VHF transmitters are restricted to line of sight, meaning ranges of 10 to 50 miles, depending upon the terrain, the effects of buildings and the transmitter power.

Because of the limited range it is necessary to retune on a long journey, and it may be better for those habitually travelling long distances or living in areas of poor provision of transmitters to use an AM radio working on medium/long wavebands.

When conditions are poor, interference can arise, and some of the suppression devices described previously fall off in performance at very high frequencies unless specifically designed for the VHF band. Available suppression devices include reactive HT cable, resistive distributor caps, screened plug caps, screened leads and resistive spark plugs.

For VHF/FM receiver installation the following points should be particularly noted:

(a) Earthing of the receiver chassis and the aerial mounting is important. Use a separate earthing wire at the radio, and scrape paint away at the aerial mounting.
(b) If possible, use a good quality roof aerial to obtain maximum height and distance from interference generating devices on the vehicle.
(c) Use of a high quality aerial downlead is important, since losses in cheap cable can be significant.
(d) The polarisation of FM transmissions may be horizontal, vertical, circular or slanted. Because of this the optimum mounting angle is at 45° to the vehicle roof.

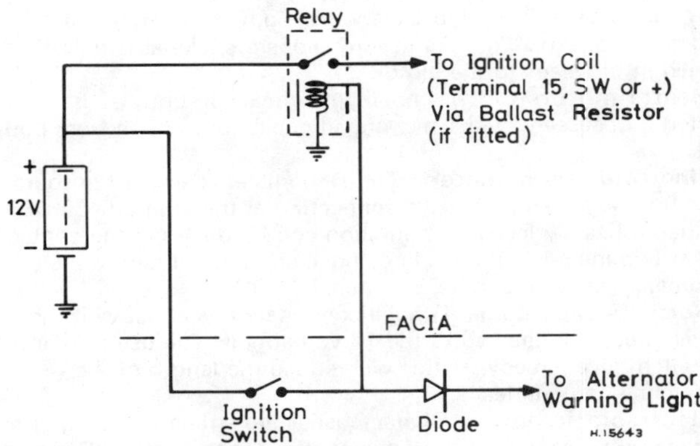

Fig. 13.146 Use of ignition coil relay to suppress case breakthrough

Citizens' Band radio (CB)

In the UK, CB transmitter/receivers work within the 27 MHz and 934 MHz bands, using the FM mode. At present interest is concentrated on 27 MHz where the design and manufacture of equipment is less difficult. Maximum transmitted power is 4 watts, and 40 channels spaced 10 kHz apart within the range 27.60125 to 27.99125 MHz are available.

Aerials are the key to effective transmission and reception. Regulations limit the aerial length to 1.65 metres including the loading coil and any associated circuitry, so tuning the aerial is necessary to obtain optimum results. The choice of a CB aerial is dependent on whether it is to be permanently installed or removable, and the performance will hinge on correct tuning and the location point on the vehicle. Common practice is to clip the aerial to the roof gutter or to employ wing mounting where the aerial can be rapidly unscrewed. An alternative is to use the boot rim to render the aerial theftproof, but a popular solution is to use the 'magmount' – a type of mounting having a strong magnetic base clamping to the vehicle at any point, usually the roof.

Aerial location determines the signal distribution for both transmission and reception, but it is wise to choose a point away from the engine compartment to minimise interference from vehicle electrical equipment.

The aerial is subject to considerable wind and acceleration forces. Cheaper units will whip backwards and forwards and in so doing will alter the relationship with the metal surface of the vehicle with which it forms a ground plane aerial system. The radiation pattern will change correspondingly, giving rise to break-up of both incoming and outgoing signals.

Interference problems on the vehicle carrying CB equipment fall into two categories:

(a) Interference to nearby TV and radio receivers when transmitting.
(b) Interference to CB set reception due to electrical equipment on the vehicle.

Problems of break-through to TV and radio are not frequent, but can be difficult to solve. Mostly trouble is not detected or reported because the vehicle is moving and the symptoms rapidly disappear at the TV/radio receiver, but when the CB set is used as a base station any trouble with nearby receivers will soon result in a complaint.

It must not be assumed by the CB operator that his equipment is faultless, for much depends upon the design. Harmonics (that is, multiples) of 27 MHz may be transmitted unknowingly and these can fall into other user's bands. Where trouble of this nature occurs, low pass filters in the aerial or supply leads can help, and should be fitted in base station aerials as a matter of course. In stubborn cases it may be necessary to call for assistance from the licensing authority, or, if possible, to have the equipment checked by the manufacturers.

Interference received on the CB set from the vehicle equipment is, fortunately, not usually a severe problem. The precautions outlined previously for radio/cassette units apply, but there are some extra points worth noting.

It is common practice to use a slide-mount on CB equipment enabling the set to be easily removed for use as a base station, for example. Care must be taken that the slide mount fittings are properly earthed and that first class connection occurs between the set and slide-mount.

Vehicle manufacturers in the UK are required to provide suppression of electrical equipment to cover 40 to 250 MHz to protect TV and VHF radio bands. Such suppression appears to be adequately effective at 27 MHz, but suppression of individual items such as alternators/dynamos, clocks, stabilisers, flashers, wiper motors, etc, may still be necessary. The suppression capacitors and chokes available from auto-electrical suppliers for entertainment receivers will usually give the required results with CB equipment.

Other vehicle radio transmitters

Besides CB radio already mentioned, a considerable increase in the use of transceivers (ie combined transmitter and receiver units) has taken place in the last decade. Previously this type of equipment was fitted mainly to military, fire, ambulance and police vehicles, but a large business radio and radio telephone usage has developed.

Generally the suppression techniques described previously will suffice, with only a few difficult cases arising. Suppression is carried out to satisfy the 'receive mode', but care must be taken to use heavy duty chokes in the equipment supply cables since the loading on 'transmit' is relatively high.

13 Bodywork and fittings

Boot lid (Solara) – removal and refitting

1 Support the boot lid in its open position and place cloths under the front corners to protect the paintwork.

2 Using a pencil, mark the position of the hinges on the boot lid.

3 Remove the bolts and lift the boot lid from the car.

4 If necessary, unbolt the boot lock and slide out the cage nuts. Extract the washer and remove the private lock. Remove the rubber buffers, badges, and clips. Unbolt the striker if necessary.

5 Refitting is a reversal of removal, but make sure that the boot lid is positioned centrally in the body aperture. If necessary, loosen the hinge bolts and adjust the boot lid, then tighten the bolts.

Boot lid spring (Solara) – removal and refitting

6 Support the boot lid in the open position.

7 Attach a length of strong wire to the rear hook of the spring, then pull the spring from the bracket.

8 Withdraw the spring and sheath (photo).

9 Refitting is a reversal of removal.

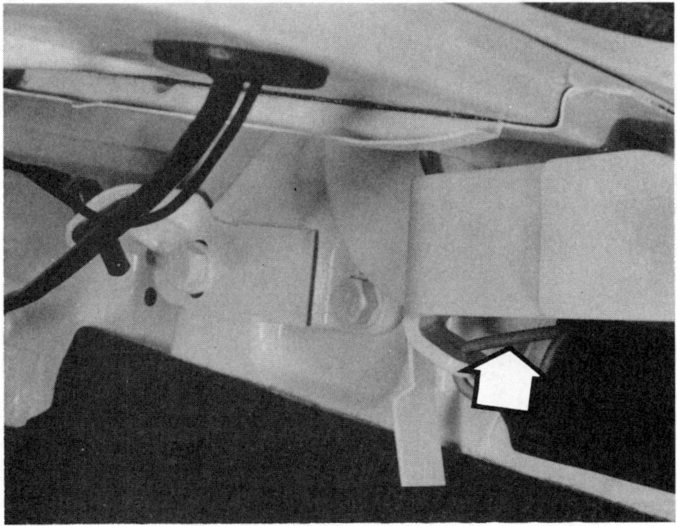

13.8 Solara luggage boot spring (arrowed)

Boot lid hinge (Solara) – removal and refitting

10 Support the boot lid in its open position and place cloths under the front corners to protect the paintwork.

11 Using a pencil, mark the position of the hinge to be removed on the boot lid and body.

12 Attach a length of strong wire to the rear hook of the spring, then pull the spring from the bracket.

13 Unbolt and remove the hinge from the boot lid and body.

14 Refitting is a reversal of removal.

Console – removal and refitting

15 Remove both front seats as follows. Move the seat rearwards and remove the Allen screws from the seat rail. Move the seat forwards and remove the Allen screws from the rear. Withdraw the seat and recover the spacers.

16 Remove the radio (if fitted) or alternatively remove the console-to-switch panel screws.

17 Remove the carpet side trim in front of the console and remove the two front mounting screws.

18 Extract the grip pin from the gear lever gaiter.

19 Disengage the console from the rear duct, then withdraw the console over the handbrake lever and gear lever.

20 If necessary remove the rear ducts, seal, console base, gaiter and upper console (photos).

21 Refitting is a reversal of removal (photo).

Heater unit (commencing A-Series) – description

22 As from A-Series models a vacuum capsule is located on the side of the heater unit in order to operate the lower airflow flap.

23 Removal and refitting procedures are similar to those described in Chapter 12, but in addition it is necessary to disconnect the vacuum supply pipe.

Remote door mirror – removal and refitting

24 Remove the screw and withdraw the outer adjuster knob and washer (photo).

25 Remove the remaining adjuster knob.

26 Remove the screw and withdraw the cover plate (photo).

27 Unscrew the notched nut and withdraw the mirror (photo).

28 Refitting is a reversal of removal.

Sun roof – maintenance, removal and refitting

29 A sun roof is available as a factory-fitted option on all models. Maintenance consists of keeping the guide rails clean, and ensuring that the drain tube outlets (in the front and rear wheel arches) are kept clear. Particular attention should be paid to this latter point if under-sealing or other underbody anti-rust treatment is carried out.

30 To remove the sun roof from the car, first open it as far as it will go. Lever the retaining stud out of one runner, prise the front of the runner free and pull the runner off the guide rail. Repeat the operation on the other runner.

31 Slowly close the sun roof, at the same time lifting it, until the front supports are clear of the guide rails.

32 From outside the car, lift the front of the sun roof and draw it forwards until the rear supports can be freed from the guide rails. Take care not to scratch the car paintwork. The sun roof can now be lifted away from the car.

33 Refitting is the reverse of the removal procedure, but note the following points:

(a) Fit the runners to the guide rails with the tip uppermost
(b) If new retaining studs are used, trim off any excess under the runners
(c) If dismantling has taken place, slacken the front side support screws on each side (after withdrawing the inner panel if necessary). With the sun roof open by about 2 in (50 mm), push the side supports outwards as far as possible, then retighten the screws.

Sun roof – dismantling and reassembly

34 With the sun roof removed from the car as described above, commence dismantling by removing the inner panel.

13.20A Gearlever gaiter screw

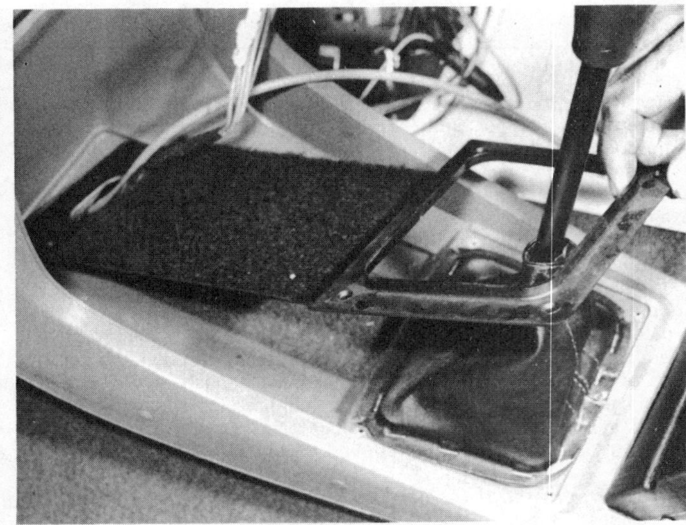
13.20B Removing centre console base

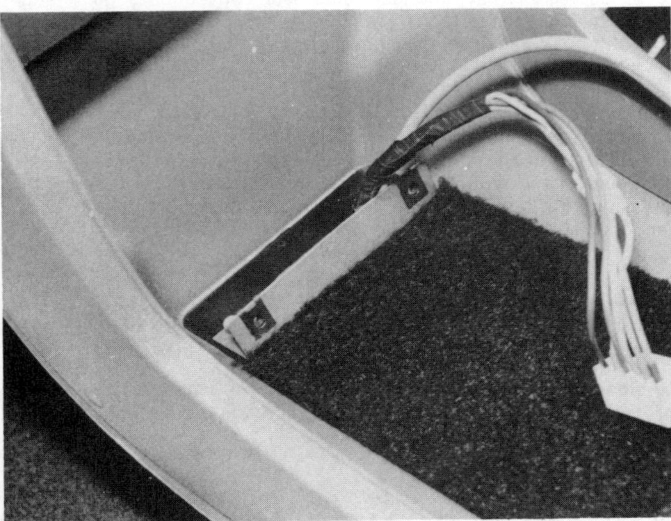
13.21 Centre console wiring harness correctly routed

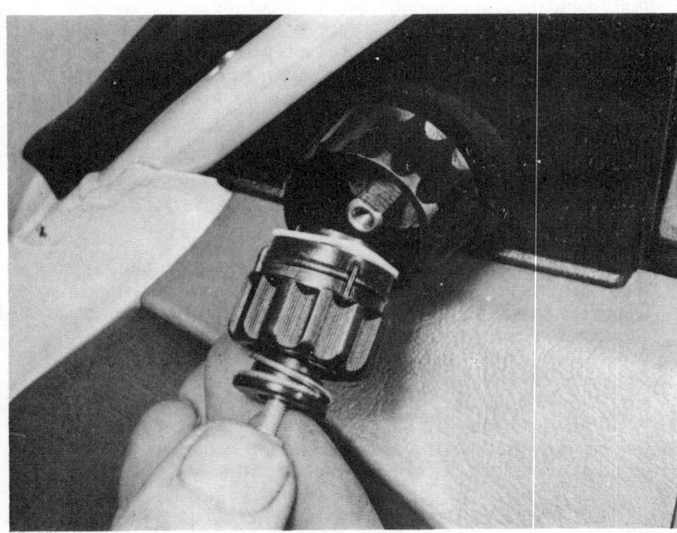
13.24 Removing door mirror adjuster knob

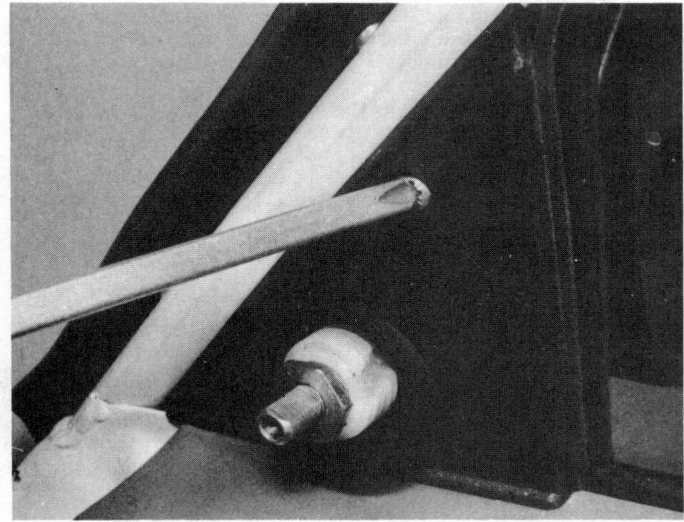
13.26 Removing door mirror cover plate

13.27 Door mirror notched retaining nut

35 Dismantling is now self-explanatory with reference to Fig. 13.147. Remote components in the order in which they are listed in the key.

36 Reassembly is the reverse of the dismantling procedure, but note the following points:

- *(a) Grease the ends of the latch before refitting*
- *(b) Grease both ends of the connecting rods, and make sure that the dowel is properly engaged with the control lever*
- *(c) Refit the anti-rattle springs with their open ends to the rear and their curved sections towards the panel*
- *(d) Do not refit the inner panel until the front side supports have been adjusted as described in paragraph 33*
- *(e) Use a suitable adhesive to stick back any displaced lining*

Radiator grille – removal and refitting

37 Using a small screwdriver or thin rod, push out the peg in the centre of the grille upper retaining catch (photo). Recover the peg as it is released from the catch.

38 Tip the grille forward at the top and lift it up to disengage the lower lugs from their locations (photo). Remove the grille from the car.

39 Refitting is the reverse sequence of removal.

Interior rear view mirror – removal and refitting

40 This is bonded to the windscreen glass.

41 The mirror can be removed using a hot air gun, or by pulling a thin nylon cord back and forth between the glass and the mounting patch.

42 New mirrors are supplied with an adhesive patch, but to refix an old mirror, clean old adhesive from the glass and mirror and bond with a special glass bonding agent.

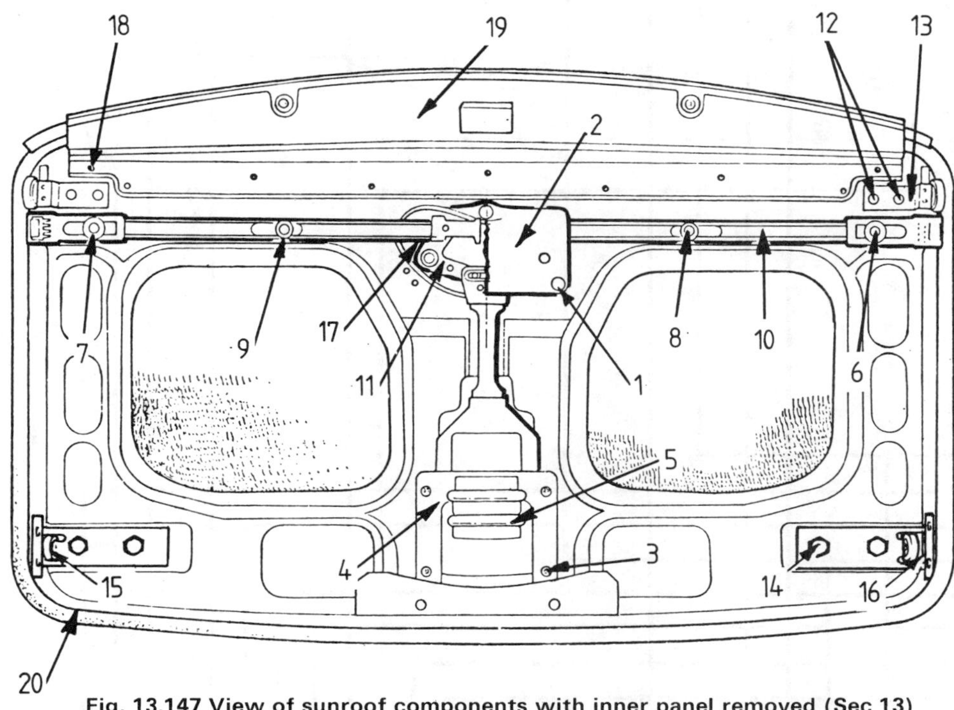

Fig. 13.147 View of sunroof components with inner panel removed (Sec 13)

1 Lock cover screws
2 Lock cover
3 Latch housing screws
4 Latch housing
5 Latch
6 Connecting rod outer fixing screws
7 Connecting rod outer guide
8 Inner guide fixing nuts
9 Inner guide
10 Connecting rod
11 Control lever
12 Rear support screws
13 Rear support
14 Front support screws
15 Front support
16 Anti-rattle spring
17 Lock cover bracket
18 Gutter screws
19 Gutter
20 Weatherstrip

13.37 Pushing out radiator grille catch plug

13.38 Radiator grille lower fixing lugs

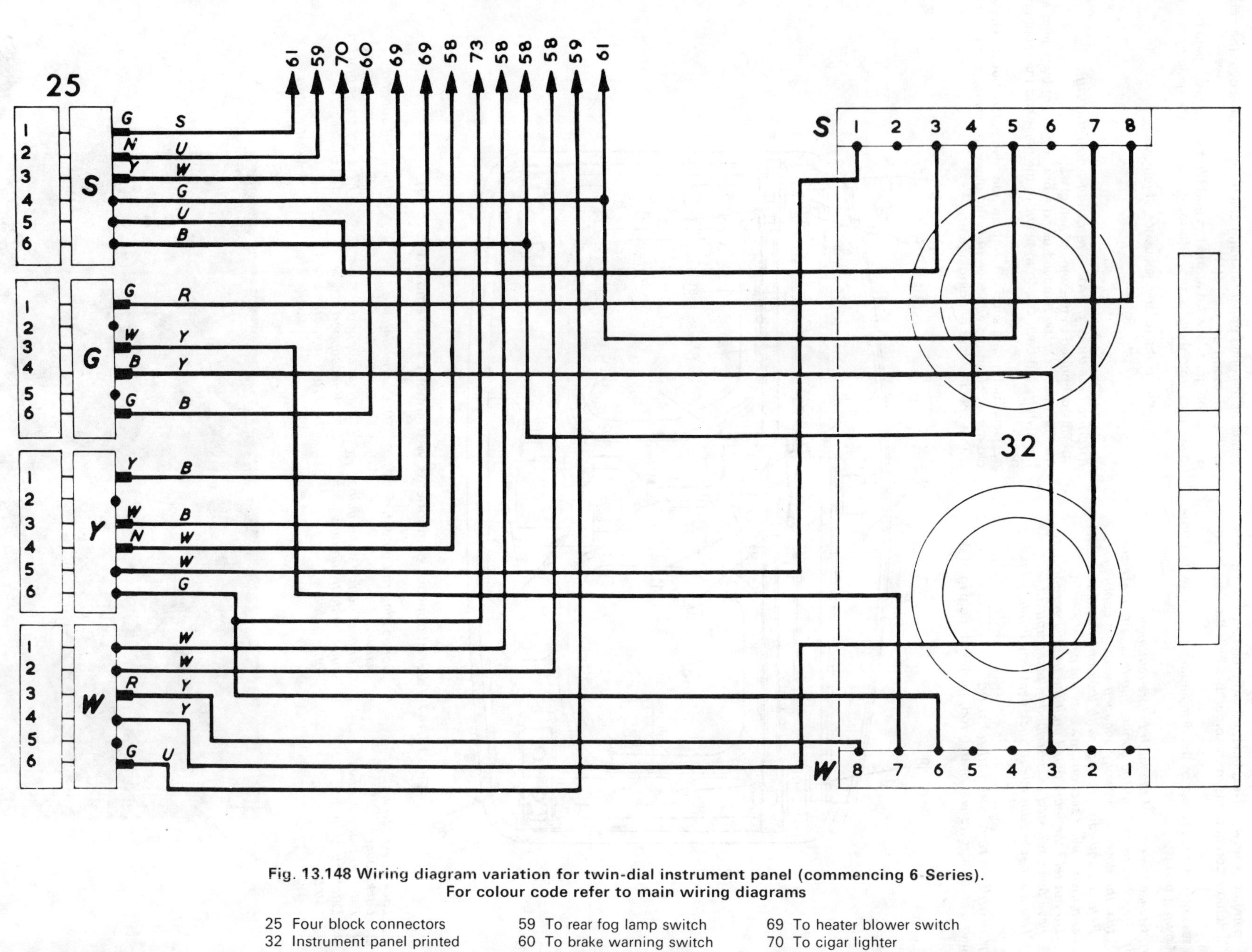

Fig. 13.148 Wiring diagram variation for twin-dial instrument panel (commencing 6 Series). For colour code refer to main wiring diagrams

25 Four block connectors
32 Instrument panel printed circuit
58 To hazard warning switch
59 To rear fog lamp switch
60 To brake warning switch
61 To heated rear window switch
69 To heater blower switch
70 To cigar lighter
73 To switch illumination lamp

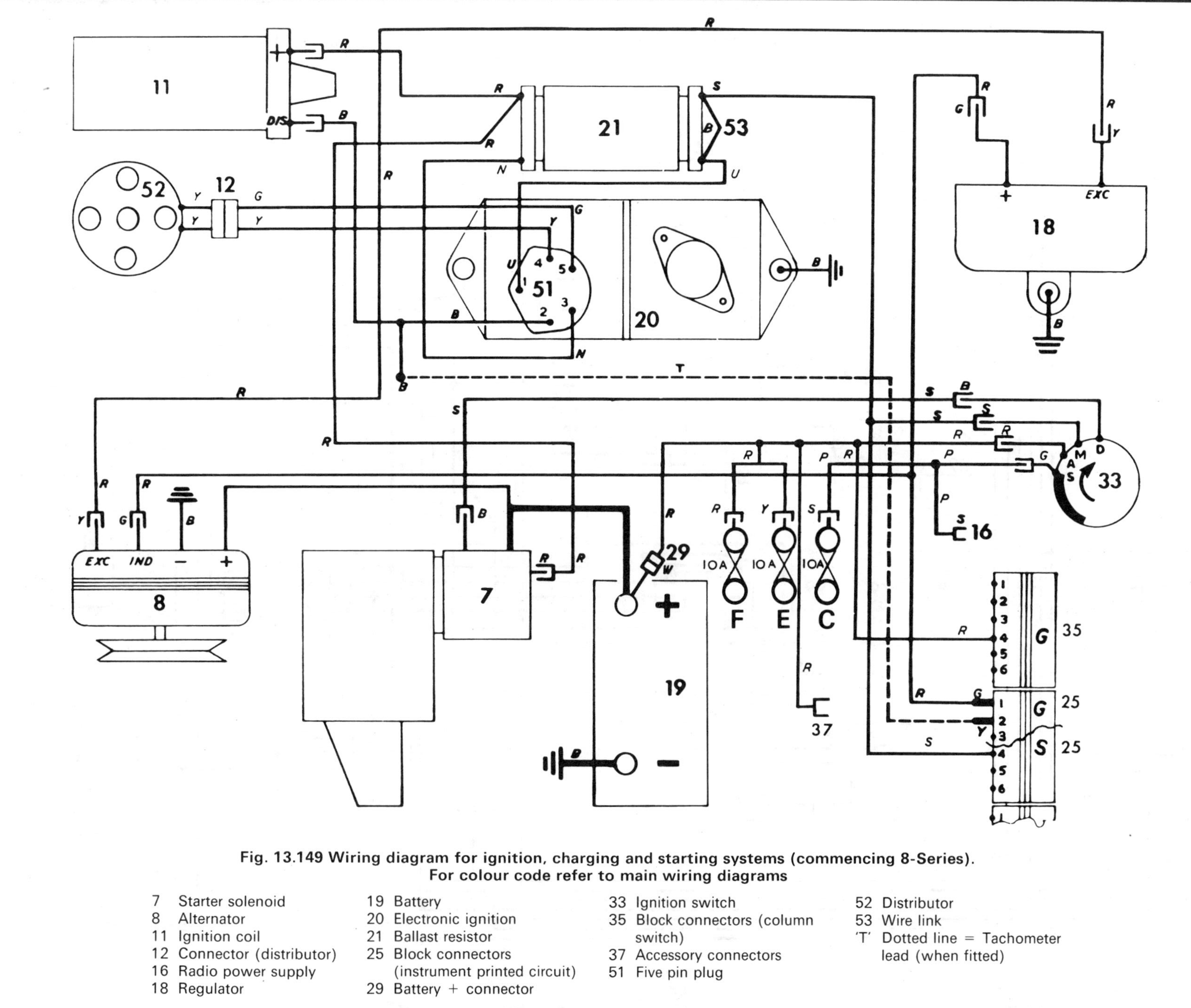

Fig. 13.149 Wiring diagram for ignition, charging and starting systems (commencing 8-Series). For colour code refer to main wiring diagrams

7 Starter solenoid
8 Alternator
11 Ignition coil
12 Connector (distributor)
16 Radio power supply
18 Regulator
19 Battery
20 Electronic ignition
21 Ballast resistor
25 Block connectors (instrument printed circuit)
29 Battery + connector
33 Ignition switch
35 Block connectors (column switch)
37 Accessory connectors
51 Five pin plug
52 Distributor
53 Wire link
'T' Dotted line = Tachometer lead (when fitted)

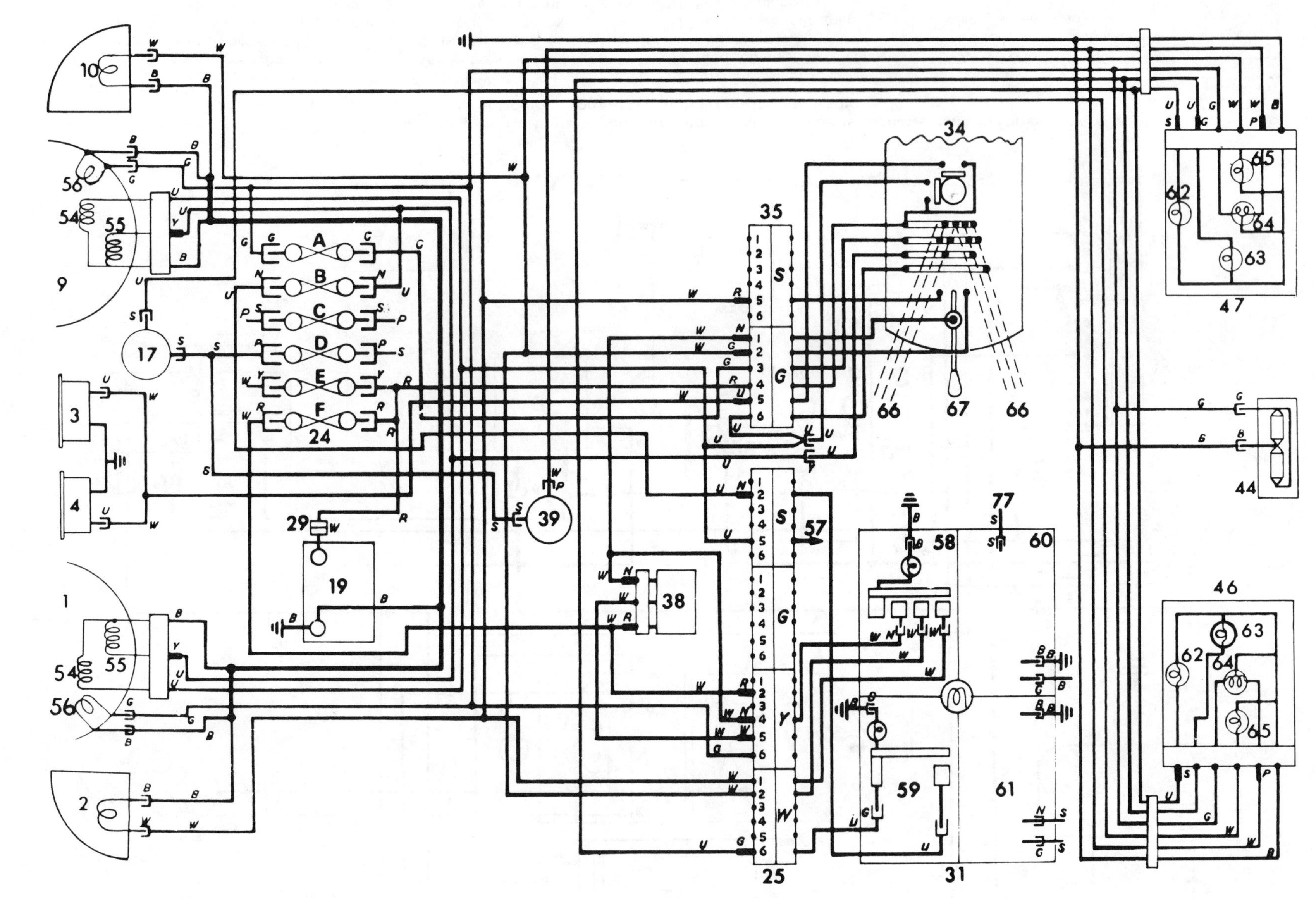

Fig. 13.150 Wiring diagram for exterior lights, horns and indicators (commencing 8-Series)

Key to Fig. 13.150 Wiring diagram for exterior lights, horns and indicators (commencing 8-Series)

1 LH headlamp
2 LH indicator
3 Horn
4 Second horn (if fitted)
9 RH headlamp
10 RH indicator
17 Reverse lamp switch
19 Battery
24 Fuses
25 Block connectors (instrument printed circuit)
29 Battery + connector
31 Switch unit assembly
34 Combination switch
35 Block connectors (column switch)
38 Flasher unit
39 Stop-lamp switch
44 Number plate lamp
46 Tail lamp assembly, LH
47 Tail lamp assembly, RH
54 Main beam filament
56 Sidelamp filament
57 To main beam warning light
58 Hazard warning switch
59 Rear foglamps switch
60 Brake pad wear warning light and test switch
61 Heated rear window switch
62 Reverse lamps
63 Foglamps
64 Stoptail lamps
65 Rear indicator lamps
66 Lighting switch control
67 Indicator switch control
77 From ignition switch M terminal

Colour code

B Black
N Brown
U Blue
G Green
P Purple
R Red
S Slate
Y Yellow
W White

Key to Fig. 13.151 Wiring diagram for instrument panel printed circuit, switches and auxiliaries (commencing 8-Series). For colour code refer to Fig. 13.150

5 Radiator fan switch
6 Radiator fan
13 Water temperature sender
14 Oil warning switch
15 Oil pressure sender
19 Battery positive supply
22 Brake pressure differential warning actuator
23 Screenwash pump
25 Block connectors (instrument printed circuit)
26 Heated rear window relay
27 Heater blower fan
28 Screen wiper motor
30 Courtesy switch (LH side)
32 Instrument panel printed circuit
34 Combination switch
36 Handbrake warning switch
37* Accessory connectors
40 Courtesy switch (RH side)
41 Front interior lamp
42 Fuel gauge tank unit
43* Luggage compartment lamp
45* Luggage compartment lamp switch
48* Rear interior lamp
49* Glovebox lamp
50* Glovebox lamp switch
58 Hazard warning switch
59 Rear fog lamp switch
60 Brake pad wear warning light and test switch
61 Heated rear window switch
68* Instrument panel rheostat
69 Heater blower switch
70 Cigar lighter
71 Wash/wipe control
72 Centre panel lamp
73 Switch illumination lamp
74 From ignition switch S terminal
76 Heated rear window
77 From ignition switch M terminal
78 Choke switch
79* Headlamp wash relay
80* Headlamp wash relay
81* Headlamp wiper motors
82* Rear window wiper motor switch
83* Rear window wiper motor
84* Rear window washer pump
85 Brake pad wear indicators
* *When fitted*

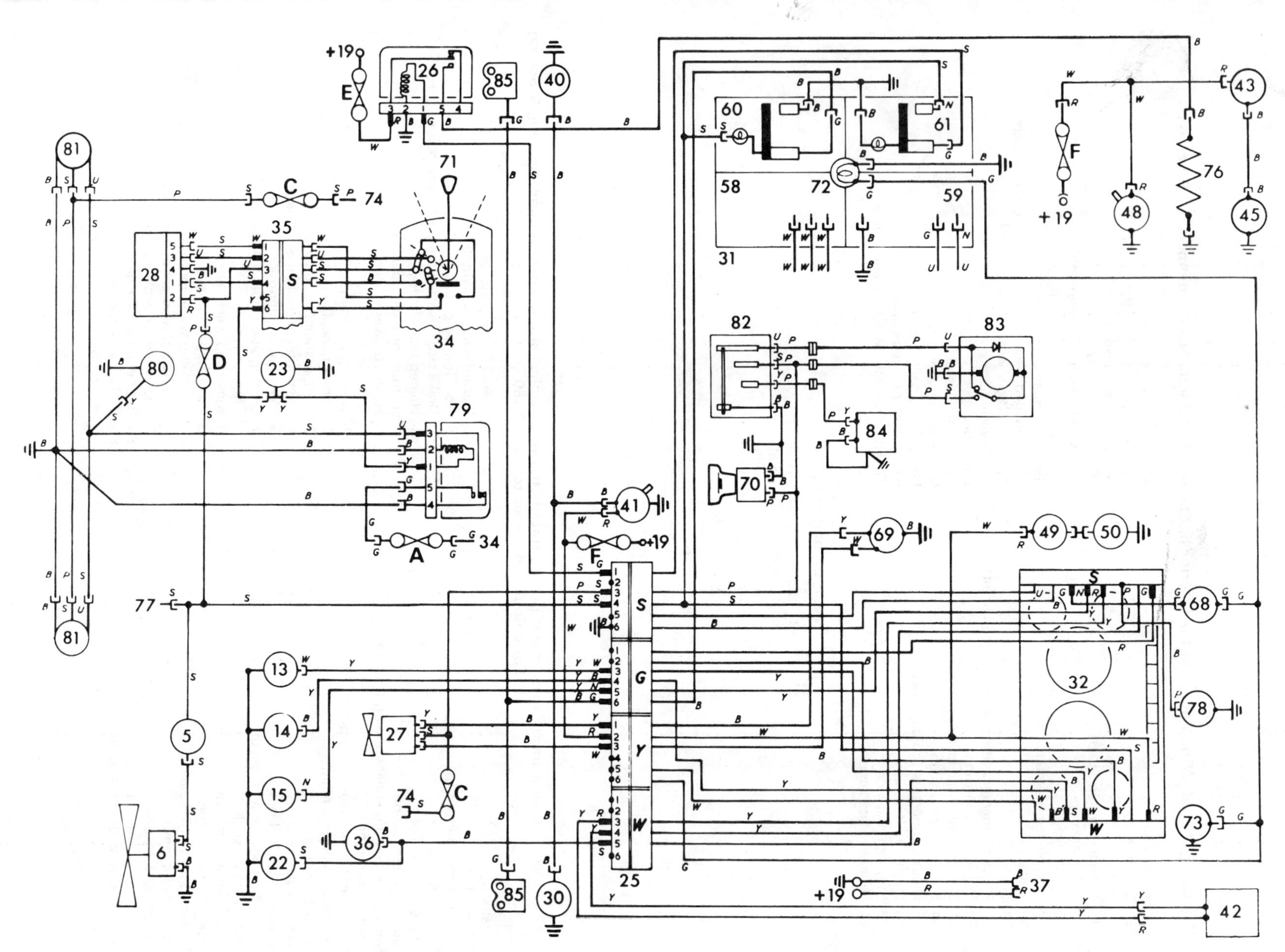

Fig. 13.151 Wiring diagram for instrument panel printed circuit, switches and auxiliaries (commencing 8-Series)

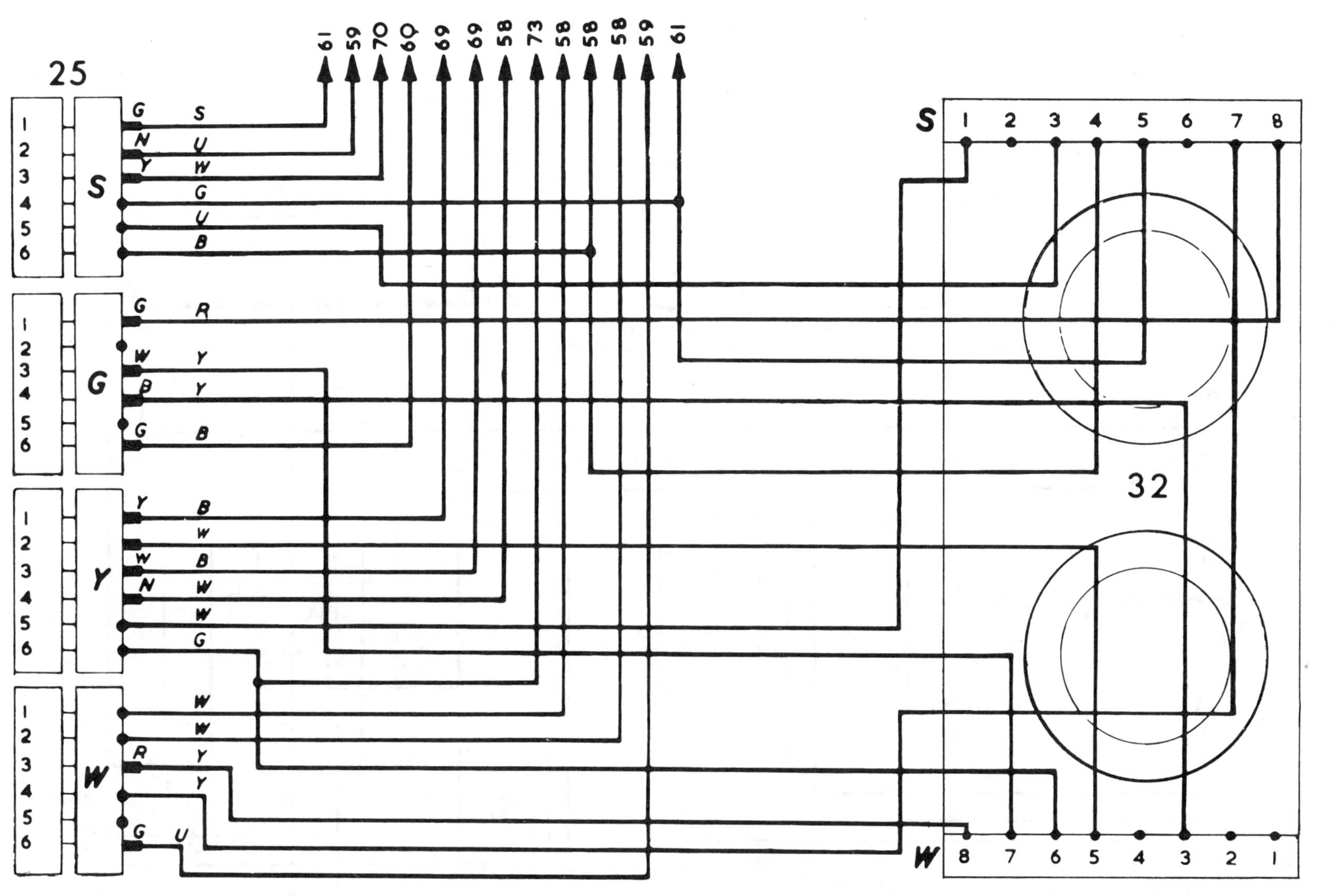

Fig. 13.152 Wiring diagram for twin-dial instrument panel (commencing 8-Series). For colour code refer to Fig. 13.150

25 Block connectors (instrument printed circuit)
32 Instrument panel printed circuit
58 To hazard warning switch
59 To rear foglamps switch
60 To brake pad wear warning light and test switch
61 To heated rear window switch
69 To heater blower switch
70 To cigar lighter
73 To switch illumination lamp

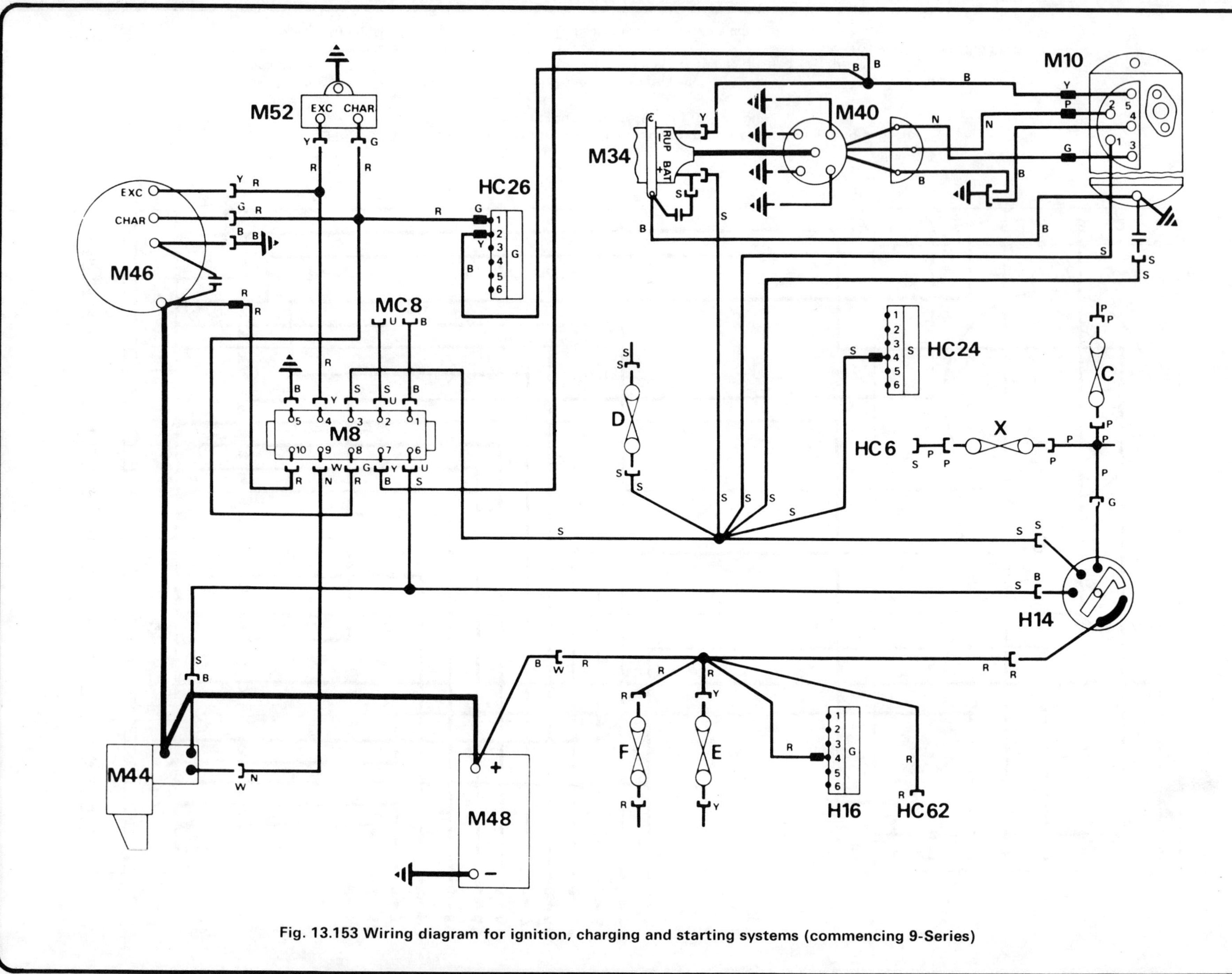

Fig. 13.153 Wiring diagram for ignition, charging and starting systems (commencing 9-Series)

Key to Fig. 13.153 Wiring diagram for ignition, charging and starting systems (commencing 9-Series). For colour code refer to Fig. 13.150

MC8 Connections to automatic transmission
H14 Ignition switch
H16 Connector, green, column switch
HC6 Radio connection
HC24 Connector, slate, instrument panel
HC26 Connector, green, instrument panel
HC62 Accessory socket, supply terminal
X In-line fuse, radio supply
M8 Diagnostic connector
M10 Control unit, electronic ignition
M34 Ignition coil
M40 Distributor
M44 Starter motor
M46 Alternator
M48 Battery
M52 Alternator regulator

Key to Fig. 13.154 Wiring diagram for lights, horns and indicators (commencing 9-Series). For colour code see Fig. 13.150

H30 Glovebox lamp and switch
H66 Front interior lamp
HC24 Connector, slate, instrument panel
HC26 Connector, green, instrument panel
HC28 Connector, yellow, instrument panel
HC30 Connector, white, instrument panel
M2 Horns
M4 Right-hand front direction indicator
M5 Left-hand front direction indicator
M16 Sidelamps
M32 Fuse unit
M32 Reverse lamp switch
M48 Battery
M59 Headlamp dip beams
M60 Headlamp main beams
T1 Warning lamp, side and tail
T2 Warning lamp, headlamp main beam
T5 Warning lamp, direction indicators
T9 Illumination, instrument panel
C1 Tail lamps
C4 Luggage compartment lamp
C8 Rear interior lamp
C10 Right-hand rear direction indicator
C11 Left-hand rear direction indicator
C14 Number plate lamp
C16 Switch, luggage compartment lamp
C17 Reverse lamps
C18 Stop-lamps
C19 Rear foglamps
H8 Door courtesy switches
H12 Direction indicator flasher unit
H13 Hazard warning
H14 Ignition switch
H16 Column switch
H18 Stop-lamp switch
H22 Switch illumination
H24 Ashtray and heater control illumination
H26 Illumination rheostat
H28 Rear foglamp switch

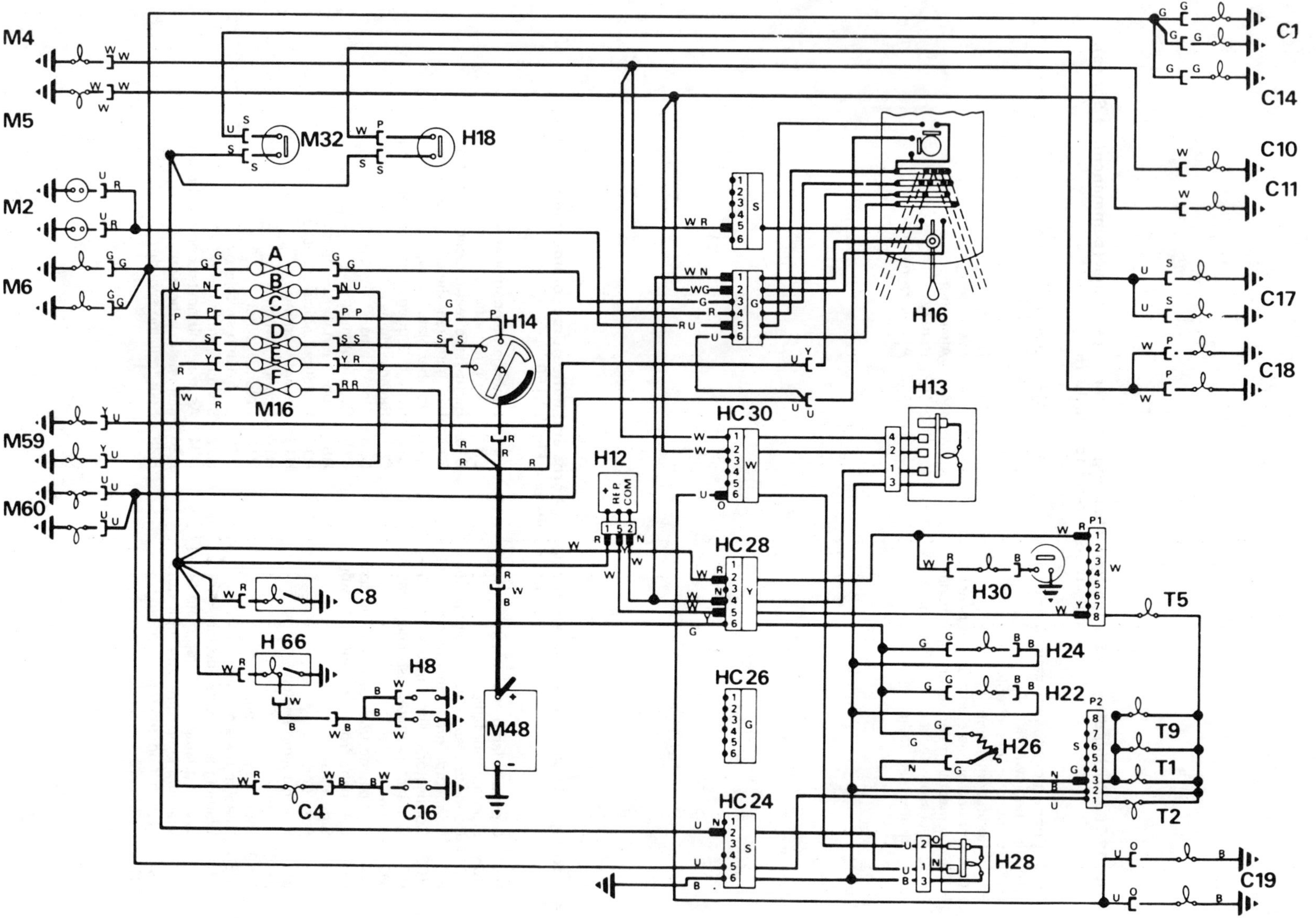

Fig. 13.154 Wiring diagram for lights, horns and indicators (commencing 9-Series)

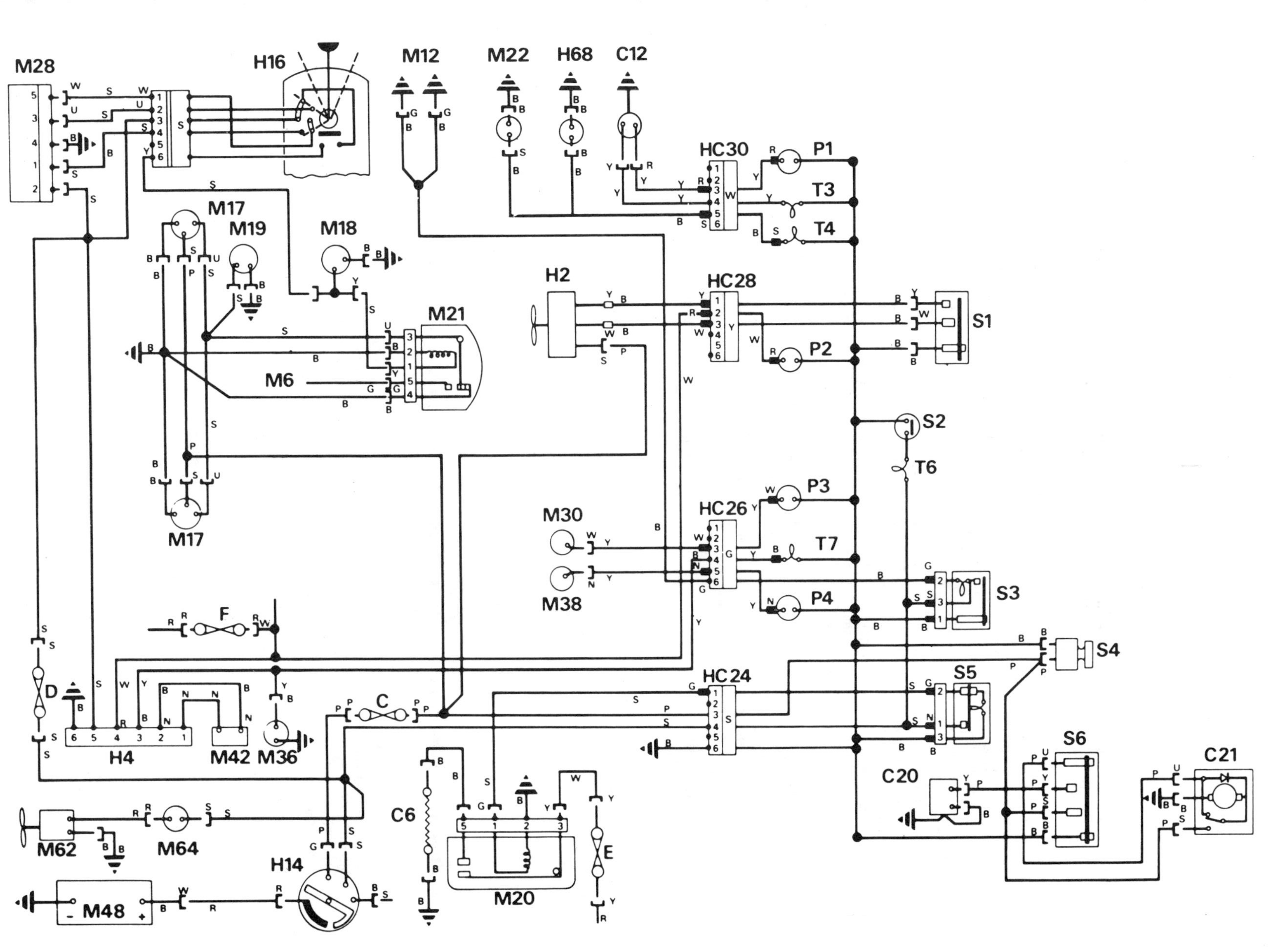

Fig. 13.155 Wiring diagram for six-dial instrument panel, switches and auxiliaries (commencing 9-Series)

Key to Fig. 13.155 Wiring diagram for six-dial instrument panel, switches and auxiliaries (commencing 9-Series). For colour code refer to Fig. 13.150

C6 Element, heated rear window
C12 Fuel tank gauge unit
C20 Rear window washer pump
C21 Rear window wiper motor
H2 Heater blower motor
H4 Control box, engine oil level indicator
H14 Ignition switch
H16 Column switch
H68 Switch, handbrake warning light
HC24 Connector, slate, instrument panel
HC26 Connector, green, instrument panel
HC28 Connector, yellow, instrument panel
HC30 Connector, white, instrument panel
M6 Connection from sidelamps
M12 Brake pad wear indicators
M17 Headlamp wiper motors
M18 Windscreen washer pump
M19 Headlamp washer pump
M20 Relay, heated rear window
M21 Headlamp wash/wipe relay
M22 Brake fluid level indicator
M28 Windscreen wiper motor
M30 Sender unit, water temperature gauge
M36 Sender unit, oil warning light
M38 Sender unit, oil pressure gauge
M42 Dipstick, engine oil levél indicator
M48 Battery
M62 Electric fan motor, engine cooling
M64 Switch, electric cooling fan
P1 Fuel gauge
P2 Clock
P3 Water temperature gauge
P4 Oil pressure gauge
S1 Switch, heater blower motor
S2 Switch, choke control
S3 Test switch, brake pad wear indicator
S4 Cigarette lighter
S5 Switch, heated rear window
S6 Switch, rear window wiper motor
T3 Warning light, low fuel level
T4 Warning light, handbrake and low brake fluid level
T6 Warning light, choke control
T7 Warning light, low oil pressure and low oil level

Key to Fig. 13.156 Wiring diagam for twin-dial instrument panel (commencing 9-Series). For colour code refer to Fig. 13.150

A Connectors, instrument panel
B Instrument panel printed circuit
H13 Hazard warning switch
H22 Switch illumination
H28 Switch, rear foglamps
S1 Switch, rear foglamps
S2 Switch, choke control
S3 Test switch, brake pad wear indicator
S4 Cigarette lighter
S5 Switch, heated rear window

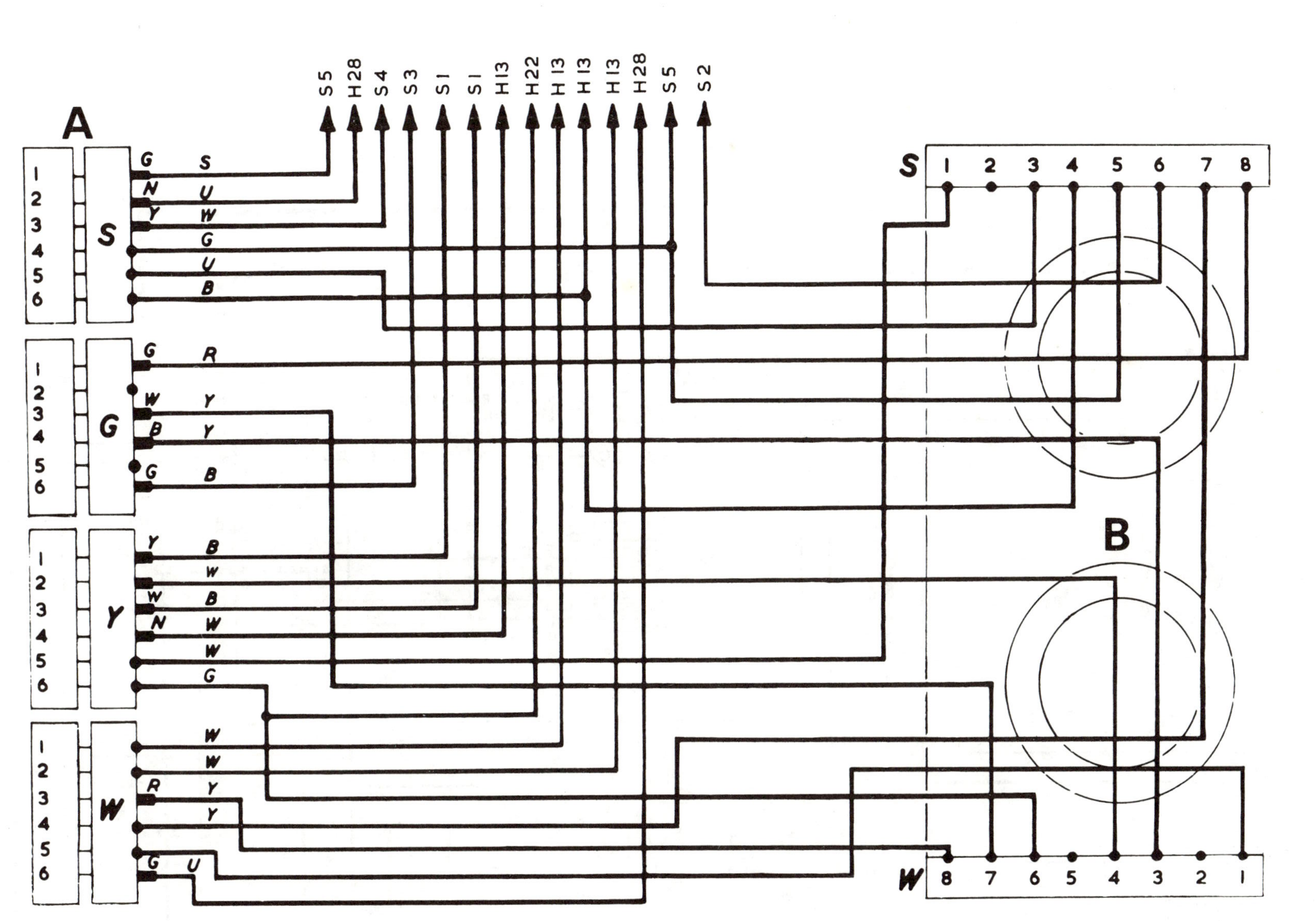

Fig. 13.156 Wiring diagram for twin-dial instrument panel (commencing 9-Series)

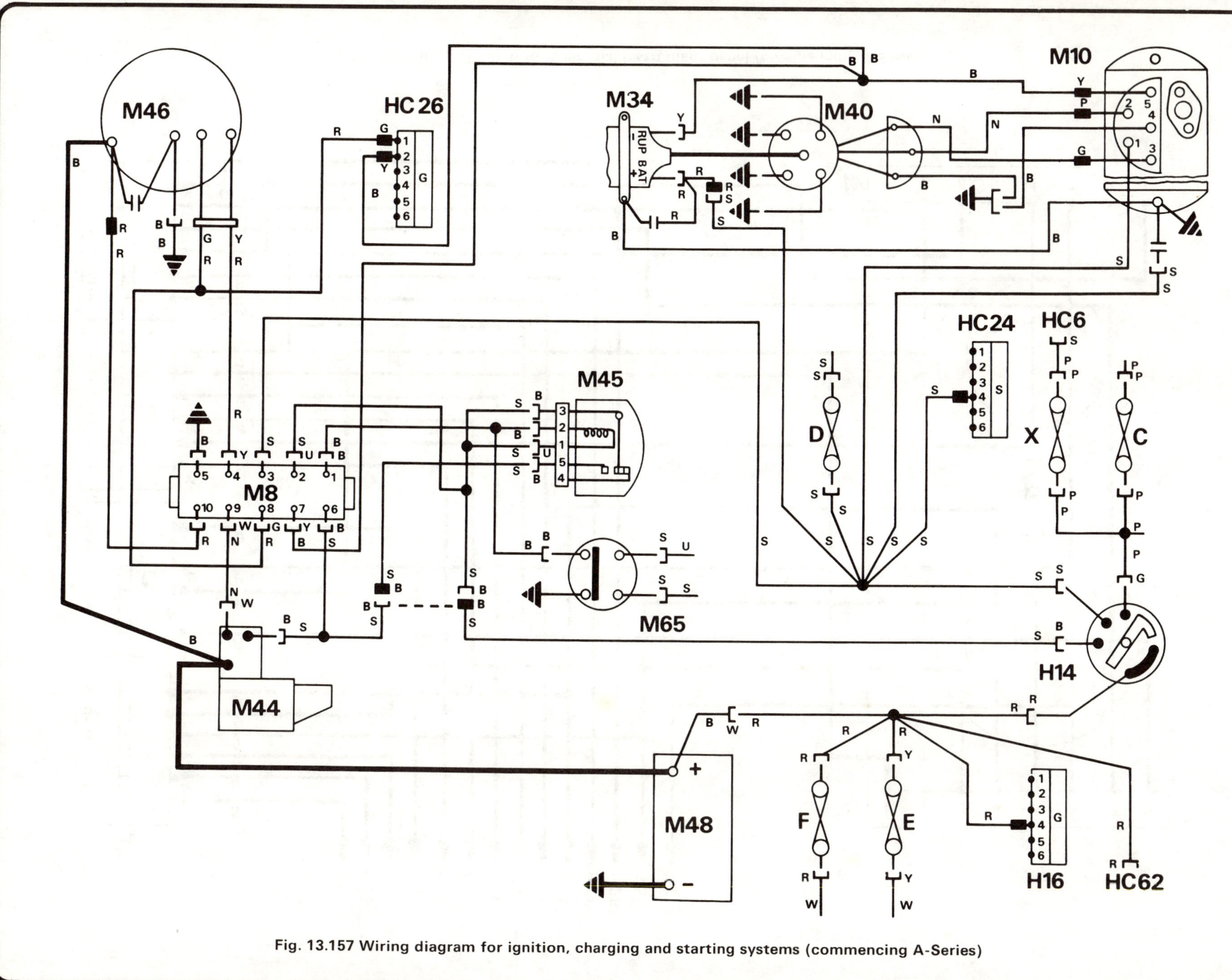

Fig. 13.157 Wiring diagram for ignition, charging and starting systems (commencing A-Series)

Key to Fig. 13.157 Wiring diagram for ignition, charging and starting systems (commencing A-Series). For colour code refer to Fig. 13.150

H14 Ignition switch
H16 Connector, green, column switch
HC6 Radio connection
HC24 Connector, slate, instrument panel
HC26 Connector, green, instrument panel
HC62 Accessory socket, supply terminal
X In-line fuse, radio supply
M8 Diagnostic connector
M10 Control unit, electronic ignition
M34 Ignition coil
M40 Distributor
M44 Starter motor
M45 Starter inhibitor relay, automatic transmission
M46 Alternator
M48 Battery
M52 Alternator regulator
M65 Inhibitor switch, automatic transmission

Key to Fig. 13.158 Wiring diagram for lights, horns and indicators, six-dial instrument panel (commencing A-Series). For colour code refer to Fig. 13.150

C1 Tail lamps
C4 Luggage compartment lamp
C8 Rear interior lamp
C10 Right-hand rear direction indicator
C11 Left-hand rear direction indicator
C14 Number plate lamps
C16 Switch, luggage compartment lamp
C17 Reverse lamps
C18 Stop-lamps
C19 Rear foglamps
H7 Heater control illumination
H8 Door courtesy switches
H12 Direction indicator flasher unit
H13 Hazard warning switch
H14 Ignition switch
H15 Illumination, automatic transmission control
H16 Column switch
H18 Stop-lamp switch
H22 Switch illumination
H24 Ashtray illumination
H26 Illumination rheostat
H28 Rear foglamp switch
H30 Glovebox lamp and switch
H66 Front interior lamp
HC24 Connector, slate, instrument panel
HC26 Connector, green, instrument panel
HC28 Connector, yellow, instrument panel
HC30 Connector, white, instrument panel
M2 Horns
M4 Left-hand front direction indicator
M5 Right-hand front direction indicator
M6 Sidelamps
M16 Fuse unit
M32 Reverse lamp switch
M48 Battery
M59 Headlamp dip beams
M60 Headlamp main beams
PC1 Printed circuit connector, white
PC2 Printed circuit connector, slate
T1 Warning lamp, side and tail
T2 Warning lamp, headlamp main beam
T5 Warning lamp, direction indicators
T9 Illumination, instrument panel

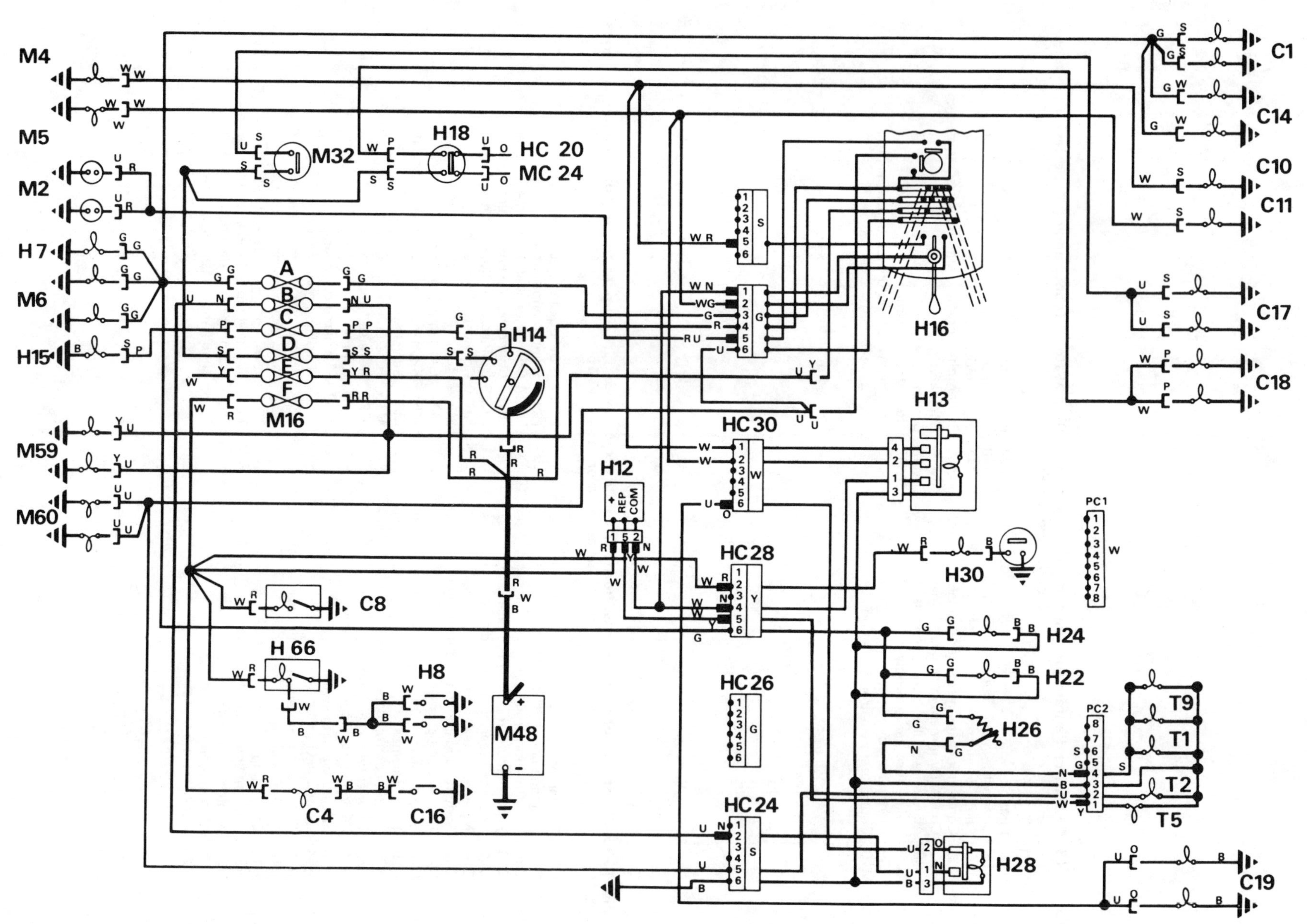

Fig. 13.158 Wiring diagram for lights, horns and indicators, six-dial instrument panel (commencing A-Series)

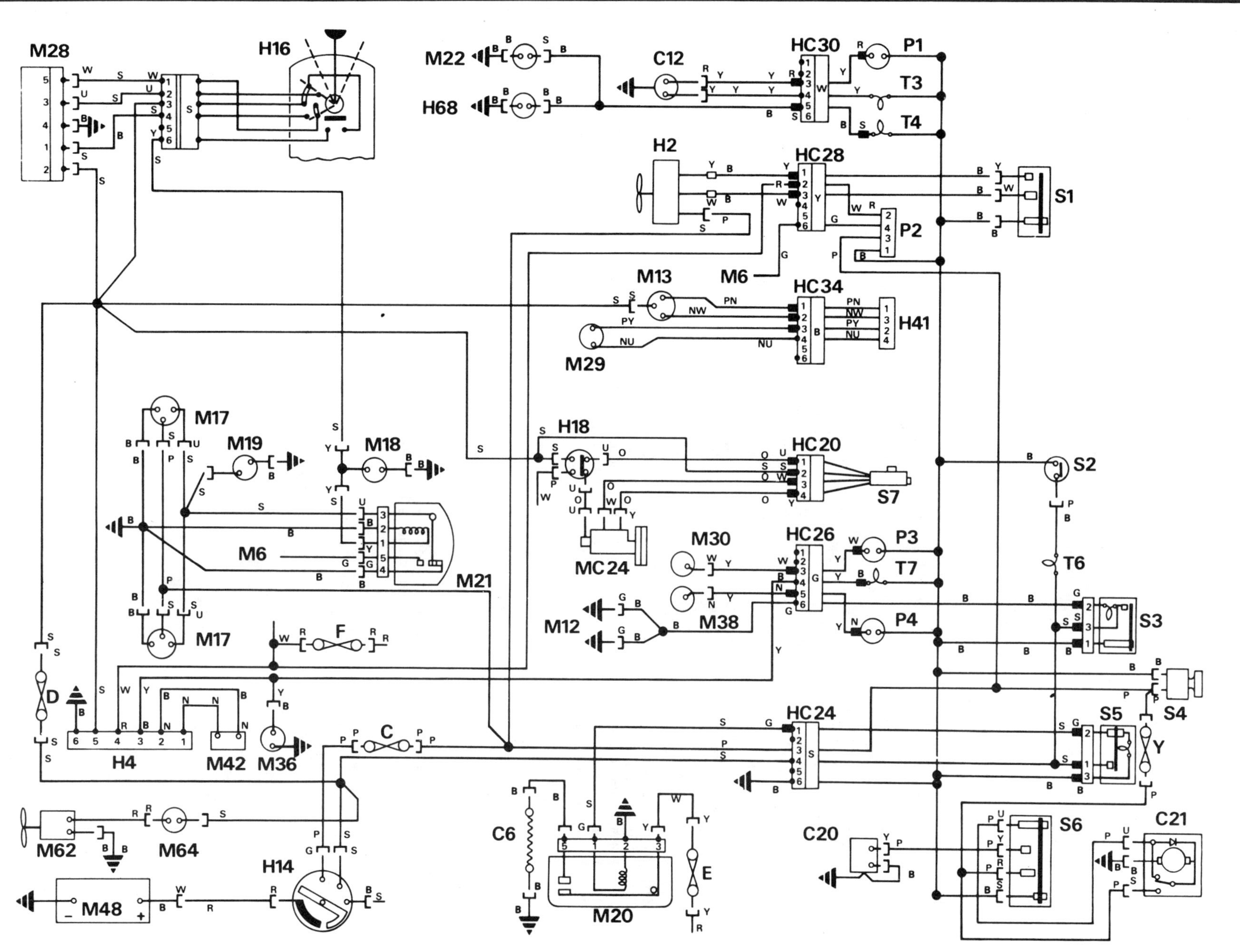

Fig. 13.159 Wiring diagram for six-dial instrument panel, switches and auxiliaries (commencing A-Series)

Key to Fig. 13.159 Wiring diagram for six-dial instrument panel switches and auxiliaries (commencing A-Series). For colour code refer to Fig. 13.150

C6 Element, heated rear window
C12 Fuel tank gauge unit
C20 Rear window washer pump
C21 Rear window wiper motor
H2 Heater blower motor
H4 Control box, engine oil level indicator
H14 Ignition switch
H16 Column switch
H18 Stop-lamp switch
H41 Connector, trip computer
H68 Switch, handbrake warning light
HC20 Connector, speed control switch
HC24 Connector, slate, instrument panel
HC26 Connector, green, instrument panel
HC28 Connector, yellow, instrument panel
HC30 Connector, white, instrument panel
HC32 Connector, black, instrument panel
M6 Connection from sidelamps
M12 Brake pad wear indicators
M13 Fuel flow meter
M17 Headlamp wiper motors
M18 Windscreen washer pump
M19 Headlamp washer pump
M20 Relay, heated rear window
M21 Headlamp wash/wipe relay
M22 Brake fluid level indicator
M28 Windscreen wiper motor
M29 Speed sensor
M30 Sender unit, water temperature gauge
M36 Sender unit, oil warning light
M38 Sender unit, oil pressure gauge
M42 Dipstick, engine oil level indicator
M48 Battery
M62 Electric fan motor, engine cooling
M64 Switch, electric cooling fan
MC24 Servo, speed control
P1 Fuel gauge
P2 Connector, digital clock or trip computer
P3 Water temperature gauge
P4 Oil pressure gauge
S1 Switch, heater blower motor
S2 Switch, choke control
S3 Test switch, brake pad wear indicator
S4 Cigarette lighter
S5 Switch, rear window wiper motor
S7 Switch, speed control
T3 Warning light, low fuel level
T4 Warning light, handbrake and low brake fluid level
T6 Warning light, choke control
T7 Warning light, low oil pressure and low oil level
Y In-line fuse, tailgate wiper

Key to Fig. 13.160 Wiring diagram for four-dial instrument panel (commencing A-Series). For colour code refer to Fig. 13.150

A Connectors, instrument panel
H13 Hazard warning switch
H22 Switch illumination
H28 Switch, rear fog lamps
S1 Switch, heater blower motor
S3 Test switch, brake pad wear indicator
S4 Cigarette lighter
S5 Switch, heated rear window

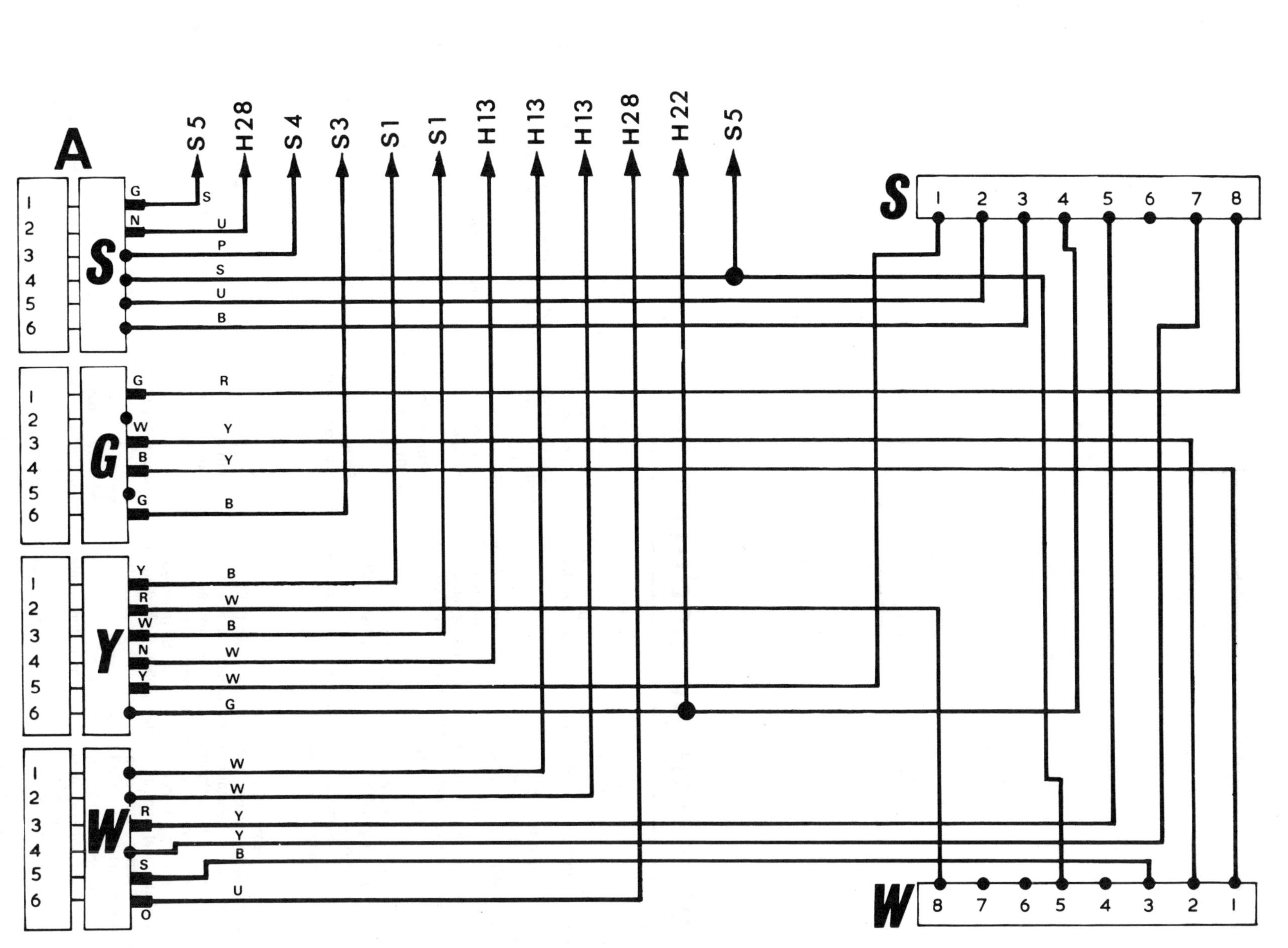

Fig. 13.160 Wiring diagram for four-dial instrument panel (commencing A-Series)

General repair procedures

Whenever servicing, repair or overhaul work is carried out on the car or its components, it is necessary to observe the following procedures and instructions. This will assist in carrying out the operation efficiently and to a professional standard of workmanship.

Joint mating faces and gaskets

Where a gasket is used between the mating faces of two components, ensure that it is renewed on reassembly, and fit it dry unless otherwise stated in the repair procedure. Make sure that the mating faces are clean and dry with all traces of old gasket removed. When cleaning a joint face, use a tool which is not likely to score or damage the face, and remove any burrs or nicks with an oilstone or fine file.

Make sure that tapped holes are cleaned with a pipe cleaner, and keep them free of jointing compound if this is being used unless specifically instructed otherwise.

Ensure that all orifices, channels or pipes are clear and blow through them, preferably using compressed air.

Oil seals

Whenever an oil seal is removed from its working location, either individually or as part of an assembly, it should be renewed.

The very fine sealing lip of the seal is easily damaged and will not seal if the surface it contacts is not completely clean and free from scratches, nicks or grooves. If the original sealing surface of the component cannot be restored, the component should be renewed.

Protect the lips of the seal from any surface which may damage them in the course of fitting. Use tape or a conical sleeve where possible. Lubricate the seal lips with oil before fitting and, on dual lipped seals, fill the space between the lips with grease.

Unless otherwise stated, oil seals must be fitted with their sealing lips toward the lubricant to be sealed.

Use a tubular drift or block of wood of the appropriate size to install the seal and, if the seal housing is shouldered, drive the seal down to the shoulder. If the seal housing is unshouldered, the seal should be fitted with its face flush with the housing top face.

Screw threads and fastenings

Always ensure that a blind tapped hole is completely free from oil, grease, water or other fluid before installing the bolt or stud. Failure to do this could cause the housing to crack due to the hydraulic action of the bolt or stud as it is screwed in.

When tightening a castellated nut to accept a split pin, tighten the nut to the specified torque, where applicable, and then tighten further to the next split pin hole. Never slacken the nut to align a split pin hole unless stated in the repair procedure.

When checking or retightening a nut or bolt to a specified torque setting, slacken the nut or bolt by a quarter of a turn, and then retighten to the specified setting.

Locknuts, locktabs and washers

Any fastening which will rotate against a component or housing in the course of tightening should always have a washer between it and the relevant component or housing.

Spring or split washers should always be renewed when they are used to lock a critical component such as a big-end bearing retaining nut or bolt.

Locktabs which are folded over to retain a nut or bolt should always be renewed.

Self-locking nuts can be reused in non-critical areas, providing resistance can be felt when the locking portion passes over the bolt or stud thread.

Split pins must always be replaced with new ones of the correct size for the hole.

Special tools

Some repair procedures in this manual entail the use of special tools such as a press, two or three-legged pullers, spring compressors etc. Wherever possible, suitable readily available alternatives to the manufacturer's special tools are described, and are shown in use. In some instances, where no alternative is possible, it has been necessary to resort to the use of a manufacturer's tool and this has been done for reasons of safety as well as the efficient completion of the repair operation. Unless you are highly skilled and have a thorough understanding of the procedure described, never attempt to bypass the use of any special tool when the procedure described specifies its use. Not only is there a very great risk of personal injury, but expensive damage could be caused to the components involved.

Safety first!

Professional motor mechanics are trained in safe working procedures. However enthusiastic you may be about getting on with the job in hand, do take the time to ensure that your safety is not put at risk. A moment's lack of attention can result in an accident, as can failure to observe certain elementary precautions.

There will always be new ways of having accidents, and the following points do not pretend to be a comprehensive list of all dangers; they are intended rather to make you aware of the risks and to encourage a safety-conscious approach to all work you carry out on your vehicle.

Essential DOs and DON'Ts

DON'T rely on a single jack when working underneath the vehicle. Always use reliable additional means of support, such as axle stands, securely placed under a part of the vehicle that you know will not give way.

DON'T attempt to loosen or tighten high-torque nuts (e.g. wheel hub nuts) while the vehicle is on a jack; it may be pulled off.

DON'T start the engine without first ascertaining that the transmission is in neutral (or 'Park' where applicable) and the parking brake applied.

DON'T suddenly remove the filler cap from a hot cooling system – cover it with a cloth and release the pressure gradually first, or you may get scalded by escaping coolant.

DON'T attempt to drain oil until you are sure it has cooled sufficiently to avoid scalding you.

DON'T grasp any part of the engine, exhaust or catalytic converter without first ascertaining that it is sufficiently cool to avoid burning you.

DON'T allow brake fluid or antifreeze to contact vehicle paintwork.

DON'T syphon toxic liquids such as fuel, brake fluid or antifreeze by mouth, or allow them to remain on your skin.

DON'T inhale dust – it may be injurious to health (see *Asbestos* below).

DON'T allow any spilt oil or grease to remain on the floor – wipe it up straight away, before someone slips on it.

DON'T use ill-fitting spanners or other tools which may slip and cause injury.

DON'T attempt to lift a heavy component which may be beyond your capability – get assistance.

DON'T rush to finish a job, or take unverified short cuts.

DON'T allow children or animals in or around an unattended vehicle.

DO wear eye protection when using power tools such as drill, sander, bench grinder etc, and when working under the vehicle.

DO use a barrier cream on your hands prior to undertaking dirty jobs – it will protect your skin from infection as well as making the dirt easier to remove afterwards; but make sure your hands aren't left slippery.

DO keep loose clothing (cuffs, tie etc) and long hair well out of the way of moving mechanical parts.

DO remove rings, wristwatch etc, before working on the vehicle – especially the electrical system.

DO ensure that any lifting tackle used has a safe working load rating adequate for the job.

DO keep your work area tidy – it is only too easy to fall over articles left lying around.

DO get someone to check periodically that all is well, when working alone on the vehicle.

DO carry out work in a logical sequence and check that everything is correctly assembled and tightened afterwards.

DO remember that your vehicle's safety affects that of yourself and others. If in doubt on any point, get specialist advice.

IF, in spite of following these precautions, you are unfortunate enough to injure yourself, seek medical attention as soon as possible.

Asbestos

Certain friction, insulating, sealing, and other products – such as brake linings, brake bands, clutch linings, torque converters, gaskets, etc – contain asbestos. *Extreme care must be taken to avoid inhalation of dust from such products since it is hazardous to health.* If in doubt, assume that they *do* contain asbestos.

Fire

Remember at all times that petrol (gasoline) is highly flammable. Never smoke, or have any kind of naked flame around, when working on the vehicle. But the risk does not end there – a spark caused by an electrical short-circuit, by two metal surfaces contacting each other, by careless use of tools, or even by static electricity built up in your body under certain conditions, can ignite petrol vapour, which in a confined space is highly explosive.

Always disconnect the battery earth (ground) terminal before working on any part of the fuel or electrical system, and never risk spilling fuel on to a hot engine or exhaust.

It is recommended that a fire extinguisher of a type suitable for fuel and electrical fires is kept handy in the garage or workplace at all times. Never try to extinguish a fuel or electrical fire with water.

Fumes

Certain fumes are highly toxic and can quickly cause unconsciousness and even death if inhaled to any extent. Petrol (gasoline) vapour comes into this category, as do the vapours from certain solvents such as trichloroethylene. Any draining or pouring of such volatile fluids should be done in a well ventilated area.

When using cleaning fluids and solvents, read the instructions carefully. Never use materials from unmarked containers – they may give off poisonous vapours.

Never run the engine of a motor vehicle in an enclosed space such as a garage. Exhaust fumes contain carbon monoxide which is extremely poisonous; if you need to run the engine, always do so in the open air or at least have the rear of the vehicle outside the workplace.

If you are fortunate enough to have the use of an inspection pit, never drain or pour petrol, and never run the engine, while the vehicle is standing over it; the fumes, being heavier than air, will concentrate in the pit with possibly lethal results.

The battery

Never cause a spark, or allow a naked light, near the vehicle's battery. It will normally be giving off a certain amount of hydrogen gas, which is highly explosive.

Always disconnect the battery earth (ground) terminal before working on the fuel or electrical systems.

If possible, loosen the filler plugs or cover when charging the battery from an external source. Do not charge at an excessive rate or the battery may burst.

Take care when topping up and when carrying the battery. The acid electrolyte, even when diluted, is very corrosive and should not be allowed to contact the eyes or skin.

If you ever need to prepare electrolyte yourself, always add the acid slowly to the water, and never the other way round. Protect against splashes by wearing rubber gloves and goggles.

When jump starting a car using a booster battery, for negative earth (ground) vehicles, connect the jump leads in the following sequence: First connect one jump lead between the positive (+) terminals of the two batteries. Then connect the other jump lead first to the negative (–) terminal of the booster battery, and then to a good earthing (ground) point on the vehicle to be started, at least 18 in (45 cm) from the battery if possible. Ensure that hands and jump leads are clear of any moving parts, and that the two vehicles do not touch. Disconnect the leads in the reverse order.

Mains electricity

When using an electric power tool, inspection light etc, which works from the mains, always ensure that the appliance is correctly connected to its plug and that, where necessary, it is properly earthed (grounded). Do not use such appliances in damp conditions and, again, beware of creating a spark or applying excessive heat in the vicinity of fuel or fuel vapour.

Ignition HT voltage

A severe electric shock can result from touching certain parts of the ignition system, such as the HT leads, when the engine is running or being cranked, particularly if components are damp or the insulation is defective. Where an electronic ignition system is fitted, the HT voltage is much higher and could prove fatal.

Fault diagnosis

Introduction

The vehicle owner who does his or her own maintenance according to the recommended schedules should not have to use this section of the manual very often. Modern component reliability is such that, provided those items subject to wear or deterioration are inspected or renewed at the specified intervals, sudden failure is comparatively rare. Faults do not usually just happen as a result of sudden failure, but develop over a period of time. Major mechanical failures in particular are usually preceded by characteristic symptoms over hundreds or even thousands of miles. Those components which do occasionally fail without warning are often small and easily carried in the vehicle.

With any fault finding, the first step is to decide where to begin investigations. Sometimes this is obvious, but on other occasions a little detective work will be necessary. The owner who makes half a dozen haphazard adjustments or replacements may be successful in curing a fault (or its symptoms), but he will be none the wiser if the fault recurs and he may well have spent more time and money than was necessary. A calm and logical approach will be found to be more satisfactory in the long run. Always take into account any warning signs or abnormalities that may have been noticed in the period preceding the fault – power loss, high or low gauge readings, unusual noises or smells, etc – and remember that failure of components such as fuses or spark plugs may only be pointers to some underlying fault.

The pages which follow here are intended to help in cases of failure to start or breakdown on the road. There is also a Fault Diagnosis Section at the end of each Chapter which should be consulted if the preliminary checks prove unfruitful. Whatever the fault, certain basic principles apply. These are as follows:

Verify the fault. This is simply a matter of being sure that you know what the symptoms are before starting work. This is particularly important if you are investigating a fault for someone else who may not have described it very accurately.

Don't overlook the obvious. For example, if the vehicle won't start, is there petrol in the tank? (Don't take anyone else's word on this particular point, and don't trust the fuel gauge either!) If an electrical fault is indicated, look for loose or broken wires before digging out the test gear.

Cure the disease, not the symptom. Substituting a flat battery with a fully charged one will get you off the hard shoulder, but if the underlying cause is not attended to, the new battery will go the same way. Similarly, changing oil-fouled spark plugs for a new set will get you moving again, but remember that the reason for the fouling (if it wasn't simply an incorrect grade of plug) will have to be established and corrected.

Don't take anything for granted. Particularly, don't forget that a 'new' component may itself be defective (especially if it's been rattling round in the boot for months), and don't leave components out of a fault diagnosis sequence just because they are new or recently fitted. When you do finally diagnose a difficult fault, you'll probably realise that all the evidence was there from the start.

Electrical faults

Electrical faults can be more puzzling than straightforward mechanical failures, but they are no less susceptible to logical analysis if the basic principles of operation are understood. Vehicle electrical wiring exists in extremely unfavourable conditions – heat, vibration and chemical attack – and the first things to look for are loose or corroded connections and broken or chafed wires, especially where the wires pass through holes in the bodywork or are subject to vibration.

All metal-bodied vehicles in current production have one pole of the battery 'earthed', ie connected to the vehicle bodywork, and in nearly all modern vehicles it is the negative (–) terminal. The various electrical components – motors, bulb holders etc – are also connected to earth, either by means of a lead or directly by their mountings. Electric current flows through the component and then back to the battery via the bodywork. If the component mounting is loose or corroded, or if a good path back to the battery is not available, the circuit will be incomplete and malfunction will result. The engine and/or gearbox are also earthed by means of flexible metal straps to the body or subframe; if these straps are loose or missing, starter motor, generator and ignition trouble may result.

Assuming the earth return to be satisfactory, electrical faults will be due either to component malfunction or to defects in the current supply. Individual components are dealt with in Chapter 11. If supply wires are broken or cracked internally this results in an open-circuit, and the easiest way to check for this is to bypass the suspect wire temporarily with a length of wire having a crocodile clip or suitable connector at each end. Alternatively, a 12V test lamp can be used to verify the presence of supply voltage at various points along the wire and the break can be thus isolated.

If a bare portion of a live wire touches the bodywork or other earthed metal part, the electricity will take the low-resistance path thus formed back to the battery: this is known as a short-circuit. Hopefully a short-circuit will blow a fuse, but otherwise it may cause burning of the insulation (and possibly further short-circuits) or even a fire. This is why it is inadvisable to bypass persistently blowing fuses with silver foil or wire.

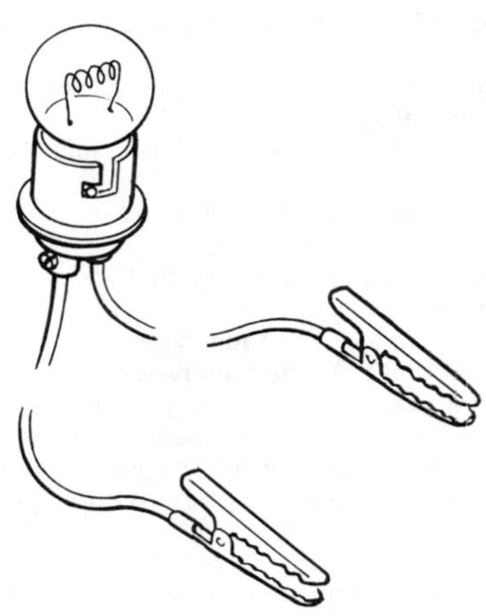

A simple test lamp is useful for checking electrical bulbs

Spares and tool kit

Most vehicles are supplied only with sufficient tools for wheel changing; the *Maintenance and minor repair* tool kit detailed in *Tools and working facilities*, with the addition of a hammer, is probably sufficient for those repairs that most motorists would consider attempting at the roadside. In addition a few items which can be fitted without too much trouble in the event of a breakdown should be carried. Experience and available space will modify the list below, but the following may save having to call on professional assistance:

Spark plugs, clean and correctly gapped
HT lead and plug cap – long enough to reach the plug furthest from the distributor
Distributor rotor
Drivebelt(s) – emergency type may suffice
Spare fuses
Set of principal light bulbs
Tin of radiator sealer and hose bandage
Exhaust bandage
Roll of insulating tape
Length of soft iron wire
Length of electrical flex
Torch or inspection lamp (can double as test lamp)
Battery jump leads
Tow-rope
Ignition waterproofing aerosol
Litre of engine oil
Sealed can of hydraulic fluid
Emergency windscreen
Hose clips
Tube of filler paste

If spare fuel is carried, a can designed for the purpose should be used to minimise risks of leakage and collision damage. A first aid kit and a warning triangle, whilst not at present compulsory in the UK, are obviously sensible items to carry in addition to the above.

When touring abroad it may be advisable to carry additional spares which, even if you cannot fit them yourself, could save having to wait while parts are obtained. The items below may be worth considering:

Clutch and throttle cables
Cylinder head gasket
Alternator brushes
Fuel pump repair kit
Tyre valve core

One of the motoring organisations will be able to advise on availability of fuel etc in foreign countries.

Engine will not start

Engine fails to turn when starter operated

Flat battery (recharge, use jump leads, or push start)
Battery terminals loose or corroded
Battery earth to body defective
Engine earth strap loose or broken
Starter motor (or solenoid) wiring loose or broken
Automatic transmission selector in wrong position, or inhibitor switch faulty
Ignition/starter switch faulty
Major mechanical failure (seizure)
Starter or solenoid internal fault (see Chapter 11)

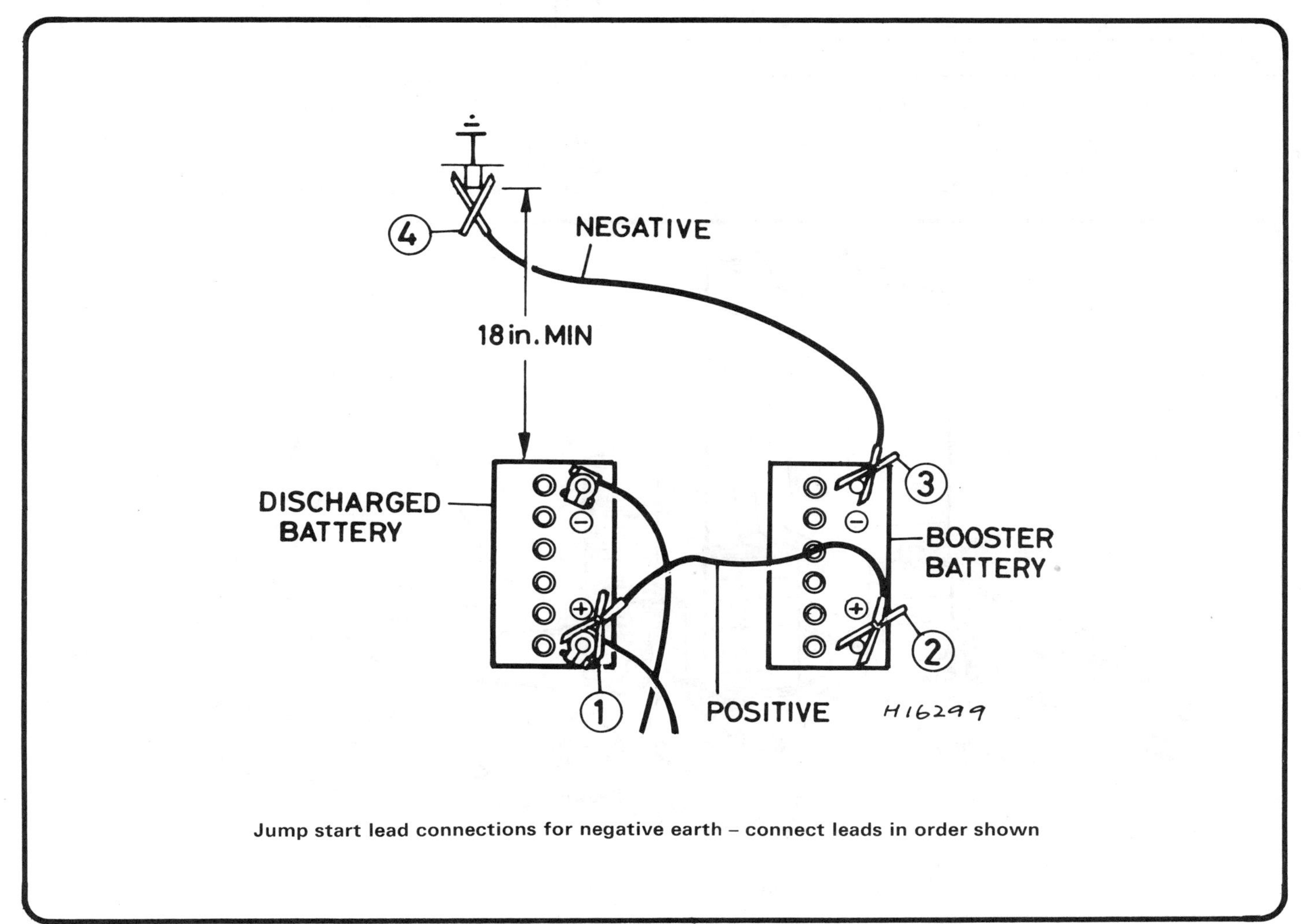

Jump start lead connections for negative earth – connect leads in order shown

Starter motor turns engine slowly

Partially discharged battery (recharge, use jump leads, or push start)
Battery terminals loose or corroded
Battery earth to body defective
Engine earth strap loose
Starter motor (or solenoid) wiring loose
Starter motor internal fault (see Chapter 11)

Starter motor spins without turning engine

Flat battery
Starter motor pinion sticking on sleeve
Flywheel gear teeth damaged or worn
Starter motor mounting bolts loose

Engine turns normally but fails to start

Damp or dirty HT leads and distributor cap (crank engine and check for spark)
No fuel in tank (check for delivery at carburettor)
Excessive choke (hot engine) or insufficient choke (cold engine)
Fouled or incorrectly gapped spark plugs (remove, clean and regap)
Other ignition system fault (see Chapter 4)
Other fuel system fault (see Chapter 3)
Poor compression (see Chapter 1)
Major mechanical failure (eg camshaft drive)

Engine fires but will not run

Insufficient choke (cold engine)
Air leaks at carburettor or inlet manifold
Fuel starvation (see Chapter 3)
Ballast resistor defective, or other ignition fault (see Chapter 4)

Engine cuts out and will not restart

Engine cuts out suddenly – ignition fault

Loose or disconnected LT wires
Wet HT leads or distributor cap (after traversing water splash)
Coil or condenser failure (check for spark)
Other ignition fault (see Chapter 4)

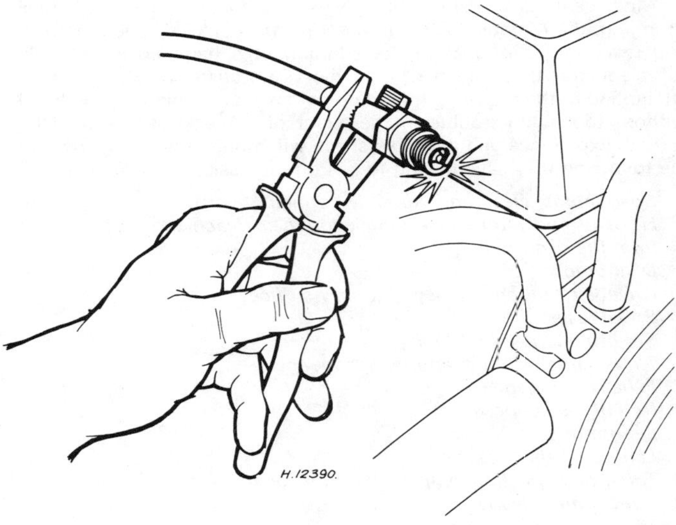

Crank engine and check for spark. Note use of insulated tool to hold plug lead

Engine misfires before cutting out – fuel fault

Fuel tank empty
Fuel pump defective or filter blocked (check for delivery)
Fuel tank filler vent blocked (suction will be evident on releasing cap)
Carburettor needle valve sticking
Carburettor jets blocked (fuel contaminated)
Other fuel system fault (see Chapter 3)

Engine cuts out – other causes

Serious overheating
Major mechanical failure (eg camshaft drive)

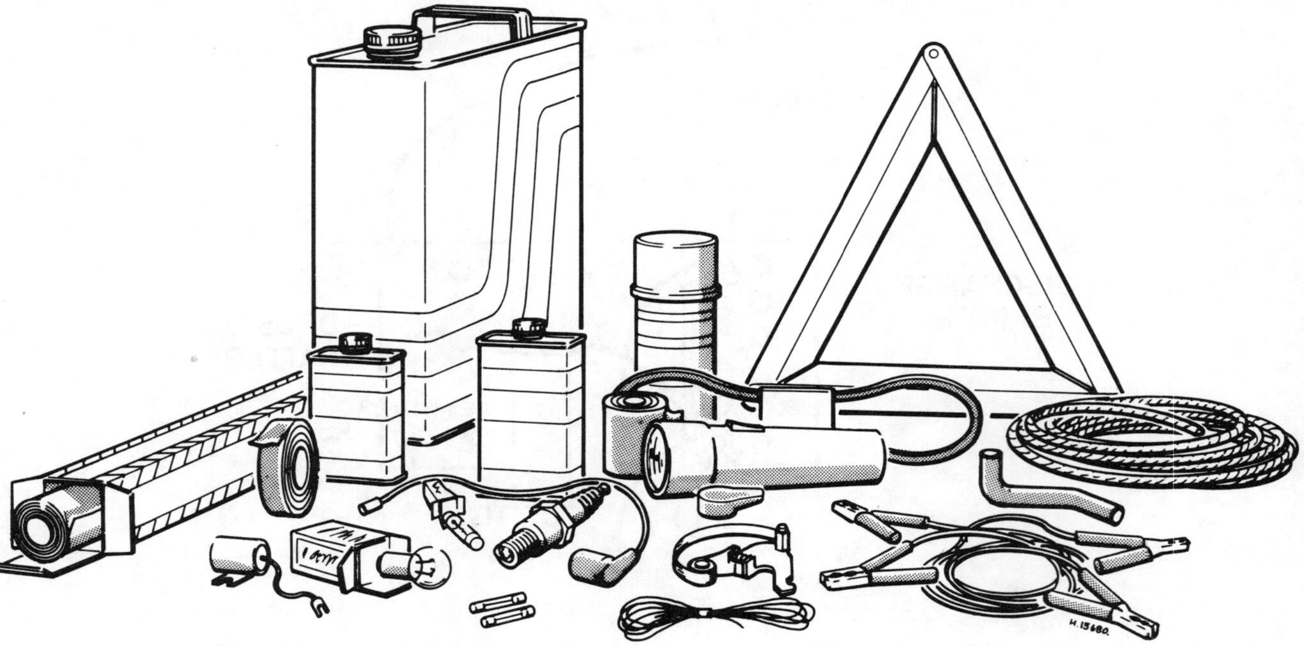

Carrying a few spares may save you a long walk!

Engine overheats

Ignition (no-charge) warning light illuminated
Slack or broken drivebelt – retension or renew (Chapter 2)

Ignition warning light not illuminated
Coolant loss due to internal or external leakage (see Chapter 2)
Thermostat defective
Low oil level
Brakes binding
Radiator clogged externally or internally
Electric cooling fan not operating correctly
Engine waterways clogged
Ignition timing incorrect or automatic advance malfunctioning
Mixture too weak

Note: *Do not add cold water to an overheated engine or damage may result*

Low engine oil pressure

Gauge reads low or warning light illuminated with engine running
Oil level low or incorrect grade
Defective gauge or sender unit
Wire to sender unit earthed
Engine overheating
Oil filter clogged or bypass valve defective
Oil pressure relief valve defective
Oil pick-up strainer clogged
Oil pump worn or mountings loose
Worn main or big-end bearings

Note: *Low oil pressure in a high-mileage engine at tickover is not necessarily a cause for concern. Sudden pressure loss at speed is far more significant. In any event, check the gauge or warning light sender before condemning the engine.*

Engine noises

Pre-ignition (pinking) on acceleration
Incorrect grade of fuel
Ignition timing incorrect
Distributor faulty or worn
Worn or maladjusted carburettor
Excessive carbon build-up in engine

Whistling or wheezing noises
Leaking vacuum hose
Leaking carburettor or manifold gasket
Blowing head gasket

Tapping or rattling
Incorrect valve clearances
Worn valve gear
Worn timing chain
Broken piston ring (ticking noise)

Knocking or thumping
Unintentional mechanical contact (eg fan blades)
Worn drivebelt
Peripheral component fault (generator, water pump etc)
Worn big-end bearings (regular heavy knocking, perhaps less under load)
Worn main bearings (rumbling and knocking, perhaps worsening under load)
Piston slap (most noticeable when cold)

Conversion factors

Length (distance)						
Inches (in)	X	25.4	= Millimetres (mm)	X	0.0394	= Inches (in)
Feet (ft)	X	0.305	= Metres (m)	X	3.281	= Feet (ft)
Miles	X	1.609	= Kilometres (km)	X	0.621	= Miles
Volume (capacity)						
Cubic inches (cu in; in^3)	X	16.387	= Cubic centimetres (cc; cm^3)	X	0.061	= Cubic inches (cu in; in^3)
Imperial pints (Imp pt)	X	0.568	= Litres (l)	X	1.76	= Imperial pints (Imp pt)
Imperial quarts (Imp qt)	X	1.137	= Litres (l)	X	0.88	= Imperial quarts (Imp qt)
Imperial quarts (Imp qt)	X	1.201	= US quarts (US qt)	X	0.833	= Imperial quarts (Imp qt)
US quarts (US qt)	X	0.946	= Litres (l)	X	1.057	= US quarts (US qt)
Imperial gallons (Imp gal)	X	4.546	= Litres (l)	X	0.22	= Imperial gallons (Imp gal)
Imperial gallons (Imp gal)	X	1.201	= US gallons (US gal)	X	0.833	= Imperial gallons (Imp gal)
US gallons (US gal)	X	3.785	= Litres (l)	X	0.264	= US gallons (US gal)
Mass (weight)						
Ounces (oz)	X	28.35	= Grams (g)	X	0.035	= Ounces (oz)
Pounds (lb)	X	0.454	= Kilograms (kg)	X	2.205	= Pounds (lb)
Force						
Ounces-force (ozf; oz)	X	0.278	= Newtons (N)	X	3.6	= Ounces-force (ozf; oz)
Pounds-force (lbf; lb)	X	4.448	= Newtons (N)	X	0.225	= Pounds-force (lbf; lb)
Newtons (N)	X	0.1	= Kilograms-force (kgf; kg)	X	9.81	= Newtons (N)
Pressure						
Pounds-force per square inch (psi; lbf/in^2; lb/in^2)	X	0.070	= Kilograms-force per square centimetre (kgf/cm^2; kg/cm^2)	X	14.223	= Pounds-force per square inch (psi; lbf/in^2; lb/in^2)
Pounds-force per square inch (psi; lbf/in^2; lb/in^2)	X	0.068	= Atmospheres (atm)	X	14.696	= Pounds-force per square inch (psi; lbf/in^2; lb/in^2)
Pounds-force per square inch (psi; lbf/in^2; lb/in^2)	X	0.069	= Bars	X	14.5	= Pounds-force per square inch (psi; lbf/in^2; lb/in^2)
Pounds-force per square inch (psi; lbf/in^2; lb/in^2)	X	6.895	= Kilopascals (kPa)	X	0.145	= Pounds-force per square inch (psi; lbf/in^2; lb/in^2)
Kilopascals (kPa)	X	0.01	= Kilograms-force per square centimetre (kgf/cm^2; kg/cm^2)	X	98.1	= Kilopascals (kPa)
Torque (moment of force)						
Pounds-force inches (lbf in; lb in)	X	1.152	= Kilograms-force centimetre (kgf cm; kg cm)	X	0.868	= Pounds-force inches (lbf in; lb in)
Pounds-force inches (lbf in; lb in)	X	0.113	= Newton metres (Nm)	X	8.85	= Pounds-force inches (lbf in; lb in)
Pounds-force inches (lbf in; lb in)	X	0.083	= Pounds-force feet (lbf ft; lb ft)	X	12	= Pounds-force inches (lbf in; lb in)
Pounds-force feet (lbf ft; lb ft)	X	0.138	= Kilograms-force metres (kgf m; kg m)	X	7.233	= Pounds-force feet (lbf ft; lb ft)
Pounds-force feet (lbf ft; lb ft)	X	1.356	= Newton metres (Nm)	X	0.738	= Pounds-force feet (lbf ft; lb ft)
Newton metres (Nm)	X	0.102	= Kilograms-force metres (kgf m; kg m)	X	9.804	= Newton metres (Nm)
Power						
Horsepower (hp)	X	745.7	= Watts (W)	X	0.0013	= Horsepower (hp)
Velocity (speed)						
Miles per hour (miles/hr; mph)	X	1.609	= Kilometres per hour (km/hr; kph)	X	0.621	= Miles per hour (miles/hr; mph)
Fuel consumption*						
Miles per gallon, Imperial (mpg)	X	0.354	= Kilometres per litre (km/l)	X	2.825	= Miles per gallon, Imperial (mpg)
Miles per gallon, US (mpg)	X	0.425	= Kilometres per litre (km/l)	X	2.352	= Miles per gallon, US (mpg)

Temperature

Degrees Fahrenheit = (°C x 1.8) + 32

Degrees Celsius (Degrees Centigrade; °C) = (°F - 32) x 0.56

**It is common practice to convert from miles per gallon (mpg) to litres/100 kilometres (l/100km), where mpg (Imperial) x l/100 km = 282 and mpg (US) x l/100 km = 235*

Index

D

E

F

Printed by
J H Haynes & Co Ltd
Sparkford Nr Yeovil
Somerset BA22 7JJ England